Chemical Synthesis of Hormones, Pheromones and Other Bioregulators

Postgraduate Chemistry Series

The *Postgraduate Chemistry Series* of advanced textbooks is designed to provide a broad understanding of selected growth areas of chemistry at postgraduate student and research level. Volumes concentrate on material in advance of a normal undergraduate text, although the relevant background to a subject is included. Key discoveries and trends in current research are highlighted, and volumes are extensively referenced and cross-referenced. Detailed and effective indexes are an important feature of the series. In some universities, the series will also serve as a valuable reference for final year honours students.

Editorial Board

Titles in the Series

Practical Biotransformations: A Beginner's Guide
Gideon Grogan

Photochemistry of Organic Compounds: From Concepts to Practice
Petr Klán and Jakob Wirz

Catalysis in Asymmetric Synthesis, 2nd Edition
Vittorio Caprio and Jonathan Williams

Reaction Mechanisms in Organic Synthesis
Rakesh Parashar

Stoichiometric Asymmetric Synthesis
Mark Rizzacasa and Michael Perkins

Organic Synthesis using Transition Metals
Rod Bates

Organic Synthesis with Carbohydrates
Geert-Jan Boons and Karl J. Hale

Protecting Groups in Organic Synthesis
James R. Hanson

Chemical Synthesis of Hormones, Pheromones and Other Bioregulators

KENJI MORI

Emeritus Professor, The University of Tokyo, Japan

A John Wiley and Sons, Ltd., Publication

Library of Congress Cataloging-in-Publication Data

Mori, K. (Kenji), 1935–
   Chemical synthesis of hormones, pheromones, and other bioregulators / Kenji Mori.
      p. ; cm.
   Includes bibliographical references and index.
   ISBN 978-0-470-69724-5 (cloth)—ISBN 978-0-470-69723-8 (pbk.) 1. Bioorganic
chemistry. 2. Biomolecules–Synthesis. 3. Hormones–Synthesis. 4. Pheromones–
Synthesis. 5. Plant hormones–Synthesis. I. Title.
   [DNLM: 1. Hormones–chemical synthesis. 2. Biological Factors–chemical
synthesis. 3. Pheromones–chemical synthesis. WK 102 M854c 2010]
   QD415.M67 2010
   571.7′4–dc22
                                                              2010013112

A catalog record for this book is available from the British Library.

ISBN: 978-0-470-69724-5 (h/b) 978-0-470-69723-8 (p/b)

Set in 10/12pt Times-Roman by Laserwords Private Limited, Chennai, India
Printed in Great Britain by Antony Rowe Ltd, Chippenham, Wiltshire

Contents

Preface

There are numerous kinds of small and biofunctional molecules. Hormones, pheromones and other bioregulators such as antibiotics and antifeedants are important small molecules for organisms. This book summarizes the chemical synthesis of over 170 of these small molecules, which have been synthesized by my research group in Tokyo since 1959.

In preparing this book, I was careful to make my presentation simple and effective by using as many schemes as possible. When the description is insufficient for your purpose, you can refer to the original full papers, all of which are listed at the end of each chapter. Of course, full papers contain experimental details. Accordingly, this book is a source to assess the applicability and usefulness of many synthetic reactions for your own work.

The second notable feature of this book is the fact that all the materials are taken from my personal experience as a chemist. You will be able to know my thoughts in choosing a target molecule, planning its synthesis and evaluating its biological functions in cooperation with biologists. Biofunctional molecules of agricultural interest are treated extensively in this book. I believe that agriculture is as important as medicine for our survival.

This book describes my personal scientific history. I hope you will enjoy it. By knowing the past achievements, you will attain your own insights for planning a better synthesis in future.

"The truth will make you free." (John 8:32)

Kenji Mori
Tokyo, 2010

Abbreviations

Ac	acetyl
AIBN	2,2′-azobisisobutyronitrile
9-BBN	9-borabicyclo[3.3.1]nonane
Bn	benzyl
Boc	t-butoxycarbonyl
Bu	butyl
Bz	benzoyl
CAN	ceric ammonium nitrate
Cbz	benzyloxycarbonyl
CD	circular dichroism
CSA	camphorsulfonic acid
DABCO	1,4-diazabicyclo[2.2.2]octane
DAST	N,N-Diethylaminosulfur trifluoride
DBN	1,5-diazabicyclo[4.3.0]non-5-ene
DBU	1,8-diazabicyclo[5.4.0]undec-7-ene
DCC	$N,N′$-dicyclohexylcarbodiimide
DDQ	2,3-dichloro-5,6-dicyano-1,4-benzoquinone
DEAD	diethyl azodicarboxylate
DHP	3,4-dihydro-2H-pyrane
DIAD	diisopropyl azodicarboxylate
DIBAL-H	diisobutylaluminum hydride
DMAP	4-N,N-dimethylaminopyridine
DME	1,2-dimethoxyethane
DMF	N,N-dimethylformamide
DMP	Dess−Martin periodinane [1,1,1-tris(acetyloxy)-1,1-dihydro-1,2-benziodoxol-3-(1H)-one]
DMSO	dimethyl sulfoxide
DNB	3,5-dinitrobenzoyl
EE	2-ethoxyethyl
Ee	enantiomeric excess
Eq	molar equivalent
Et	ethyl
Fmoc	9-fluorenylmethoxycarbonyl
GLC	gas-liquid chromatography
HLADH	horse liver alcohol dehydrogenase
HMDS	1,1,1,3,3,3-hexamethyldisilazane
HMPA	hexamethylphosphoric triamide
HOBt	1-hydroxybenzotriazole
HPLC	high-performance liquid chromatography
Im	1-imidazolyl or imidazole

IR	infrared
LDA	lithium diisopropylamide
MCPBA	*m*-chloroperbenzoic acid
Me	methyl
MEM	2-methoxyethoxymethyl
MOM	methoxymethyl
MPLC	medium-pressure liquid chromatography
MS	molecular sieves or mass spectrum
Ms	methanesulfonyl
MTPA	α-methoxy-α-trifluoromethylphenylacetyl
NBS	*N*-bromosuccinimide
NCS	*N*-chlorosuccinimide
NIS	*N*-iodosuccinimide
NMO	*N*-methylmorpholine *N*-oxide
NMR	nuclear magnetic resonance
PCC	pyridinium chlorochromate
PDC	pyridinium dichromate
Ph	phenyl
Piv	pivaloyl (= trimethylacetyl)
PLE	pig-liver esterase
PMB	*p*-methoxybenzyl
PPL	pig-pancreatic lipase
PPTS	pyridinium *p*-toluenesulfonate
Pr	propyl
Pyr	pyridine
TBAF	tetra(*n*-butyl)ammonium fluoride
TBS	*t*-butyldimethylsilyl
TBDPS	*t*-butyldiphenylsilyl
Tf	triflyl (= trifluoromethanesulfonyl)
TFA	trifluoroacetic acid
THF	tetrahydrofuran
THP	tetrahydropyran-2-yl
TIPS	triisopropylsilyl
TLC	thin-layer chromatography
TMEDA	*N,N,N′,N′*-tetramethylethylenediamine
TMS	trimethylsilyl
TPAP	tetra(*n*-propyl)ammonium perruthenate
Tr	trityl (= triphenylmethyl)
Triton B	*N*-benzyltrimethylammonium hydroxide
Ts	tosyl (= *p*-toluenesulfonyl)

1

Introduction—Biofunctional Molecules and Organic Synthesis

Synthetic organic chemistry is a discipline different from biology. The former, however, can be a very useful tool to solve problems in biology. This chapter explains the reason why organic synthesis is useful in the studies of biofunctional molecules, and also details the ideas and techniques employed in the synthesis of biofunctional molecules.

1.1 What are biofunctional molecules?

Biofunctional molecules are those compounds that control such characteristics of organisms as differentiation, growth, metamorphosis, homeostasis, aggregation and reproduction. Both small molecules and macromolecules are used as biofunctional molecules. Chemical synthesis of small biofunctional molecules will be the subject of this book, because I have been engaged in the chemical synthesis of small biofunctional molecules (molecular weight less than 1000) for half a century.

Biofunctional natural products with low molecular weights are classified as shown in Table 1.1. Chemical studies on vitamins, hormones and antibiotics started in the first half of the 20th century, while those on semiochemicals began in the middle of the 20th century. This book treats the chemical synthesis of hormones, pheromones and other bioregulators such as allelochemicals.

1.2 Developmental stages of studies on biofunctional molecules

Let us first consider the process by which the investigation of a biofunctional molecule develops. As shown in Figure 1.1, the careful observation of a biological phenomenon together with speculation on the cause of that phenomenon make up the first step in the discovery of a biofunctional molecule. In the studies on the plant-growth hormone gibberellin, the first step was the observation in Japan in 1898 that the infection of rice seedlings by fungus *Gibberella fujikuroi* causes elongation of the seedlings to bring about the so-called "bakanaé" (= foolish seedlings)[1] disease, a destructive pest that reduces the yield of rice in Asia.[1]

Chemical Synthesis of Hormones, Pheromones and Other Bioregulators   Kenji Mori
© 2010 John Wiley & Sons, Ltd

Table 1.1   *Classification of biofunctional molecules*

Name	Definition
Vitamins	Biofunctional molecules that are taken in as food constituents; being essential to the proper nourishment of the organism. Derived from *vita* (L.) = life + amine
Hormones	Biofunctional molecules that are secreted and pass into the target organ of the same individual.Derived from *horman* (Gk.) = stir up
Antibiotics	Biofunctional molecules mainly of microbial origin that kill other micro-organisms. Derived from *anti* (Gk.) = against + *bios* (Gk.) = made of life
Semiochemicals	Biofunctional molecules that spread information between individuals. (They are also called signal substances.) Derived from *semio* (Gk.) = sign
(a) Pheromones	Biofunctional molecules that are used for communication between individuals within the same species. Derived from *pherein* (Gk.) = to carry + *horman* (Gk.) = stir up
(b) Allelochemicals	Biofunctional molecules that are used for communication between individuals belonging to different species. Derived from *allelon* (Gk.) = of each other
(1) Allomones	Biofunctional molecules that evoke advantageous reactions for their producers. Derived from *allos* (Gk.) = other
(2) Kairomones	Biofunctional molecules that evoke advantageous reactions for their receivers. Derived from *kairo* (Gk.) = opportune
(3) Synomones	Biofunctional molecules that evoke advantageous reactions for both their producers and receivers. Derived from *syn* (Gk.) = together with

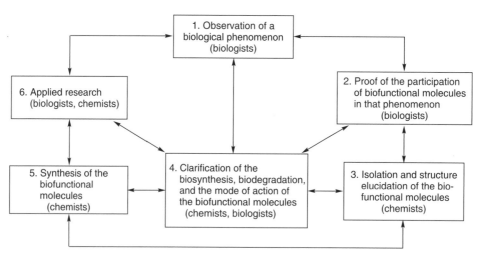

Figure 1.1   *Developmental stages of studies on biofunctional molecules: each stage is mutually interrelated with other stages*

The second stage of the research is to prove the participation of a biofuntional molecule in that specific phenomenon. In gibberellin research, Kurosawa proved that small biofunctional molecules produced by *G. fujikuroi* caused the elongation of rice seedlings.[2]

Then, the third and crucial stage comes: the isolation and structure elucidation of the biofunctional molecules responsible for the phenomenon. In 1938, Yabuta and Sumiki isolated the plant hormone

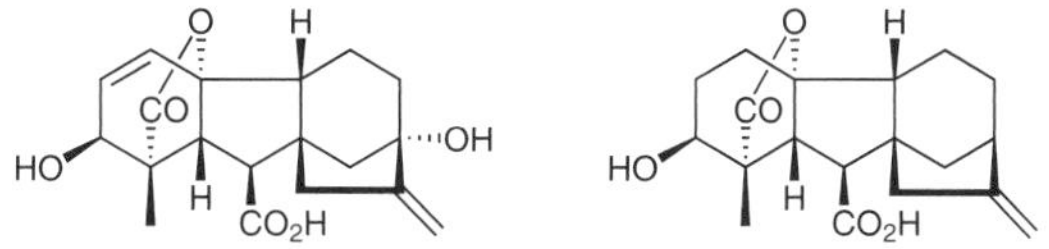

Figure 1.2 *Structures of the gibberellins*

gibberellins as crude crystals, which elongated rice seedlings.[3] The correct gross structure of gibberellin A$_3$ (**1**, Figure 1.2) was proposed by Cross *et al.* in 1959.[4] With the established structure of a biofunctional molecule, one can proceed with further chemical or biological research. Chemists and biologists begin to clarify the biosynthesis, biodegradation and the mode of action of that biofunctional molecule. On the other hand, synthetic chemists attempt the synthesis of that compound. In the case of the gibberellins, their synthesis was undertaken by many groups, culminating in total synthesis by Nagata,[5] Corey,[6] Mander,[7,8] Yamada and Nagaoka,[9] Ihara and Toyota,[10] and others. Mori's relay synthesis of (±)-gibberellin A$_4$ (**2**) in 1969 was an early success in this area.[11] As to the biosynthesis, biodegradation and mode of action of the gibberellins, chemists and biologist have been involved for many years.[12]

Finally, application of a particular biofunctional molecule in agriculture and other bioindustries or health care and medicine is the practical goal of the research. Chemists will synthesize many analogs and derivatives, and biologists will evaluate their biological effects. If one can find a useful compound, it will be commercialized for practical application. For example, gibberellin A$_3$ (**1**) is used in Japan to produce seedless grapes.

1.3 Small amounts of the samples are now sufficient for the elucidation of the structures of biofunctional molecules

Thanks to the development of microanalytical techniques and efficient separation methods, it is now possible to determine the structure of a biofunctional molecule with less than 1 mg of the material. In Table 1.2, examples are given with regard to the amounts of the samples employed for the structure elucidation of biofunctional molecules.

When Butenandt *et al.* studied bombykol (**3**), the female sex pheromone of the silkworm moth (*Bombyx mori*) in 1961, they isolated 12 mg of the crystalline 4-(*p*-nitrophenylazo)benzoate of bombykol (**3**) from half a million pheromone glands of the female silkworm moth obtained from more than a million silkworm cocoons bought in Germany, Italy and Japan.[13] A highly recommendable account of the reflection on the study of bombykol was published by Hecker and Butenandt.[14] More recent examples[15–19] in Table 1.2 show that the structures have been clarified even with microgram quantities.

A unique example of structural identification of a biofunctional molecule was reported recently by Hughson and coworkers in 2002.[19] Autoinducer-2 (AI-2) is a universal signal for interspecies communication (quorum sensing) in bacteria, which allows bacteria to coordinate gene expression. The structure of AI-2 remained elusive until 2002, when the X-ray crystallographic analysis of AI-2 sensor protein (Lux P) in a complex with AI-2 was successfully carried out. As shown in Table 1.2, the bound ligand AI-2 was a furanosyl borate diester **8**.[19] In this particular case, the structure of a biofunctional molecule could be elucidated even without isolating it.

Table 1.2 *Amounts of the samples employed for the structure elucidation of some biofunctional molecules*

Researchers and year of the work	Name of compound	Structure[a]	Amounts of sample (mg)
Butenandt et al.[13] 1961	Bombykol (pheromone of *Bombyx mori*)	**3**	12 (as a derivative)
Röller et al.[15] 1967	Juvenile hormone I (from *Hyalophora cecropia*)	**4**	0.3
Persoons et al.[16] 1976	Periplanone-B (pheromone of *Periplaneta americana*)	**5**	0.2
Oliver et al.[17] 1992	Pheromone of *Biprorulus bibax*	**6**	0.075
Wakamura et al.[18] 2001	Posticlure (pheromone of *Orgyia postica*)	**7**	0.01
Hughson et al.[19] 2002	AI-2 (bacterial quorum-sensing signal)	**8**	trace

[a]Except for **8**, the stereostructures including *cis/trans*-isomerism were determined later by synthesis.

1.4 Why must biofunctional molecules be synthesized?

One may think that there is no need for the synthesis of biofunctional molecules, because they always exist in organisms, and can be extracted and isolated. This is quite untrue, because the amounts of biofunctional molecules in organisms are usually very small. Due to their extremely low concentrations effective in organisms, biofunctional molecules can be isolated only in very small amounts, as shown in Table 1.2. It is therefore impossible to isolate hormones, pheromones and other bioregulators in gram quantities.

The limited availability of biofunctional molecules often makes it difficult to determine their precise stereostructures at the time of isolation. Accordingly, synthesis has become important as a tool to determine the structures of biofunctional molecules unambiguously. Advances in analytical techniques enabled chemists to propose the structures of biofunctional molecules even when they are available in extremely small quantities. Because of that, synthesis has become even more important than ever.

When we want to use biofunctional molecules practically in agriculture or medicine, of course we have to provide them in quantity. Organic synthesis is a method of choice for their large-scale production together with biotransformation and fermentation. More importantly, organic synthesis can provide useful compounds with better utility than the natural products themselves. Therefore, we must synthesize biofunctional molecules. Only after sufficient supply of the materials, biologists can examine their bioactivities in full depth. Biologists are always waiting for the cooperation and service of synthetic chemists.

1.5 How can we synthesize biofunctional molecules?

1.5.1 What is synthesis?

Synthesis is a process by which we can convert a simple compound **A** to a more complicated compound **B**. For that purpose we must employ an appropriate chemical reaction **C**. Synthesis is therefore a function involving the three parameters **A**, **B**, and **C**. There are three kinds of synthetic studies.

(a) Synthesis with a fixed target molecule **B**. Synthesis of natural products is the typical example in this category. Many of the syntheses in this book belong to this category.
(b) Synthesis with a fixed starting material **A**. In industries, it is always necessary to think about a clever new use of cheap starting materials in-house.
(c) Synthesis as achieved by using a particular reaction **C**. Discoverers of new reactions usually attempt to determine the scope and limitations of their new reactions by applying them to the synthesis of a certain target molecule. People in academia quite often work along this line.

   Of course there is another type of synthesis. That is synthesis of any kinds of target molecules from any kinds of starting materials to generate compounds with the desired physical or biological properties. This is the way popular in materials science and medicinal chemistry.

1.5.2 What kind of consideration is necessary before starting a synthesis?

In synthesizing biofunctional molecules, one must select a target molecule. There are a number of criteria for selecting a target. The impact and significance of the synthetic achievement must be taken into consideration. What kind of target can be regarded as the one with a great impact? It quite often depends on personal taste. It may happen that a target is chemically interesting but biologically nonsense or vice versa. So, like paintings, one and the same synthetic achievement can be highly appreciated by a certain fraction of chemists, while it can be disputed by others. Accordingly, the choice of a target molecule reflects the taste of the chemist who works with that target. There is no other way than to choose one's favorite molecule as the target. However, it happens that someone requests you to synthesize a sample for him or her. In my experience, such a request quite often brings about an interesting result. Chemists should be flexible to respond to others.

   Next, one must choose the starting material(s). All the synthesis starts from readily available commercial products. It is therefore important for a chemist to look at catalogs of big reagent manufactures like Aldrich. By reading catalogs we can have knowledge on the prices of many possible starting materials. It is also important to be familiar with the industrial intermediates in the chemical and pharmaceutical industries. We may be able to obtain such intermediates by the courtesy of people in these industries.

   Then, it is the time to make a gross plan of the synthesis. One must decide the key reaction to be employed. In the case of an enantioselective synthesis, the timing of introducing the required asymmetry correctly is always of great importance. Of course, one must think about each of the steps, and the order must be fixed by which each step is to be executed. There are many possible synthetic routes for a biofunctional molecule. At the beginning it is not so easy to devise the best route. Through experimentation one can determine realistic routes. If a certain step does not go as expected, one must reconsider and modify the synthetic route. Finally, one can make the target molecule. In many cases, if one can dream it, one can make it.

1.5.3 Synthon

According to Corey, *synthon* is defined as structural units within a molecule that are related to possible synthetic operations.[20] Some people recently regard a synthon as a synonym of a building block. This is different from Corey's original usage.

So as to understand the concept of synthon, let us analyse the synthesis of keto ester **A** in Figure 1.3. There are many different ways to disconnect the C–C bond in **A**, and eight structural units (a)–(h) are conceivable as possible synthons. Disconnection of a target molecule to possible synthons is called **retrosynthetic analysis**. If there is a reaction to connect the possible synthons to build up **A**, then we can select realistic synthons. In the case of **A**, (d) and (e) are two synthons, which can be connected by employing the Michael addition.

Retrosynthetic analysis is the basic operation in planning organic synthesis. Knowledge on synthetic reactions and insight to determine a useful reaction are of basic importance in planning synthesis. The higher the number of synthons in a target molecule, the more difficult it is to synthesize. Corey emphasizes in depth the importance of retrosynthetic analysis.[21]

1.5.4 Molecular symmetry and synthesis

Recognition of explicit or implicit symmetry in a target molecule often simplifies its synthetic plan. A classic and well-known example, as shown in Figure 1.4, is the synthesis of (±)-usnic acid (**9**) by Barton *et al.*[22] A lichen constituent usnic acid is dissymmetric at a glance. But by employing oxidative coupling reaction of phenols, usnic acid (**9**) can be dissected into a single starting material.

Our synthesis of magnosalicin (**10**), a medicinally active principle from a plant *Magnolia salicifolia*, was achieved in a single step, as shown in Figure 1.5, considering the symmetry of the molecule.[23] The details of this synthesis will be discussed later in Section 2.4.2.

1.5.5 Criteria for 'A Good Synthesis'

It sometimes happens that over twenty or thirty different syntheses are published for a single biofunctional molecule, because so many different synthetic routes are possible for that target. Then, how can we regard

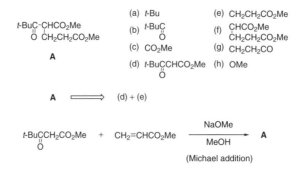

Figure 1.3 *Retrosynthetic analysis: disconnection of **A** to give two synthons (d) and (e)*

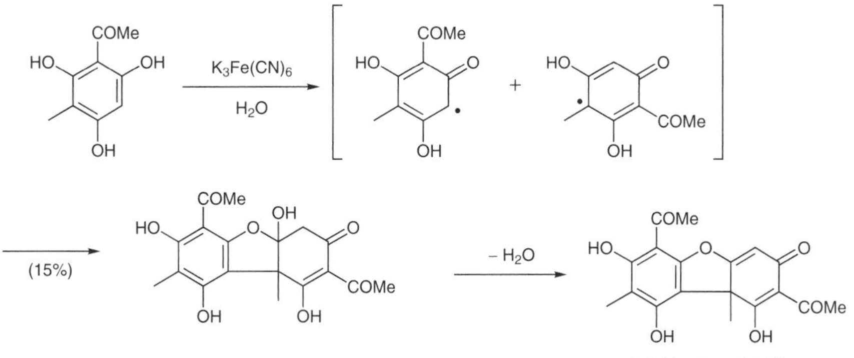

Figure 1.4　*Synthesis of (±)-usnic acid by Barton et al. Modified by permission of Shokabo Publishing Co., Ltd*

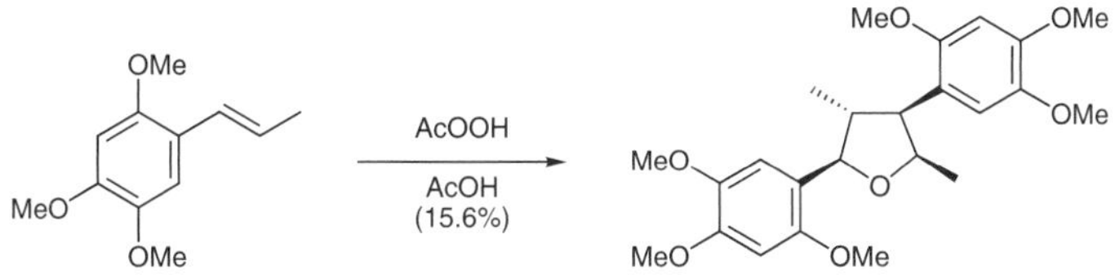

Figure 1.5　*Synthesis of (±)-magnosalicin. Modified by permission of Shokabo Publishing Co., Ltd*

a single one of the syntheses as superior to others? This is a good question, just like it is difficult to judge a painting, a composition or a novel to be better than others. Personal tastes of a researcher as a designer are always reflected on his or her achievements. The following three points, however, are the prerequisites for a particular synthetic plan to be judged as a good one.

(a) Each step of the synthesis proceeds with a good yield. Highly efficient regio- and stereoselective reactions must be employed.
(b) As to the pivotal step in a synthetic route, there should be available an alternative method to achieve that transformation. Otherwise, the synthesis may come to a dead end, and a graduate student as a practitioner may not be able to get his or her Ph.D. degree.
(c) The simpler the synthetic route, the better the synthesis. This is my own view.

In order to achieve an efficient synthesis, a convergent route is always preferred to a linear route. In the case of a linear route, as shown in the left part of Figure 1.6, the starting material **A** will be converted to the final product **ABCDEFGH** through seven steps in a linear sequence. With this linear route, the overall yield of the final product will be 48% even in the case in which the yield of each step is 90%. Usually, the yield of each of the seven steps may be 70%. Then, the overall yield will drop to only 8%. If the yield of each step is 50%, the overall yield will result in a miserable figure of only 1%. As shown in the right part of Figure 1.6, a convergent route generates blocks like **AB**, **CD**, **EF** and **GH**. Then, these will be connected twice to give the final product. The necessary steps to complete this convergent synthesis

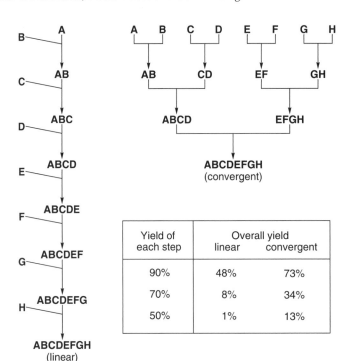

Yield of each step	Overall yield	
	linear	convergent
90%	48%	73%
70%	8%	34%
50%	1%	13%

Figure 1.6 *Linear synthesis versus convergent synthesis. Reprinted with permission of Shokabo Publishing Co., Ltd*

are the same seven steps. But the overall yield of the convergent route can be surprisingly better than what can be realized by the linear route. Thus, when each step can give the next product in 90% yield, the overall yield will be as high as 73%. Even in the case when the yield of each step is 50%, the overall yield of the convergent route will remain as still acceptable 13%. It is clear that a convergent synthesis is more efficient than a linear synthesis.

Recognition of possible synthons in a target molecule is the most important factor to make its synthesis simple or complicated. Let us compare the efficiency and simplicity of three different syntheses of (8*E*,10*E*)-8,10-dodecadien-1-ol (**11**), the female-produced sex pheromone of the codling moth, *Cydia pomonella*. This moth is a notorious pest of apple orchards.

Figure 1.7 summarizes the synthesis of **11** reported by Descoins and Henrick in 1972.[24] In their retrosynthetic analysis, they dissected **11** into two parts by breaking the C–C bond between C-5 and C-6. Cyclopropyl bromide was converted to 3,5-heptadienyl bromide (**A**), while tetrahydropyran furnished another building block **B**. Coupling of the Grignard reagent **C** derived from **B** with **A** in the presence of lithium tetrachlorocuprate was followed by the removal of the THP protective group to give **11**. The pure pheromone **11** was found to be crystalline, and could be purified by recrystallization. This synthesis is convergent. But the two starting materials are expensive, and the synthetic route is not short enough for economical manufacturing of **11**.

Mori's synthesis of **11** in 1974 was also convergent, as shown in Figure 1.8.[25] In this synthesis, the C–C bond at between C-6 and C-7 was disconnected to enable the use of two cheap C_6 starting materials, sorbic acid and hexane-1,6-diol. The former was converted to the building block **A**, while the latter was

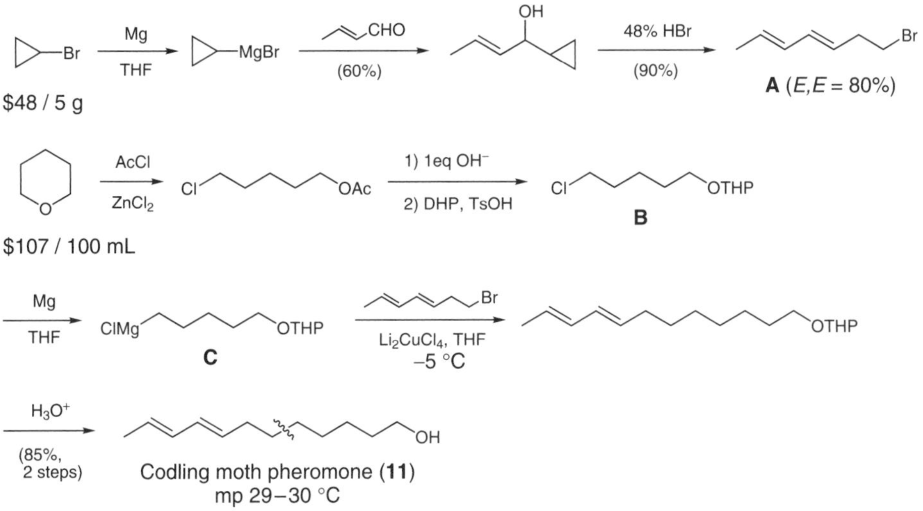

Figure 1.7 Synthesis of the sex pheromone of the codling moth by Descoins and Henrick. Modified by permission of Shokabo Publishing Co., Ltd

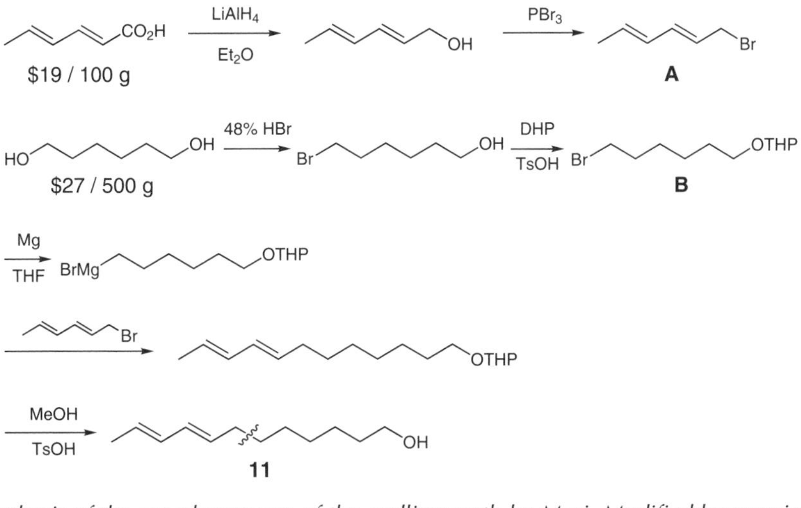

Figure 1.8 Synthesis of the sex pheromone of the codling moth by Mori. Modified by permission of Shokabo Publishing Co., Ltd

converted to another building block **B**. Coupling of the Grignard reagent prepared from **B** with **A** yielded **11** after deprotection.

Further improvement of Mori's synthesis was reported by Henrick in 1977 (Figure 1.9).[26] Coupling of (2*E*, 4*E*)-2,4-hexadienyl acetate (**A**) with the Grignard reagent **B** yielded the desired and crystalline product **11** in 60–70% overall yield based on (2*E*, 4*E*)-2,4-hexadien-1-ol. This convergent synthesis was employed

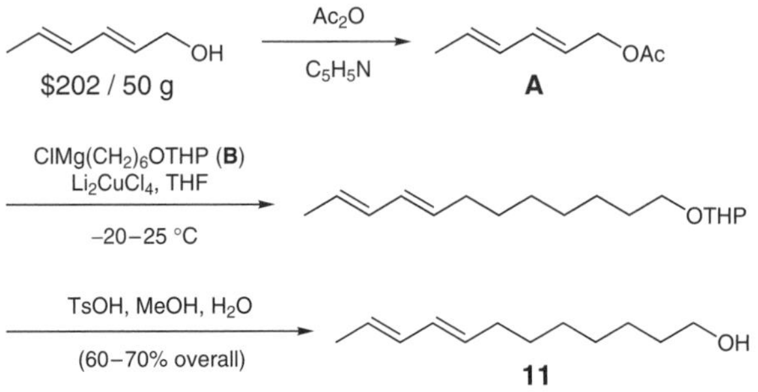

Figure 1.9   *Synthesis of the sex pheromone of the codling moth by Henrick. Modified by permission of Shokabo Publishing Co., Ltd*

for the commercial production of **11**. Improvement of a synthetic manufacturing process, so-called "process chemistry", is important in realizing the practical use of biofunctional molecules.

1.6   What kind of knowledge and techniques are necessary to synthesize biofunctional molecules?

Of course, knowledge in synthetic reactions and techniques in organic experiments including purification methods are essential requirements to execute synthesis of biofunctional molecules. However, two additional things must be learned.

1.6.1   Stereochemistry and reactivity

Many biofunctional molecules are chiral and nonracemic. It is therefore important to know the relationship between stereochemistry and reactivity. Let us examine the following two examples.

We first think about the different reactivities of axial and equatorial isomers through the examples shown in Figure 1.10. Esters **12** and **13** are called methyl 4-epidehydroabietate (**12**) and methyl dehydroabietate (**13**). Dehydroabietic acid can be esterified with methanol and sulfuric acid to give **13**, while 4-epidehydroabietic acid cannot be esterified under the same conditions. Ester **12** can be prepared only through methylation with diazomethane. Although ester **13** can be hydrolysed with sodium hydroxide in methanol, the stereoisomer **12** cannot be hydrolysed under the same conditions. This reduced reactivity

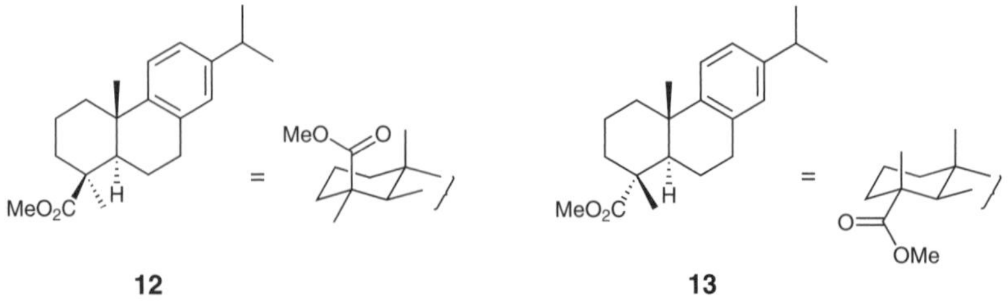

Figure 1.10   *Reactivity of the two stereoisomers of resin acid methyl ester*

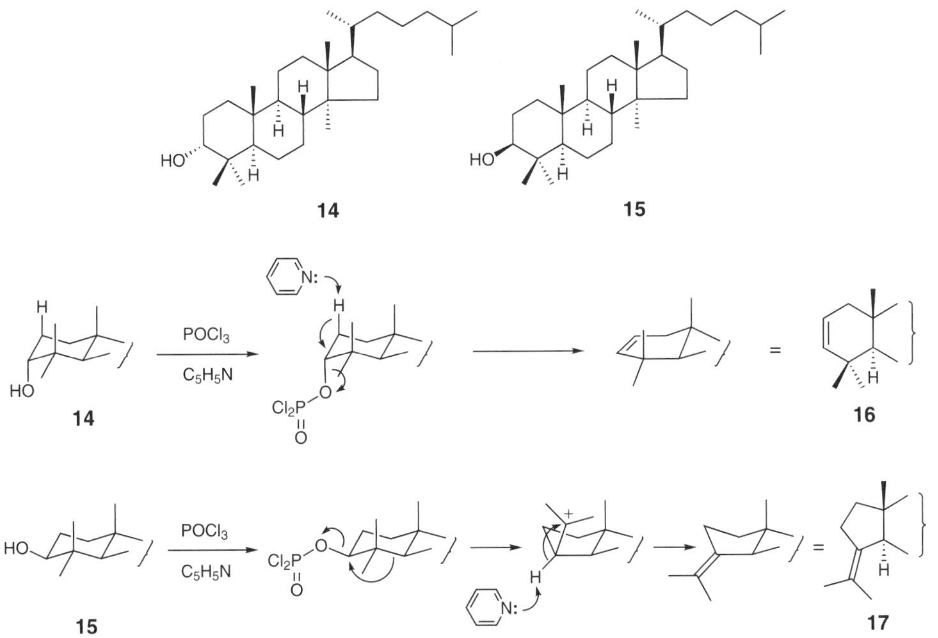

Figure 1.11 *Reactivity of the two stereoisomers of triterpene alcohol. Modified by permission of Shokabo Publishing Co., Ltd*

of **12** originates from the steric congestion around the ester group, because it is in axial orientation. The equatorial counterpart **13** is more reactive, because the ester group of **13** is not sterically hindered.

Figure 1.11 shows the second example. Triterpene alcohol **14** with an axial hydroxy group can be dehydrated smoothly by treatment with phosphorus oxychloride in pyridine to give **16** through conventional E2 elimination mechanism. However, dehydration of the equatorial alcohol **15** leads to a rearrangement product **17** through the mechanism as shown in Figure 1.11. This type of simple stereochemical knowledge is very useful in synthetic planning.

1.6.2 Stereochemistry and analytical methods

Knowledge on analytical methods is very important for quick elucidation of the structures of synthetic intermediates. It is also very important for the unambiguous identification of the final synthetic product with the natural product. Two examples will be given here.

The first example illustrates the importance of NMR spectroscopy in modern organic synthesis. Commercial NMR spectrometers manufactured by Varian Associates in the USA became available to chemical communities in late 1950s to early 1960s. In Japan, the first NMR spectrometer that became available was Varian V4300C operating at 56.4 MHz. In 1960 I synthesized (±)-lactone **18** by the route shown in Figure 1.12.[27] It is worthwhile for you to think about the mechanisms of conversion of **A** to **B** and that of **C** to **E** via **D**.

The racemic lactone (±)-**18** was obtained as a pure and crystalline compound, and its relative stereochemistry had to be determined. I thought, in 1960, ^{1}H NMR analysis to be the most appropriate method to solve the problem, because it had already been known to use a vicinal coupling constant for the stereochemical studies of cyclohexane compounds including terpenoids and steroids. Figures 1.13(a) and (b)

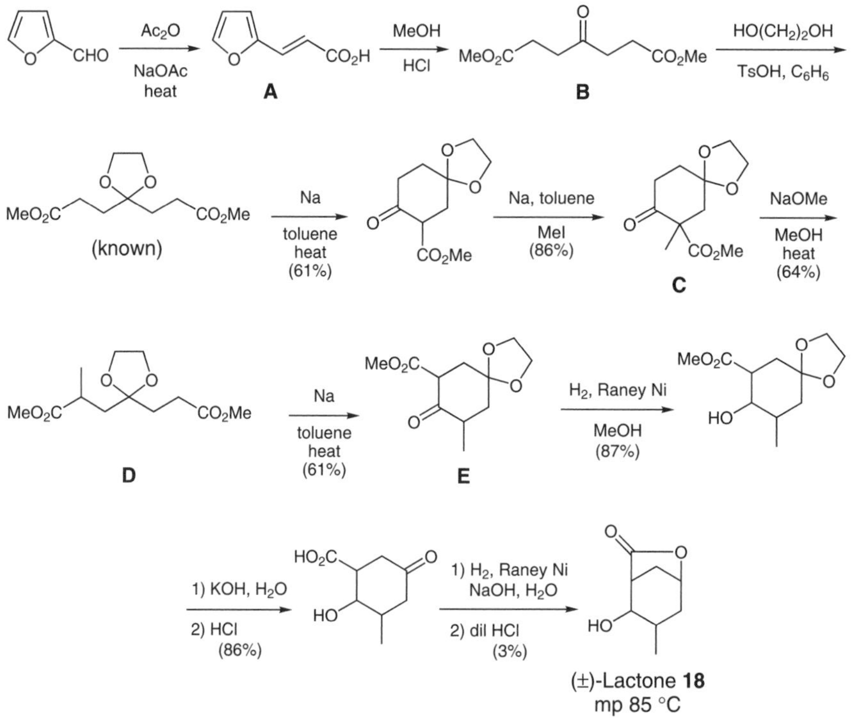

Figure 1.12   *Synthesis of (±)-lactone 18. Modified by permission of Shokabo Publishing Co., Ltd*

show two ^{1}H NMR spectra of (±)-**18**, one (a) measured at 56.4 MHz in 1960 and the other (b) measured at 400 MHz in 2008. The ^{1}H NMR spectrum of the tosylate of (±)-**18** is also shown in Figure 1.13 (c).

I was able to deduce the relative stereochemistry of (±)-**18** by examining its ^{1}H NMR spectrum measured in 1960. I noticed the presence of a signal at $\delta = 3.54$ (1H, dd, $J = 3$, 10 Hz) due to the CHOH proton.

Figure 1.14 shows the stereoformulas **18A**–**18D** of the stereoisomers of (±)-**18**. The formula shows one of the two enantiomers. The Newman projections depicted in the middle row show the stereochemical relationships between the substituents at C-2 and C-3, while the Newman projections in the bottom row indicate the situations at C-2 and C-1. The Karplus equation, as follows, is known to correlate the magnitude of the coupling constant J with the dihedral angle ϕ.

$$J = \begin{cases} 8.5\cos^2\phi - 0.28 & 0° \le \phi \le 90° \\ 9.5\cos^2\phi - 0.28 & 90° \le \phi \le 180° \end{cases}$$

Let us first examine the projections in the middle row. In the case of **18A**, the dihedral angle ϕ between the bonds C–Ha and C–Hb is 180°, which demands $J = 9$ Hz according to the Karplus equation. In the cases of **18B**, **18C** and **18D**, that dihedral angle ϕ is 60°. Therefore, in these three cases, the magnitude of J will be 1.8 Hz according to the Karplus equation. As you can see from the two Newman projections in the bottom row, the dihedral angle ϕ between the bonds C–Ha and C–Hc is fixed at 60°, which demands $J = 1.8$ Hz. Accordingly, only in the case of the stereoisomer **18A**, we can observe a large constant $J = 9$ Hz. The ^{1}H NMR spectrum of (±)-**18** as shown in Figure 1.13(a) shows the

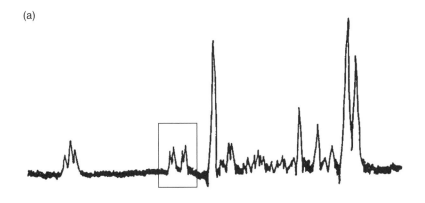

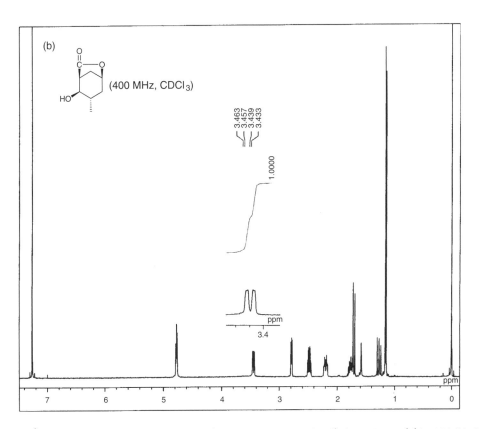

Figure 1.13   ¹H NMR spectra of (±)-**18** measured at (a) 56.4 MHz in CHCl₃ in 1960, and (b) 400 MHz in CDCl₃ in 2008. ¹H NMR spectrum of the corresponding tosylate is also shown in (c) at 400 MHz in CDCl₃ in 2008

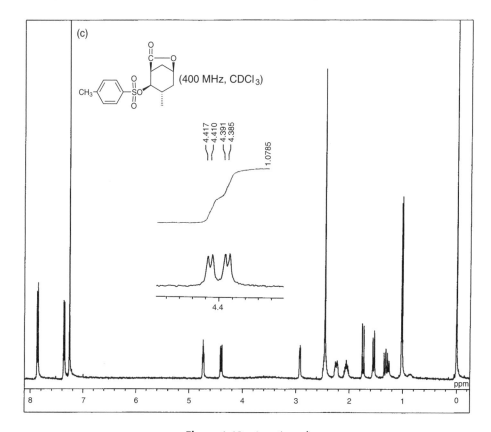

Figure 1.13   *(continued)*

magnitude of the coupling constant J_{HaHb} as 10 Hz. Therefore, the relative configuration of $(\pm)$-**18** was determined unambiguously as **18A**.[27] NMR spectroscopy is useful and important as a tool to determine the stereostructures of the synthetic products.

The second example shows the usefulness of X-ray crystallographic analysis in modern organic synthesis. Advances in computer science made this technique a routinely useful one. In 1986 I was interested in clarifying the steric course of the oxidation reaction as shown in Figure 1.15.[23] The reaction had been studied by Schmauder *et al.*, and the product reported as **21**.[28] Because the product could be recrystallized from isopropyl alcohol as a beautiful monoclinic with mp 96–97 °C, its X-ray crystallographic analysis was carried out. Figure 1.16 shows the molecular structure of the product. It was not **21** but **20**. This information was used immediately for the synthesis of a very similar and bioactive lignan called magnosalicin (**10**, see Section 2.4.2).[23]

There are many other useful analytical methods. Chromatographic methods such as gas chromatography (GC) and high-performance liquid chromatography (HPLC) are used daily for identification and estimation of the purity of a synthetic product. Chiroptical methods, such as circular dichroism (CD) spectroscopy, are also important especially in studying the relationships between absolute configuration and bioactivity of biofunctional molecules. In later chapters I will give some examples of application of CD spectroscopy in enantioselective synthesis.

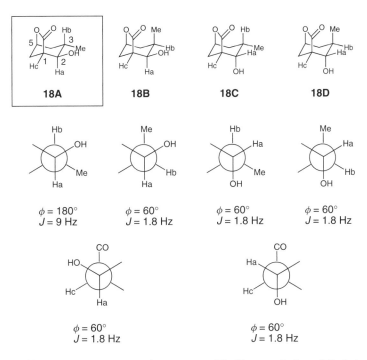

Figure 1.14 *Four possible stereoisomers* ***A***–***D*** *of* (±)-***18***. *Modified by permission of Shokabo Publishing Co., Ltd*

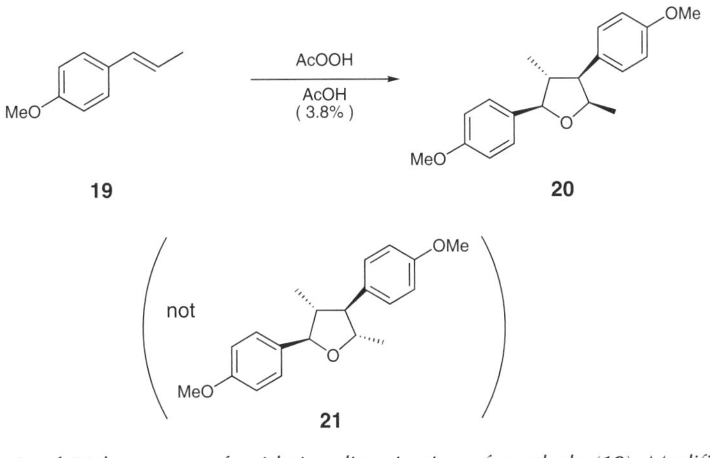

Figure 1.15 *Synthesis of* ***20*** *by means of oxidative dimerization of anethole* (***19***). *Modified by permission of Shokabo Publishing Co., Ltd*

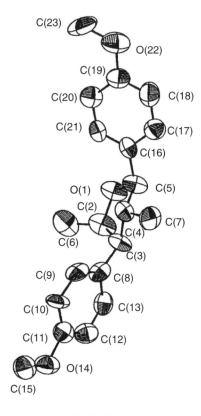

Figure 1.16   *Structure **20** as clarified by X-ray crystallographic analysis*

References

1. Tamura, S. In *Gibberellins*, Takahashi, N.; Phinney, B.O.; MacMillan, J. eds., Springer; New York, 1991, pp. 1–8.
2. Kurosawa, E. *Nat. Hist. Soc. Formosa* **1926**, *16*, 213–227.
3. Yabuta, T.; Sumiki, Y. *J. Agric. Chem. Soc. Jpn*. **1938**, *14*, 1526.
4. Cross, B.E.; Grove, J.F.; MacMillan, J.; Moffatt, J.S.; Mulholland, T.P.C.; Seaton, J.C.; Sheppard, N. *Proc. Chem. Soc*. **1959**, 302–303.
5. Nagata, W.; Wakabayashi, T.; Narisada, M.; Hayase, Y.; Kamata, S. *J. Am. Chem. Soc*. **1971**, *93*, 5740–5758.
6. (a) Corey, E.J.; Danheiser, R.L.; Chandrasekaran, S.; Siret, P.; Keck, G.E.; Gras, J.-L. *J. Am. Chem. Soc*. **1978**, *100*, 8031–8034. (b) Corey, E.J.; Danheiser, R.L.; Chandrasekaran, S.; Keck, G.E.; Gopalan, B.; Larsen, S.D.; Siret, P.; Gras, J.-L. *J. Am. Chem. Soc*. **1978**, *100*, 8034–8036.
7. Lombardo, L.; Mander, L.N.; Turner, J.V. *J. Am. Chem. Soc*. **1980**, *102*, 6626–6628.
8. Mander, L.N. *Chem. Rev*. **1992**, *92*, 573–612.
9. Nagaoka, H.; Shimano, M.; Yamada, Y. *Tetrahedron Lett*. **1989**, *30*, 971–974.
10. Toyota, M.; Yokota, M.; Ihara, M. *J. Am. Chem. Soc*. **2001**, *123*, 1856–1861.
11. Mori, K.; Shiozaki, M.; Itaya, N.; Matsui, M.; Sumiki, Y. *Tetrahedron* **1969**, *25*, 1293–1321.
12. Murofushi, N. *et al*. in *Comprehensive Natural Products Chemistry* Vol. *8*, Mori, K. Vol. ed., Elsevier; Oxford, 1999, pp. 35–57.
13. Butenandt, A.; Beckmann, R.; Hecker, E. *Hoppe-Seyler's Z. Physiol. Chem*. **1961**, *324*, 71–83.

14. Hecker, E.; Butenandt, A. in *Techniques in Pheromone Research*, Hummel, H.E.; Miller, T.A., eds., Springer; New York, 1984, pp. 1–44.
15. Röller, H.; Dahm, K.H.; Sweeley, C.C.: Trost, B.M. *Angew. Chem. Int. Ed.* **1967**, *6*, 179–180.
16. Persoons, C.J.; Verwiel, P.E.J.; Ritter, F.J.; Talman, E.; Nooijen, P.J.F.; Nooijen, W.J. *Tetrahedron Lett.* **1976**, 2055–2058.
17. Oliver, J.E.; Aldrich, J.R.; Lusby, W.R.; Waters, R.M.; James, D.G. *Tetrahedron Lett.* **1992**, *33*, 891–894.
18. Wakamura, S.; Arakaki, N.; Yamamoto, M.; Hiradate, S.; Yasui, H.; Yasuda, T.; Ando, T. *Tetrahedron Lett.* **2001**, *42*, 687–689.
19. Chen, X.; Schauder, S.; Potier, N.; Van Dorsselae, A.; Pelczer, I.; Bassler, B.L.; Hughson, F.M. *Nature* **2002**, *415*, 545–549.
20. Corey, E.J. *Pure Appl. Chem.* **1967**, *14*, 19–37.
21. Corey, E.J.; Cheng, X.-M. *The Logic of Chemical Synthesis*, Wiley; New York, 1989, pp. 1–436.
22. Barton, D.H.R.; Deflorin, A.M.; Edwards, O.E. *J. Chem. Soc.* **1956**, 530–534.
23. Mori, K.; Komatsu, M.; Kido, M.; Nakagawa, K. *Tetrahedron* **1986**, *42*, 523–528.
24. Descoins, C.; Henrick, C.A. *Tetrahedron Lett.* **1972**, 2999–3002.
25. Mori, K. *Tetrahedron* **1974**, *30*, 3807–3810.
26. Henrick, C.A. *Tetrahedron* **1977**, *33*, 1845–1889.
27. Mori, K.; Matsui, M.; Sumiki, Y. *Agric. Biol. Chem.* **1961**, *25*, 902–906.
28. Schmauder, H.-P.; Groger, D.; Lohmann, D.; Gruner, H.; Foken, H.; Zschunke, A. *Pharmazie* **1979**, *34*, 22–25.

2

Synthesis of Phytohormones, Phytoalexins and Other Biofunctional Molecules of Plant Origin

Since ancient times, mankind has utilized higher plants as the sources of foods, clothes, houses, medicinals and so on, which are indispensable for mankind's existence on this planet. Chemists became interested in clarifying the physiology and ecology of higher plants. Studies on natural products of plant origin have prospered for these two centuries, and the field is now called phytochemistry. Chemical studies on physiology of higher plants clarified the roles of various phytohormones in higher plants. Semiochemicals and phytoalexins are important biofunctional molecules in plant ecology. In this chapter we will learn the synthetic chemistry of biofunctional molecules related to higher plants.

2.1 Phytohormones

2.1.1 What are phytohormones?

Phytohormones are those compounds that regulate and control the physiological phenomena of higher plants such as growth, maturation, abscission and dormancy. They are biosynthesized at a special part of a plant, then migrate to the places where they control the various physiological phenomena. Seven major phytohormones are shown in Figure 2.1.

(a) Indole-3-acetic acid. Kögl in the Netherlands was the first organic chemist who investigated phytohormones from the chemical point of view. In 1934 he isolated from human urine a compound that promotes the plant growth, and named it hetroauxin.[1] Heteroauxin was identified as indole-3-acetic acid, which had been synthesized by Majima and Hoshino in Japan in 1925.[2] Indole-3-actic acid is now called auxin, and one of the important phytohormones to promote the growth of higher plants.

(b) Gibberellins. Gibberellins were discovered and named in 1938 by Yabuta and Sumiki in Japan.[3] Gibberellin A_3 is the major one of the gibberellins commercially manufactured by fermentation. Gibberellins promote plant growth. They were first isolated as fungal metabolites, and then shown to be phytohormones.

Chemical Synthesis of Hormones, Pheromones and Other Bioregulators   Kenji Mori
© 2010 John Wiley & Sons, Ltd

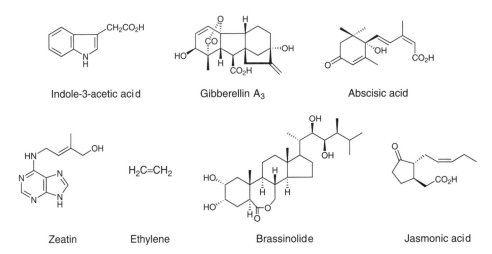

Figure 2.1   *Structures of major phytohormones*

(c)   Abscisic acid. In 1963 Addicott in the USA together with a Japanese chemist Ohkuma isolated 9 mg of abscisic acid from 300 kg of cotton fruits.[4] This is a phytohormone that promotes abscission and dormacy of higher plants.

(d)   Zeatin. In 1964 Letham in Australia isolated and identified zeatin as a hormone to promote cell division and growth.[5] He isolated this adenine derivative from immature seeds of maize.

(e)   Ethylene. In 1934 ethylene was detected in the volatiles produced by apple. It is a phytohormone to promote maturation of fruits. This simplest hormone is used practically to promote the maturation of immature apples and bananas.

(f)   Brassinolide. In 1979 Grove and coworkers in the USA isolated 4 mg of brassinolide from 40 kg of the pollen of the rape.[6] This hormone promotes the elongation and growth of plant cells.

(g)   Jasmonic acid. In 1982 Ueda and Kato in Japan isolated jasmonic acid as a senescence-promoting substance in plants.[7] This acid and related compounds are widely distributed among plants.

Now I will describe my own synthetic works on phytohormones.

2.1.2   Gibberellins

2.1.2.1   *Structures of gibberellins*

The gibberellins were first isolated in 1938 in the building in which I was educated and started my career as a chemist—Department of Agricultural Chemistry at the University of Tokyo. In 1959, I was a new Ph.D. student in the Laboratory of Organic Chemistry of that Department. My thesis adviser Professor Matsui told me, "Let us synthesize the gibberellins." I hesitated for a moment, because the structure of the gibberellins was not yet elucidated in 1959. Without a structure no one can synthesize the gibberellins. He then continued, "If you think it to be a too difficult task, I have another research subject." Hearing his words, I immediately decided to work on the gibberellins. As a young student I was brave enough to face the challenge. It took nine years of my life to synthesize some members of the gibberellins.

Figure 2.2 shows the structures of the gibberellins and the numbering system of the carbon framework, *ent*-gibberellane. In 1956 Cross in the UK proposed structure **A** with C-19 → C-2 γ-lactone ring for

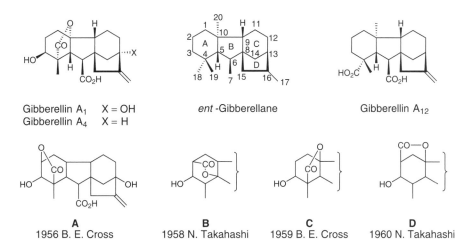

Figure 2.2   *Structures of the gibberellins. Modified by permission of Shokabo Publishing Co., Ltd*

gibberellin A_1.[8] Takahashi in Japan, however, proposed a different structure **B** with a δ-lactone ring for gibberellin A_1.[9] Cross revised his structure in 1959 to propose **C** with a C-19 → C-10 γ-lactone ring as the correct structure for gibberellin A_1.[10] Takahashi *et al.* then thought **D** with a γ-lactone ring as a possible structure of gibberellin A_1, although they did not publish this idea but presented it orally at a conference.

What Professor Matsui asked me in 1959 was to synthesize ring A model compounds, and compare their spectral properties with those of the gibberellins. IR spectra of the gibberellins show their lactone carbonyl absorption ($\nu_{C=O}$) at 1746–1770 cm^{-1} when measured as nujol pastes. The carbonyl absorption of γ-lactones appears at 1770 cm^{-1}, while that of δ-lactones is known to appear at 1740 cm^{-1}. Thus, the $\nu_{C=O}$ values of the gibberellins are just in between those of γ- and δ-lactones. In the case of ^{1}H NMR analysis, if the structure **A** or **C** is correct, the protons of the methyl group attached to C-4 will show a 3H singlet, while in the cases of **B** and **D**, that protons will appear as a 3H doublet. Professor Matsui and I believed that the synthesis and spectroscopic analysis of ring A model compounds would enable us to deduce the correct structure of the ring A of the gibberellins.

Figure 2.3 shows the structures of racemic lactones synthesized by myself as the model compounds of the ring A of the gibberellins.[11,12] Infrared absorptions due to the lactonic carbonyl group of each of the lactones are also shown in the figure. Infrared spectra of these lactones were measured as nujol pastes and also as solutions in dioxane. 1,4-Dioxane was the solvent that could dissolve the gibberellins, while chloroform could not.

γ-Lactones without a hydroxy group such as (±)-**1**, (±)-**2** and (±)-**12** showed their carbonyl absorptions ($\nu_{C=O}$) at 1768–1780 cm^{-1} as typical γ-lactones, while δ-lactones absorbed at 1730–1750 cm^{-1} regardless of the presence or absence of a hydroxy group. However, in the cases of γ-lactones with a hydroxy group such as (±)-**3**, (±)-**4**, (±)-**5**, (±)-**10** and (±)-**11**, they showed their carbonyl absorptions at 1755, 1745, 1768, 1742 and 1760 cm^{-1}, respectively, when measured as nujol pastes. These values were not so different from those of the δ-lactones. If the IR spectra of these hydroxylated γ-lactones were measured as dioxane solutions, their carbonyl absorptions were observed at 1782, 1780, 1778, 1778 and 1778 cm^{-1}, respectively. These values were in good accord with their γ-lactone structure. In the crystalline state as nujol pastes, there exists a hydrogen bond between the hydroxy and the carbonyl groups, shifting the carbonyl absorption to a lower wave number region. In a dioxane solution, however, there is no intermolecular hydrogen

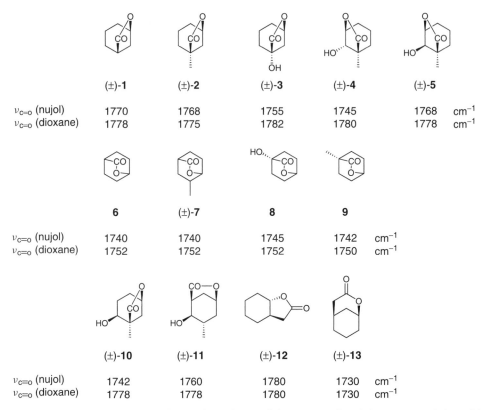

	(±)-1	(±)-2	(±)-3	(±)-4	(±)-5	
$\nu_{C=O}$ (nujol)	1770	1768	1755	1745	1768	cm^{-1}
$\nu_{C=O}$ (dioxane)	1778	1775	1782	1780	1778	cm^{-1}

	6	(±)-7	8	9	
$\nu_{C=O}$ (nujol)	1740	1740	1745	1742	cm^{-1}
$\nu_{C=O}$ (dioxane)	1752	1752	1752	1750	cm^{-1}

	(±)-10	(±)-11	(±)-12	(±)-13	
$\nu_{C=O}$ (nujol)	1742	1760	1780	1730	cm^{-1}
$\nu_{C=O}$ (dioxane)	1778	1778	1780	1730	cm^{-1}

Figure 2.3 *Structures of lactones synthesized as the model compounds of the ring A of the gibberellins. Reprinted with permission of Shokabo Publishing Co., Ltd*

Table 2.1 *IR data of the synthetic ring A model lactones and the gibberellins*

	γ-lactones	δ-lactones	Gibberellins
$\nu_{C=O}$ (nujol) cm^{-1}	1742–1780	1730–1745	1746–1770
$\nu_{C=O}$ (dioxane) cm^{-1}	1775–1782	1730–1752	1777–1786

bonding, and the carbonyl absorption therefore appears at the normal position for γ-lactones. Gibberellins, as dioxane solutions, absorb at $\nu_{C=O} = 1777$–1780 cm^{-1}, which is the normal position for γ-lactones. Therefore, gibberellins must possess γ-lactone structure as shown in **A**, **C** and **D** of Figure 2.2. These IR spectral comparisons are summarized in Table 2.1.[11,12]

Subsequently, the ^{1}H NMR spectra of the model compounds were compared with those of the gibberellins. A 3H-singlet was observed in the cases of (±)-**2**, (±)-**4**, (±)-**5**, **9** and (±)-**10**, while a 3H-doublet was observed in the cases of (±)-**7** and (±)-**11**. Because the methyl group at C-4 of the gibberellins showed a 3H-singlet, the structure of the ring A could be neither **B** nor **D** in Figure 2.2. We therefore concluded the formula **C** of Cross to be correct, and proceeded to the next stage of gibberellin synthesis. A lesson learned through the study of ring A models was the importance of conditions for the measurement with analytical instruments. In the present case of IR spectroscopy, measurement as a solution or a nujol paste brought about a significant difference in $v_{C=O}$ values, which caused ambiguity in proposing a possible structure of the gibberellins.

2.1.2.2 Synthesis of (±)-epigibberic acid

Gibberellins and their degradation products possess characteristic tetracyclic carbon frameworks. I synthesized (±)-**14**,[13] (±)-**15**,[14] (±)-**16**,[15] and (±)-**17**[16] (Figure 2.4). In this section I will detail the synthesis of (±)-epigibberic acid (**18**), completed in October 1962, just a month before my wedding ceremony with my wife Keiko (*nee* Suzuki).[17,18]

(+)-Epigibberic acid (**18**) was obtained by Cross in 1954 by treating gibberellin A_3 with hot mineral acid. Its absolute configuration at C-9 is the same as that of the gibberellins. I therefore surmised that a synthesis of (±)-epigibberic acid (**18**) might be extended to the synthesis of the gibberellins. My idea as a young chemist with almost no experience was to synthesize the tetracyclic gibberellane system one by one, starting from the ring A. Figure 2.5 summarizes the synthesis of the rings A and B part of (±)-**18**.[19]

o-Xylene (**A**) was converted to anhydride **B** as shown in Figure 2.5. Intramolecular Friedel–Crafts acylation of **B** followed by esterification yielded bicyclic keto ester **C**. Methoxycarbonylation of **C** with sodium amide and dimethyl carbonate afforded β-keto ester **D**. In 1961 I had to prepare dimethyl carbonate by myself from phosgene and methanol, because it was commercially unavailable. Alkylation of **D** with methyl bromoacetate and sodium methoxide was followed by hydrolysis, decarboxylation and esterification to give diester **E** as crystals. Use of two lachrymatory substances, *o*-xylyl bromide and methyl bromoacetate, often made my neighbors as well as myself unhappy, but this process was confirmed to be an efficient and reproducible one by House in the USA.[20] Loewenthal in Israel reported another synthesis of **E**, in which he described it as an oil.[21]

Conversion of **E** to (±)-epigibberic acid (**18**) is shown in Figure 2.6. I attempted to construct the ring C of **F** by Robinson annulation reaction. Annulation of **E** with methyl isopropenyl ketone was first attempted in the presence of sodium amide in diethyl ether, which was the original condition as reported by Robinson in 1930s. The reaction, however, afforded a messy, brown-colored mixture of products, whose outlook was just like soil humic acid. One of my friends, a soil chemist, said to me, "Are you synthesizing soil organic matters?" Then, I repeated the reaction under much harsher conditions employing potassium *t*-butoxide.

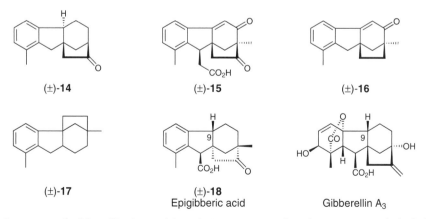

Figure 2.4 *Structures of gibberellin A_3 and its relatives. Reprinted with permission of Shokabo Publishing Co., Ltd*

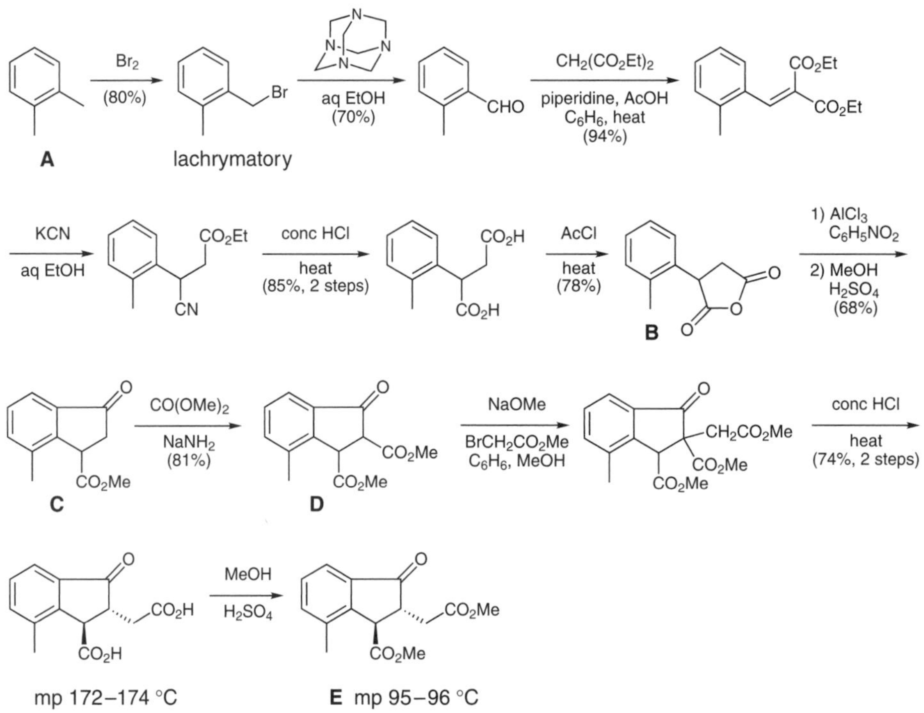

Figure 2.5 *Synthesis of (±)-epigibberic acid (1). Reprinted with permission of Shokabo Publishing Co., Ltd*

It was in vain. Only by using milder conditions, such as sodium methoxide in methanol, was the reaction successful. The product was a half-ester **F**, which must have been generated via lactone **L**.[cf. 21,22] Quite often a reaction can be successful only under mild conditions. This was a good lesson for me.

Alkaline hydrolysis of **F** furnished the corresponding diacid, whose treatment with acetic anhydride caused inversion of configuration of the carboxyl group on ring B to give anhydride **G**. This steric inversion was necessary to make the later hydrogenation step (**I** to **J**) stereoselective. The anhydride **G** was treated with boron trifluoride etherate to effect intramolecular acylation, yielding **H** after methylation. Subsequent protection of the carbonyl group in ring D proceeded in poor yield due to a side reaction (cleavage of the D-ring with generation of an ester **M**). The desired product **I** was hydrogenated over Raney nickel. Interaction of the less-hindered β-side of **I** with the catalyst resulted in the addition of hydrogen from the β-side only to give **J**. Oxidation of **J** yielded keto ester **K**. Wolff–Kishner reduction of **K** brought about reduction of the carbonyl group to a methylene group and also concomitant epimerization of the carboxyl group. Finally, acidification of the reaction mixture with dilute HCl afforded (±)-epigibberic acid (**18**) as crystals. Loewenthal of Technion, Haifa, completed his synthesis of (±)-gibberic acid slightly earlier in 1962 than we did, but I was happy to have been able to publish our preliminary communication about synthesis of (±)-**18** in the same year of 1962.

2.1.2.3   *Partial synthesis of gibberellin C*

Figure 2.7 shows my idea at the time when I finished the synthesis of (±)-epigibberic acid to convert it to gibberellin A₄. A phenolic compound **A** can be prepared from (±)-epigibberic acid (**18**). Reduction of the

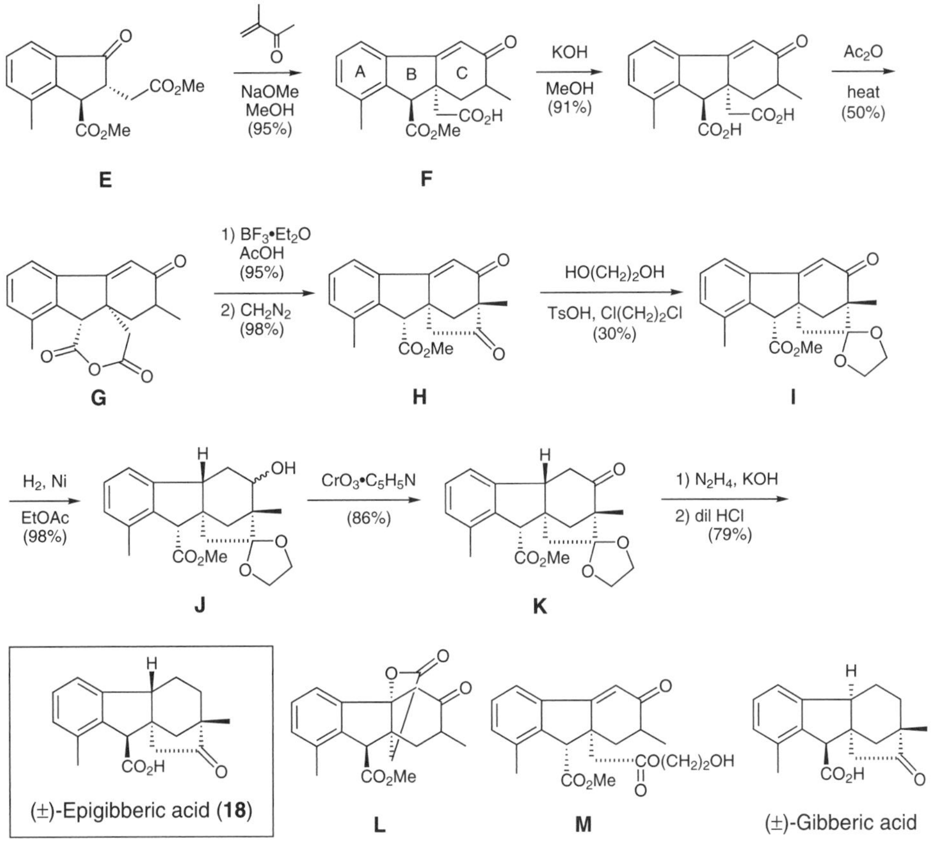

Figure 2.6 *Synthesis of (±)-epigibberic acid (2). Modified by permission of Shokabo Publishing Co., Ltd*

aromatic A-ring of **A** to give **B** must be feasible. Subsequent one-carbon addition at C-4 of **B** followed by functional group transformation will allow the synthesis of gibberellin C (**19**). This keto acid **19** was first obtained by Yabuta and Sumiki by acid treatment of gibberellin A_1 (see Figure 2.2), named gibberellin C, and showed about 1% of the plant-growth-promoting activity of gibberellin A_3. Acid-generated carbocation of gibberellin A_1 at C-16 gives **19** via Wagner–Meerwein rearrangement. This rearrangement can readily be understood by inspecting a molecular model of gibberellin A_1.

Availability of gibberellin A_3 as a commercial product of Kyowa Fermentation Industry Co. in Japan made me speculate that the phenol **A** in Figure 2.7 might be prepared by degradation of gibberellin A_3. If so, **A** would serve as a useful relay compound in our gibberellin synthesis. As shown in Figure 2.8, treatment of gibberellin A_3 with hot mineral acid generates gibberic acid and epigibberic acid via aromatization of ring A and Wagner–Meerwein rearrangement involving rings C and D. Keto lactone **A**, which could be prepared readily from gibberellin A_3 by methylation followed by MnO_2 oxidation, might give phenol **C** after acid treatment and methylation. My experimental result, however, was different. The product was a dienone ester **B**, as proved by its UV, IR and ^{1}H NMR data. The structure **B** made me examine the possibility of introducing a carboxy group at C-4 of **B**. This one-carbon addition might enable us to use **B** as the starting material for a partial synthesis of gibberellin C.

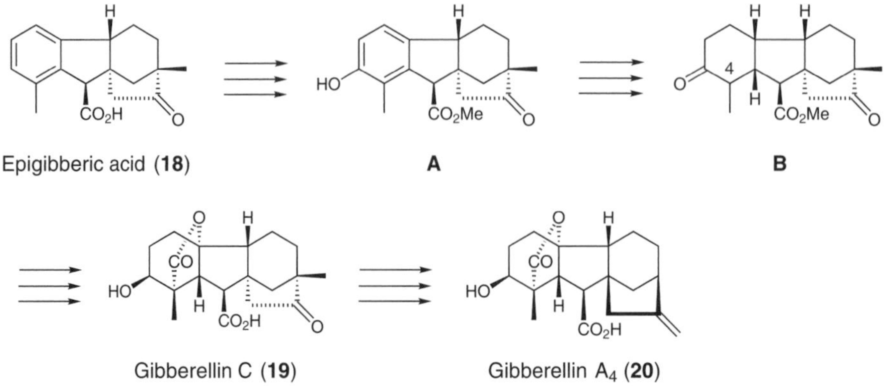

Epigibberic acid (**18**)   **A**   **B**

Gibberellin C (**19**)   Gibberellin A$_4$ (**20**)

Figure 2.7 *A plan for converting epigibberic acid to gibberellin A$_4$. Reprinted with permission of Shokabo Publishing Co., Ltd*

The diketo ester **B** was treated with ethylene glycol and *p*-toluenesulfonic acid to give monoacetal **D**, which served as the substrate for one-carbon introduction at C-4. Carboxylation of **D** was attempted under various different conditions employing bases and carboxylating agents available in 1964. As shown in Figure 2.8, a combination of triphenylmethylsodium and carbon dioxide gave a positive result. The resulting acid was methylated, and the product was treated with dilute hydrochloric acid to remove the ethylene acetal protective group on ring D. The desired product **E** was isolated after chromatographic purification (by preparative GC in initial experiments and by conventional SiO$_2$ chromatography in later experiments) as crystals. Subsequent reprotection of the ring D-ketone afforded **F**, whose reduction with sodium borohydride was followed by catalytic hydrogenation to give **G**. The lactone in ring A was constructed by treating **G** with dilute sulfuric acid. After methylation of the ring B carboxyl group, **H** was obtained. Treatment of **H** with dilute base caused epimerization at C-3 by way of retroaldol/aldol mechanism, yielding **I**.[23] Finally, acid hydrolysis of **I** furnished gibberellin C (**19**).[24,25]

Gibberellin C with plant-growth-promoting activity was thus synthesized from **B**. Preparation of **B**, however, depended on not synthesis but degradation of gibberellin A$_3$. The present synthesis of gibberellin C was therefore a partial synthesis, not a total synthesis. In the past, there were many partial syntheses of complicated natural products owing to our incapability to achieve de novo synthesis of the intermediates (relay compounds). Nowadays, there exist only very few partial syntheses, and almost all the syntheses are total syntheses. Since 1964, I concentrated my efforts to synthesize **B** from epigibberic acid, and also to connect all the intermediates from xylene to gibberellin A$_4$ on a single line.

2.1.2.4 *Formal total synthesis of gibberellin A$_4$*

In order to convert epigibberic acid (**18**) into the diketo ester **B**, the aromatic ring A of **18** must be hydrogenated into a cyclohexane ring. After extensive experimentation,[26] the conversion could be achieved as shown in Figure 2.9.[25,27,28] Classical electrophilic nitration of methyl epigibberate was followed by reduction and diazotization, and the resultant diazonium salt was hydrolysed to give phenol **A**. Hydrogenation of the phenol ring over RhO$_2$/PtO$_2$ in acetic acid was followed by oxidation to give diketo ester **C**, which gave unsaturated diketo ester **D** after 4 steps.

The diketo ester **D** was then converted to **B**. In the course of this work, both **C** and **D** could be provided by degradation of commercially available gibberellin A$_3$. The relay synthesis of gibberellin A$_4$ was completed in December 1967 just before Professor Yabuta's 80th birthday. Since the conversion of

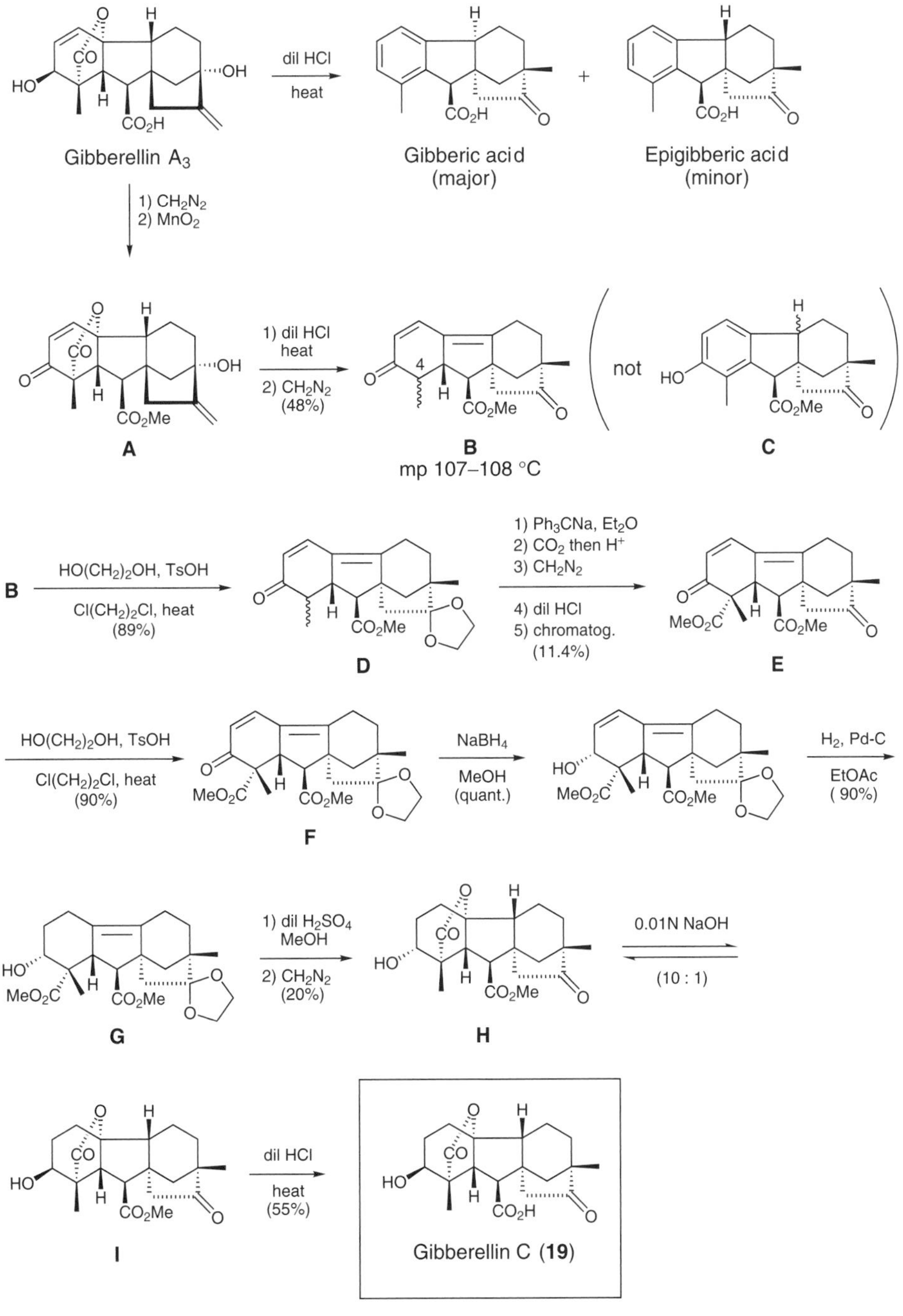

Figure 2.8 *Partial synthesis of gibberellin C. Modified by permission of Shokabo Publishing Co., Ltd*

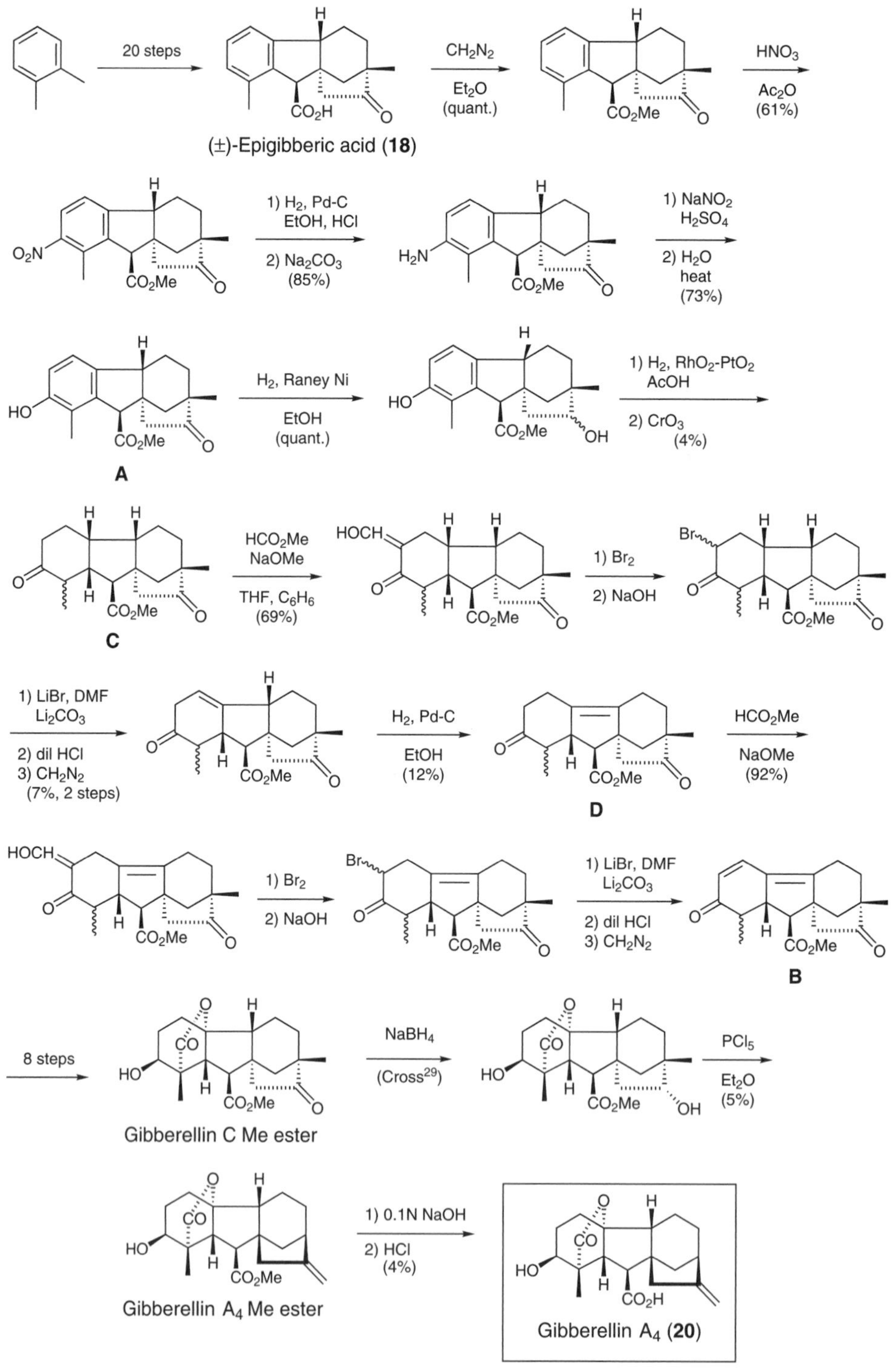

Figure 2.9   *Synthesis of gibberellin A₄. Modified by permission of Shokabo Publishing Co., Ltd*

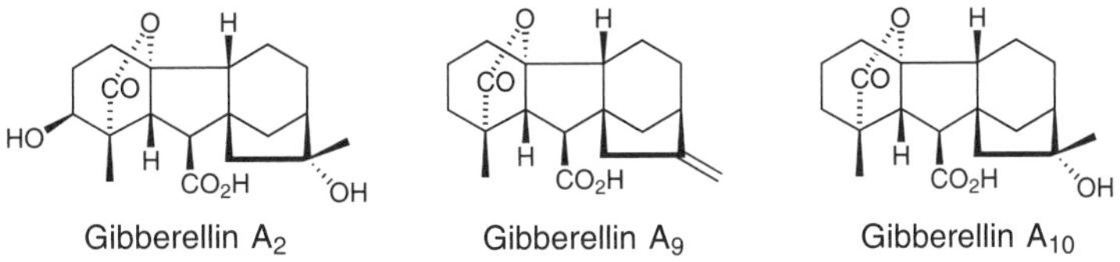

Gibberellin A$_2$ Gibberellin A$_9$ Gibberellin A$_{10}$

Figure 2.9   *(continued)*

gibberelin C methyl ester to gibberellin A$_4$ methyl ester had already been accomplished by Cross *et al.*,[29] the final step that was done by my own hands was the hydrolysis of gibberellin A$_4$ methyl ester to gibberellin A$_4$ (**20**). Previously reported conversion of gibberellin A$_4$ to gibberellins A$_2$, A$_9$ and A$_{10}$ made the present synthesis of gibberellin A$_4$ as the formal total synthesis of the latter, too. In July 1968, just before writing up the full paper of this work, I was promoted to Associate Professor.

When I finished the gibberellin synthesis in 1968, Professor K. Arima, a famous Professor of Microbiology in our Department, said to me. "Congratulations, Dr. Mori, on the completion of the gibberellin synthesis. But you spent 9 years of your life to do it. Don't forget that the fungus *Gibberella fujikuroi* makes the gibberellins within a couple of days." This criticism made me think that we chemists can be respected by biologists only when we synthesize those compounds that are difficult to prepare by biological systems.

2.1.2.5   *Other works related to gibberellins*

There is a bridgehead hydroxy group in gibberellin A$_3$ at C-13 between rings C and D, while gibberellin A$_4$ does not possess it. I wanted to synthesize this type of compounds with a bridgehead hydroxy group at C-13. (+)-Epiallogibberic acid (**21**) was obtained by heating gibberellin A$_3$ with hydrazine hydrate, and possesses a hydroxy group at C-13. I chose **21** as my target, and synthesized (−)-epiallogibberic acid (**21′**), as shown in Figure 2.10.[30−32]

The key step was the reductive rearrangement of diketone **A** to hydroxy ketone **B** with a hydroxy group at C-13. A possible intermediate leading to **B** was thought to be a strained diol **A*** generated from **A** by reductive C−C bond formation presumably via radical anion. Cleavage of the cyclopropane ring of **A*** gave three products **B**, **C** and **D**, which were isolated in the yields depicted in Figure 2.10.

Gibberellin A$_{12}$ (**22**) is the simplest one among all the C$_{20}$-gibberellins. Its synthesis was achieved as shown in Figure 2.11.[33,34] As to the synthesis of the key intermediate (±)-**A**, it will be detailed in Figure 2.15 in connection with the synthesis of a diterpene called steviol. Our synthesis of gibberellin A$_{12}$ (**22**) was designed on the basis of its biosynthetic pathway, formally yielding (±)-gibberellin A$_{12}$. Unfortunately, conversion of (±)-**B** to (±)-**C** proceeded in a miserable yield of only 4.5%, and therefore the efficiency of synthesis was extremely low. Conversion of **E** to **F** had been done by Galt and Hanson,[35] and **F** was methylenated by us to give **G**. Cross and Norton converted **G** into gibberellin A$_{12}$.[36] Since we did not resolve the racemic intermediates, our synthesis was a formal total synthesis of (±)-gibberellin A$_{12}$.

A number of research groups worked on the total synthesis of gibberellins. The works of W. Nagata, E.J. Corey and L.N. Mander were reviewed by Mander.[37] Other notable syntheses were Nagaoka's work on (±)-gibberellin A$_3$[38] and Toyota and Ihara's work on gibberellin A$_{12}$.[39]

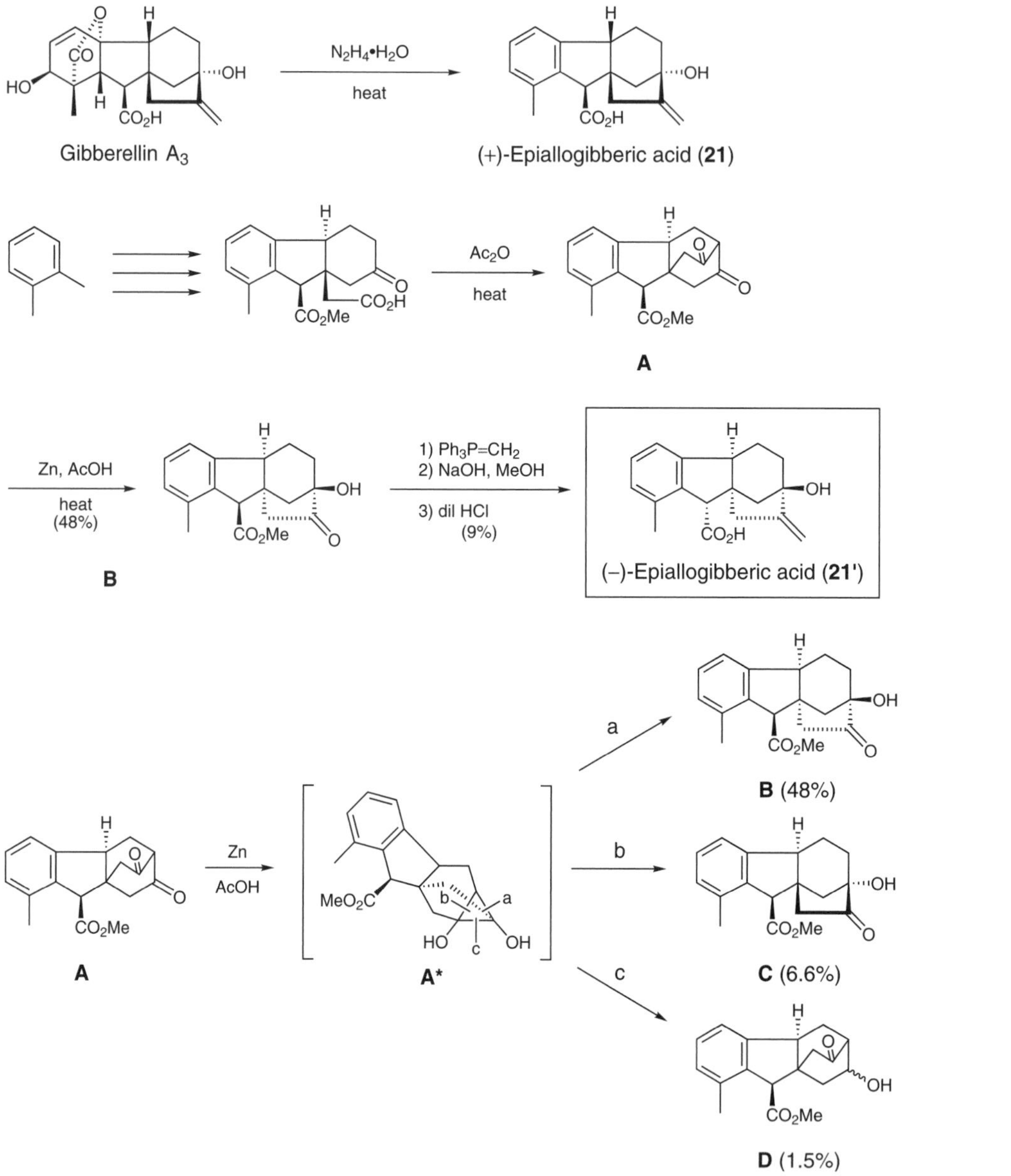

Figure 2.10   *Synthesis of epiallogibberic acid. Modified by permission of Shokabo Publishing Co., Ltd*

2.1.3   Diterpenes related to gibberellins

2.1.3.1   *Structures of diterpenes related to gibberellins*

Gibberellins belong to tetracyclic diterpenoids. (−)-Kaur-16-en-19-oic acid (**23**) and (−)-kaur-16-en-19-ol (**24**) were first isolated in 1963 by Jefferies and coworkers from Australian shrub *Ricinocarpus stylosus*,

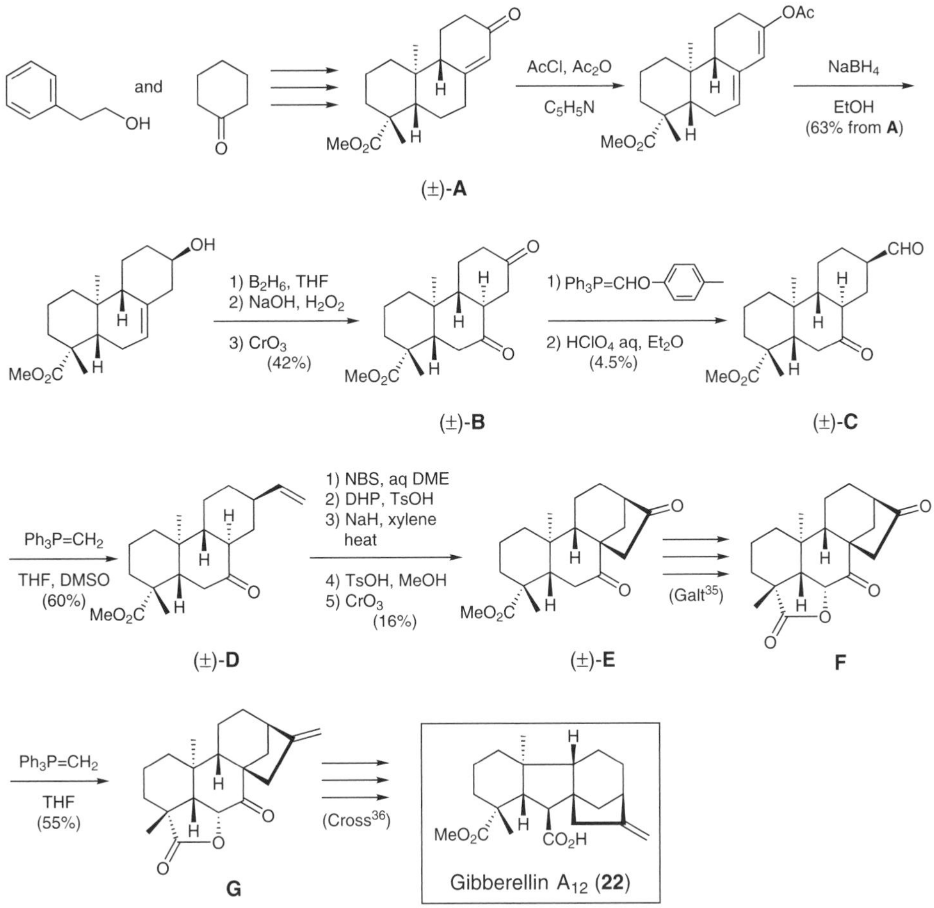

Figure 2.11 *Synthesis of gibberellin A_{12}. Reprinted with permission of Shokabo Publishing Co., Ltd*

and later isolated from barley and also from the culture broth of *Gibberella fujikuroi*. Their plant-growth-promoting activity was discovered in 1966 by Katsumi, and they were shown to be the intermediates in gibberellin biosynthesis.

Quite independently from gibberellin research, stevioside was known since 1931 as a sweet glycoside from *Stevia rebaudiana*, a plant of Compositae. It is about 300 times sweeter than sucrose, and has been used by people in Paraguay as a sweetener. It is now used world-wide as a low-calorie sweetener. Mosettig clarified its structure, in 1963, as depicted in Figure 2.12. In the same year of 1963, Ruddat found its aglycone, steviol (**25**), to be bioactive as a plant-growth promoter. Steviol (**25**), like gibberellins A_1 and A_3, possesses a bridgehead hydroxy group at C-13. I therefore started my synthetic works on these tetracyclic diterpenes. There is a good review on the synthesis of tri- and tetracyclic diterpenoids.[40]

2.1.3.2 Synthesis of (±)-kaur-16-en-19-oic acid and (±)-kaur-16-en-19-ol

I worked for two years (1964–1966) to synthesize (±)-kaur-16-en-19-oic acid (**23**) and (±)-kaur-16-en-19-ol (**24**), as shown in Figures 2.13 and 2.14.[41–43]

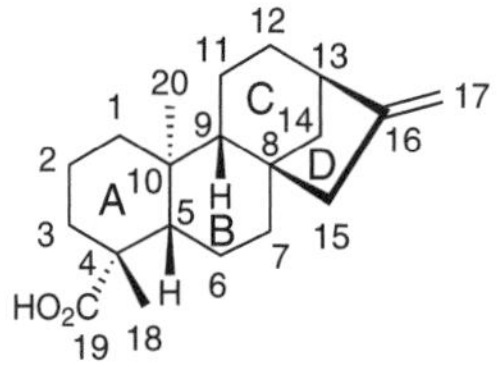

(−)-Kaur-16-en-19-oic acid **(23)** 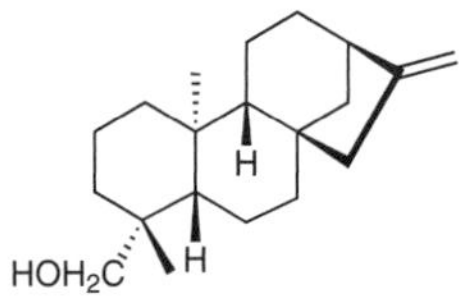   (−)-Kaur-16-en-19-ol **(24)**

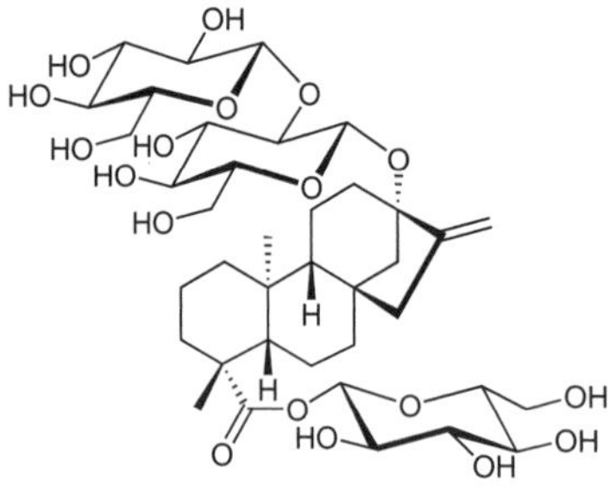

Stevioside 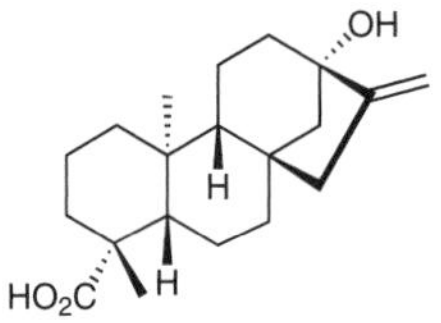   Steviol **(25)**

Figure 2.12   *Structures of diterpenes related to gibberellins. Reprinted with permission of Shokabo Publishing Co., Ltd*

Methyl acroylacetate (Nazarov's ester, **A** in Figure 2.13) was one of the starting materials. This ester **A** was prepared from ethyl acrylate, ethanol and methyl acetoacetate via methyl 5-ethoxy-3-oxopentanoate. Robinson annulation between **A** and (±)-**B** to give **C** was the first key reaction. Tetralone (±)-**B** was prepared from naphthalene. Methylation of (±)-**C** with methyl iodide and potassium *t*-butoxide stereoselectively gave (±)-**D**, whose carbonyl group at C-3 and the double bond at C-5 were reduced to furnish (±)-**E**. Catalytic hydrogenation of (±)-**F** over Raney nickel was followed by chromic-acid oxidation of the resulting alcohol to give keto ester (±)-**G**. Intramolecular aldol reaction of keto aldehyde (±)-**H** afforded tetracyclic aldol (±)-**I**, which was oxidized to give nicely crystalline diketo ester (±)-**J**. Wittig methylenation of (±)-**J** proceeded regioselectively to give (±)-**K**. Finally, Wolff–Kishner reduction of (±)-**K** afforded 58 mg of (±)-kaur-16-en-19-oic acid **(23)**.[41,43] I then slightly modified the synthetic route, and prepared 218 mg of (±)-kaur-16-en-19-ol **(24)**, as shown in Figure 2.14.[42,43]

Plant-growth-promoting activity of synthetic (±)-**24** was compared with that of natural (−)-**24** by Professor M. Katsumi, using dwarf maize d$_5$. The synthetic (±)-**24** was exactly half as active as the natural (−)-**24**.[43] This result indicated that unnatural (+)-**24** showed no bioactivity at all. A half of the synthetic (±)-**24** was biologically useless after such a lengthy and multistep synthetic transformations. This was a very good lesson given to me. Through this experience I later devoted myself to enantioselective syntheses of bioactive natural products so that I could avoid the wasteful synthesis of inactive enantiomers.

2.1.3.3   *Synthesis of (±)-steviol*

It took seven years (1964–1970) to complete the synthesis of (±)-steviol **(25)**.[44–48] A plausible method had to be developed for the construction of the C/D ring system with a bridgehead hydroxy group at C-13. In addition, there was a need to develop an efficient method to supply sufficient amounts of intermediates so that we could finally reach the target (±)-**25**.

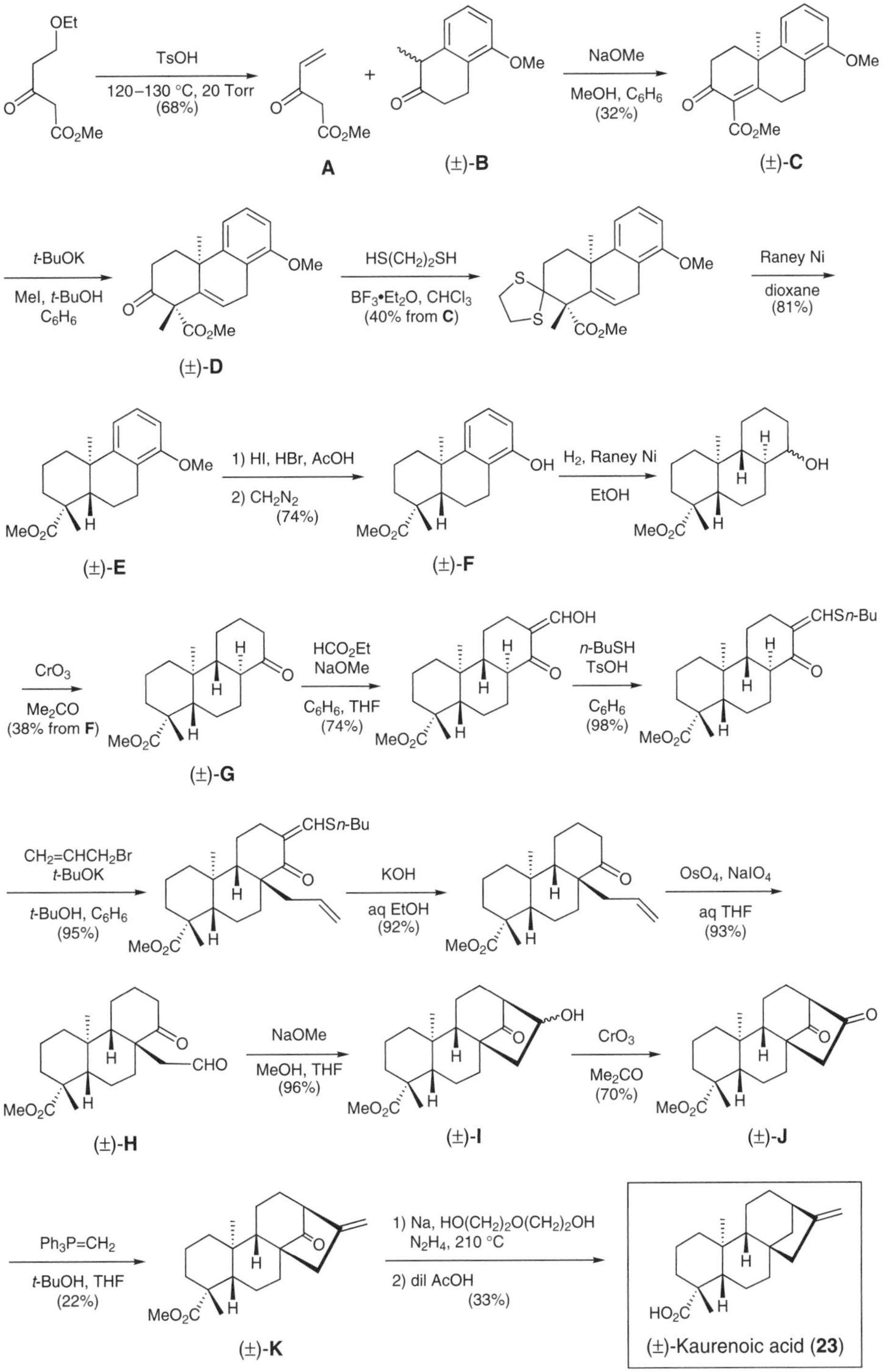

Figure 2.13   *Synthesis of (±)-kaur-16-en-19-oic acid and (±)-kaur-16-en-19-ol (1). Reprinted with permission of Shokabo Publishing Co., Ltd*

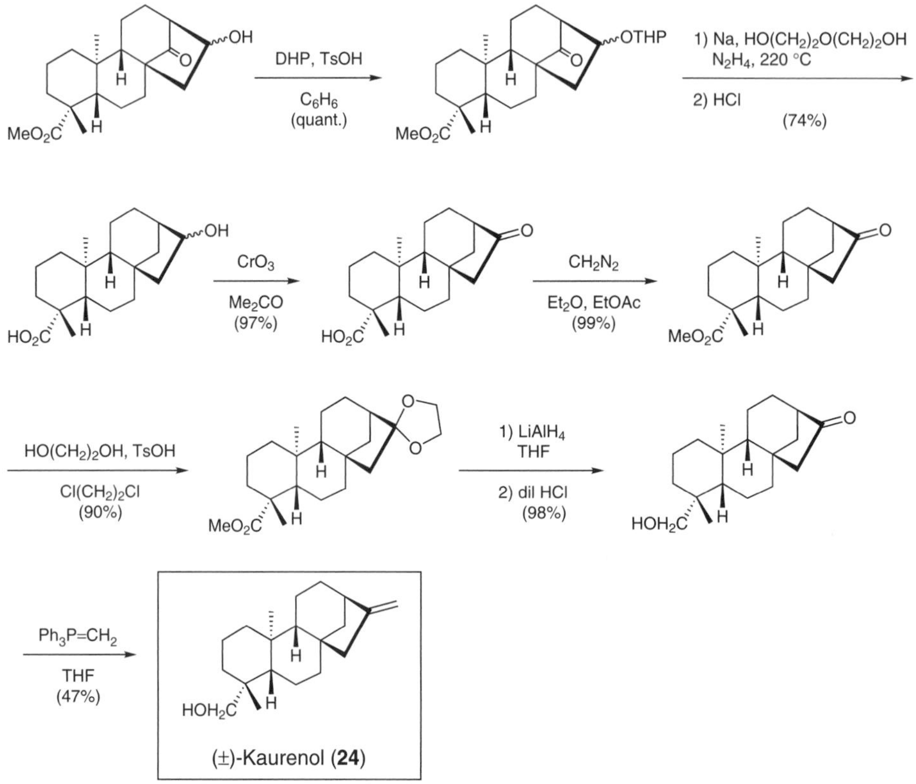

Figure 2.14 *Synthesis of (±)-kaur-16-en-19-oic acid and (±)-kaur-16-en-19-ol (2). Reprinted with permission of Shokabo Publishing Co., Ltd*

In 1965 I synthesized (±)-**L** in Figure 2.15, which was known as the ozonolysis product of steviol methyl ester.[44] A large-scale preparative method for (±)-**H** was published in 1966.[45] Unlike the contemporary way of natural products synthesis, there was no sophisticated method like transition-metal-catalysed reactions. I therefore started my synthesis of (±)-**L** in a considerably large scale. My experimental records indicate that 745 g of (±)-**C** was secured starting from 795 g of (±)-**B**. The low-yielding pivotal step was cyclization of (±)-**D** to give crystalline (±)-desoxypodocarpic acid (**E**) in only 23% yield. Chromic-acid oxidation of (±)-**E** methyl ester afforded (±)-**F**, whose nitration was regioselective to give (±)-**G**. Conventional functional group manipulation of (±)-**G** gave (±)-**H**, which was subjected to Birch reduction to give (±)-**I**. Claisen rearrangement of allyl vinyl ether (±)-**J** gave (±)-**K**, which was oxidized to keto acid (±)-**L**. Synthetic (±)-**L** was identified with the ozonolysis product of steviol methyl ester by spectral comparisons.[44,46]

Figure 2.16 summarizes our synthesis of (±)-steviol (**25**).[47,48] The unsaturated aldehyde (±)-**A** [obtained by Claisen rearrangement of (±)-**J** in Figure 2.15] was converted to (±)-**B**, whose hydroboration-oxidation afforded (±)-**C** in 62% yield, together with the minor product (±)-**K** (13%). Acid treatment of (±)-**C** effected hydrolysis of the acetal and intramolecular aldol reaction to give aldol (±)-**D** as an epimeric mixture. Chromic-acid oxidation of (±)-**D** afforded diketone (±)-**E**. Treatment of (±)-**E** with zinc amalgam and hydrochloric acid gave the desired hydroxy ketone (±)-**G** (41%) as the major product together with its stereoisomer (±)-**F** (19%). Wittig methylenation of (±)-**G** was followed by conversion of ester (±)-**H** to acid, giving (±)-steviol (**25**). Synthesis of (±)-steviol starting from (±)-**C** was also achieved via (±)-**I** and

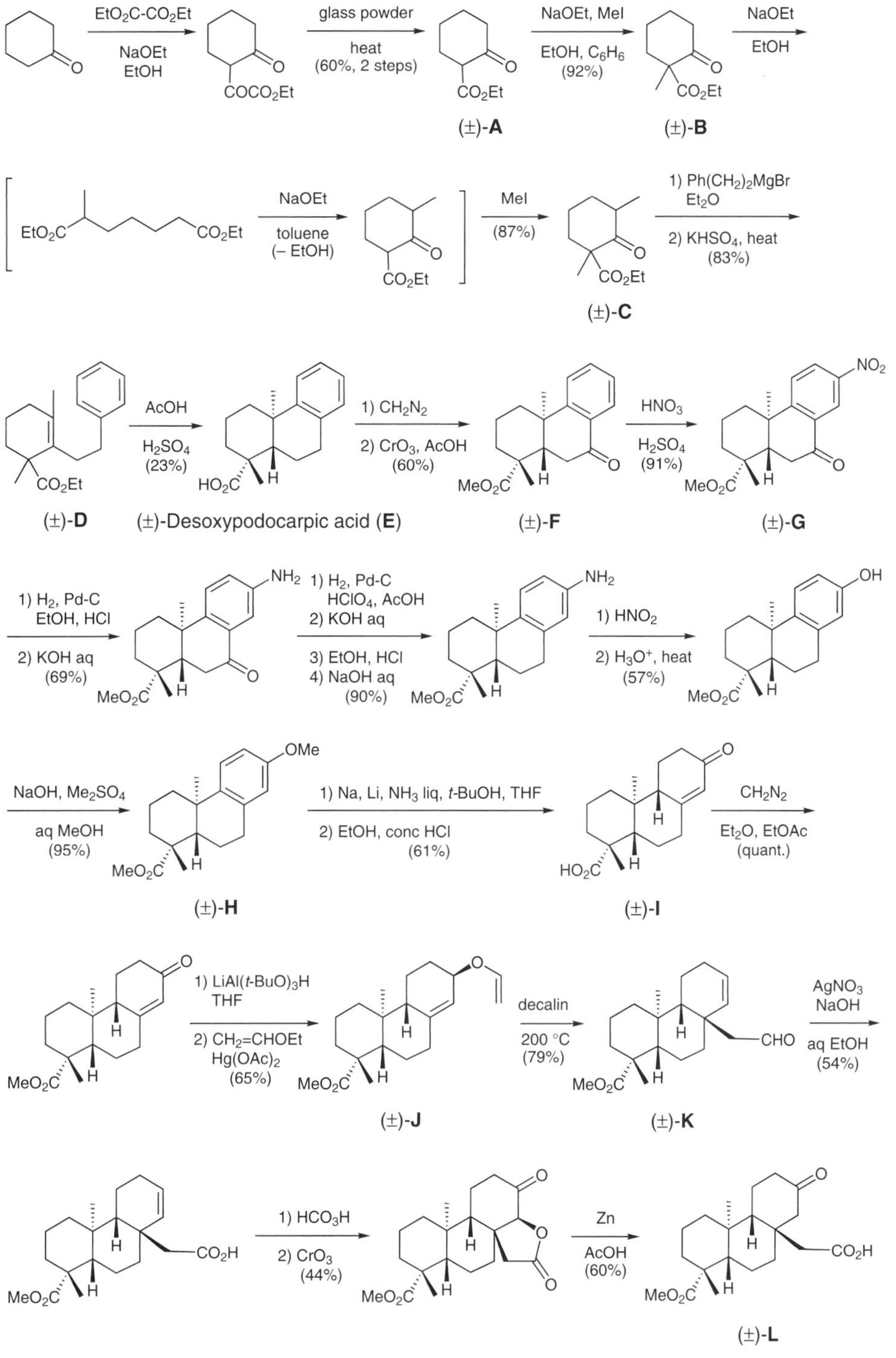

Figure 2.15   *Synthesis of (±)-steviol (1). Modified by permission of Shokabo Publishing Co., Ltd*

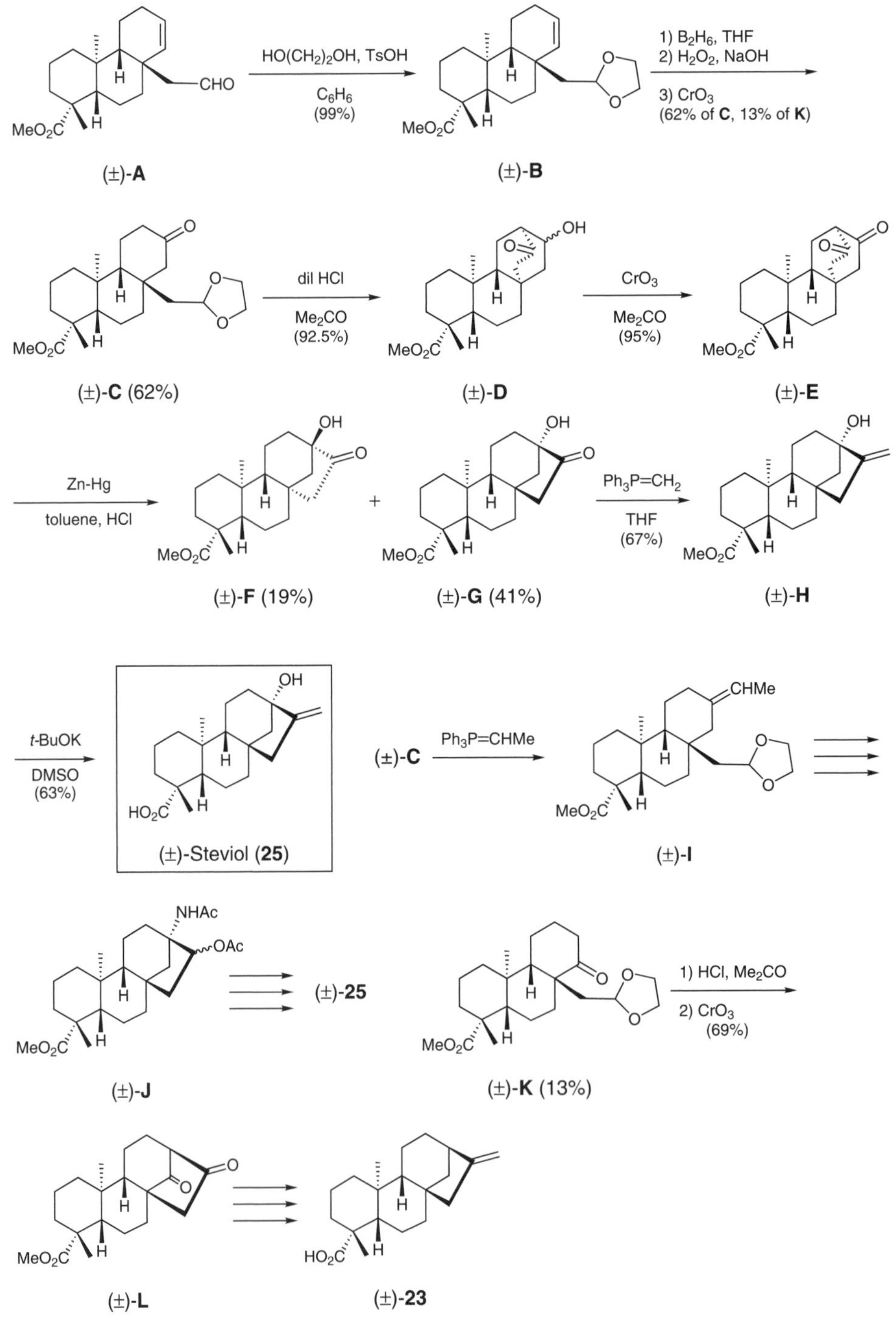

Figure 2.16 *Synthesis of (±)-steviol (2). Reprinted with permission of Shokabo Publishing Co., Ltd*

$(\pm)$-**J**.[46] $(\pm)$-Kaur-16-en-19-oic acid (**23**) was synthesized from the minor product $(\pm)$-**K** of hydroboration-oxidation via $(\pm)$-**L**.[46]

2.1.3.4 Synthesis of (±)-tripterifordin

Tripterifordin (**26**) is an anti-HIV diterpene isolated in 1992 from a Chinese plant *Tripterygium wilfordii* by Lee and coworkers. Although it is not a plant hormone, its structural similarity with gibberellins attracted me to synthesize it.

In 1966 I attempted the synthesis of the δ-lactone system of ring A of gibberellin A$_{15}$. As a part of that work, I synthesized pentacyclic keto lactone $(\pm)$-**C** (Figure 2.17) by starting from $(\pm)$-**A**, whose synthesis was shown in Figure 2.14.[49] Amide $(\pm)$-**B**, obtained from $(\pm)$-**A**, was subjected to photolysis in the presence of iodine and lead tetraacetate in benzene. Subsequent manipulation yielded $(\pm)$-**C** in 20% yield.[50] The crystals of $(\pm)$-**C** were kept in my sample box, and were taken out after 26 years. Methylenation of **C** was followed by hydration of the resulting olefin to give $(\pm)$-tripterifordin (**26**).[51] A lesson learned through this work was the benefit of keeping the synthetic samples in a good shape so that we can use them even after many years.

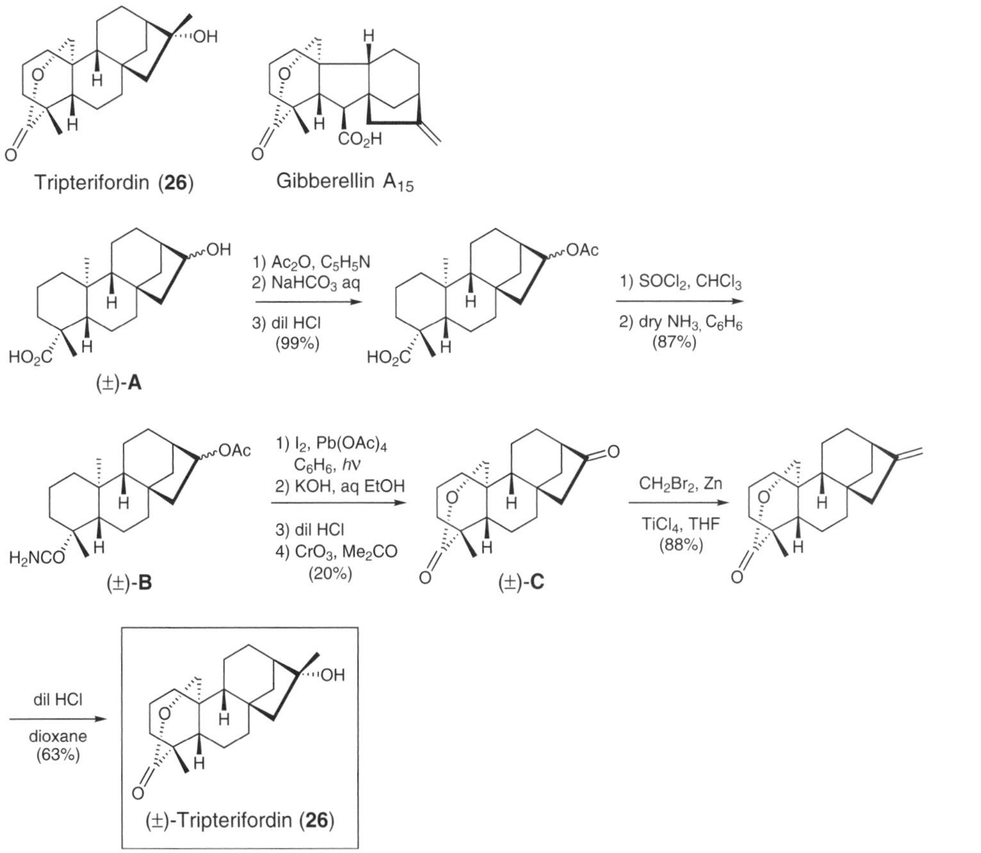

Figure 2.17 *Synthesis of (±)-tripterifordin. Reprinted with permission of Shokabo Publishing Co., Ltd*

2.1.4   Abscisic acid and its relatives—synthesis of optically active compounds

2.1.4.1   *Stereochemical problems on abscisic acid*

Because naturally occurring (+)-abscisic acid (**27**) did not give crystals good enough for X-ray crystallographic analysis, it was rather difficult to determine its absolute configuration.

Mills's empirical rule was known to correlate the absolute configuration of 2-cyclohexen-1-ols with their magnitudes of specific rotations. Thus, (*S*)-**A** of Figure 2.18 is more levorotatory than (*R*)-**B**, and their molar rotations differ very much ($\Delta[M]_D = 100–400°$). In 1967, Cornforth proposed the absolute configuration of (+)-abscisic acid (**27**) as *R* on the basis of Mills's rule.[52] Reduction of methyl (+)-absciscate (**C**) with sodium borohydride gave two hydroxy esters **D** and **E**, which could be separated by preparative TLC. The ester **E** was identified with (±)-**E** obtained by reduction of (±)-**F** basing on comparisons by TLC and MS. Because (±)-**E** was prepared from (±)-**F**, its two hydroxy groups must be in *cis*-relationship. Accordingly, **E** must be the *cis*-dihydroxy ester, while **D** must be the *trans* one. The optically active **E** derived from (+)-abscisic acid showed $[M]_D = +390°$, while **D** showed $[M]_D = +990°$. Since **E** with *cis*-hydroxy groups was more levorotatory, its secondary hydroxy group must be β-oriented, and the absolute configuration of (+)-abscisic acid (**27**) was claimed as *R*.[52] Initially, this proposal was accepted by everyone, because J.W. Cornforth was a Nobel Prize winner.

In 1970, however, Burden and Taylor reported the following experimental results, which were in conflict with Cornforth's proposal.[53] They oxidized a carotenoid violaxanthin (**A** in Figure 2.19), and obtained xanthoxin (**B**), which was a strong plant-growth retardant like (+)-abscisic acid. They also observed that further oxidation of **B** gave (+)-*trans*-abscisic acid.[53] The absolute configuration of violaxanthin (**A**) was known as depicted on the basis of an X-ray analysis of a related compound. Then, *trans*-abscisic acid obtained by oxidation of violaxanthin must be with *S*-configuration showing negative rotation. In reality (+)-*trans*-abscisic acid was obtained. Accordingly, Burden and Taylor challenged Cornforth, and proposed *S*-configuration to (+)-abscisic acid. Mills' rule is an empirical rule and it may not be applicable to **D** and **E** of Figure 2.18. Any empirical rule sometimes shows this kind of limitation.

Ryback, an associate of Cornforth, solved the problem by determining the absolute configuration of a degradation product of the naturally occurring (+)-abscisic acid as shown in the bottom of Figure 2.19.[54] He first converted (+)-abscisic acid to **C**. Ozonolysis of **C** yielded **D** after further oxidation and methylation. Then, **D** was synthesized from (*S*)-malic acid. The synthetic **D** showed the same optical rotation as that of **D** obtained from (+)-abscisic acid. Therefore, *S*-configuration was assigned to (+)-abscisic acid.[54]

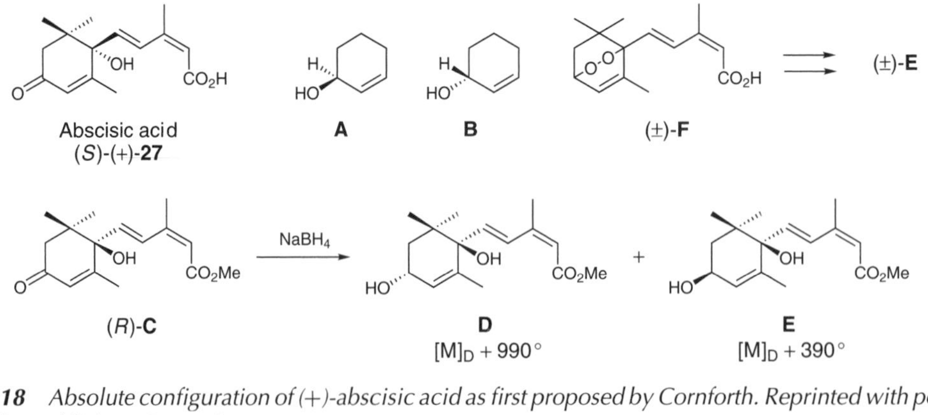

Figure 2.18   *Absolute configuration of (+)-abscisic acid as first proposed by Cornforth. Reprinted with permission of Shokabo Publishing Co., Ltd*

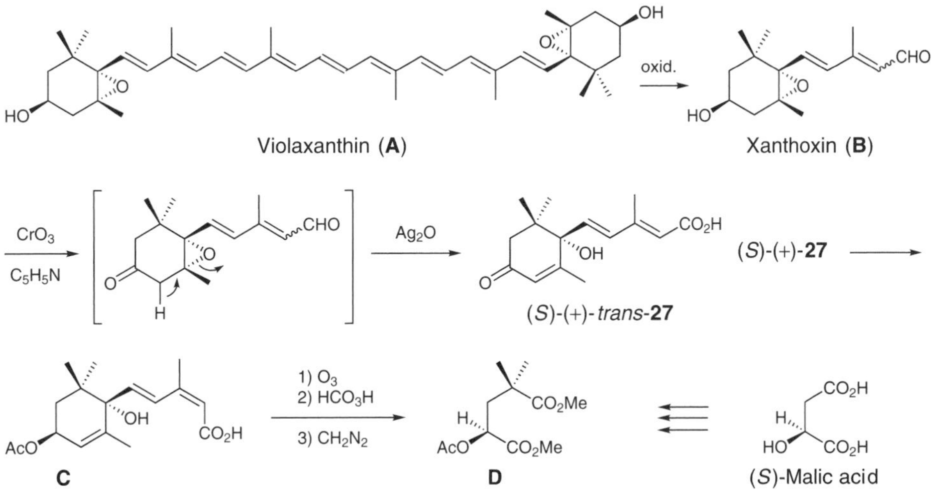

Figure 2.19 *Experiments by Burden and Ryback to clarify the absolute configuration of (+)-abscisic acid. Reprinted with permission of Shokabo Publishing Co., Ltd*

Almost at the same time Harada reported the *S*-configuration of (+)-abscisic acid on the basis of a theoretical calculation employing his exciton-chirality method.[55]

2.1.4.2 *Synthetic approaches to assign the S-configuration to (+)-abscisic acid*

Three different groups in Japan and the USA used synthetic methods to clarify the absolute configuration of (+)-abscisic acid (**27**). Oritani and Yamashita were the first to reach the correct conclusion,[56] and Koreeda *et al.* soon reached the same conclusion.[57] I will describe my own solution to this problem.[58–60] The key compound in my work was ketone **A** of Figure 2.20, which was synthesized in an optically active form. Ketone **A** was then converted to optically active grasshopper ketone (**28**). This allenic ketone **28** was isolated as a defense substance of the North American grasshopper to repel other insects, and its absolute configuration had been established by X-ray analysis. This correlation would determine the absolute configuration of the starting ketone **A** beyond doubt. Subsequently, the ketone **A** was converted via **G** to (*S*)-(+)-dehydrovomifoliol (**29**), which had been converted to (*S*)-(+)-abscisic acid (**27**) by Masamune and coworkers.[61] Therefore, my synthesis correlates abscisic acid (**27**) with grasshopper ketone (**28**), and it firmly establishes the absolute configuration of (+)-abscisic acid (**27**) on the basis of the X-ray analysis of grasshopper ketone (**28**).

My idea to employ ketone **A** as the key compound came out by chance. Ketone (±)-**D** was obtained by hydrogenation of diketone **B** over platinum oxide in methanol. A Swiss patent claimed the hydrogenation product to be **C**. It was untrue, and the actual product was hydroxy ketone (±)-**D**. The reason why **D** is generated is still unclear to me. Optical resolution of (±)-**D** could be achieved by converting (±)-**D** into its ester **E** with 3β-acetoxyetienic acid. The ester **E** was highly crystalline and could be purified by chromatography and recrystallization. Reduction of **E** with lithium tri(*t*-butoxy)aluminum hydride stereoselectively generated axial alcohol **F**. The axial hydroxy group of **F** was protected as the corresponding THP ether, which was reduced with lithium aluminum hydride. Oxidation of the resulting alcohol yielded the ketone **A**, which showed a positive Cotton effect in its CD spectrum. Since the ketone **A** could be converted to the natural enantiomer of grasshopper ketone (**28**), its absolute configuration must be as

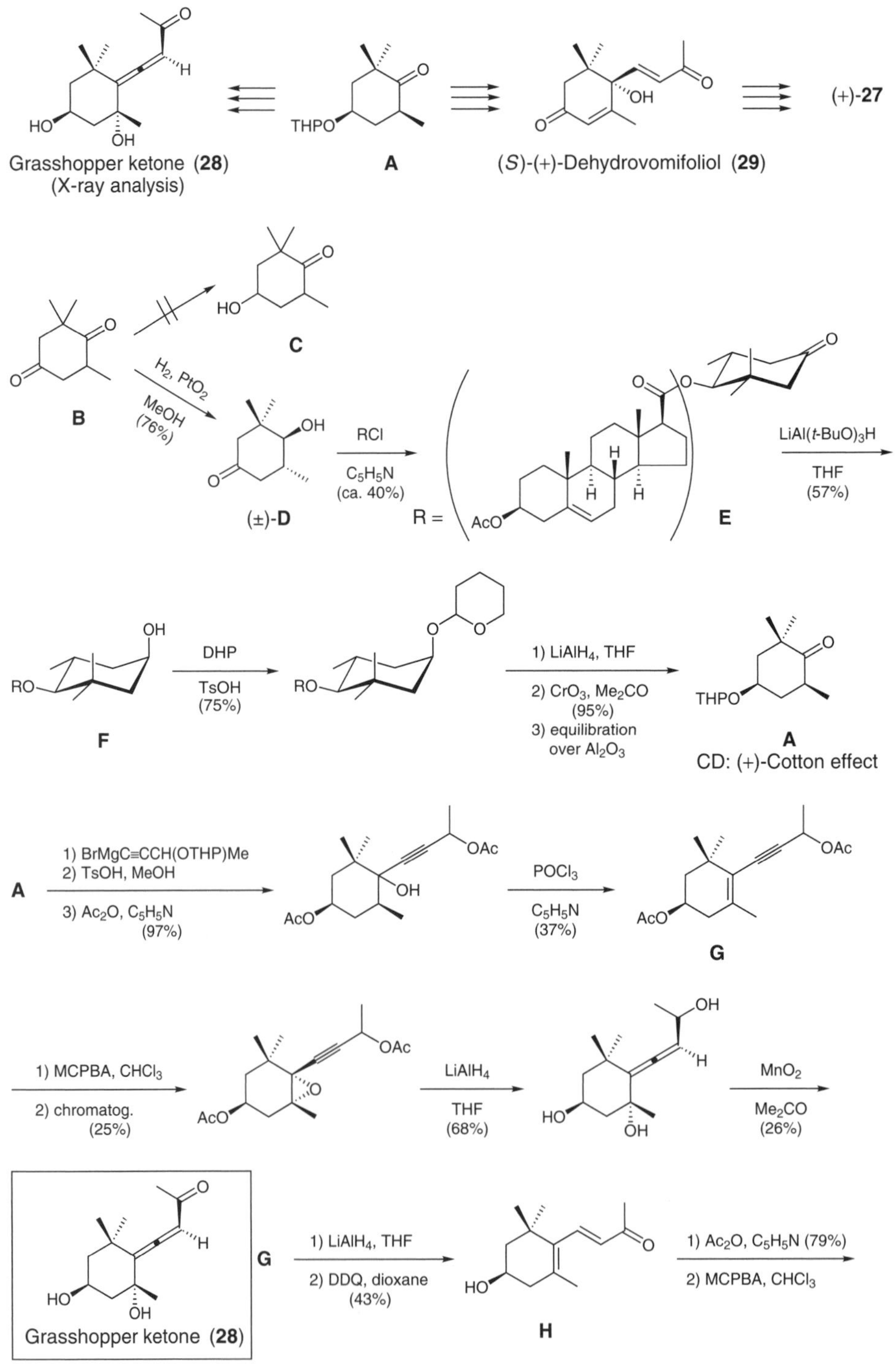

Figure 2.20 *Synthesis of the enantiomers of dehydrovomifoliol. Reprinted with permission of Shokabo Publishing Co., Ltd*

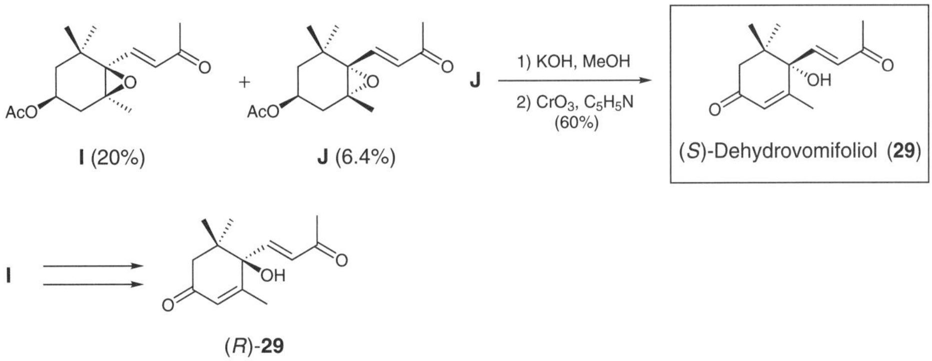

Figure 2.20 *(continued)*

depicted in **A**. Then, **G**, an intermediate on the way to **28**, was converted to **H**. The corresponding acetate was epoxidized with *m*-chloroperbenzoic acid to give epoxides **I** and **J**. The epoxide **J** was converted to (*S*)-dehydrovomifoliol (**29**), while **I** yielded (*R*)-**29**. Accordingly, the absolute configuration of (+)-abscisic acid (**27**) was confirmed to be *S*.[60]

2.1.5 Brassinosteroids

2.1.5.1 Structures of brassinosteroids

In 1979, Grove and coworkers isolated 4 mg of crystalline brassinolide from 40 kg of the pollen of rape plant *Brassica napus*, and determined its structure as shown in **30** (Figure 2.21) by X-ray analysis.[6]

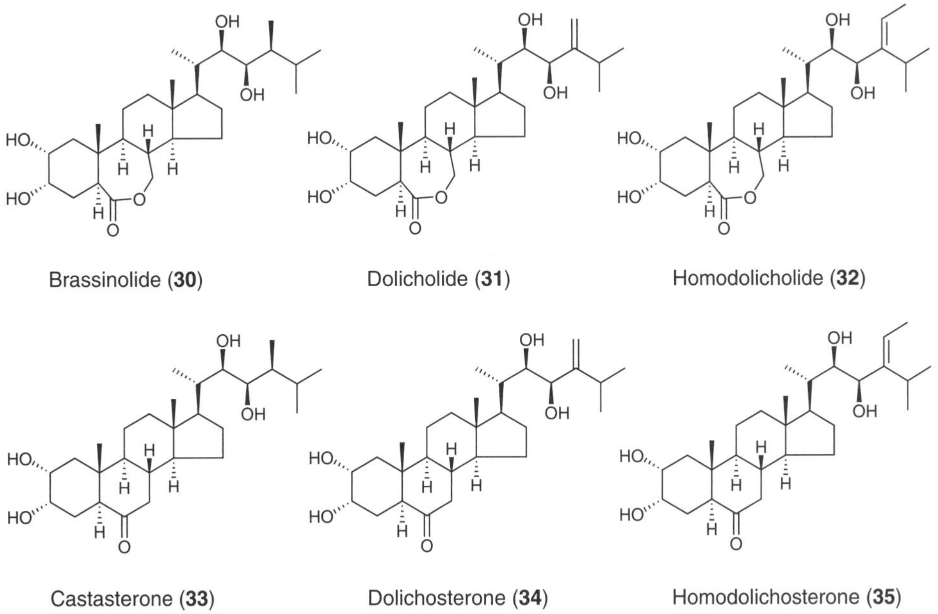

Figure 2.21 *Major members of the naturally occurring brassinosteroids. Reprinted with permission of Shokabo Publishing Co., Ltd*

The first synthesis of brassinolide (**30**) was achieved by Fung and Siddall in 1980,[62] and also by Ikekawa and coworkers in the same year.[63] Since then, many plant-growth promoters related to **30** with steroidal structures have been isolated from plants (Figure 2.21). They are called brassinosteroids. The early phase of our synthetic works on brassinosteroids has been reviewed.[64]

2.1.5.2   *Synthesis of (22S,23S,24S)-28-homobrassinolide*

In late 1979, Professor Marumo of Nagoya University requested me to synthesize brassinolide or its analog as soon as possible. I therefore began the synthesis of (22S,23S,24S)-28-homobrassinolide (**36**) by myself as shown in Figure 2.22.[65] Stigmasterol, an abundant phytosterol of soybeans, was employed as the starting material, and the product **36** showed plant-growth-promoting activity, although it was about 1% of brassinolide itself. Hydroxylation of **A** to give **B** proceeded in a stepwise manner. The double bond in ring A was hydroxylated quickly, and that in the side-chain could be hydroxylated only very sluggishly to give tetraol **B** in 52% yield after four days. The absolute configuration of the two hydroxy-bearing carbons in the side-chain was 22S,23S as determined by X-ray analysis of the corresponding bis-acetonide. Brassinolide (**30**) possesses 22R,23R-stereochemistry.

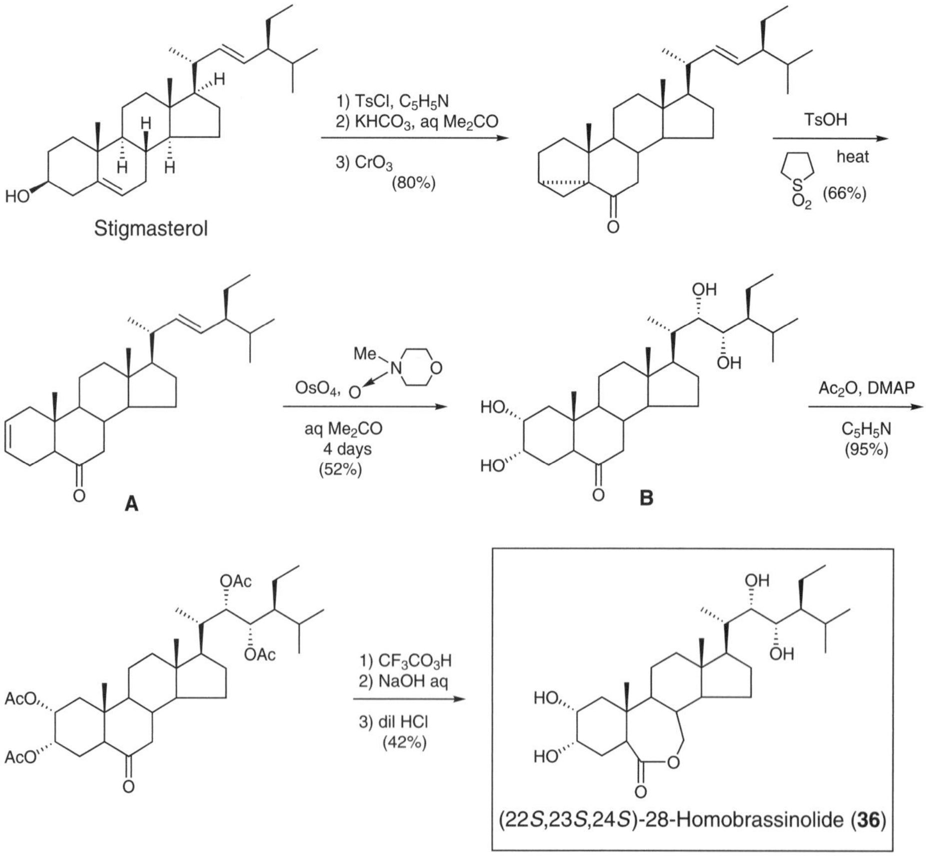

Figure 2.22   *Synthesis of (22S,23S,24S)-28-homobrassinolide. Reprinted with permission of Shokabo Publishing Co., Ltd*

2.1.5.3 Synthesis of brassinolide and castasterone

In 1982, we reported two different syntheses of brassinolide (**30**) from stigmasterol.[66,67] The second synthesis,[67] whose full paper was published in 1984,[68] was later extended to a successful synthesis of 30 g of brassinolide in 7.0% overall yield from stigmasterol (Figure 2.23).[69–71]

The key step in this synthesis was the cleavage of epoxide **C** with trimethylaluminum and *n*-butyllithium to give castasterone (**33**, a naturally occurring brassinosteroid) in a regio- and stereoselective manner after deprotection.[68] Another notable step was the conversion of ketone **A** to ethylene acetal **B** by treatment with acetone ethylene acetal and *p*-toluenesulfonic acid. If 2-butanone ethylene acetal, a more conventional acetalization reagent, was used instead of acetone ethylene acetal, then the product was not the crystalline **B** but an oily and diastereomeric mixture of the acetals generated by exchange reaction at C-2 and C-3, liberating acetone and incorporating 2-butanone, in addition to the acetal formation at C-6. This type of possible exchange reactions are quite often overlooked at the planning stage, rendering synthesis complicated by formation of unwanted byproducts. Aburatani noticed the present exchange reaction by checking the MS of the oily product that had been assumed to be **B**.[69]

2.1.5.4 Synthesis of dolicholide and dolichosterone

In 1982 Yokota *et al.* isolated 160 μg of dolicholide (**31**) and 50 μg of dolichosterone (**34**) from 34 kg of the immature seeds of a bean, *Dolichos lablab*.[72] So as to confirm the proposed structures, we synthesized these brassinosteroids as shown in Figure 2.24.[68,73,74] In this case, too, we used organoaluminum chemistry and epoxide cleavage as the key reactions.

Stigmasterol was converted to 6,6-ethylenedioxy aldehyde **A**, which yielded lactonic alcohol **B**. Oxidation of **B** gave lactonic aldehyde **C**. Organoaluminum reagent **D** was added to **C** to furnish allylic alcohol **E** stereoselectively. Oxidation of **E** with *m*-chloroperbenzoic acid provided **F**. When this epoxide **F** was heated with aluminum isopropoxide, the epoxy ring was cleaved to give dolicholide (**31**) after removing the acetonide protective group with dilute acetic acid.[68] Reaction of **A** with **D** gave **G**, which eventually afforded dolichosterone (**34**).[68]

2.1.5.5 Synthesis of homodolicholide and homodolichosterone

The amounts of homodolicholide (**32**)[75] and homodolichosterone (**35**)[76] isolated from *Dolichos lablab* were only 12 μg and 20 μg, respectively. It was therefore necessary to synthesize them in substantial amounts so as to evaluate their plant-growth-promoting activity. We synthesized these two brassinosteroids, as shown in Figure 2.25.[68]

The key steps were the same as those used for the synthesis of dolicholide (**31**) and dolichosterone (**34**): (i) chain elongation by means of organoaluminum chemistry, (ii) stereoselective epoxidation of the resulting allylic alcohol, and (iii) regio- and stereoselective cleavage of the epoxy ring.

Because homodolicholide (**32**) and homodolichosterone (**35**) possess the same carbon skeleton as that of stigmasterol, their alternative syntheses were achieved, utilizing all the carbon atoms of stigmasterol.[77–79]

2.1.5.6 Synthesis of 25-methyldolichosterone, 25-methyl-2,3-diepidolichosterone, 25-methylcastasterone and 25-methylbrassinolide

In 1987 Kim *et al.* isolated 25-methyldolichosterone (**37**, Figure 2.26) from immature seeds of *Phaseolus vulgaris*, and found it to be about one order of magnitude more bioactive plant-growth regulator than

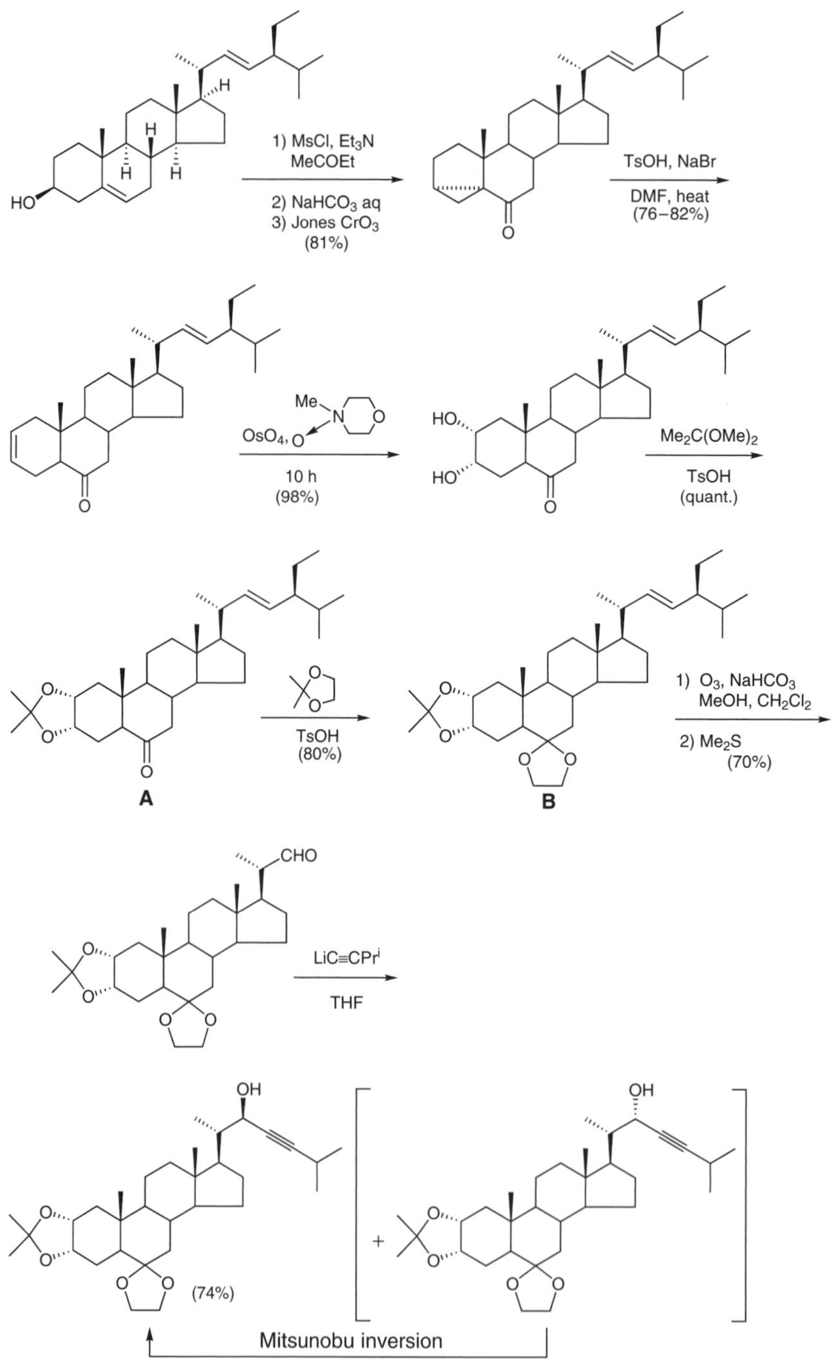

Figure 2.23 *Synthesis of brassinolide and castasterone. Reprinted with permission of Shokabo Publishing Co., Ltd*

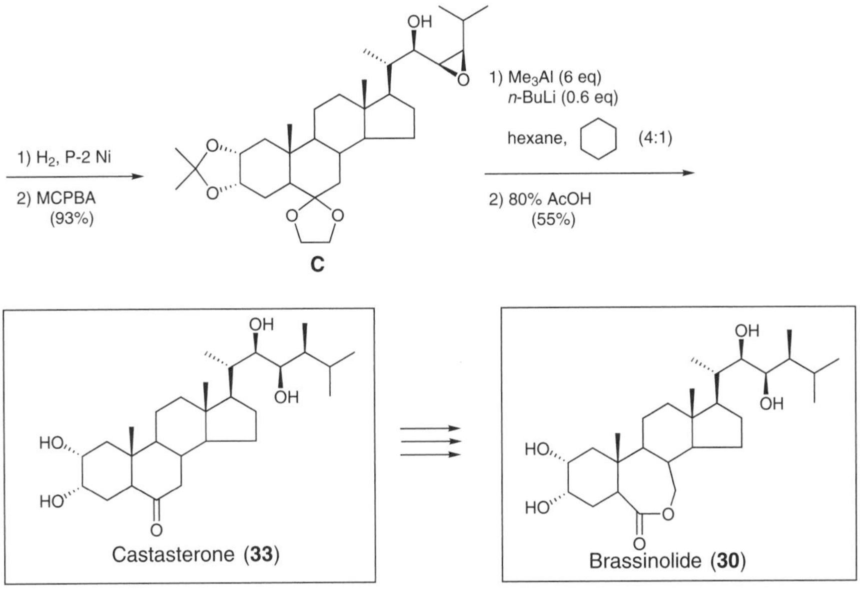

Figure 2.23   *(continued)*

dolichosterone itself.[80] In search of a more potent plant-growth regulator, we synthesized four 25-methylate brassinosteroids, as shown in Figure 2.26.[81]

As a result, 25-methylbrassinolide (**40**) was found to be more bioactive than the natural brassinolide. Later, 25-methyl-2,3-diepidolichosterone (**38**) was isolated from *Phaseolus vulgaris*. 25-Methyldolichosterone (**37**) and 25-methylcastasterone (**39**) were also synthesized.[81]

In our synthesis of brassinosteroids, organoaluminum chemistry was proved to be useful in constructing the side-chain double bond with the desired geometry, and the stereoselective cleavage of epoxides was shown to be an excellent method to generate the correct configuration of the side-chain.

Brassinosteroids promote plant growth to increase the dry weight. Their practical use will continue to be an interesting problem in agronomy. Nakatani and coworkers discovered the inhibitory activity of both castasterone (**33**) and (22*S*,23*S*,24*S*)-28-homobrassinolide (**36**) against the insect moulting hormone, 20-hydroxyecdysone.[82] The biological significance of this inhibitory activity is not clarified yet.

2.1.6   Phyllanthrinolactone, a leaf-closing factor

The phenomenon of nyctinasty or "plant sleep" has been recorded since the ancient time of Alexander the Great.[83] For example, the pinnate leaves of a large tamarind tree (*Tamarindus indica*) fold together at night as if the tree sleeps.[83] In 1995 Yamamura and coworkers isolated 3.1 mg of phyllanthrinolactone (**41**, Figure 2.27) from 19.2 kg of the fresh nyctinastic plant *Phyllanthus urinaria* as its leaf-closing factor.[84] It is bioactive only for that plant in the daytime at a very low concentration of 1×10^{-7} M.

Because the absolute configuration of the aglycone part of phyllanthrinolactone was unknown, we carried out its synthesis as shown in Figure 2.27.[85,86] The racemate of the aglycone part (±)-**H** was synthesized

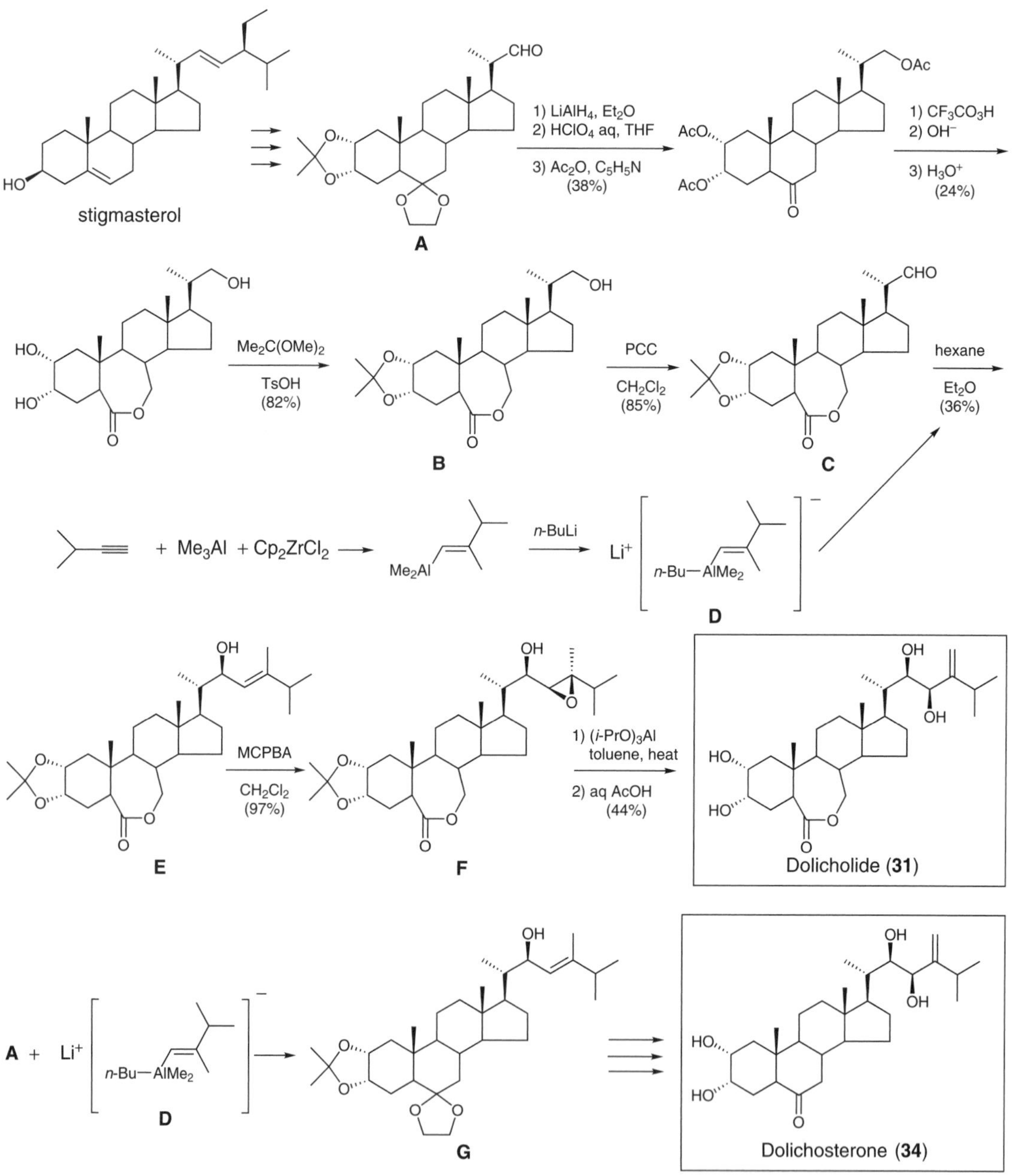

Figure 2.24 *Synthesis of dolicholide and dolichosterone. Reprinted with permission of Shokabo Publishing Co., Ltd*

from (±)-**B**, which was obtained by cycloaddition of dichloroketene to **A**. Although epoxidation of (±)-**C** afforded a diastereomeric mixture of (±)-**D** and (±)-**E**, the major and crystalline isomer turned out to be the desired (±)-**D**, while (±)-**E** was an oil. Hydroxylactone (±)-**F** was converted to the desired stereoisomer (±)-**H** via ketolactone (±)-**G**. Glucosidation of (±)-**H** with 2,3,4,6-tetra-*O*-acetyl-α-D-glucopyranosyl

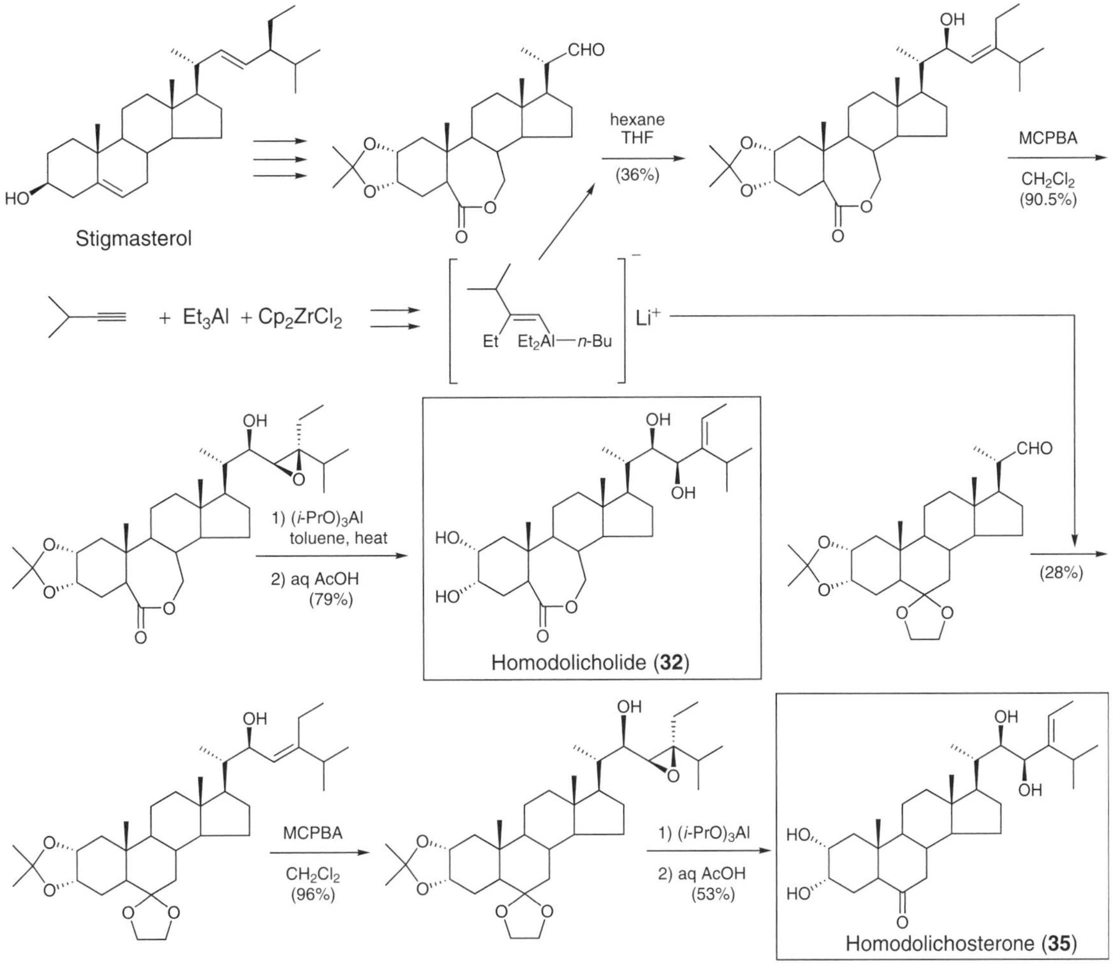

Figure 2.25 *Synthesis of homodolicholide and homodolichosterone. Reprinted with permission of Shokabo Publishing Co., Ltd*

bromide yielded a mixture of (±)-**I** and **J** and **K**, which was separated by silica-gel chromatography. Fortunately, one of the two glucosides obtained in 15.6% yield was crystalline, and its structure could be solved by X-ray crystallographic analysis, as depicted in **K**.

The tetraacetates **J** and **K** were deacetylated by treatment with potassium cyanide in methanol to give **41** and **41″**, respectively. As byproducts, **41′** and **41‴** were also obtained. All the glucosides **41**, **41′**, **41″** and **41‴** were amorphous. The acetate **J** of phyllanthrinolactone (**41**) was also amorphous. We were lucky to prepare **K**, whose structure could be solved by X-ray analysis. Accordingly, the structure of phyllanthrinolactone was firmly established as **41**, based on the X-ray analysis of **K**.

The leaf-closing activity of **41**, **41′**, **41″** and **41‴** was bioassayed against the leaves of *Phyllanthus urinaria*. Only **41** was bioactive at the concentrations of $10^{-3}, 10^{-4}$, and 10^{-5} g/L. In this case, chirality plays an important role in biological recognition.

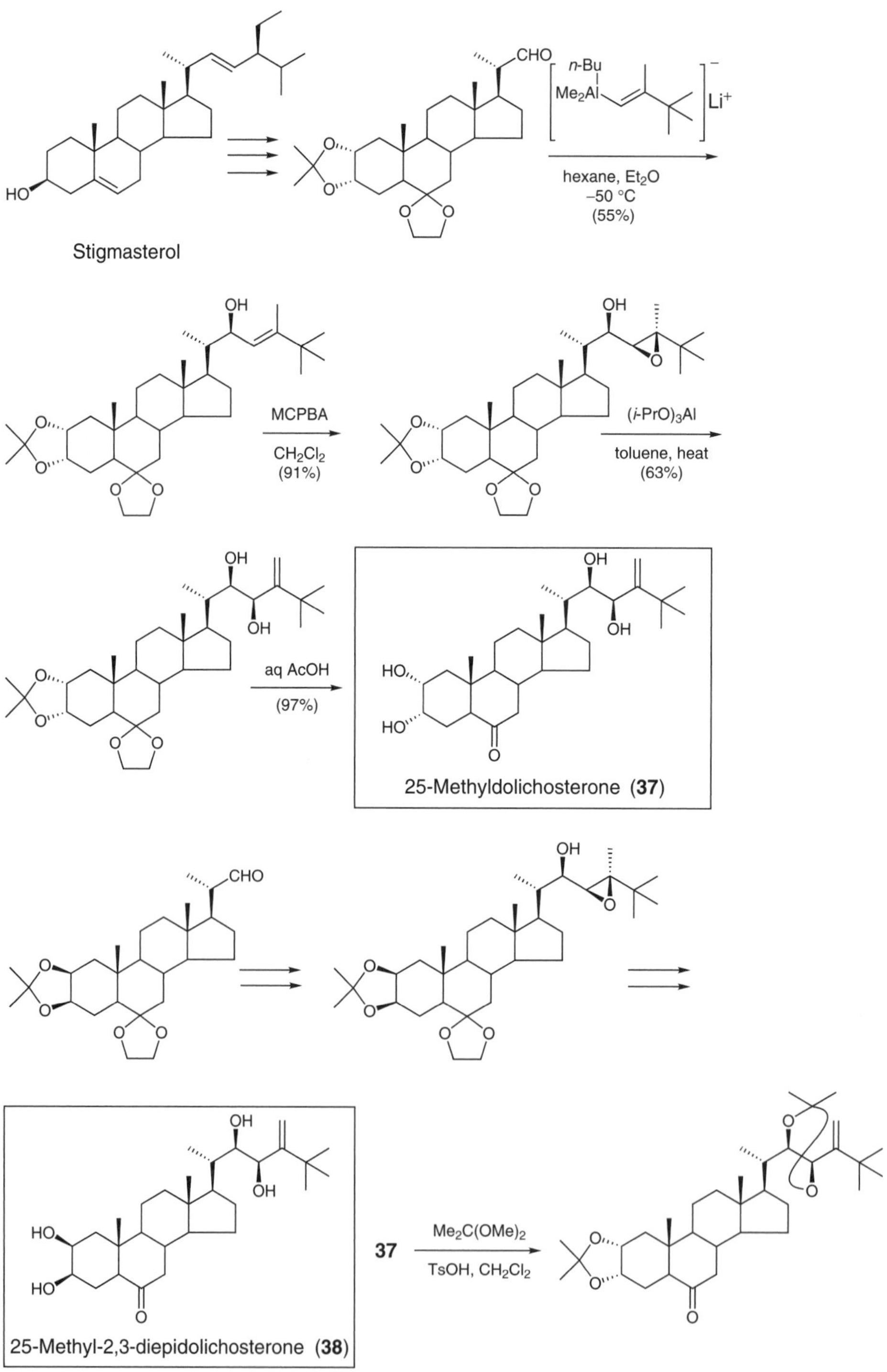

Figure 2.26   *Synthesis of 25-methylbrassinosteroids. Modified by permission of Shokabo Publishing Co., Ltd*

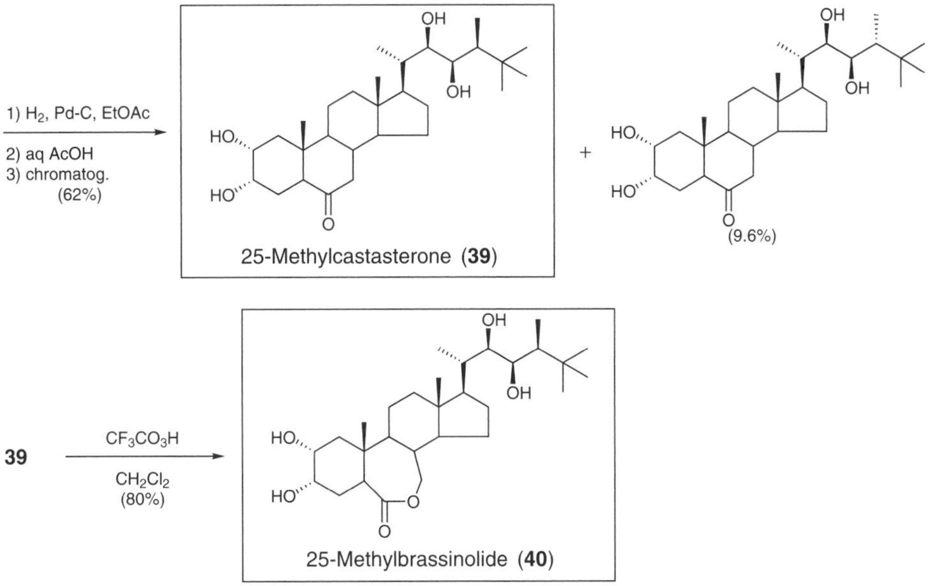

1) H_2, Pd-C, EtOAc

2) aq AcOH
3) chromatog.
(62%)

25-Methylcastasterone (**39**)

+

(9.6%)

39 → CF$_3$CO$_3$H / CH$_2$Cl$_2$ (80%)

25-Methylbrassinolide (**40**)

Figure 2.26   *(continued)*

2.2   Phytoalexins

2.2.1   What are phytoalexins?

Upon attack by pathogenic micro-organisms, higher plants are known to produce antimicrobial compounds as defense substances. These allelochemicals are called phytoalexins. I chose several phytoalexins as my synthetic targets. Synthesis of some of the phytoalexins will be discussed in this section.

2.2.2   Synthesis of pisatin

(+)-Pisatin (**42**, Figure 2.28) was isolated as an antifungal phytoalexin by Cruickshank and Perrin in 1960[87] from the pods of the garden peas (*Pisum sativum*), which has been inoculated with fungal spores. Its structure was proposed in 1962,[88] and its absolute configuration was determined in 1980.[89] We became interested in synthesizing both the enantiomers of pisatin, and accomplished the synthesis of the enantiomers of pterocarpin (**43**) in 1988 as a model study.[90] Then, in 1989, the enantiomers of pisatin (**42**) were synthesized, as shown in Figure 2.28.[91]

The known isoflavone **B**[92] was synthesized from sesamol (**A**) as depicted. The carbonyl group of **B** was reduced to give a mixture of (±)-**C** and (±)-**D**, whose dehydration afforded alkene **E**. This was oxidized with osmium tetroxide to give (±)-**F**. Optical resolution of (±)-**F** could be achieved by HPLC separation of the diastereomers **G** and **H** prepared by acylation of (±)-**F** with (+)-camphor-10-sulfonyl chloride. When **G** was treated with TBAF both the camphor-10-sulfonyl and TBS protective groups were removed to give **I**. The triol **I** was acylated with 1.1 eq. of trifluoromethanesulfonyl chloride (TfCl) to give monotriflate **J**. Treatment of **J** with sodium carbonate effected ring closure to give (+)-pisatin (**42**). Similarly, **H** yielded unnatural (−)-pisatin.

(+)-Pisatin (**42**) was shown to be an active antifungal agent, and (−)-pisatin also showed weak bioactivity. In this case, even the unnatural enantiomer (−)-**42** was bioactive, as tested by Dr. G. Russell in

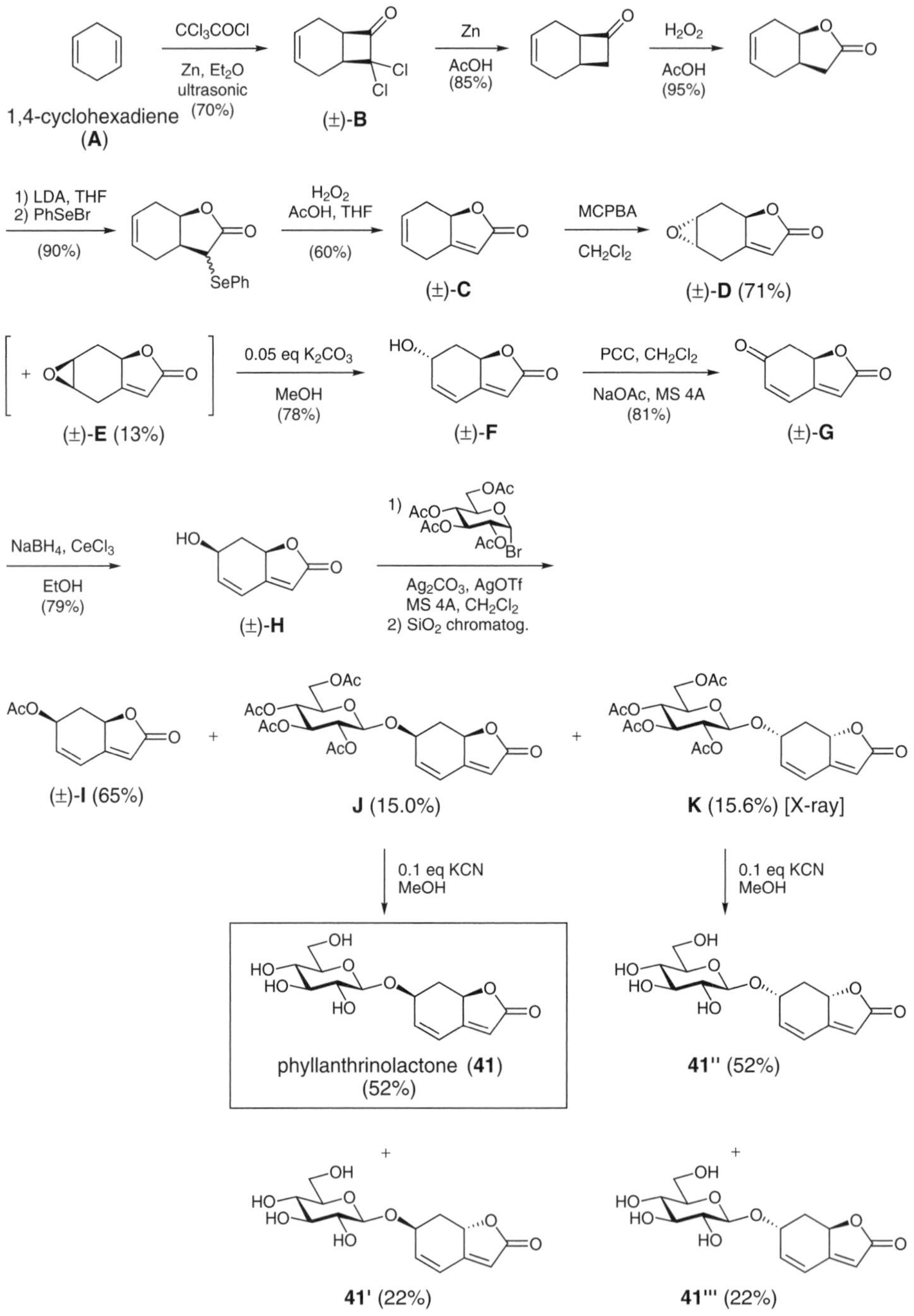

Figure 2.27 *Synthesis of phyllanthrinolactone*

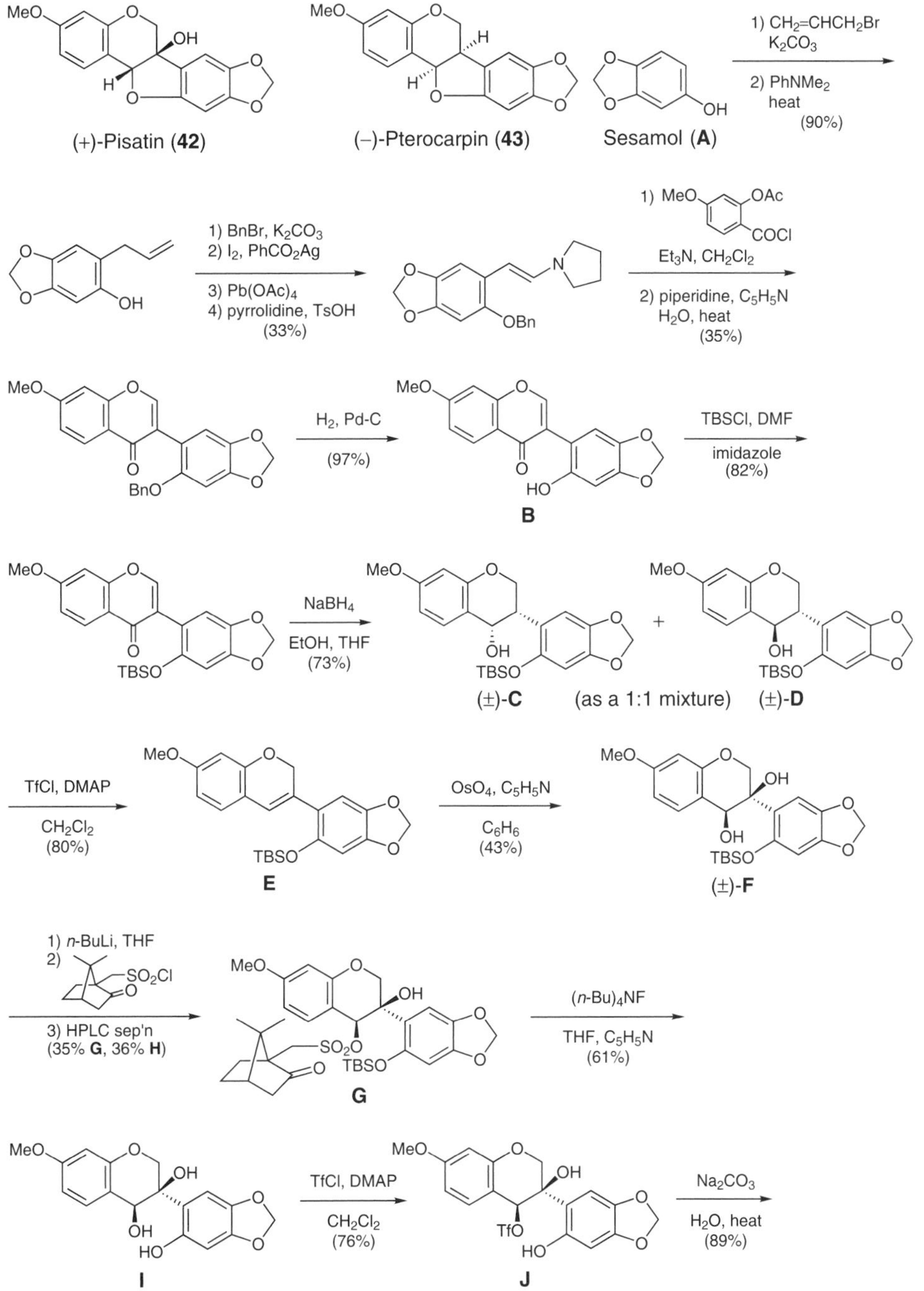

Figure 2.28 *Synthesis of the enantiomers of pisatin. Modified by permission of Shokabo Publishing Co., Ltd*

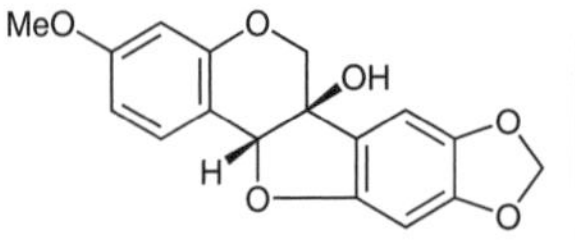

(+)-Pisatin (**42**)

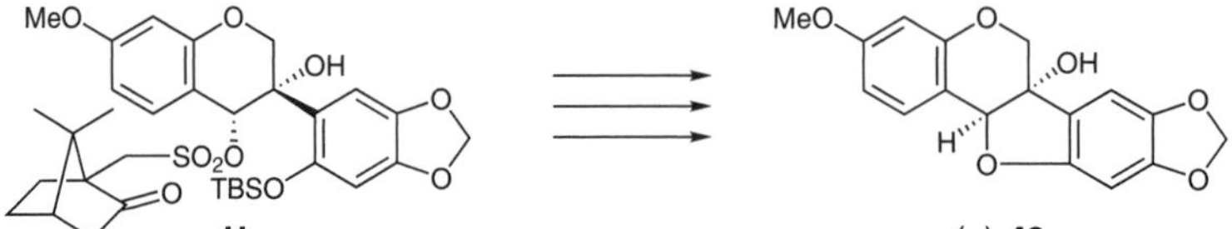

Figure 2.28 (continued)

New Zealand. It must be added that the bioactivity of (+)-**42** was far weaker than those of commercial fungicides.

2.2.3 Synthesis of 2-(4-hydroxyphenyl)naphthalene-1,8-dicarboxylic anhydride

Anthracnose caused by infection with a fungus *Colletotrichum musae* is a common disease of banana, *Musa acuminata*. The unripe green fruit of banana shows resistance to the growth of fungal hyphae, and the pathogen is quiescent until the fruit ripens. Phytoalexins produced by the unripe fruit were isolated by Hirai *et al.*, and 2-(4-hydroxyphenyl)naphthalene-1,8-dicarboxylic anhydride (**44**, Figure 2.29) was identified as the major and the strongest component of the phytoalexins.[93]

Figure 2.29 shows the synthesis of **44** from acenaphthene.[94] Suzuki–Miyaura coupling of 3-bromoacenaphthene (**A**) with 4-methoxyphenylboronic acid (**B**) was the key step to give **C** in 86% yield.

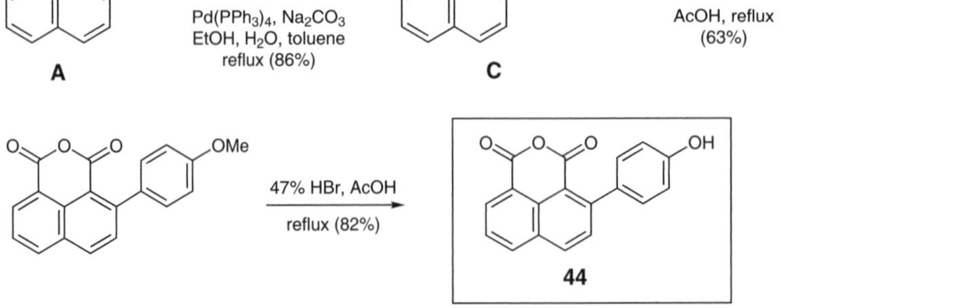

Figure 2.29 *Synthesis of 2-(4-hydroxyphenyl)naphthalene-1,8-dicarboxylic anhydride, a banana phytoalexin*

Oxidation and demethylation of **C** yielded **44**. Phenolic compounds are frequently produced by plants as phytoalexins.

2.2.4 Synthesis of oryzalexins

Rice blast disease as caused by a pathogenic fungus *Magnaporthe grisea* (old name: *Pyricularia oryzae*) is the most disastrous disease against rice plants, *Oryza sativa*. In 1983 Akatsuka and coworkers isolated three new diterpenes, oryzalexins A (**45**, Figure 2.30), B (**46**) and C (**47**) as phytoalexins produced by rice plants after infection with *M. grisea*.[95]

The amounts of phytoalexins produced by rice plants are limited. For example, 50 kg of rice leaves after fungal infection afforded 1.5 mg of oryzalexin A (**45**).[95] We became interested in synthesizing both the enantiomers of oryzalexins so as to estimate their antifungal activities.

Our synthetic plan is shown in Figure 2.30. The immediate synthetic precursor of oryzalexins A, B and C would be (±)-**A**. The precursor (±)-**A** would be synthesized from (±)-**B**, which would be prepared from 1,6-dihydroxynaphthalene (**C**).[96] Optical resolution of (±)-**A** would be feasible to give both the enantiomers of oryzalexins.

The synthesis of (±)-3-hydroxyisopimara-8(14),15-diene (**A** in Figure 2.30 and **H** in Figure 2.31) is shown in Figure 2.31. 1,6-Dihydroxynaphthalene (**A**) was converted to (±)-**B**, which also served as an intermediate in the synthesis of (±)-kaur-16-en-19-oic acid (cf. Figure 2.13). Hydrogenation of the tricyclic intermediate (±)-**C** was not stereoselective, giving a 2:1 mixture of the desired (±)-**D** and undesired (±)-**E**. This mixture was further manipulated without separation. When the mixture of the stereoisomeric phenols was hydrogenated over Raney nickel, only the undesired stereoisomer afforded a hydrogenolysis product. After chromic-acid oxidation, the mixture was purified by chromatography to give (±)-**F** as crystals, while (±)-**G** was obtained as an oil.

Conversion of (±)-**F** to (±)-**H** [=(±)-3β-hydroxyisopimara-8(14),15-diene] could be executed in the usual manner. The stereoisomers (±)-**H** and (±)-**I** were separable by chromatography over silica gel impregnated with silver nitrate. The overall yield of (±)-**H** was 8.5% (15 steps) based on (±)-**C**.

Figure 2.32 summarizes the optical resolution of (±)-**A** and subsequent syntheses of the enantiomers of oryzalexins A, B and C from the resolved (+)-**A** and (−)-**A**. Esterification of (±)-**A** with (−)-camphanic

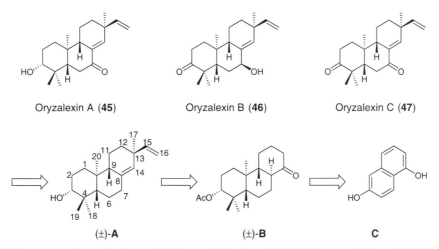

Figure 2.30 Structures of oryzalexins and their synthetic plan. Modified by permission of Shokabo Publishing Co., Ltd

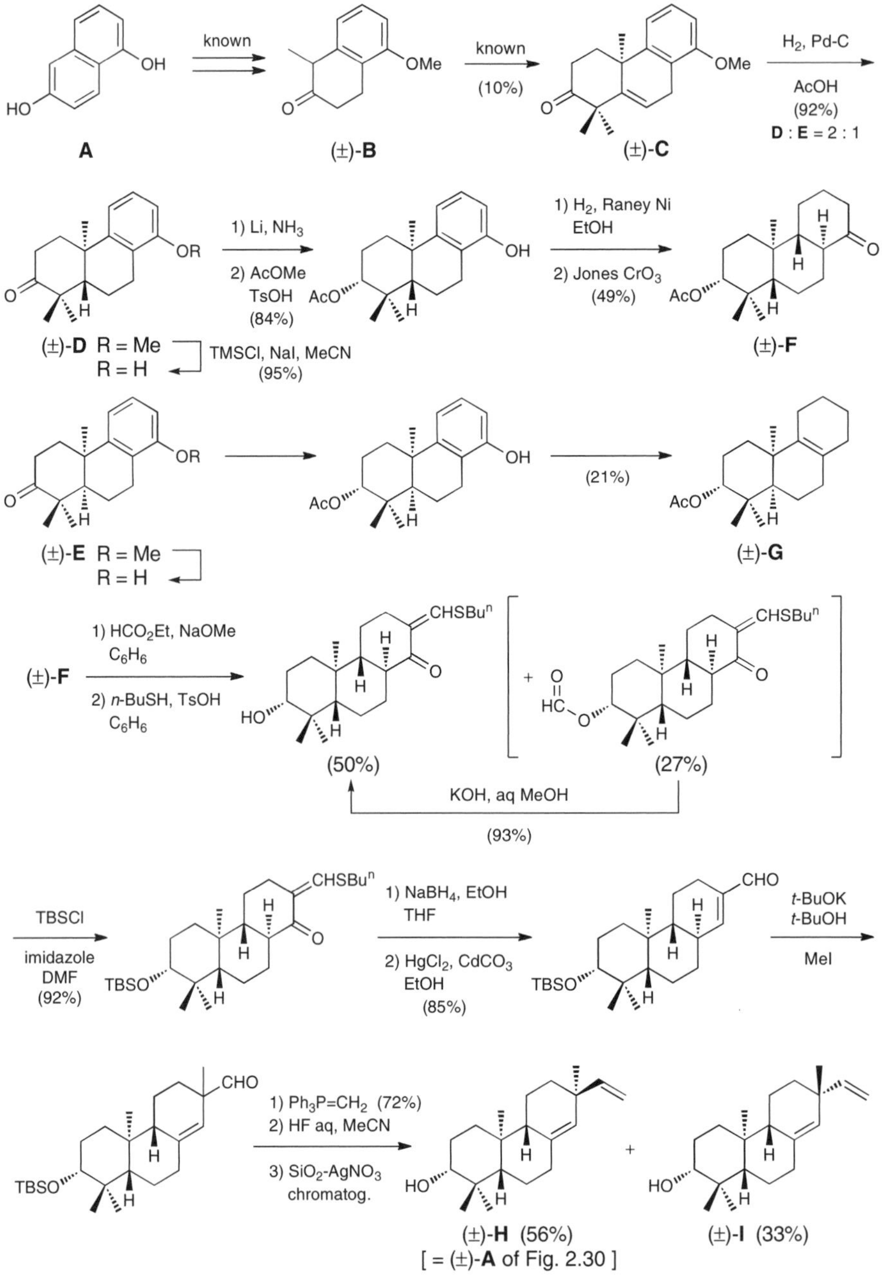

Figure 2.31 *Synthesis of oryzalexins (1). Modified by permission of Shokabo Publishing Co., Ltd*

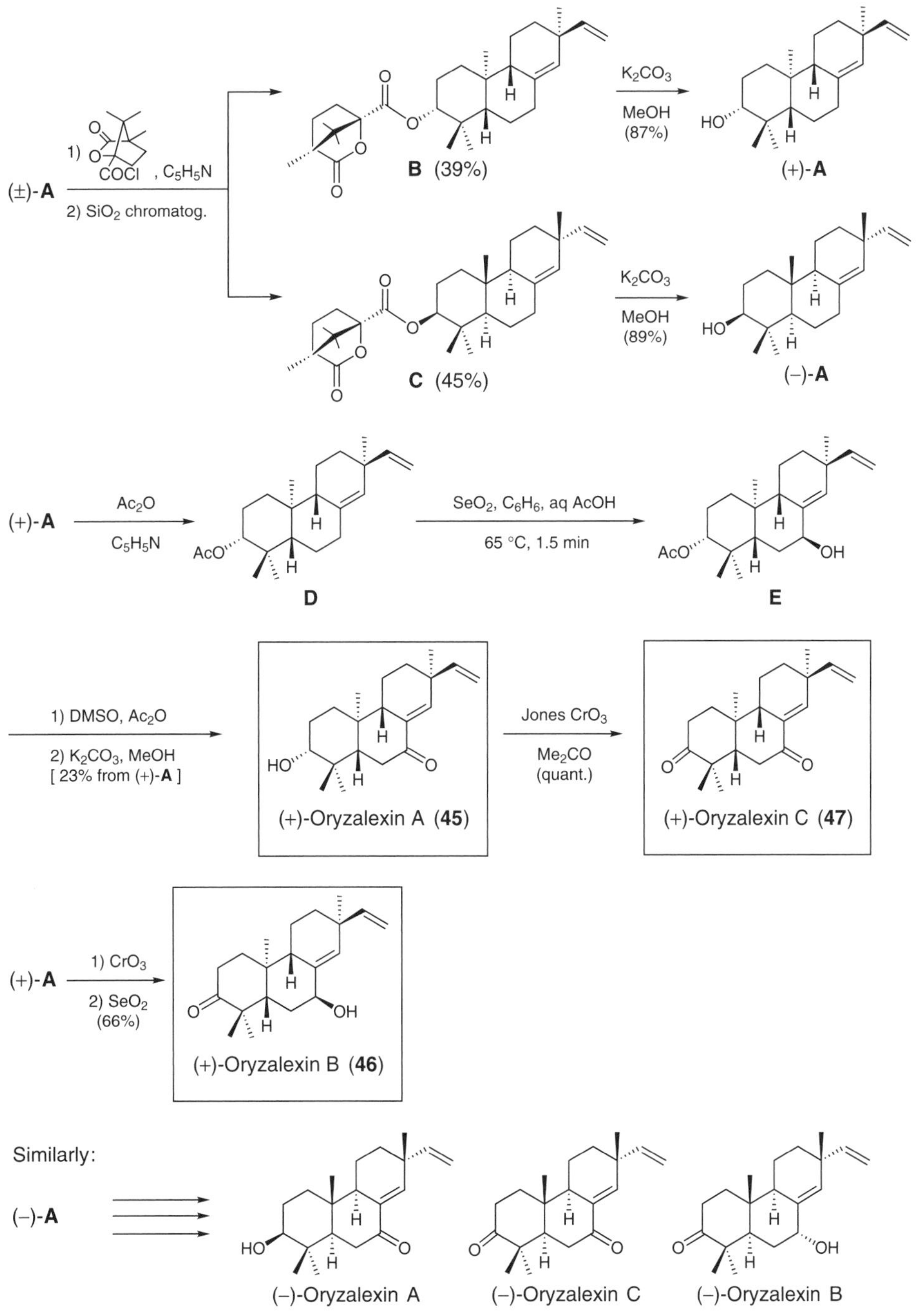

Figure 2.32 *Synthesis of oryzalexins (2). Modified by permission of Shokabo Publishing Co., Ltd*

chloride yielded a diastereomeric mixture of esters **B** and **C**, which were separated by silica-gel chromatography. Treatment of **B** with potassium carbonate in methanol removed (−)-camphanic acid to give (+)-**A**, while **C** furnished (−)-**A**. Further functional group transformation of (+)-**A** afforded (+)-oryzalexin A (**45**), (+)-oryzalexin B (**46**) and (+)-oryzalexin C (**47**). Similarly, (−)-**A** was converted to unnatural (−)-enantiomers of oryzalexins A, B and C. A noteworthy step in this functional group transformation was hydroxylation of **D** to give alcohol **E**. After extensive trials, the hydroxylation could be achieved by treatment of **D** with selenium dioxide in benzene/acetic acid/water at 65 °C for 1.5 min. Careful optimization of a difficult step is the responsibility of a practitioner in organic synthesis.

Antifungal activities of the synthetic samples were examined by Prof. Akatsuka to confirm the strong bioactivities of (+)-**45**, (+)-**46** and (+)-**47** against *M. grisea*. The unnatural enantiomers were only weakly antifungal.

2.2.5 Synthesis of phytocassanes

In late 1990s Koga and coworkers reported phytocassanes A−E (Figure 2.33) as new diterpene phytoalexins produced by rice plants infected with such disastrous fungi as *Magnaporthe grisea, Rhizoctonia solani*, and *Phytophtora infestans*.[97,98] Their cassane diterpene structures were proposed as shown in Figure 2.33 on the basis of extensive spectroscopic analysis. Although their CD spectra were recorded, their absolute configuration remained unknown.

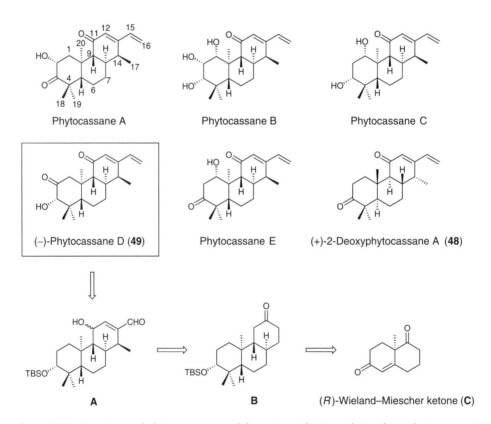

Figure 2.33 *Structures of phytocassanes and the retrosynthetic analysis of (−)-phytocassane D*

I became interested in clarifying the absolute configuration of phytocassanes so as to know whether they, like gibberellins and oryzalexins, belong to the *ent*-diterpene series or not. We first synthesized (+)-2-deoxyphytocassane A (**48**), and measured its CD spectrum.[99] The result strongly suggested that phytocassanes belong to the *ent*-diterpene series. In order to confirm the absolute configuration of phyto-cassanes definitely, (−)-phytocassane D (**49**) was synthesized, basing on its retrosynthetic analysis as shown in Figure 2.33. The synthesis was achieved as summarized in Figures 2.34 and 2.35.[100] The synthetic (−)-**49** showed the CD spectrum identical to that of the natural product, verifying its *ent*-diterpene stereochemistry.

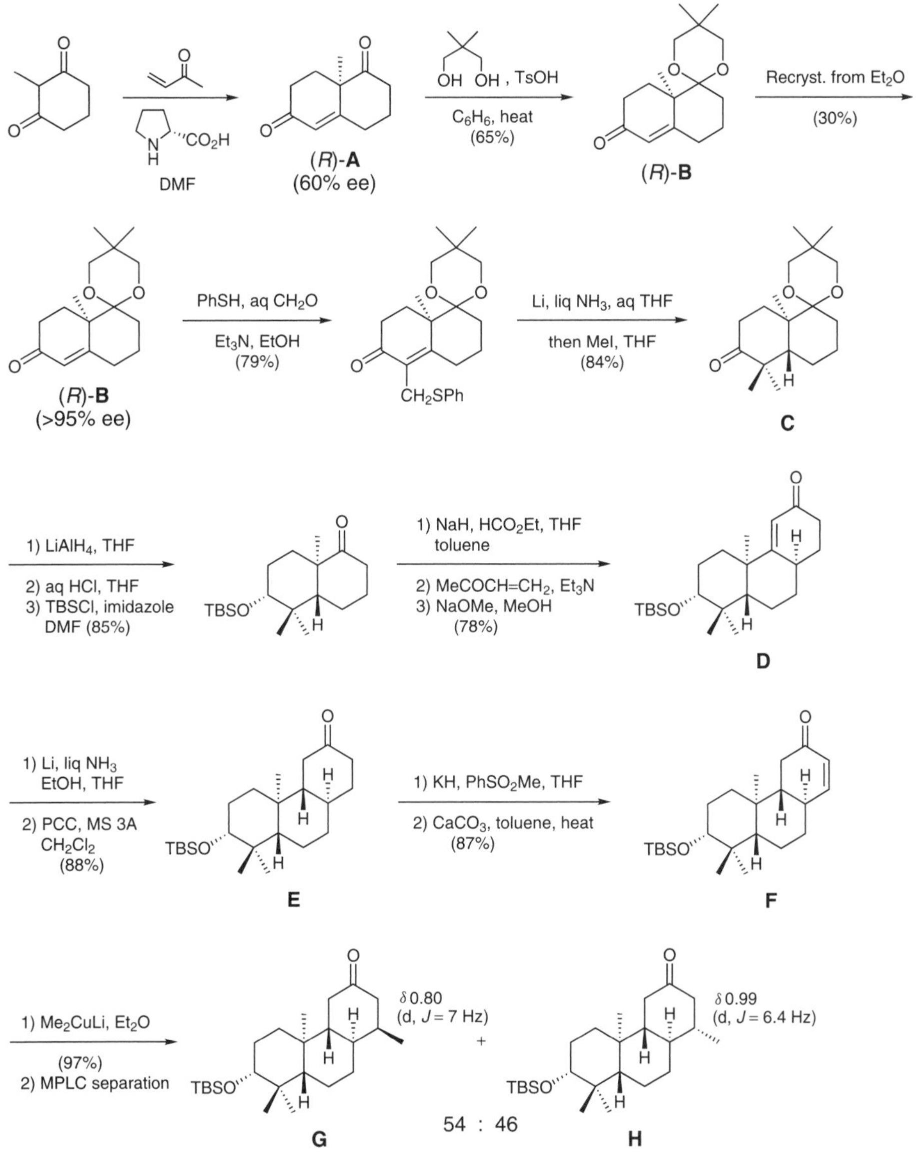

Figure 2.34   *Synthesis of (−)-phytocassane D (1)*

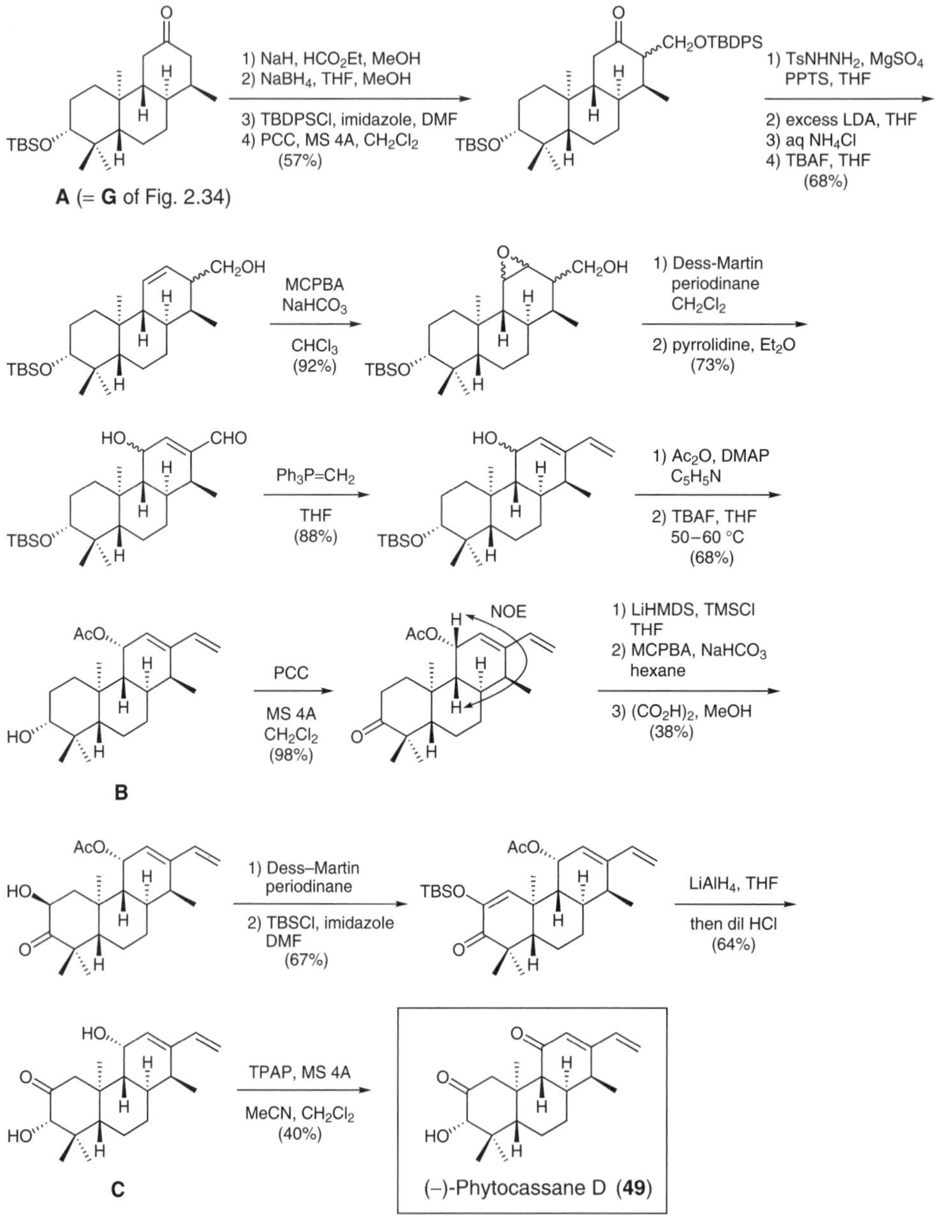

Figure 2.35   *Synthesis of (−)-phytocassane D (2)*

The earlier part of the synthesis of (−)-phytocassane D (**49**) is shown in Figure 2.34. The starting material was (*R*)-Wieland–Miescher ketone (**A**) prepared by the well-known organocatalytic Robinson annulation reaction. To enhance the enantiomeric purity of **A**, the corresponding acetal **B** was recrystallized from ether to give (*R*)-**B** of >95% ee. This was converted to **C**. Then, the second Robinson annulation furnished **D**, which yielded the isomeric ketone **F** via **E**. Methylation of **F** with lithium dimethylcuprate was nonstereoselective, giving the desired **G** and the undesired **H** in a ratio of 54:46. Fortunately, they were separable by MPLC.

Conversion of the tricyclic ketone **A** to (−)-phytocassane D (**49**) is summarized in Figure 2.35. First, the ring C was modified to put all the required functionalities except the oxidation level at C-11 affording **B**. Then, the ring A was modified to give **C**, whose mild oxidation furnished (−)-phytocassane D (**49**). Its IR, NMR and CD spectra were identical to those of the natural product. Comparison of the CD data of all the phytocassanes led to the conclusion that they belong to *ent*-diterpenes like oryzalexins and gibberellins.[100]

2.3 Plant allelochemicals

Higher plants interact with other organisms such as animals including insects, other plants, and micro-organisms by means of allelochemicals. Some of them are advantageous for higher plants as the producers. Then, these allelochemicals are called allomones. In other cases, allelochemicals are beneficial to receivers such as other plants, animals and micro-organisms. They are called kairomones. In this section two types of compounds will be discussed that work as kairomones.

2.3.1 Synthesis of glycinoeclepin A

Cyst nematodes are well known as serious pests of many crops, and their extermination is an important agricultural problem. They generally have a limited host range, and their specificity is thought to be based on response to chemical hatching stimulus secreted by the host plants as kairomone. In 1985 Masamune *et al.* isolated a degraded triterpene, glycinoeclepin A (**50**, Figure 2.36), as potent hatching stimulus for

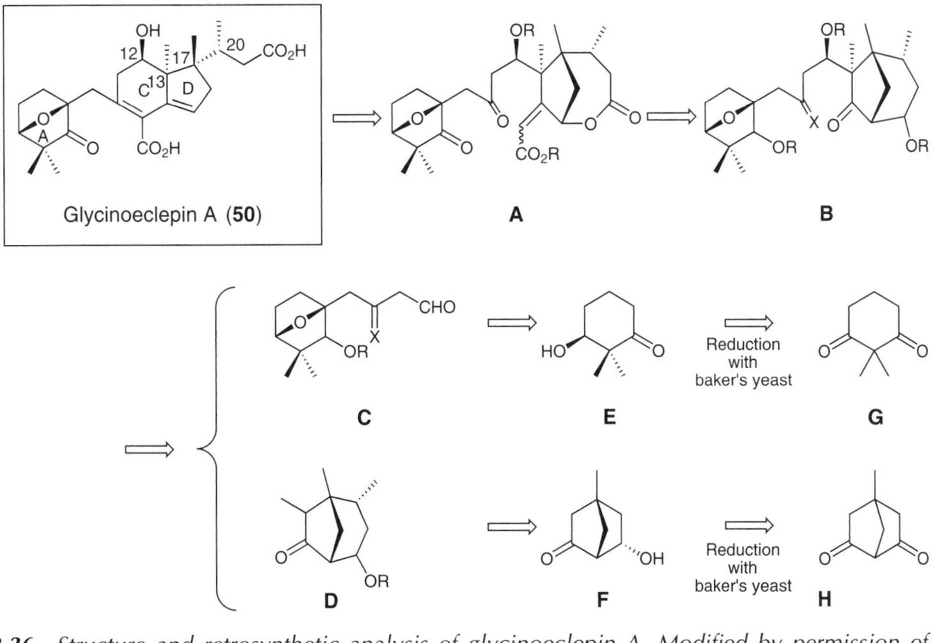

Figure 2.36 *Structure and retrosynthetic analysis of glycinoeclepin A. Modified by permission of Shokabo Publishing Co., Ltd*

the soybean cyst nematode (*Heterodera glycines*) from the dried root of the kidney bean (*Phaseolus vulgaris*).[101] The isolated amount of **50** was 1.25 mg as its *bis-p*-bromophenacyl ester from 1 ton of the dried bean roots cultivated on 10 ha of bean field, and the structure was determined by X-ray analysis of the *bis-p*-bromophenacyl ester.[101]

Glycinoeclepin A (**50**) shows very strong hatch-stimulating activity for the soybean cyst nematode at the concentration of 10^{-12}–10^{-13} g/mL. We therefore became interested in synthesizing it according to the retrosynthetic analysis as shown in Figure 2.36.[102,103] The starting materials are hydroxy ketones **E**[104] and **F**, which are the reduction products of the symmetrical diketones **G** and **H** with baker's yeast. There are many biochemical reactions that are useful in organic synthesis owing to their high enantioselectivity.[105] The two building blocks **C** and **D** would be derived from **E** and **F**, respectively, and then coupled by aldol reaction to give **B**. The intermediate **B** would yield glycinoeclepin A (**50**) via **A**. This plan was realized as a convergent and efficient synthesis of **50**.

Figure 2.37 summarizes the synthesis of the left half part **D** of the target **50**. Olefinic alcohol **B**, derived from (*S*)-hydroxy ketone **A**, was treated with *N*-iodosuccinimide to give iodo ether **C**. Further manipulation of **C** yielded aldehyde **D** to be employed for the key aldol condensation.

Synthesis of the right half part **E** of the target **50** is shown in Figure 2.38. Reduction of **A** was achieved successfully by using a large amount of baker's yeast to give **B** after acetylation. Ring expansion of **C** to give **D** was successful only after extensive experimentation to optimize the reaction conditions, and **D** was converted to **E**, the partner for the aldol reaction.

The subsequent synthetic steps leading to glycinoeclepin A (**50**) are shown in Figure 2.39. Successful aldol reaction of **A** with **B** was achieved by employing the zinc enolate derived from **B** to give **C**, which was so unstable that the product could be isolated only after esterification to give **D**. Treatment of **D** with sodium hydride yielded **E**, whose functional group transformation afforded **F**, the pivotal intermediate for

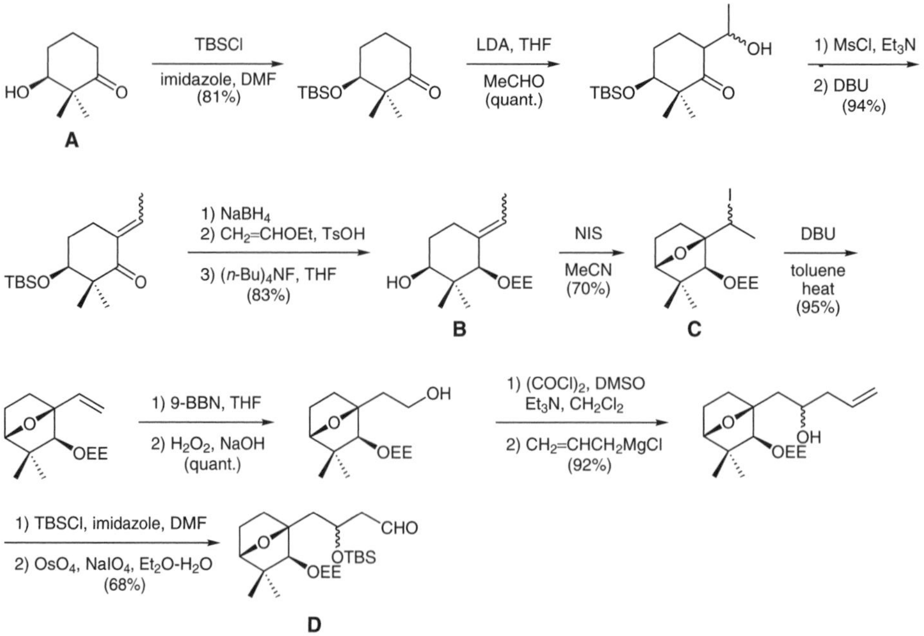

Figure 2.37 *Synthesis of the left half of glycinoeclepin A. Modified by permission of Shokabo Publishing Co., Ltd*

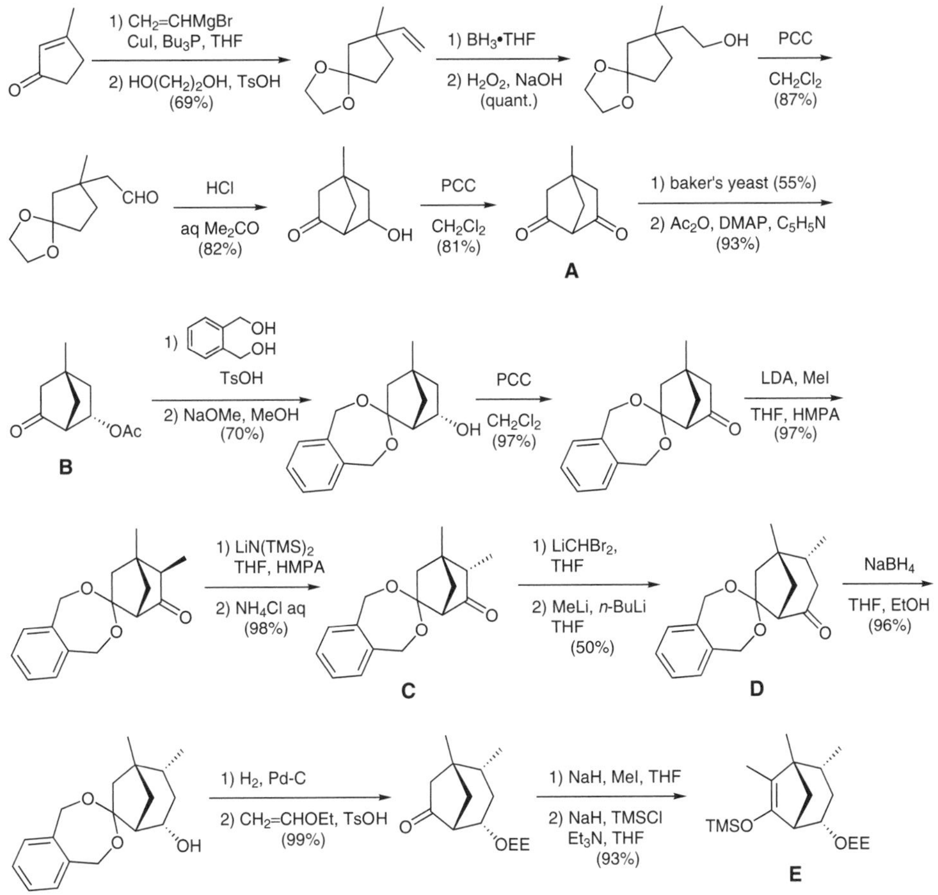

Figure 2.38 *Synthesis of the right half of glycinoeclepin A. Modified by permission of Shokabo Publishing Co., Ltd*

the C-ring formation. We chose lithium dimethylcuprate as the electron source [Cu(I)→Cu(II)], and **F** gave **G** with the desired ring system and the side-chain. A plausible mechanism for this reaction is shown at the bottom of Figure 2.39. Further functional group manipulation of **G** finally furnished 220 mg of glycinoeclepin A (**50**) as needles, mp 120–121.5 °C. The overall yield of **50** was 4.2% (32 steps) based on ketol **E** or 4.4% (31 steps) based on ketol **F** in Figure 2.36.

Our synthetic glycinoeclepin A (**50**) showed almost the same or slightly stronger hatch-stimulating activity in vitro as the natural product itself. However, it showed no nematocidal effect in laboratory pots and in a soybean field, as tested by Sumitomo Chemical Co. and Novartis. Glycinoeclepin A (**50**) seems to be adsorbed readily by soil particles or decomposed very quickly by soil micro-organisms. It works as a hatching stimulus only when it was secreted by bean roots to effect nematode eggs located in the extreme vicinity to the roots. Practical use of this semiochemical in agriculture is not so simple and needs further study.

2.3.2 Synthesis of strigolactones

Parasitic weeds such as witchweeds (*Striga* spp.) and broomrapes (*Orobanche* spp.) are known to cause severe yield losses in grains and legumes in Africa, Asia and the USA. The seeds of such weeds remain

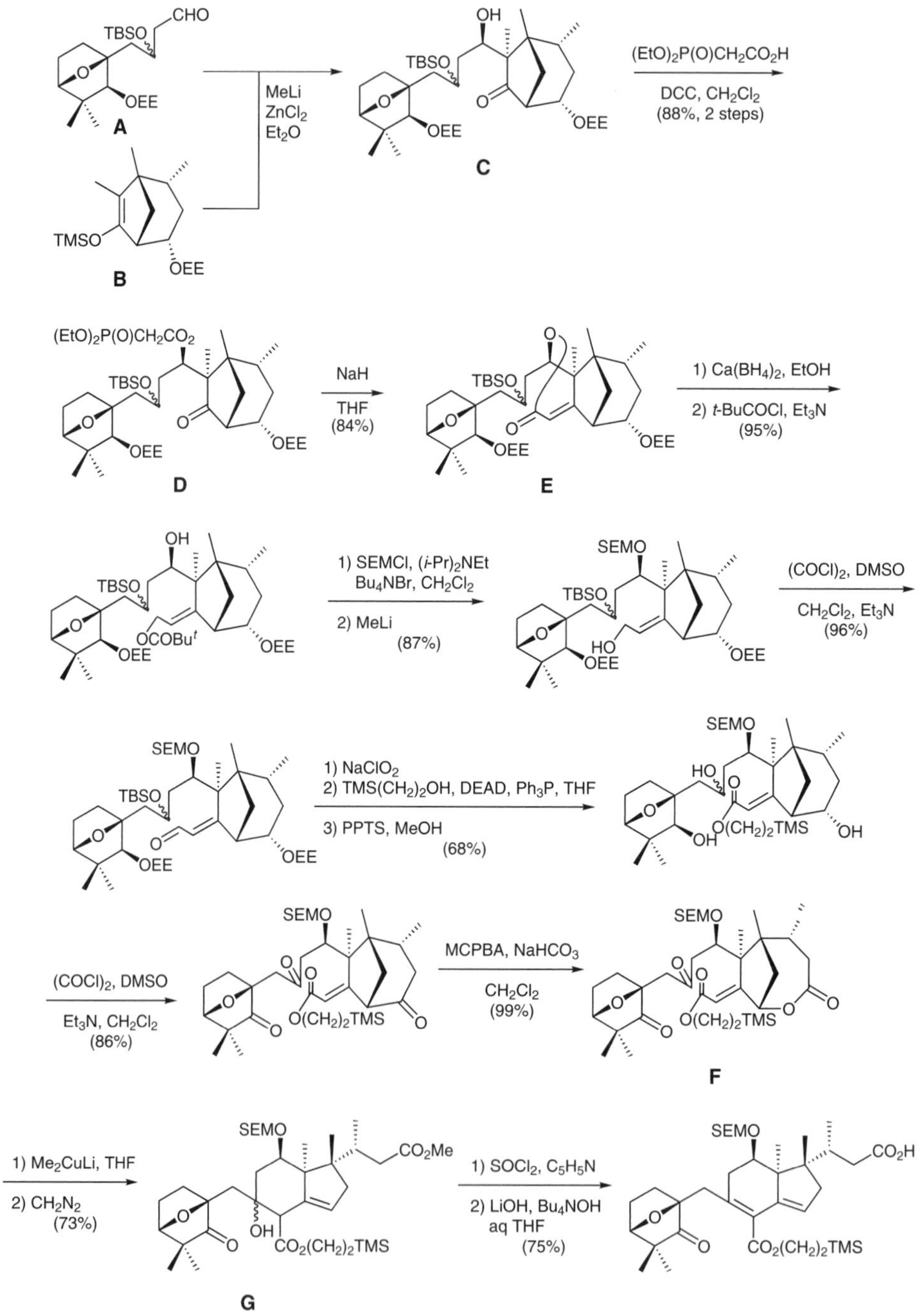

Figure 2.39 *Synthesis of glycinoeclepin A. Modified by permission of Shokabo Publishing Co., Ltd*

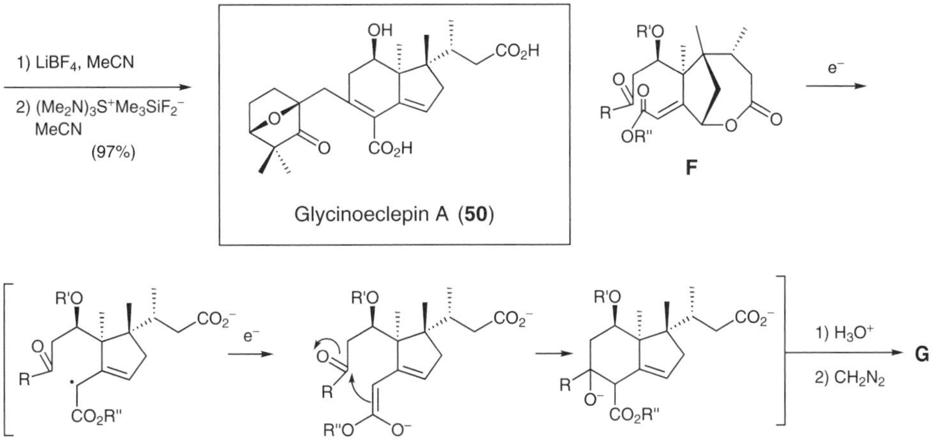

Figure 2.39 *(continued)*

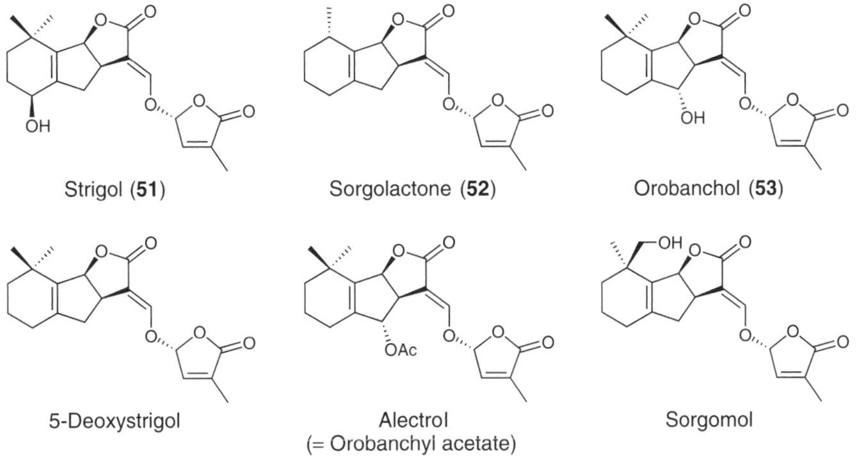

Figure 2.40 *Structures of strigolactones*

dormant in soil until exudates from their host plants induce germination. Several active principles of the exudates as shown in Figure 2.40 have been isolated to date, and are known under the general name "strigolactones." It recently has become clear that strigolactones are important signals for host recognition not only for parasitic weeds but also for symbiotic arbuscular mycorrhizal fungi. Furthermore, strigolactones are regarded as phytohormones.

Strigol (**51**) was the first strigolactone to be isolated from cotton root in 1972 as a strong stimulant for the germination of parasitic weeds.[106] Two decades later in 1992, sorgolactone (**52**) was isolated from a host plant, *Sorghum bicolor*.[107] Then, in 1998, orobanchol (**53**) was isolated from red clover (*Trifolium pratense*) as a germination stimulant for clover broomrape (*Orobanche minor*).[108] 5-Deoxystrigol was isolated as a host-recognition signal for arbuscular mycorrhizal fungi, while alectrol (= orobanchyl acetate) was isolated from the root exudates of cowpea (*Vigna unguiculata*) as a germination stimulant for the parasites *Alectra* *vogelii* and *Striga gesnerioides*. Sorgomol is the latest member of strigolactones, which has been isolated from *Sorghum bicolor*.[109]

These strigolactones were isolated in only very small amounts, and their synthesis was important to establish their structures and also to supply sufficient materials for biological study. Our synthesis of sorgolactone (**52**), strigol (**51**) and orobanchol (**53**) will be discussed here.

2.3.2.1   Synthesis of sorgolactone

The amount of sorgolactone (**52**) isolated from *Sorghum bicolor* was only 5 μg at the time of isolation,[107] and its sample was no longer available. This fact justified its synthesis. We synthesized both (±)-**52**[110,112] and (+)-**52**.[111,112] Figure 2.41 summarizes our synthesis of (+)-sorgolactone (**52**).[112]

The chiral source for our synthesis of (+)-**52** was methyl (*S*)-citronellate, which was converted to acetylenic phenylselenyl ester **A**. Radical-mediated cyclization of **A** afforded **B**. Further conversion of **B** to **C** was followed by its reduction and lactonization to give lactones **D** and **E**. Formylation of **D** gave **F**, which was coupled with bromolactone (±)-**G** to give (+)-sorgolactone (**52**) and its diastereomer **52′**. Similarly, **E** yielded two additional diastereomers, **52″** and **52‴**.

Bioassays of these four stereoisomers of **52** were carried out by employing seeds of *Orobanche minor*. All stereoisomers exhibited strong activity, which decreased in the order **52‴** > **52** > **52′** = **52″**. The natural (+)-sorgolactone (**52**) was not the strongest germination stimulant.

2.3.2.2   Synthesis of strigol

Prior to our synthesis of the naturally occurring (+)-enantiomer of strigol (**51**) in gram quantities, all the attempts to prepare (+)-**51** ended up in securing less than 0.1 g of (+)-**51**. We achieved a more efficient synthesis of (+)-**51** by employing lipase-catalysed optical resolution of an intermediate as the key step (Figure 2.42).[113]

The key intermediate used for the enzymatic optical resolution was hydroxy lactone (±)-**C** prepared from citral via (±)-**A**. Treatment of (±)-**C** with vinyl acetate in the presence of lipase AK (Amano) effected asymmetric acetylation, and the recovered (+)-hydroxy lactone **C** as well as the acetate (+)-**D** were obtained. Subsequently, (+)-**C** was converted to (+)-strigol (**51**). Three stereoisomers of strigol, **51′**, **51″** and **51‴** were also synthesized.

2.3.2.3   Synthesis of orobanchol

In 1998, Yokota *et al.* isolated orobanchol (**53**) from *Orobanche minor*.[108] The scarcity of the isolated orobanchol did not allow its NMR study, and only its GC-MS data were available. Even with this limitation its structure was proposed as **A** in Figure 2.43, although the position of the hydroxy group and that of the double bond in the ring **B** were uncertain. We synthesized various stereoisomers of (±)-**A** and (±)-**53**, and executed their GC-MS comparison with orobanchol. The comparison enabled us to propose **53** as the structure of orobanchol.[114,115] Our synthesis of (+)-orobanchol (**53**) will be discussed here (Figure 2.43).[113]

(+)-Hydroxy lactone **B**, which was employed in our synthesis of (+)-strigol, was deoxygenated to give **C**. Oxidation of **C** with chromic anhydride-3,5-dimethylpyrazole complex in dichloromethane afforded the desired ketone **D** together with its regioisomer **E**. Reduction of **D** gave the desired alcohol **F** as the minor product. The major product **G**, however, could be epimerized by Mitsunobu inversion to the desired **F**. Coupling of **F** with bromolactone (±)-**H** furnished (+)-orobanchol (**53**) and its epimer **53′**.

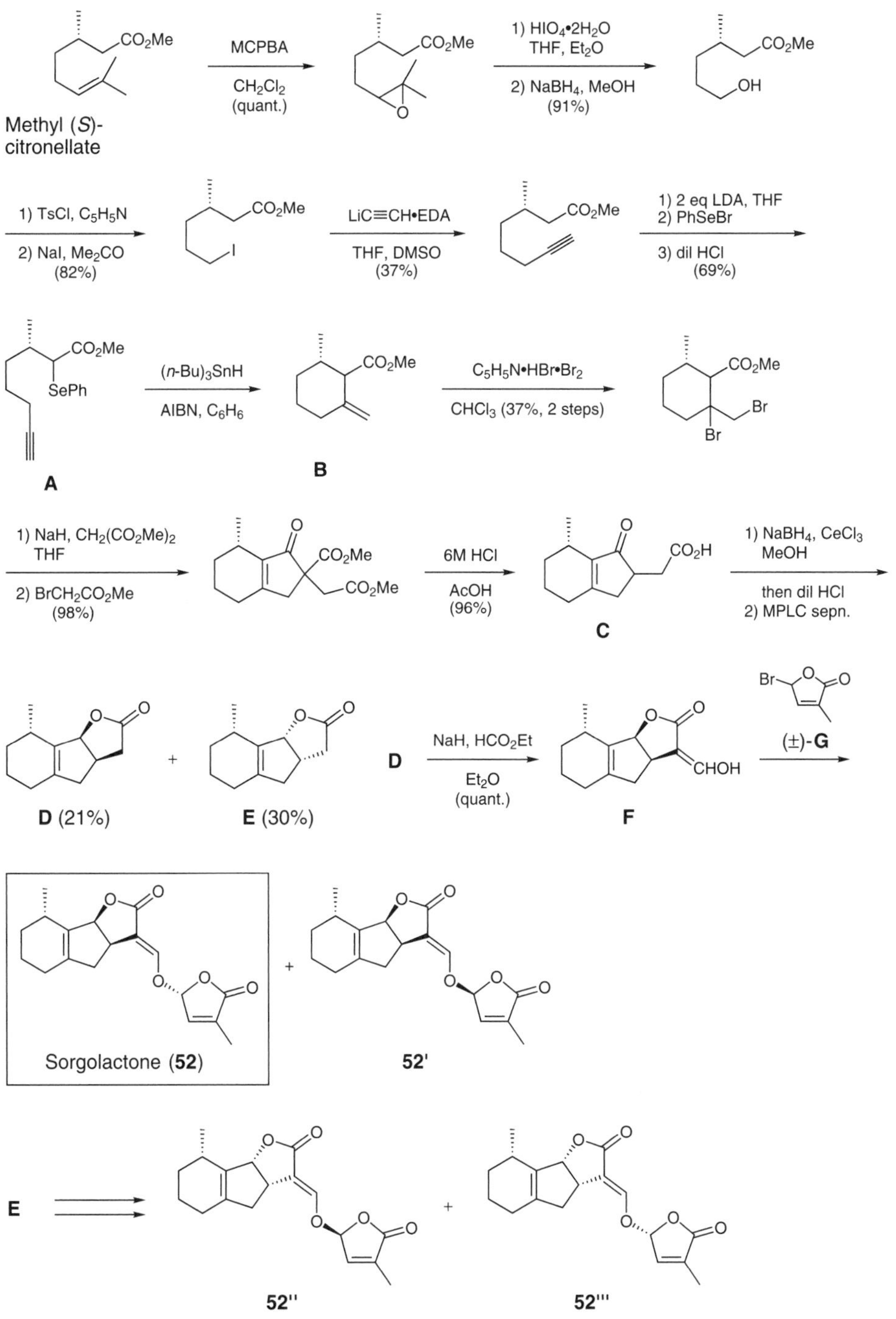

Figure 2.41   *Synthesis of sorgolactone*

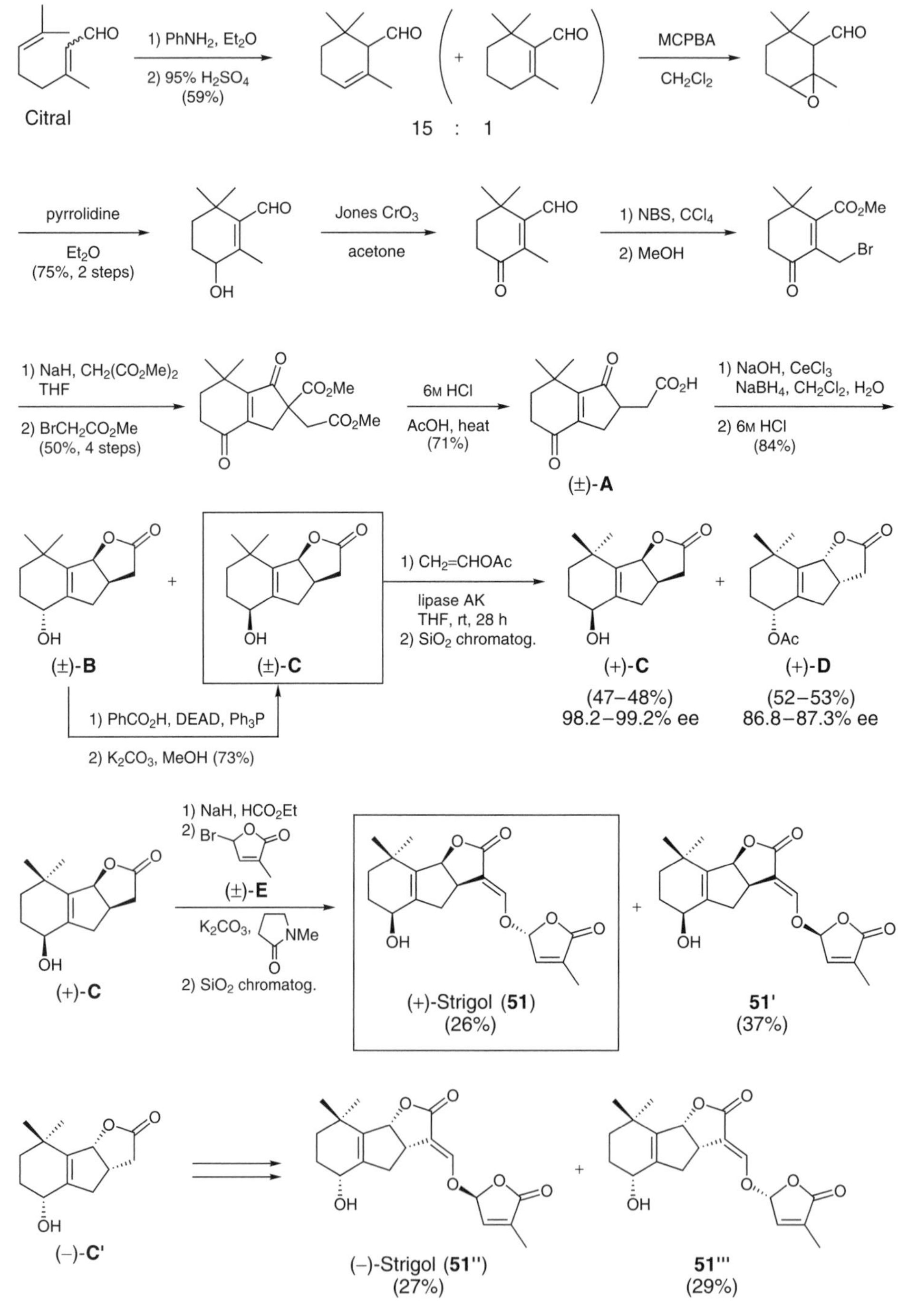

Figure 2.42 *Synthesis of strigol*

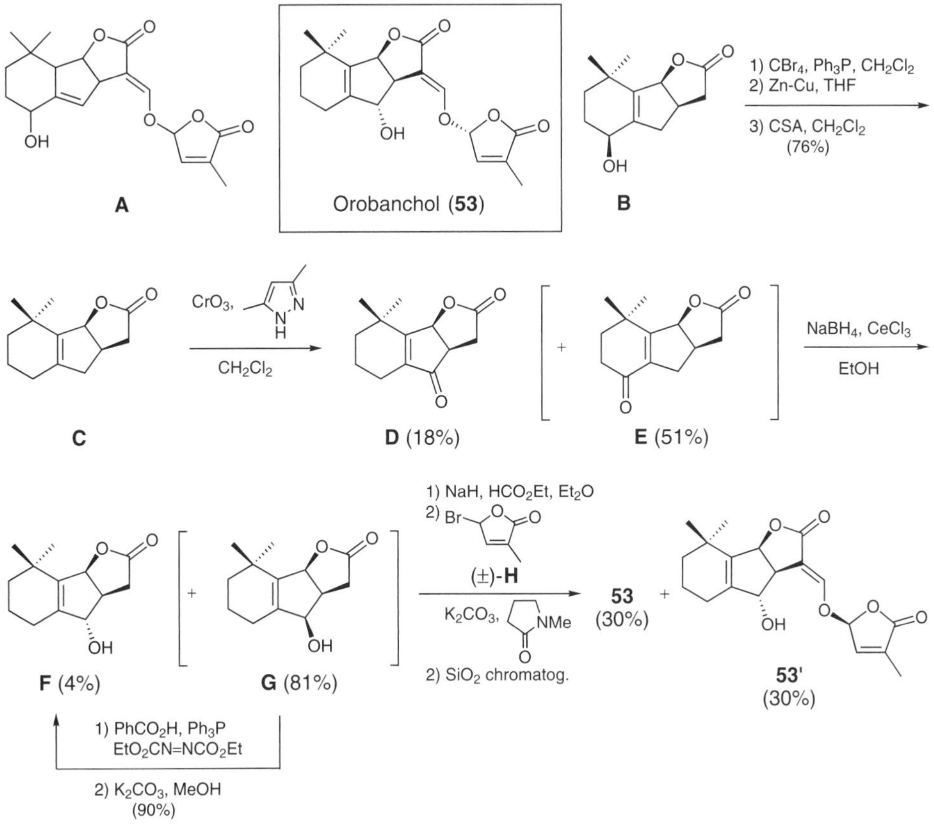

Figure 2.43 *Synthesis of orobanchol*

The trimethylsilyl ether of **53** showed a mass spectrum identical to that of the trimethylsilyl ether of the naturally occurring orobanchol.

In the study of such scarce natural products like strigolactones, synthesis played a very important role to settle their structural problems.

2.4 Other bioactive compounds of plant origin

There are many other phytochemicals that are potentially useful as medicinals, agrochemicals and food ingredients. I will discuss some of them in this section.

2.4.1 Synthesis of arnebinol

In 1983 Sankawa and coworkers isolated arnebinol (**54**) from the root of *Arnebia euchroma*.[116] Its unique *ansa*-type structure **54** as revealed by X-ray analysis together with its bioactivity as an inhibitor of prostaglandin biosynthesis prompted us to synthesize it as shown in Figure 2.44.[117,118]

Geraniol (**A**) was converted to **B**, the precursor for cyclization. Treatment of **B** with potassium carbonate effected deacetylation and cyclization to give arnebinol (**54**) together with isoarnebinol.

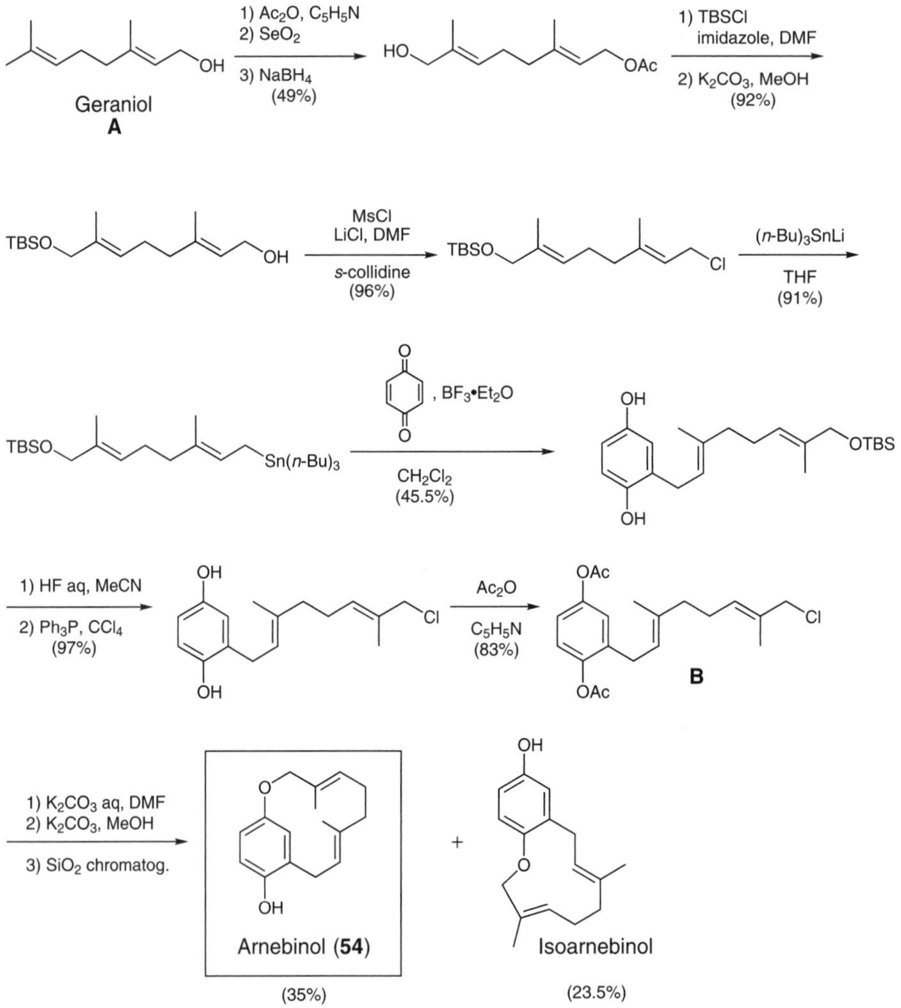

Figure 2.44 *Synthesis of arnebinol. Modified by permission of Shokabo Publishing Co., Ltd*

2.4.2   Synthesis of magnosalicin

Synthesis of (±)-magnosalicin was briefly mentioned in Figure 1.5. In 1984 Sankawa and coworkers isolated a new neolignan named magnosalicin (**55**, Figure 2.45) from *Magnolia salicifolia* as an antiallergy compound.[119] Buds of *M. salicifolia* are known as an oriental medicinal drug, which has been used for nasal allergy and nasal empyema. Magnosalicin (**55**) showed a significant inhibitory effect to histamine release from rat peritoneal mast cells. Its structure as racemic **55** was established by X-ray analysis.

The biogenetic precursor of magnosalicin (**55**), a phenylpropane dimer, might be α-asarone, as suggested by Sankawa.[119] We therefore attempted the oxidation of α-asarone with peracetic acid, and obtained (±)-magnosalicin (**55**) in a single step, as shown in Figure 2.45.[120] Although there were many stereoisomers of **55** in the worked-up mixture, only the desired (±)-**55** was crystalline, and all the other byproducts were oils. Sometimes, biogenetic consideration may simplify the synthetic process.

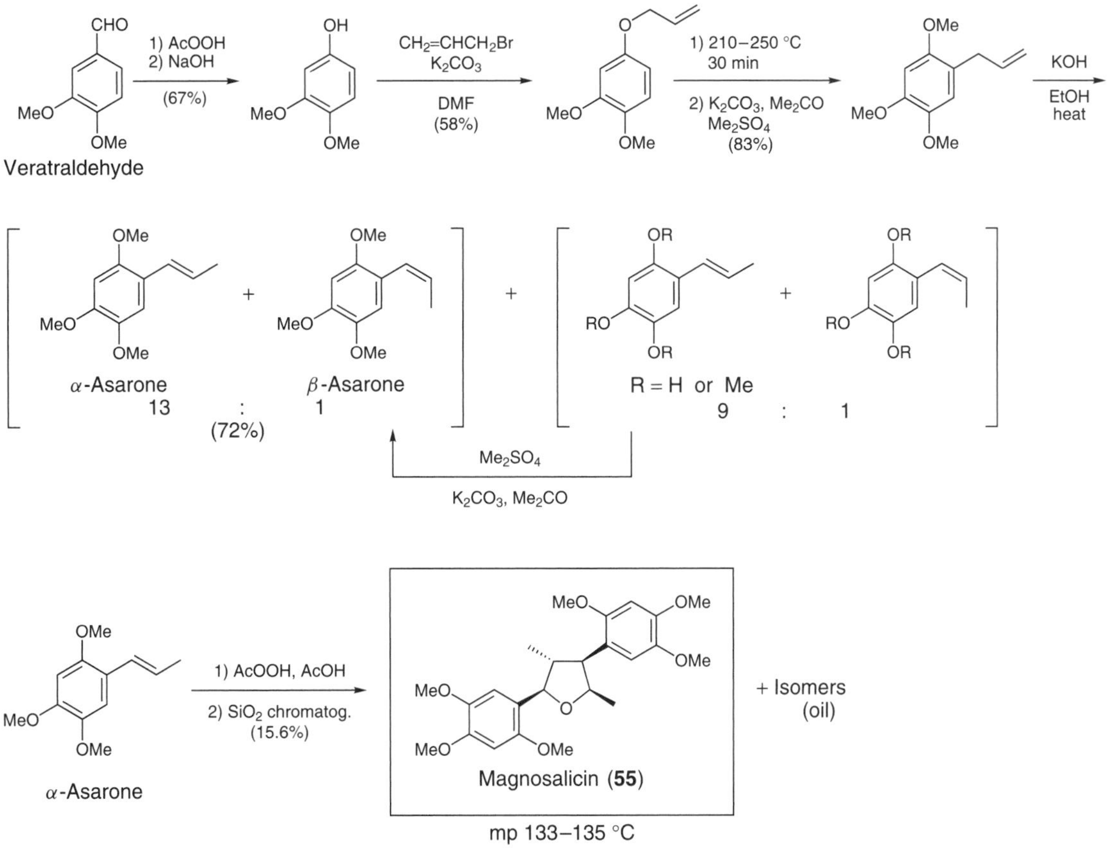

Figure 2.45 *Synthesis of magnosalicin. Modified by permission of Shokabo Publishing Co., Ltd*

2.4.3 Synthesis of hernandulcin

In 1985, Kinghorn and coworkers isolated an extremely sweet bisabolene-type sesquiterpene from an Aztec herb *Lippia dulcis*.[121] This plant was known to the Aztecs as *tzonpelic xihuitl* (sweet herb), and was described in a book written between 1570–1576 by a Spanish physician F. Hernández.[121] Kinghorn named the sesquiterpene (+)-hernandulcin, showed it to be more than 1000 times as sweet as sucrose, and elucidated its structure as **56** (Figure 2.46) including the $(6R^*, 1'R^*)$-relative stereochemistry.

In view of the intense sweetness of (+)-hernandulcin, we became interested in synthesizing all of the four possible stereoisomers of **56** so as to establish the absolute configuration of (+)-hernandulcin and also to clarify the relationship between stereochemistry and taste.[122,123] Figure 2.46 summarizes our synthesis. (R)-(+)-Limonene was epoxidized to give bisepoxide **A**. The less-hindered epoxide in the side-chain was then cleaved with isoprenylmagnesium chloride in the presence of copper(I) iodide to give **B**. Treatment of **B** with sodium phenylselenide gave **C** and **D**, which were separable by chromatography. Phenylselenide **C** was oxidized with hydrogen peroxide to afford **E**. Oxidation of **E** was followed by chromatographic purification to give $(6S, 1'S)$-(+)-hernandulcin (**56** = natural product) and $(6S, 1'R)$-(+)-*epi*-hernandulcin. Similarly, by starting from (S)-(−)-limonene, $(6R, 1'R)$-(−)-hernandulcin and its $(6R, 1'S)$-isomer were synthesized.

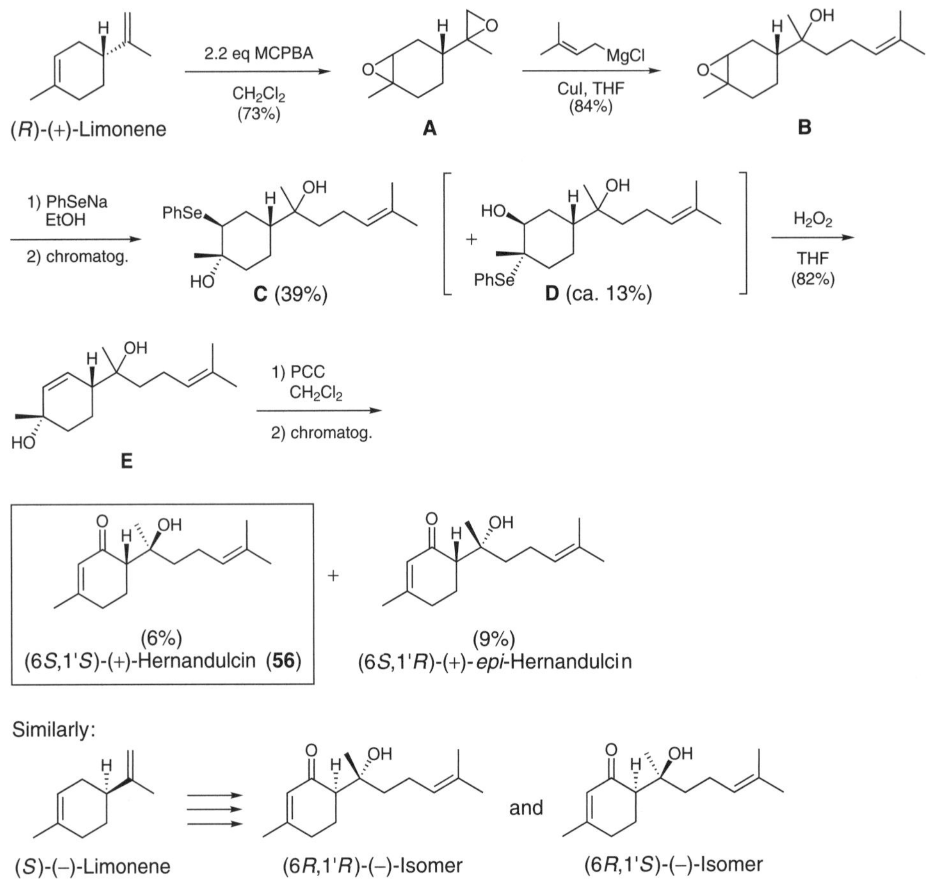

Figure 2.46   *Synthesis of hernandulcin. Modified by permission of Shokabo Publishing Co., Ltd*

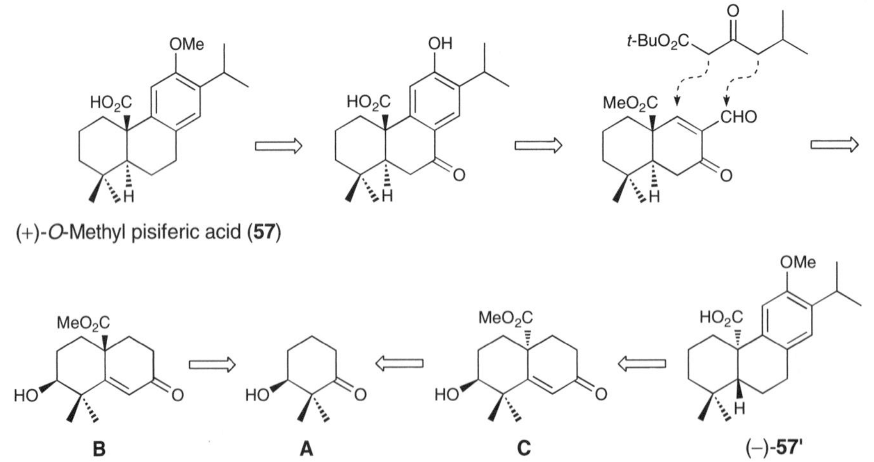

Figure 2.47   *Retrosynthetic analysis of O-methyl pisiferic acid. Modified by permission of Shokabo Publishing Co., Ltd*

The sensory test of the four isomers revealed that **56** was 1100–1200 times as sweet as sucrose. Other three stereoisomers were all bitter and somewhat pungent with no perceptible sweet taste. The naturally occurring (6*S*, 1'*S*)-(+)-hernandulcin (**56**) was the only stereoisomer with sweet taste.

2.4.4   Synthesis of *O*-methyl pisiferic acid

(+)-*O*-Methyl pisiferic acid (**57**, Figure 2.47) was first isolated in 1980 by Yatagai and Takahashi from the leaves of a Japanese tree *Chamaecyparis pisifera*.[124] Discovery of its bioactivity as a mite-growth

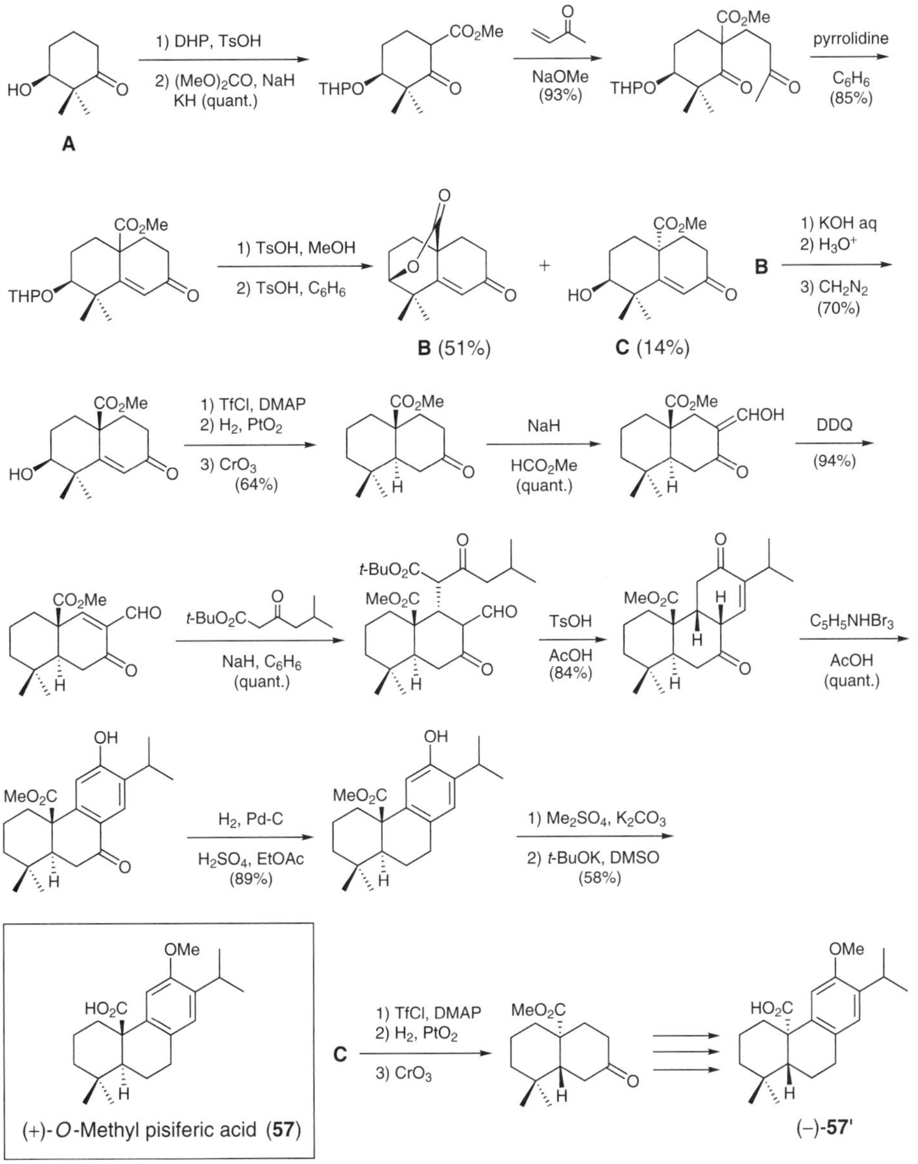

Figure 2.48   *Synthesis of O-methyl pisiferic acid. Modified by permission of Shokabo Publishing Co., Ltd*

regulator by Marumo and coworkers[125] prompted us to synthesize it. This diterpene acid **57** inhibits both the hatching and the feeding of the two-spotted spider mite, *Tetranychus urticae*, which is a serious pest to many crops.

Figure 2.47 shows our retrosynthetic analysis of *O*-methyl pisiferic acid (**57**).[126] The starting material is the same (*S*)-3-hydroxy ketone **A** as employed by us for the synthesis of glycinoeclepin A (see Figure 2.36). The ready availability of **A** by reduction of 2,2-dimethyl-1,3-cyclohexanedione with baker's yeast makes **A** a versatile starting material in terpene synthesis.[104] Conversion of **A** to two diastereomeric compounds **B** and **C** enables the synthesis of both (+)-**57** and (−)-**57′**.

Our synthesis of the enantiomers **57** and **57′** of *O*-methyl pisiferic acid is summarized in Figure 2.48.[126] Annulation of **A** to give bicyclic intermediates was not stereoselective, giving both **B** and **C**. As shown in

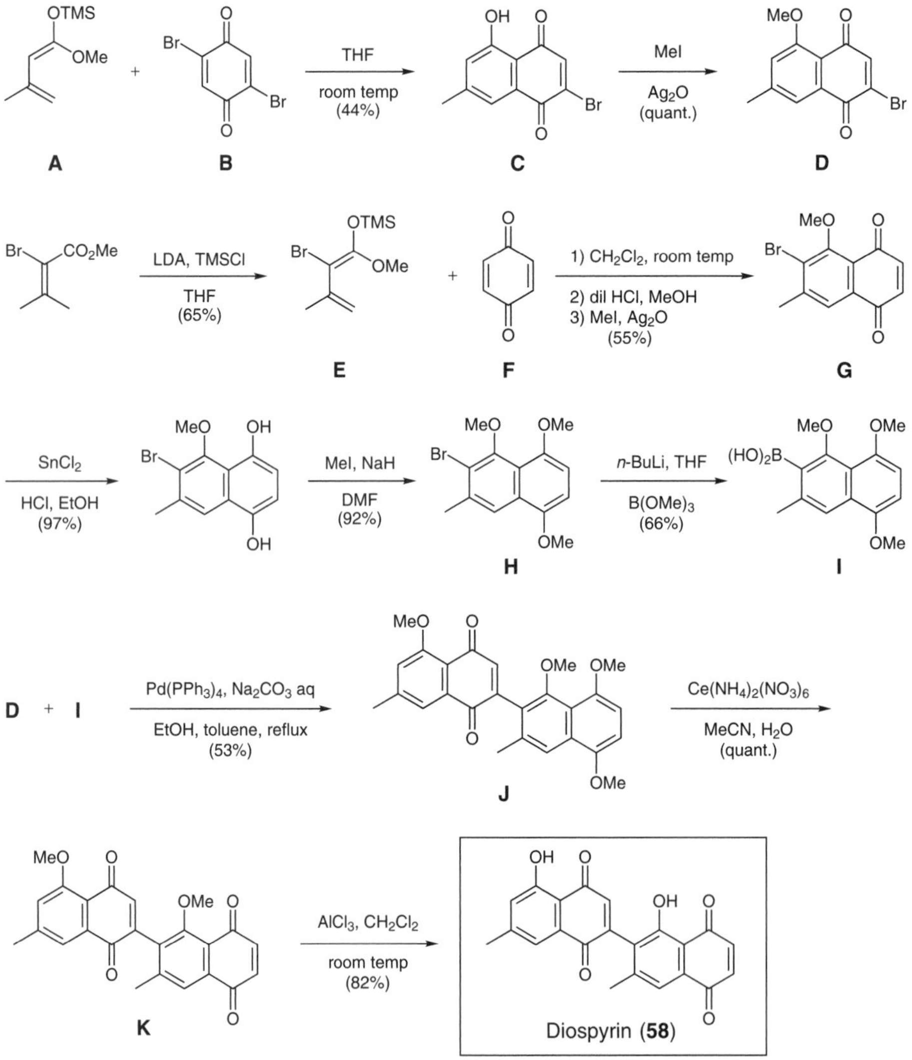

Figure 2.49 *Synthesis of diospyrin*

the figure, **B** was converted to the naturally occurring (+)-*O*-methyl pisiferic acid (**57**), while **C** furnished the unnatural (−)-**57′**. The naturally occurring (+)-**57** showed high toxicity ($LD_{50} = 62.5$ ppm) against mites (*Tetranychus urticae*), while (−)-**57′** was nontoxic even at the dosage of 250 ppm. It thus became clear that the pesticidal activity of **57** depends on its absolute configuration.

2.4.5 Synthesis of diospyrin

Diospyrin was first isolated in 1961 by Kapil and Dhar as an orange-red constituent of an Indian tree, *Diospyros montana*.[127] Its structure as **58** (Figure 2.49) was proposed by Sidhu and Pardhasaradhi in 1967.[128] Hazra and coworkers discovered potentially useful bioactivities of diospyrin such as in vitro antiplasmoidal effects against malaria, leishmaniasis and trypanosomiasis.[129] As requested by Dr. Hazra, we carried out a synthesis of diospyrin as shown in Figure 2.49.[130]

Our plan was to connect the two 7-methyljuglone units **D** and **I** by Suzuki–Miyaura coupling. Diospyrin (**58**) is reported to be optically inactive, and there is no restricted rotation around the connecting bond between C-2 and C-6′. It therefore seemed possible to connect **D** and **I** at the positions C-2 and C-6′ under conventional conditions. Diels–Alder reactions were employed to prepare **D** and **I**. Accordingly, Diels–Alder cycloaddition between **A** and **B** gave **C**, which was methylated to furnish **D**. Similarly, the adduct **G** was prepared from **E** and **F**. The naphthoquinone **G** was converted to boronic acid **I** via bromonaphthalene **H**.

Suzuki–Miyaura coupling of **D** with **I** smoothly afforded **J**, which was oxidized with ceric ammonium nitrate to give **K**. For the final demethylation of **K**, several methods were tried without success: treatment of **K** with boron tribromide, 47% hydrobromic acid, lithium chloride in DMF, trimethylsilyl iodide, and potassium thiophenolate in diethylene glycol. Finally, clean demethylation of **K** was achieved with aluminum chloride in dichloromethane at room temperature to give diospyrin (**58**) as orange plates. The proposed structure **58** was thus confirmed by its synthesis.[130]

2.4.6 Synthesis of mispyric acid

In 1999, Hecht and coworkers isolated mispyric acid (**59**, Figure 2.50), a monocyclic triperpene dicarboxylic acid, from the stem bark of an Australian plant *Mischocarpus pyriformis* as an inhibitor of DNA

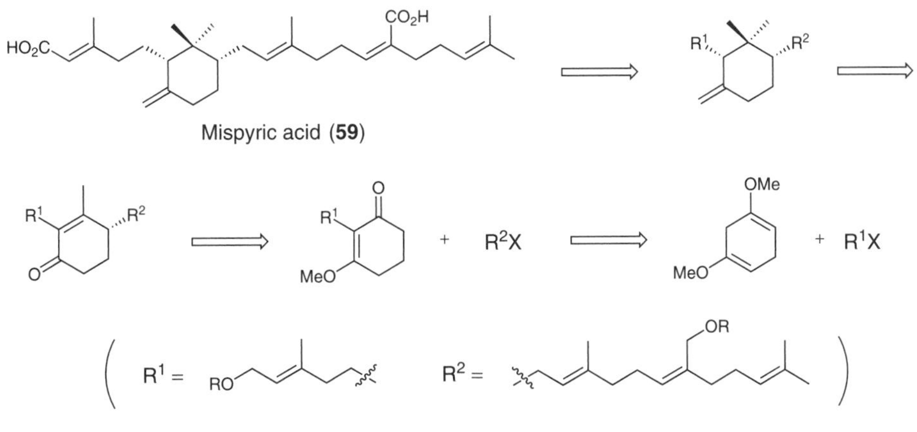

Mispyric acid (**59**)

Figure 2.50 *Retrosynthetic analysis of mispyric acid*

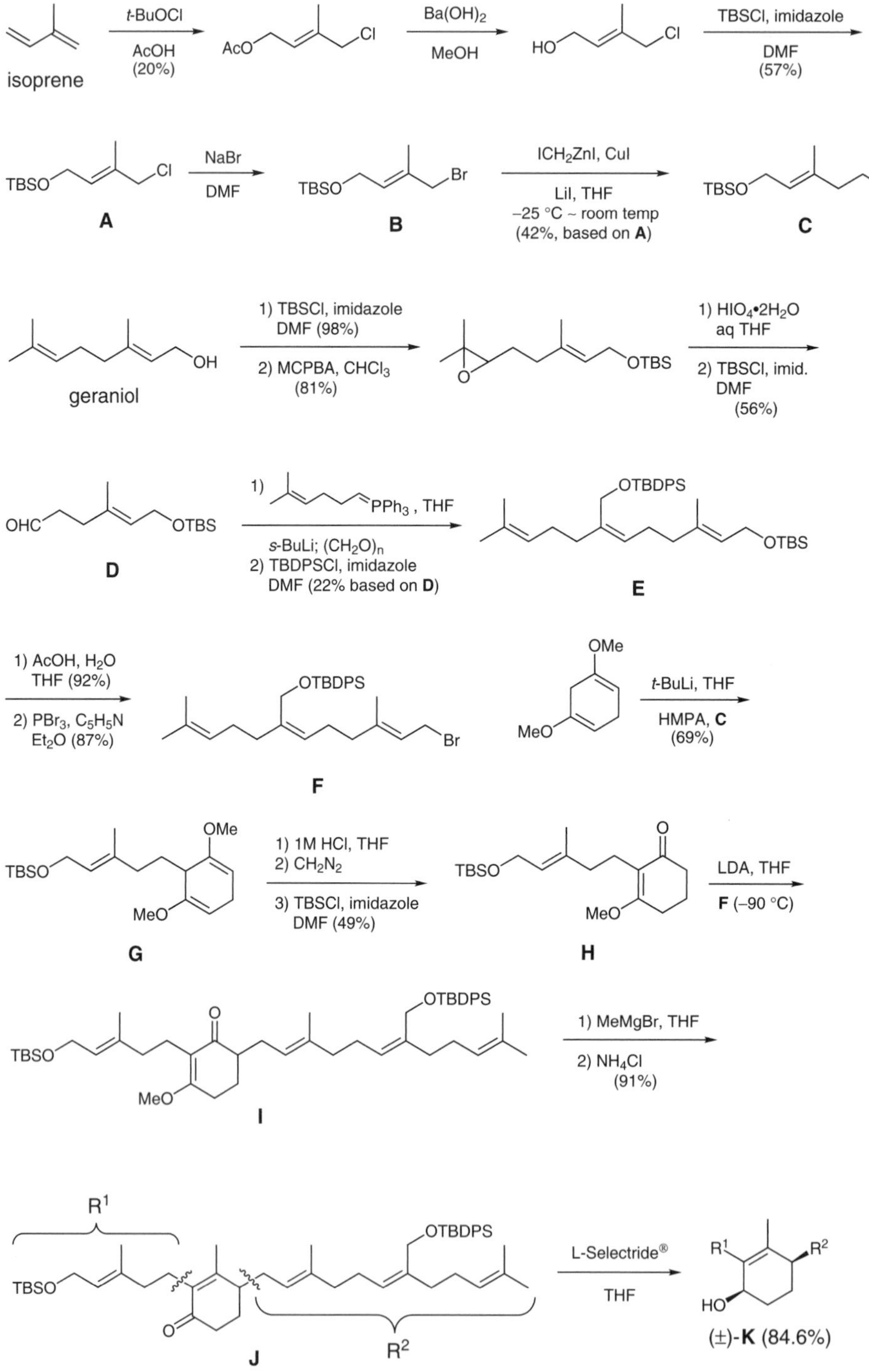

Figure 2.51   *Synthesis of mispyric acid (1)*

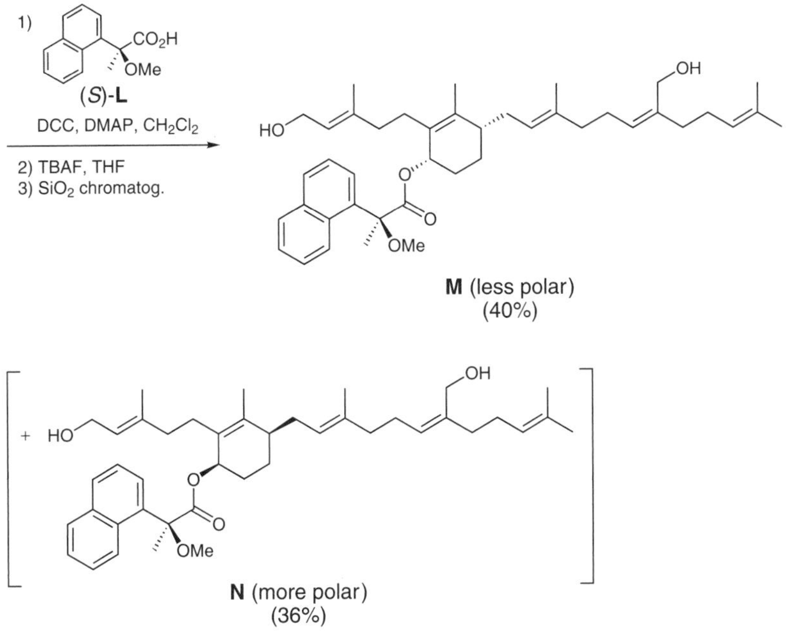

M (less polar)
(40%)

N (more polar)
(36%)

Figure 2.51   *(continued)*

polymerase β.[131] Since the absolute configuration of mispyric acid was unknown, we synthesized both the enantiomers of **59**, and (2*S*, 4*S*)-(+)-**59** was shown to be the natural product.[132,133]

Figure 2.50 shows our synthetic plan of mispyric acid (**59**). 1,5-Dimethoxy-1,4-cyclohexadiene was chosen as the starting material, and its stepwise alkylation with two different alkylating agents, R¹X and R²X, was thought to provide **59**. Of course, optical resolution of an intermediate was necessary to synthesize both the enantiomers of **59**.

The early phase of the synthesis is summarized in Figure 2.51. The first alkylating agent **C** was prepared from isoprene via **A** and **B**. Then, the second one **F** was prepared from geraniol via **D** and **E**. The first alkylation of 1,5-dimethoxy-1,4-cyclohexadiene with **C** yielded **G**, which was converted to **H**. Alkylation of **H** with **F** furnished **I**, to which was added a methyl group to give **J**. Reduction of **J** with L-Selectride® afforded alcohol (±)-**K**, which was esterified with Harada's resolving agent (*S*)-**L**.[134] The resulting esters were inseparable. Fortunately, however, diol ester **M** and **N** were separable and they were obtained in pure forms. Other attempts failed to resolve (±)-**K**.

Figure 2.52 shows the later part of the synthesis. The pure diastereomer **A** was treated with *t*-butyldimethylsilyl chloride, and the product was reduced to give alcohol **B**. Cyclopropanation of the allylic alcohol **B** furnished **C**, which was oxidized to afford bicyclic ketone **D**. Dissolving metal reduction of **D** gave **E**. At this stage, the absolute configuration of (−)-**E** was elucidated by examining its CD spectrum as shown in Figure 2.53.

In the CD spectrum of (−)-**E**, a negative Cotton effect was observed, while (+)-**E′** showed a positive Cotton effect. Accordingly, the octant projection of (−)-**E** and (+)-**E′** must be as shown in Figure 2.53, and (−)-**E** was (2*R*, 4*S*)-**E**, while (+)-**E′** was (2*S*, 4*R*)-**E′**. This is a typical example of application of the octant rule in determining the absolute configuration of substituted cyclohexanones.

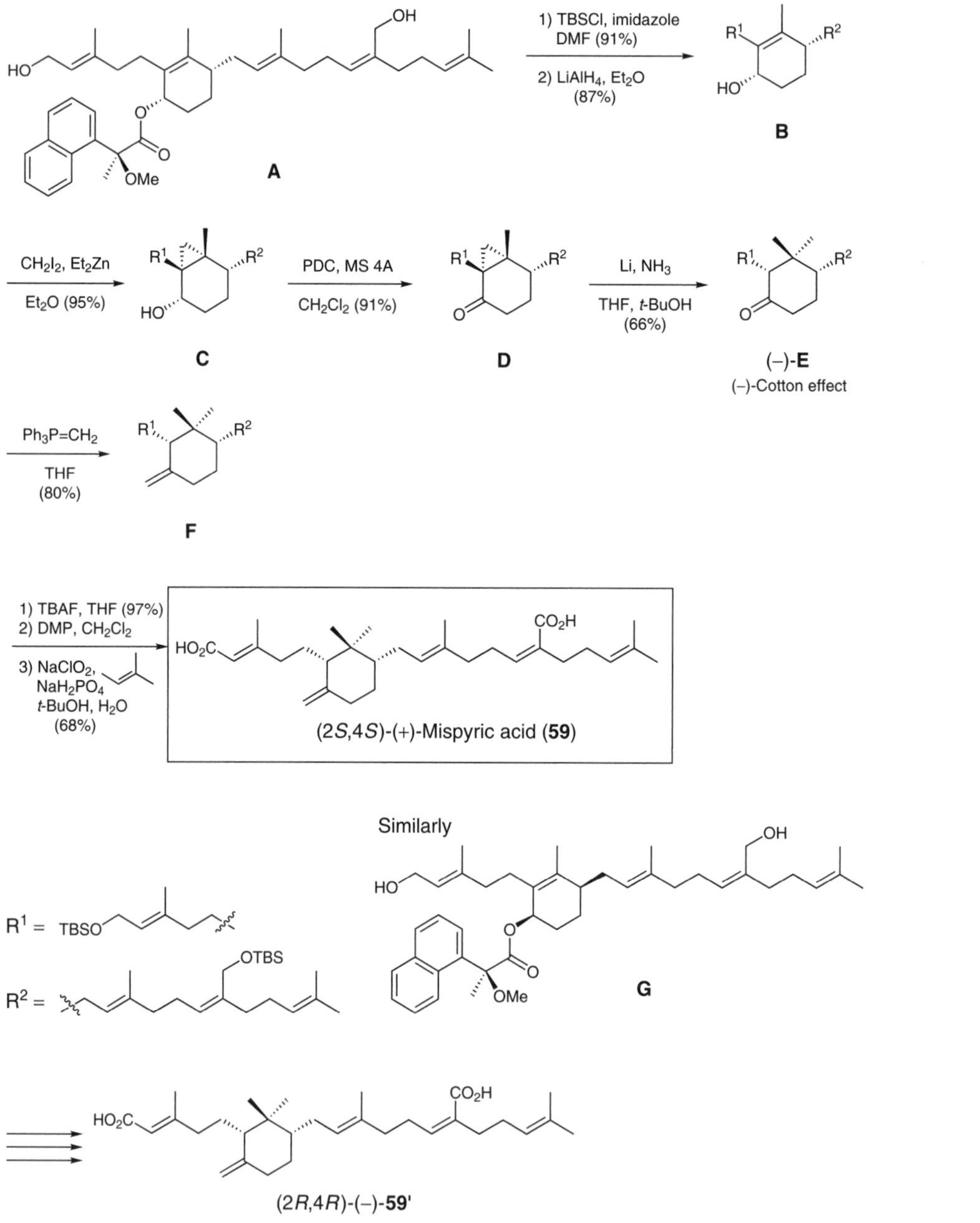

Figure 2.52   *Synthesis of mispyric acid (2)*

To continue the synthesis (Figure 2.52), (−)-ketone **E** was treated with methylene triphenylphosphorane to give **F**. Deprotection followed by oxidation of **F** afforded the naturally occurring (+)-mispyric acid (**59**), whose absolute configuration was determined as 2*S*, 4*S*. Similarly, the other diastereomer **G** obtained by optical resolution afforded unnatural (2*R*, 4*R*)-(−)-**59'**.

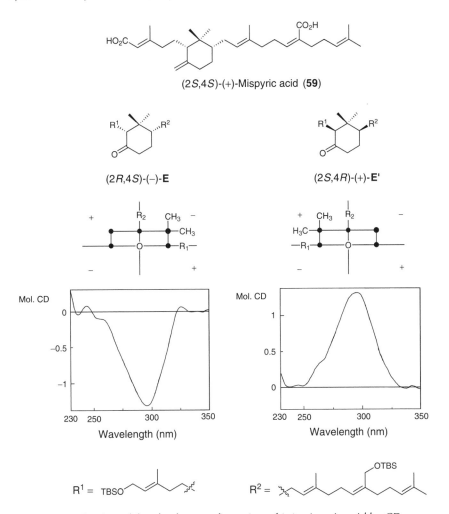

Figure 2.53   *Determination of the absolute configuration of (+)-mispyric acid by CD measurements*

In this chapter we discussed the syntheses of fifty-nine plant-related and bioactive natural products. Phytochemicals will continue to be targets for synthetic chemists due to their structural diversities and biological importance.

References

1.   Kögl, F.; Haagen-Smit, A.J.; Erxleben, H. *Hoppe-Seyler's Z. Physiol. Chem.* **1934**, *228*, 90–112.
2.   Majima, R.; Hoshino, T. *Ber. Deutsch. Chem. Ges.* **1925**, *53*, 2042–2062.
3.   Yabuta, T.; Sumiki, Y. *Nippon Nôgeikagaku Kaishi* (*J. Agric. Chem. Soc. Jpn.*) **1938**, *14*, 1526.
4.   Ohkuma, K.; Addicott, F.T.; Smith, O.E.; Thiessen, W.E. *Tetrahedron Lett.* **1965**, 2529–2535.
5.   Letham, D.S. *Life Sci.* **1963**, *2*, 569–573.
6.   Grove, M.D.; Spencer, G.F.; Rohwedder, W.K.; Mandava, N.; Worley, J.F.; Warthen, Jr., J.D.; Steffens, G.L.; Flippen-Anderson, J.L; Cook, Jr., J.C. *Nature* **1979**, *281*, 216–217.

7.   Ueda, J.; Kato, J. *Agric. Biol. Chem*. **1982**, *46*, 1975–1976.
8.   Cross, B.E.; Grove, J.F.; MacMillan, J.; Mulholland, T.P.C *Chem. Ind*. **1956**, 954–955.
9.   Takahashi, N.; Seta, Y.; Kitamura, H.; Sumiki, Y. *Bull. Agric. Chem. Soc. Jpn*. **1958**, *22*, 432–433.
10.   Cross, B.E.; Grove, J.F.; MacMillan, J.; Moffatt, J.S.; Mulholland, T.P.C; Seaton, J.C.; Sheppard, N. *Proc. Chem. Soc*. **1959**, 302–303.
11.   Mori, K.; Matsui, M.; Sumiki, Y. *Agric. Biol. Chem*. **1961**, *25*, 205–222.
12.   Mori, K.; Matsui, M.; Sumiki, Y. *Agric. Biol. Chem*. **1961**, *25*, 902–906.
13.   Mori, K.; Matsui, M.; Sumiki, Y. *Agric. Biol. Chem*. **1961**, *25*, 907–914.
14.   Mori, K.; Matsui, M.; Sumiki, Y. *Agric. Biol. Chem*. **1963**, *27*, 22–26.
15.   Mori, K.; Matsui, M.; Sumiki, Y. *Agric. Biol. Chem*. **1964**, *28*, 72–73.
16.   Mori, K.; Matsui, M.; Sumiki, Y. *Agric. Biol. Chem*. **1964**, *28*, 243–247.
17.   Mori, K.; Matsui, M.; Sumiki, Y. *Agric. Biol. Chem*. **1962**, *26*, 783–784.
18.   Mori, K.; Matsui, M.; Sumiki, Y. *Agric. Biol. Chem*. **1963**, *27*, 537–542.
19.   Mori, K.; Matsui, M.; Sumiki, Y. *Agric. Biol. Chem*. **1963**, *27*, 27–30.
20.   House, H.O.; Sauter, F.J.; Kenyon, W.G.; Riehl, J.-J. *J. Org. Chem*. **1968**, *33*, 957–961.
21.   Loewenthal, H.J.E; Malhotra, S.K. *Proc. Chem. Soc*. **1962**, 230–231.
22.   Loewenthal, H.J.E; Malhotra, S.K. *J. Chem. Soc*. **1965**, 990–994.
23.   Mori, K.; Matsui, M.; Sumiki, Y. *Agric. Biol. Chem*. **1964**, *28*, 179–183.
24.   Mori, K.; Matsui, M.; Sumiki, Y. *Tetrahedron Lett*. **1964**, 1803–1807.
25.   Mori, K.; Shiozaki, M.; Itaya, N.; Matsui, M.; Sumiki, Y. *Tetrahedron* **1969**, *25*, 1293–1321.
26.   Mori, K.; Ogawa, T.; Itaya, N.; Matsui, M.; Sumiki, Y. *Tetrahedron* **1969**, *25*, 1281–1291.
27.   Mori, K.; Shiozaki, M.; Itaya, N.; Ogawa, T.; Matsui, M.; Sumiki, Y. *Tetrahedron Lett*. **1968**, 2183–2188.
28.   Mori, K.; Shiozaki, M.; Matsui, M.; Sumiki, Y. *Proc. Jpn. Acad*. **1968**, *44*, 717–720.
29.   Cross, B.E.; Hanson, J.R.; Speake, R.N. *J. Chem. Soc*. **1965**, 3555–3563.
30.   Mori, K.; Matsui, M.; Sumiki, Y. *Tetrahedron Lett*. **1970**, 429–432.
31.   Mori, K.; Matsui, M.; Sumiki, Y. *Proc. Jpn. Acad*. **1970**, *46*, 450–452.
32.   Mori, K. *Tetrahedron* **1971**, *27*, 4907–4919.
33.   Takemoto, I.; Mori, K.; Matsui, M. *Agric. Biol. Chem*. **1976**. *40*, 251–253.
34.   Mori, K.; Takemoto, I.; Matsui, M. *Tetrahedron* **1976**, *32*, 1497–1502.
35.   Galt, R.H.B; Hanson, J.R. *J. Chem. Soc*. **1965**, 1565–1570.
36.   Cross, B.E.; Norton, K. *J. Chem. Soc*. **1965**, 1570–1572.
37.   Mander, L.N. *Nat. Prod. Rep*. **1988**, 541–579.
38.   Nagaoka, H.; Shimano, M.; Yamada, Y. *Tetrahedron Lett*. **1989**, *30*, 971–974.
39.   Toyota, M.; Yokota, M; Ihara, M. *J. Am. Chem. Soc*. **2001**, *123*, 1856–1861.
40.   Goldsmith, D. In *The Total Synthesis of Natural Products, Vol. 8*, ApSimon, J. ed., Wiley; New York, **1992**, pp 1–243.
41.   Mori, K.; Matsui, M. *Tetrahedron Lett*. **1966**, 175–180.
42.   Mori, K.; Matsui, M.; Ikekawa, N.; Sumiki, Y. *Tetrahedron Lett*. **1966**, 3395–3400.
43.   Mori, K.; Matsui, M. *Tetrahedron* **1968**, *24*, 3095–3111.
44.   Mori, K.; Matsui, M. *Tetrahedron Lett*. **1965**, 2347–2350.
45.   Mori, K.; Matsui, M. *Tetrahedron* **1966**, *22*, 879–884.
46.   Nakahara, Y.; Mori, K.; Matsui, M. *Agric. Biol. Chem*. **1971**, *35*, 918–928.
47.   Mori, K.; Nakahara, Y.; Matsui, M. *Tetrahedron Lett*. **1970**, 2411–2414.
48.   Mori, K.; Nakahara, Y.; Matsui, M. *Tetrahedron* **1972**, *28*, 3217–3226.
49.   Mori, K.; Matsui, M. *Tetrahedron Lett*. **1966**, 1633–1635.
50.   Mori, K.; Matsui, M.; Fujisawa, N. *Tetrahedron* **1968**, *24*, 3113–3125.
51.   Mori, K.; Aki, S. *Liebigs Ann. Chem*. **1993**, 97–98.
52.   Cornforth, J.W.; Draber, W.; Milborrow, B.V.; Ryback, G. *Chem. Commun*. **1967**, 114–116.
53.   Burden, R.S.; Taylor, H.F. *Tetrahedron Lett*. **1970**, 4071–4074.
54.   Ryback, G. *J. Chem. Soc., Chem. Commun*. **1972**, 1190–1191.
55.   Harada, N. *J. Am. Chem. Soc*. **1973**, *95*, 240–242.
56.   Oritani, T.; Yamashita, K. *Tetrahedron Lett*. **1972**, 2521–2524.

57. Koreeda, M.; Weiss, G; Nakanishi, K. *J. Am. Chem. Soc*. **1973**, *95*, 239–240.
58. Mori, K. *Tetrahedron Lett*. **1973**, 723–726.
59. Mori, K. *Tetrahedron Lett*. **1973**, 2635–2638.
60. Mori, K. *Tetrahedron* **1974**, *30*, 1065–1072.
61. Takasugi, M.; Anetai, M.; Katsui, N.; Masamune, T. *Chem. Lett*. **1973**, 245–248.
62. Fung, S.; Siddall, J.B. *J. Am. Chem. Soc*. **1980**, *102*, 6580–6581.
63. Ishiguro, M.; Takatsuto, S.; Morisaki, M.; Ikekawa, N. *J. Chem. Soc., Chem. Commun*. **1980**, 962–964.
64. Mori, K. *Rev. Latinoamer. Quim*. **1985**, *16*, 55–59.
65. Mori, K. *Agric. Biol. Chem*. **1980**, *44*, 1211–1212.
66. Mori, K.; Sakakibara, M.; Ichikawa, Y.; Ueda, H.; Okada, K.; Umemura, T.; Yabuta, G.; Kuwahara, S.; Kondo, M.; Minobe, M.; Sogabe, A. *Tetrahedron* **1982**, *38*, 2099–2109.
67. Sakakibara, M.; Okada, K.; Ichikawa, Y.; Mori, K. *Heterocycles* **1982**, *17*, 301–304.
68. Mori, K.; Sakakibara, M.; Okada, K. *Tetrahedron* **1984**, *40*, 1767–1781.
69. Aburatani, M.; Takeuchi, T.; Mori, K. *Agric. Biol. Chem*. **1985**, *49*, 3557–3562.
70. Aburatani, M.; Takeuchi, T.; Mori, K. *Agric. Biol. Chem*. **1986**, *50*, 3043–3047.
71. Aburatani, M.; Takeuchi, T.; Mori, K. *Synthesis* **1987**, 181–183.
72. Yokota, T.; Baba, J.; Takahashi, N. *Tetrahedron Lett*. **1982**, *23*, 4965–4966.
73. Okada, K.; Mori, K. *Agric. Biol. Chem*. **1983**, *47*, 925–926.
74. Kondo, M.; Mori, K. *Agric. Biol. Chem*. **1983**, *47*, 97–102.
75. Yokota, T.; Baba, J.; Takahashi, N. *Agric. Biol. Chem*. **1983**, *47*, 1409–1411.
76. Baba, J.; Yokota, T.; Takahashi, N. *Agric. Biol. Chem*. **1983**, *47*, 659–661.
77. Sakakibara, M.; Mori, K. *Agric. Biol. Chem*. **1983**, *47*, 1405–1406.
78. Sakakibara, M.; Mori, K. *Agric. Biol. Chem*. **1983**, *47*, 1407–1408.
79. Sakakibara, M.; Mori, K. *Agric. Biol. Chem*. **1984**, *48*, 745–752.
80. Kim, S.K.; Yokota, T.; Takahashi, N. *Agric. Biol. Chem*. **1987**, *51*, 2303–2305.
81. Mori, K.; Takeuchi, T. *Liebigs Ann. Chem*. **1988**, 815–818.
82. Hetru, C.; Roussel, J.-P.; Mori, K.; Nakatani, Y. *C.R. Acad. Sci. Paris* **1986**, Ser. II, *302*, 417–420.
83. Schildknecht, H. *Angew. Chem. Int. Ed*. **1983**, *22*, 695–710.
84. Ueda, M.; Shigemori-Suzuki, T.; Yamamura, S. *Tetrahedron Lett*. **1995**, *36*, 6267–6270.
85. Mori, K.; Audran, G.; Nakahara, Y.; Bando, M.; Kido, M. *Tetrahedron Lett*. **1997**, *38*, 575–578.
86. Audran, G.; Mori, K. *Eur. J. Org. Chem*. **1998**, 57–62.
87. Cruickshank, I.A.M; Perrin, D.R. *Nature* **1960**, *187*, 799–800.
88. Perrin, D.D.; Perrin, D.R. *J. Am. Chem. Soc*. **1962**, *84*, 1922–1925.
89. Ingham, J.L.; Markham, K.R. *Phytochemistry* **1980**, *19*, 1203–1207.
90. Mori, K.; Kisida, H. *Liebigs Ann. Chem*. **1988**, 721–723.
91. Mori, K.; Kisida, H. *Liebigs Ann. Chem*. **1989**, 35–39.
92. Uchiyama, M.; Matsui, M. *Agric. Biol. Chem*. **1967**, *31*, 1490–1498.
93. Hirai, N.; Ishida, H.; Koshimizu, K. *Phytochemisry* **1994**, *37*, 383–385.
94. Takikawa, H.; Yoshida, M.; Mori, K. *Biosci. Biotechnol. Biochem*. **1999**, *63*, 1834–1836.
95. Akatsuka, T.; Kodama, O.; Kato, H.; Kono, Y.; Takeuchi, S. *Agric. Biol. Chem*. **1983**, *47*, 445–447.
96. Mori, K.; Waku, M. *Tetrahedron* **1985**, *41*, 5653–5660.
97. Koga, J.; Shimura, M.; Oshima, K.; Ogawa, N.; Yamauchi, T.; Ogasawara, N. *Tetrahedron* **1995**, *51*, 7907–7918.
98. Koga, J.; Ogawa, N.; Yamauchi, T.; Kikuchi, M.; Ogasawara, N.; Shimura, M. *Phytochemistry* **1997**, *44*, 249–253.
99. Yajima, A.; Mori, K. *Tetrahedron Lett*. **2000**, *41*, 351–354.
100. Yajima, A.; Mori, K. *Eur. J. Org. Chem*. **2000**, 4079–4091.
101. Fukuzawa, A.; Furusaki, A.; Ikura, M.; Masamune, T. *J. Chem. Soc., Chem. Commun*. **1985**, 222–224.
102. Mori, K.; Watanabe, H. *Pure Appl. Chem*. **1989**, *61*, 543–546.
103. Watanabe, H.; Mori, K. *J. Chem. Soc., Perkin Trans. 1*, **1991**, 2919–2934.
104. Mori, K.; Mori, H. *Org. Synth. Col. Vol*. **1993**, *8*, 312–315.
105. Mori, K. *Synlett* **1995**, 1097–1109.

106. Cook, C.E.; Whichard, L.P.; Wall, M.E.; Egley, G.H.; Coggon, P.; Luhan, P.A.; McPhail, A.T. *J. Am. Chem. Soc.* **1972**, *94*, 6198–6199.

107. Hauck, C.; Müller, S.; Schildknecht, H. *J. Plant Physiol.* **1992**, *139*, 474–478.

108. Yokota, T.; Sakai, H.; Okuno, K.; Yoneyama, K.; Takeuchi, Y. *Phytochemistry* **1998**, *49*, 1967–1973.

109. Xie, X.; Yoneyama, K.; Kusumoto, D.; Yamada, Y.; Takeuchi, Y.; Sugimoto, Y.; Yoneyama, K. *Tetrahedron Lett.* **2008**, *49*, 2066–2068.

110. Mori, K.; Matsui, J.; Bando, M.; Kido, M.; Takeuchi, Y. *Tetraheron Lett.* **1997**, *38*, 2507–2510.

111. Mori, K.; Matsui, J. *Tetrahedron Lett.* **1997**, *38*, 7891–7892.

112. Matsui, J.; Bando, M.; Kido, M.; Takeuchi, Y.; Mori, K. *Eur. J. Org. Chem.* **1999**, 2183–2194.

113. Hirayama, K.; Mori, K. *Eur. J. Org. Chem.* **1999**, 2211–2217.

114. Mori, K.; Matsui, J.; Yokota, T.; Sakai H.; Bando, M.; Takeuchi, Y. *Tetrahedron Lett.* **1999**, *40*, 943–946.

115. Matsui, J.; Yokota, T.; Bando, M.; Takeuchi, Y.; Mori, K. *Eur. J. Org. Chem.* **1999**, 2201–2210.

116. Yao, X.-S.; Ebizuka, Y.; Noguchi, H.; Kiuchi, F.; Iitaka, Y.; Sankawa, U.; Seto, H. *Tetrahedron Lett.* **1983**, *24*, 2407–2410.

117. Mori, K.; Sakakibara, M.; Waku, M. *Tetrahedron Lett.* **1984**, *25*, 1085–1086.

118. Mori, K.; Waku, M.; Sakakibara, M. *Tetrahedron* **1985**, *41*, 2825–2830.

119. Tsuruga, T.; Ebizuka, Y.; Nakajima, J.; Chun, Y.-T.; Noguchi, H.; Iitaka, Y.; Sankawa, U.; Seto, H. *Tetrahedron Lett.* **1984**, *25*, 4129–4132.

120. Mori, K.; Komatsu, M.; Kido, M.; Nakagawa, K. *Tetrahedron* **1986**, *42*, 523–528.

121. Compadre, C.M.; Pezzuto, J.M.; Kinghorn, A.D.; Kamath, S.K. *Science* **1985**, *227*, 417–419.

122. Mori, K.; Kato, M. *Tetrahedron Lett.* **1986**, *27*, 981–982.

123. Mori, K.; Kato, M. *Tetrahedron* **1986**, *42*, 5895–5900.

124. Yatagai, M.; Takahashi, T. *Phytochemistry* **1980**, *19*, 1149–1151.

125. Ahn, J.-W.; Wada, K.; Marumo, S.; Tanaka, H.; Osaka, Y. *Agric. Biol. Chem.* **1984**, *48*, 2167–2169.

126. Mori, K.; Mori, H. *Tetrahedron* **1986**, *42*, 5531–5538.

127. Kapil, R.S.; Dhar, M.M. *J. Sci. Industr. Res.* **1961**, 20B, 498–500.

128. Sidhu, G.S.; Pardhasaradhi, M. *Tetrahedron Lett.* **1967**, 1313–1316.

129. Hazra, B.; Ghosh, R.; Banerjee, A.; Kirby, G.C.; Warhust, D.C.; Phillipson, J.D. *Phytotherapy Res.* **1995**, *9*, 72–74.

130. Yoshida, M.; Mori, K. *Eur. J. Org. Chem.* **2000**, 1313–1317.

131. Sun, D.-A.; Deng, J.-Z.; Starck, S.R.; Hecht, S.M. *J. Am. Chem. Soc.* **1999**, *121*, 6120–6124.

132. Imamura, Y.; Takikawa, H.; Mori, K. *Tetrahedron Lett.* **2002**, *43*, 5743–5746.

133. Imamura, Y.; Takikawa, H.; Sasaki, M.; Mori, K. *Org. Biomol. Chem.* **2004**, *2*, 2236–2244.

134. Harada, N.; Watanabe, M.; Kuwahara, S.; Sugio, A.; Kasai, Y.; Ichikawa, A. *Tetrahedron:Asymmetry* **2000**, *11*, 1249–1253.

3

Synthesis of Insect Bioregulators Other than Pheromones

Metamorphosis of insects is regulated by hormones. My synthetic works on insect juvenile hormones since 1967 finally resulted in the preparation of the pure enantiomers of juvenile hormones to examine their bioactivity. Plants and lower animals produce insect antifeedants and repellents as their defense agents against insects. These topics will be treated in this chapter.

3.1 Insect juvenile hormones

3.1.1 What are insect hormones?

Pivotal roles of insect hormones in the metamorphosis of insects were first noticed in 1917 by a Polish zoologist Kopéc. Since then a number of terpenoids, steroids and peptides were found to be insect hormones controlling the metamorphosis of insects.

(a) Juvenile hormones (JHs). In nymphs and larvae of insects, the presence of JH ensures that each moult is to another immature stage. In other words, JH makes insects maintain their current stage. The absence of JH makes the pupa moult to the adult. JH is also known to have a function in reproduction by promoting the development of ovaries. Since the discovery of JH I by Röller *et al*. in 1967, seven JHs have been identified from various different insects. Their structures are shown in Figure 3.1.

(b) Moulting hormones (Ecdysteroids). Steroid hormones collectively called ecdysteroids mediates the moulting, or ecdysis, of insects. In 1956 Butenandt *et al*. isolated 25 mg of a moulting hormone from 500 kg of male pupae of the silkworm moth, *Bombyx mori*. The hormone was named α-ecdysone, and its structure was solved by an X-ray analysis by Hoppe and Huber in 1965. β-Ecdysone (2.5 mg) was then isolated as the second component of the moulting hormone of *B. mori*.

(c) Insect neuropeptides. These are many insect neuropeptides that control the insect life. For example, 5.4 μg of prothoracicotropic hormone (PTTH) was isolated by Suzuki *et al*. from brains of 5×10^5 heads (3.75 kg) of *B. mori*. PTTH is produced at the anterior central part of the brain, and activates the prothoracic gland to produce ecdysteroids. As I have not worked on peptide synthesis, this type of hormone is not discussed here.

Chemical Synthesis of Hormones, Pheromones and Other Bioregulators   Kenji Mori
© 2010 John Wiley & Sons, Ltd

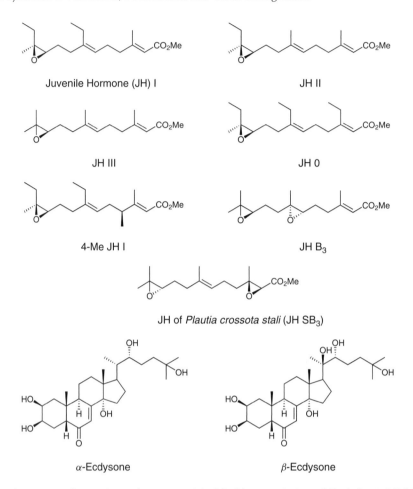

Juvenile Hormone (JH) I

JH II

JH III

JH 0

4-Me JH I

JH B₃

JH of *Plautia crossota stali* (JH SB₃)

α-Ecdysone

β-Ecdysone

Figure 3.1   *Structures of some insect hormones. Modified by permission of Shokabo Publishing Co., Ltd*

3.1.2   Synthesis of juvabione

3.1.2.1   *Isolation and identification of juvabione*

In the early 1960s, Sláma in Prague brought bugs, *Pyrrhocoris apterus*, to Boston with him so as to join the Entomological Laboratories of Harvard University, where Williams was actively pursuing his JH research. Sláma was unsuccessful in raising *P. apterus* at Harvard, because his bugs did not moult into normal adults but into supernumerary nymphs. He speculated the presence of a JH-active compound in his insect-rearing system, and found the paper used in the cages of insects to contain a JH active compound, which was tentatively called "paper factor".[1] The "paper factor" was present in American paper (but not in European or in Japanese papers) and also in balsam fir (*Abies balsamea*), from which the American paper was manufactured.

In 1966, Bowers isolated the "paper factor" from *A. balsamea*, and identified it as juvabione (**60**, Figure 3.2).[2] Administration of 5 μg of juvabione (**60**) to a fifth instar larva of *P. apterus* inhibits its normal moulting to the adult, and generates a sixth instar supernumerary nymph.[2] The acid obtained by

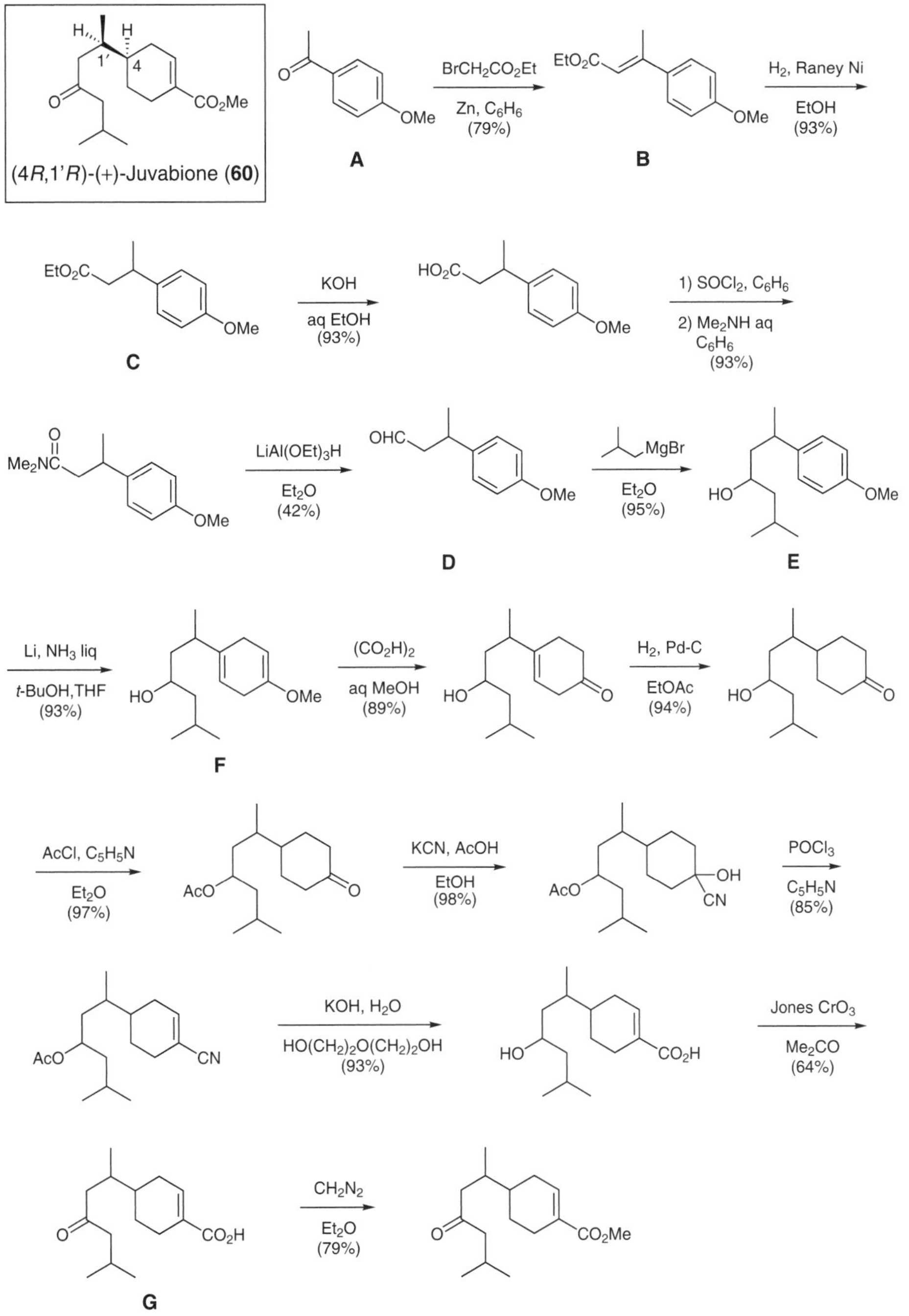

Figure 3.2 *Synthesis of (±)-juvabione. Modified by permission of Shokabo Publishing Co., Ltd*

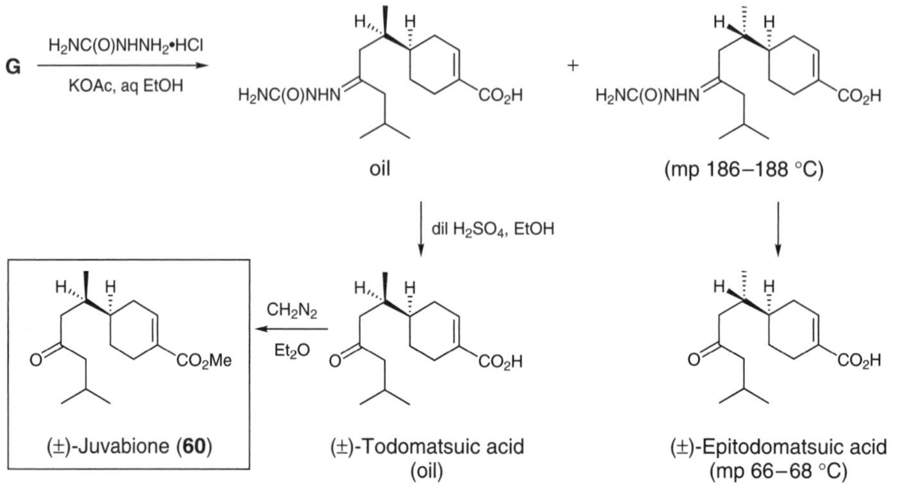

Figure 3.2   *(continued)*

hydrolysis of **60** was identified as todomatsuic acid, which had been obtained by Tutihasi in 1940 as an acid in *Abies sachalinensis*.[3] Its plain structure as a sesquiterpene acid was proposed by Momose in 1941,[4] and a synthesis of a related compound was reported by Nakazaki and Isoe.[5] I read Bowers' paper[2] in January, 1967, and decided to synthesize juvabione. I thought that it might be useful as a kind of insecticide, because it inhibited the metamorphosis. At that time I was so impressed by Rachel Carson's book "Silent Spring" that I really wanted to invent a more benign way of pest control.

3.1.2.2   Synthesis of (±)-juvabione

As shown in Figure 3.2, I synthesized 11.4 g of a stereoisomeric mixture of todomatsuic acid starting from 100 g of *p*-methoxyacetophenone (15 steps) in March, 1967.[6,7] The synthesis was straightforward, employing only basic reactions. Reformatsky reaction (**A→B**), Grignard reaction (**D→E**) and Birch reduction (**E→F**) are typical name reactions taught in every elementary organic course. Conversion of ester **C** to aldehyde **D** was carried out in four steps in 1967. It is now possible to convert **C** to **D** in a single step by low-temperature reduction with diisobutylaluminum hydride or in two steps by reduction of **C** with lithium aluminum hydride to the corresponding alcohol followed by its oxidation with pyridinium chlorochromate. Organic synthesis grows and changes every day! Since the naturally occurring (+)-todomatsuic acid was reported to give a crystalline semicarbazone,[3] the synthetic and stereoisomeric mixture of todomatsuic acid was converted to the corresponding semicarbazone, giving crystalline and oily isomers. They gave back the parent acids by acid treatment, and their methylation with diazomethane gave (±)-juvabione (**60**) and its epimer. Bowers bioassayed (±)-**60**, and found it to exhibit the JH activity against *P. apterus*. After his positive bioassay results, he sent back to me a letter of congratulations.

3.1.2.3   Synthesis of (+)-juvabione

In the early 1990s we achieved two enantioselective syntheses of (4*R*,1′*R*)-(+)-juvabione (**60**) by employing baker's yeast as an agent for asymmetric reduction.[8,9] I will describe one of them (Figure 3.3). In

1990 we found that the reduction of a prochiral diketone **A** with fermenting baker's yeast gave **B** (95% ee) in 52% yield, although it was contaminated with 8% of the isomeric hydroxy ketone **C**.[10] Fortunately, the contaminant could be removed by chromatography and recrystallization at the stage of unsaturated alcohol **D**. This alcohol **D** was oxidized to ketone **E**, whose Baeyer–Villiger oxidation and allylic rearrangement afforded lactone **F**. Methylation of **F** took place on the less-hindered β-side of the lactone ring to give **G** stereoselectively. This step fixed the desired (4R,1′R)-stereochemistry of (+)-juvabione (**60**). Subsequent transformation of **G** to **60** proceeded as shown in Figure 3.3.[8]

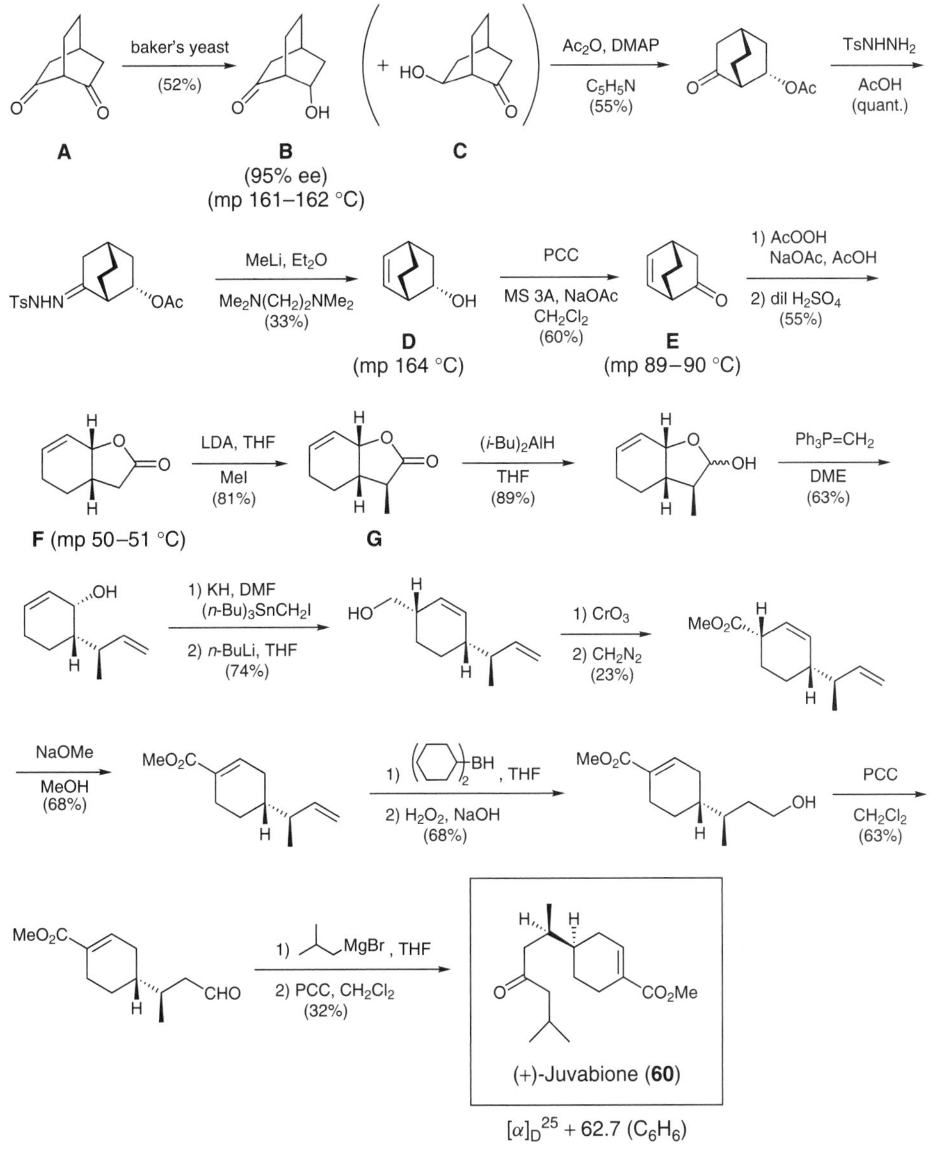

Figure 3.3   *Synthesis of (+)-juvabione. Modified by permission of Shokabo Publishing Co., Ltd*

3.1.3   Synthesis of the racemates of juvenile hormones

Since Röller's discovery of JH I in 1967,[11] I became interested in the preparation of JHs, and synthesized a stereoisomeric and racemic mixture of JH I in 1969.[12] Subsequently, we prepared a variety of JH analogs with different alkyl substituents.[13-16] One of them was almost 1000 times more active than JH I against

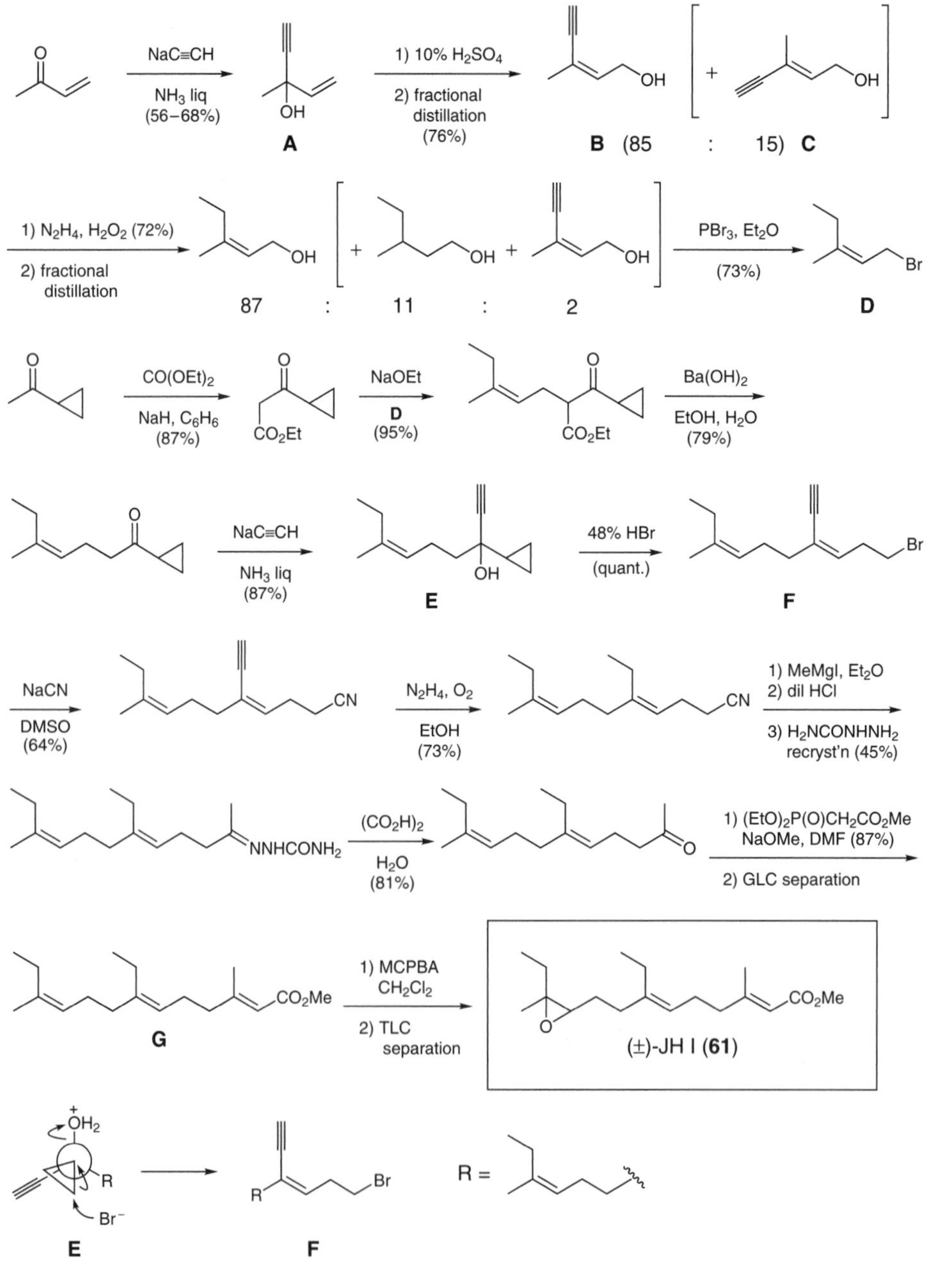

Figure 3.4   *Synthesis of (±)-JH I. Modified by permission of Shokabo Publishing Co., Ltd*

silkworm moth. This new JH analog with a terminal *n*-propyl group instead of ethyl could extend the larval period of the silkworm moth and the resulting larger larvae could produce much larger cocoons than the normal larvae. Of course larger cocoons yielded larger amounts of silk.[14,16]

My JH works enabled me to travel abroad for the first time in 1970, when a Swiss Juvenile Hormone Symposium was held in Basel. I was invited to give a lecture, and met many people including Bowers, Corey, Sörm and van Tamelen. The Symposium sponsored by CIBA-GEIGY (present Novartis) gave me the first chance to report my chemistry to an international audience,[17] and also to see the Swiss Alps.

I then achieved the synthesis of (±)-JH I and (±)-JH II. The problem was how to construct the trisub-stituted double bond in pure (*E*)- or (*Z*)-form. Our 1972 synthesis of (±)-JH I (**61**) is summarized in Figure 3.4.[18,19] Acid-catalysed allylic rearrangement of **A** was moderately selective to give (*Z*)-alkene **B** as the major isomer (**B/C** = 85:15), and **B** could be obtained in pure form by fractional distillation. The alkene **B** was used to construct the terminal (*Z*)-double bond part of the precursor **G** of (±)-JH I (**61**). Julia cleavage (**E→F**) of cyclopropyl alcohol **E** gave pure **F** with a newly generated (*E*)-double bond, which served as the (6*E*)-double bond of (±)-JH I (**61**). Stereoselectivity of the conversion of **E** to **F** can be explained by the Newman projection shown at the bottom of Figure 3.4. Apparently, the ethynyl group is less bulky than the alkyl group R, and therefore the depicted conformation is the preferred one, giving **F** with the (*Z*)-double bond. Purification of **G** was executed by preparative GLC to remove the unwanted (2*Z*)-isomer.

Stereocontrolled synthesis of (±)-JH II is shown in Figure 3.5.[20] Coupling of phenyl sulfide **A** with bromide **C**, prepared from geranyl acetate **B**, gave **D**. Then, **D** was converted to (±)-JH II (**62**).[20]

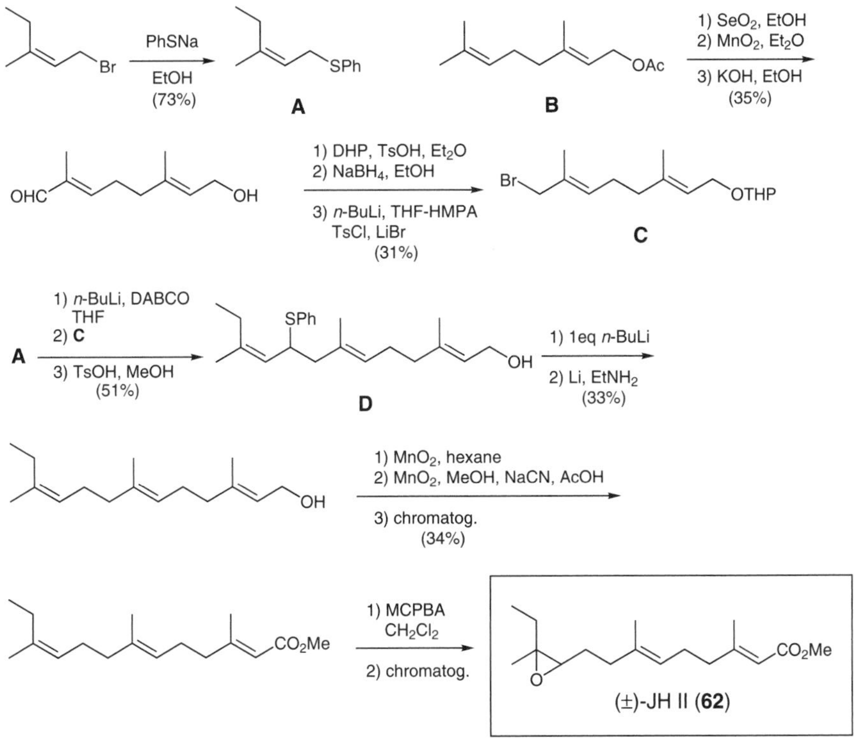

Figure 3.5 *Synthesis of (±)-JH II. Modified by permission of Shokabo Publishing Co., Ltd*

3.1.4   Synthesis of the enantiomers of juvenile hormones

3.1.4.1   *Earlier syntheses and biological studies by others*

The next challenge after the completion of the synthesis of racemic JHs was the synthesis of pure enantiomers of JHs to compare the bioactivities of natural and unnatural enantiomers of JHs. In the 1970s several groups reported the syntheses of optically active JHs. Although none of the attempts in this period could provide pure enantiomers of JHs, I summarize here the stereochemistry–bioactivity relationships reported in the 1970s and the early 1980s.

In 1971 Loew and Johnson synthesized both the natural (+)-JH I and its opposite enantiomer (−)-JH I.[21] Their bioassay showed that (+)-JH I was about 9 times more active than (−)-JH I on the wax moth, *Galleria mellonella*. On *Tenebrio molitor*, (+)-JH I was 6–8 times more active than (−)-JH I. Their synthetic (−)-JH I, however, was not enantiomerically pure, and might have contained up to 10% of (+)-JH I, which could account for most, or even all, of the observed bioactivity. Loew and Johnson therefore stated that further work must be done before a quantitative conclusion could be reached regarding the bioactivity of (−)-JH I.[21]

In 1985 Prestwich and Wawrzénczyk synthesized the enantiomers of JH I (95% ee), and bioassayed their binding affinity to the JH binding protein of the tobacco hornworm moth, *Manduca sexta*.[22] The natural (+)-JH I showed only twice the relative binding affinity as that of (−)-JH I. Despite the previous conclusion of Loew and Johnson, it was generally believed that (−)-JH I was bioactive to some extent.

In 1974 Marumo and coworkers synthesized the enantiomers of JH II by microbial asymmetric hydrolysis of the epoxy ring of (±)-JH II (prepared by Mori) with a fungus *Helminthosporium sativum*.[23] The hydrolysed diol was converted to (+)-JH II, while the epoxide remained intact was (−)-JH II. Their enantiomeric purities, however, were rather low (66–73% ee), and no definite biological data could be obtained.

The enantiomers of JH III ethyl ester were prepared in 1976 also by Marumo and coworkers by means of chromatographic separation of a JH III derivative, and their bioactivities were examined on the allatectomized fourth instar larvae of the silkworm moth, *Bombyx mori*.[24] Their (+)-JH III ethyl ester was about 50 times more active than the (−)-isomer. Schooley and coworkers reported another attempt to prepare the enantiomers of JH III by optical resolution, and their binding properties were studied with hemolymph JH binding protein of *Manduca sexta*.[25] The order of binding activity was (+)-JH III > (±)-JH III > (−)-JH III. They concluded that the binding observed for (−)-JH III [about 1/4 of that of (+)-JH III] might have been due largely to its 8% contamination with (+)-JH III.

These earlier studies on stereochemistry–JH activity relationships indicated that the synthesis of extremely pure enantiomers of JHs would allow us to determine the relationships accurately. We therefore started our enantioselective JH synthesis based on biocatalysis.

3.1.4.2   *Biocatalysis and juvenile hormone synthesis*

In 1985, we reported that reduction of a prochiral 1,3-diketone **A** (Figure 3.6) with fermenting baker's yeast (*Saccharomyces cerevisiae*) was enantioselective to give (*S*)-hydroxy ketone **B** of 98–99% ee.[26] I noticed that the Baeyer–Villiger oxidation of **B** would furnish (*S*)-hydroxylactone, a building block synthetically equivalent to the terminal epoxide moiety of (+)-JH III. This idea was used for the synthesis of (+)- and (−)-JH III in 1987.[27]

Similar reduction of a prochiral diketone **C** with another yeast, *Pichia terricola* afforded **D**, which was converted to JH I,[28] JH II,[28] JH 0[29] and 4-methyl JH I.[30] In the case of (+)-4-methyl JH I, asymmetric hydrolysis of dimethyl 3-methylglutarate with pig-liver esterase (PLE) was also employed for its

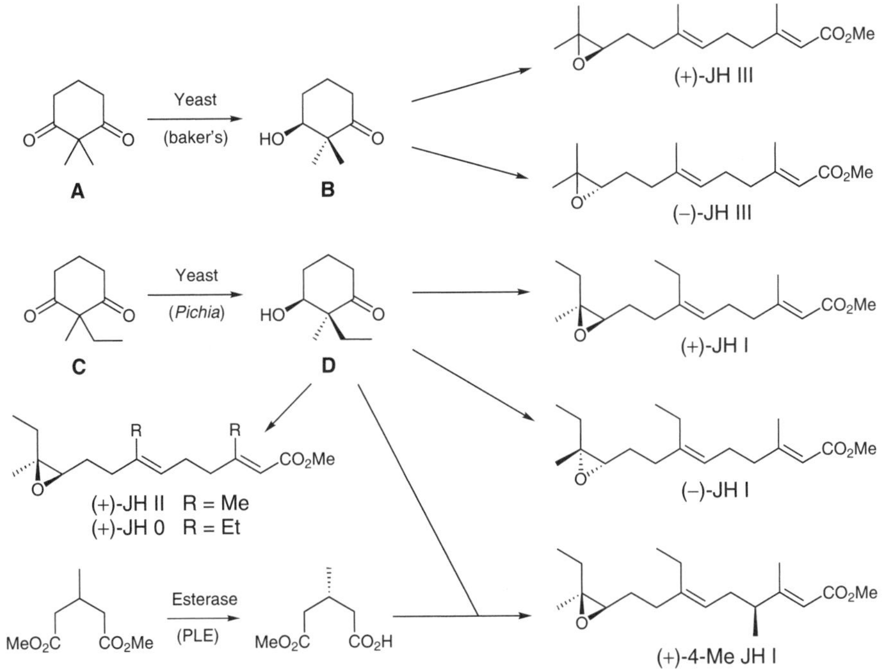

Figure 3.6 *Summary of the synthesis of enantiomerically pure JHs. Modified by permission of Shokabo Publishing Co., Ltd*

construction. As detailed below, our JH synthesis in the late 1980s to the early 1990s became possible by the use of asymmetric biocatalysis. Biocatalysis is a useful tool in asymmetric synthesis.[32–34]

3.1.4.3   Synthesis of the enantiomers of JH III

We synthesized pure enantiomers of JH III in 1987 as shown in Figure 3.7.[27] The hydroxy ketone **B**, prepared by yeast-mediated reduction of **A**[35] was acetylated to give **C**. Baeyer–Villiger oxidation of **C** with *m*-chloroperbenzoic acid gave crystalline lactone **D**, whose recrystallization furnished enantiomerically pure **D**. Reduction of **D** with lithium aluminum hydride afforded triol **E**. Acetylenic ester **F** derived from **E** was treated with sodium thiophenolate to give Michael adducts **G** and **H**. These were separable by silica-gel chromatography, and **G** was methylated to give **I**. Further elongation of the carbon chain of **I** was achieved by dianion alkylation of methyl acetoacetate with bromide **J** to give **K**. The corresponding enol phosphate **L** was methylated to give **M**. Removal of the acetonide protective group of **M** yielded dihydroxy ester **Na**. Its enantiomeric purity was confirmed as about 100% ee by HPLC analysis of the corresponding (*R*)- and (*S*)-α-methoxy-α-trifluorometylphenylacetates (MTPA esters), **Nb** and **Nc**. Finally, the hydroxy ester **Na** was converted to (*R*)-(+)-JH III (**63**) via mesylate **O**. (*S*)-(−)-JH III (**63′**) was also prepared from **Na** via **P**. Conversion of **O** to **63** proceeded with Walden inversion at C-10, while base treatment of **P** resulted in retention of configuration at C-10, giving **63′**.[27]

Our synthetic enantiomers of JH III (**63** and **63′**) were shown to be of ca.100% ee by ^{1}H NMR analysis in the presence of a chiral solvating reagent.[27] As shown in Figure 3.8, ^{1}H NMR spectra of the enantiomers of the synthetic JH III were measured at 400 MHz in the presence of (*R*)-(−)-

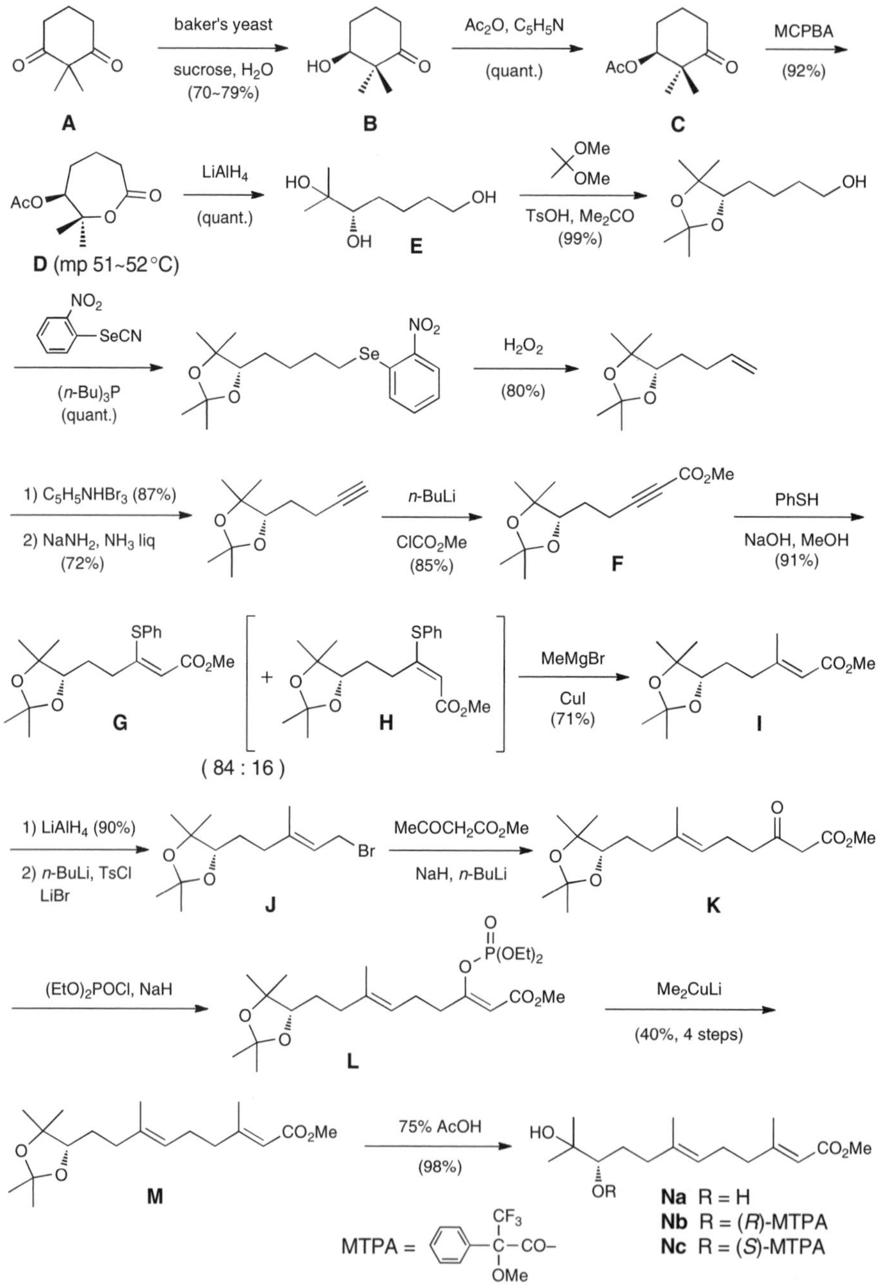

Figure 3.7 *Synthesis of the enantiomers of JH III. Modified by permission of Shokabo Publishing Co., Ltd*

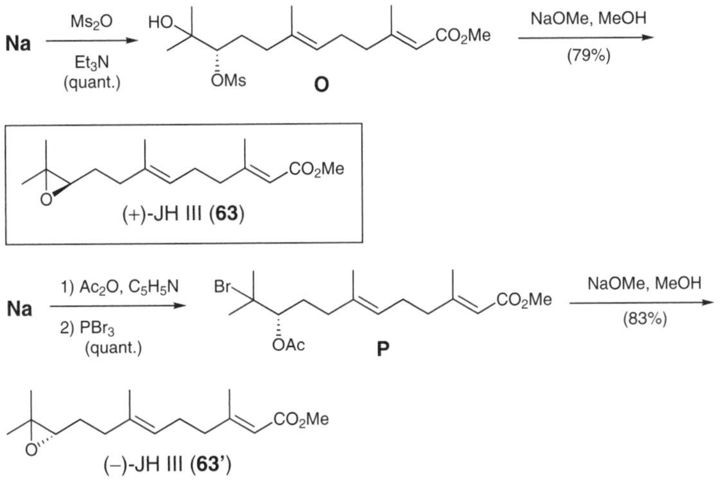

Figure 3.7   *(continued)*

2,2,2-trifluoro-1-(9-anthryl)ethanol. The proton at C-10 of JH III showed a triplet, whose δ-value was slightly different ($\Delta\delta = 0.035$ ppm) between the enantiomers. Both **63** and **63′** were enantiomerically pure.

3.1.4.4   Synthesis of the enantiomers of JH I

Figure 3.9 summarizes our synthesis of the enantiomers of JH I in 1988.[28] If we want to employ the same strategy as used for the synthesis of JH III, the synthesis of JH I demands the execution of the diastereo- and enantioselective reduction of a prochiral 1,3-diketone **A**. Unfortunately, reduction of **A** with fermenting baker's yeast was nondiastereoselective, giving both **B** and **C**. After extensive screening of yeasts, *Pichia terricola* KI 0117 donated by Kirin Brewery Co. was found to achieve highly stereoselective reduction to give >99% of **B** with 99% ee. Since both of the 3,5-dinitrobenzoates **D** and **E** were crystalline, these could be purified by recrystallization, and 100% pure **E** was secured. Conversion of **E** to **F** was achieved by methanolysis followed by acetonide formation. The enantiomeric purity of **F** was proved to be ca. 100% ee by HPLC analysis of bis-MTPA ester **G** derived from **F**.

Further steps to JH I were executed in a manner similar to the synthesis of JH III. At the stage of **Ha**, its 100% enantiomeric purity was confirmed by HPLC analysis of **Hb** and **Hc**. Subsequently, **Ha** was converted to (+)-JH I (**61**).[28] For the synthesis of unnatural (−)-JH I (**61′**), the hydroxy ketone **B** was converted to its opposite enantiomer **B′** via **I** and **J**.[31] Then, **B′** was further converted to (−)-JH I (**61′**). The enantiomeric purity of the intermediates leading to (−)-JH I was confirmed to be ca. 100% ee by HPLC analysis of **G′**.[31] Synthesis of (+)-JH II (**62**) was also achieved by starting from **B**.[28]

3.1.4.5   Bioactivities of the enantiomers of juvenile hormones

I will now tell you the story about bioassays of our enantiomers of JH III and JH I by zoologists. Both the enantiomers of JH III were bioassayed by Lanzrein and coworkers in Switzerland.[36] In the *Galleria mellonella* wax test, a local morphogenetic assay, (+)-JH III (**63**) was 5240 times more active than (−)-JH III (**63′**). In a systemic morphogenetic assay with the cockroach *Nauphoeta cinerea*, 380 times less

(+)-JH III (**63**) was necessary in order to induce detectable juvenilization (58 ng of **63** and 22 μg of **63′**). In a systemic gonadotropic assay with *N. cinerea*, 225 times less (+)-JH III was needed to induce vitellogenin synthesis in 50% of the insects. In the JH binding protein assay using the hemolymph JH binding protein of *N. cinerea*, (+)-JH III had an affinity for the JH binding protein, which was about 46 times higher than that of (−)-JH III. Thus, JH binding proteins and receptors in target tissues clearly display enantioselectivity.

When the bioactivity was assayed by topical application on allatectomized fourth instar larvae of the silkworm moth *Bombyx mori*, (+)-JH III was active at the dosage of 10–100 μg/larva, while (−)-JH III was totally inactive even at a dosage of 500 μg/larva.[37] In this assay, too, the bioresponse was highly enantioselective, in contrast to the previous data. Only after the synthesis of pure enantiomers of JH III could the enantioselective nature of the action of JH III be clarified accurately.

In the case of our synthetic enantiomers of JH I, they were bioassayed by topical application on allatectomized fourth instar larvae of *B. mori*.[37] The natural (+)-JH I (**61**) was active at the dosage of as low as 0.04 μg/larva to induce 50% larval moulting. It was about 12 000 times more active than the unnatural (−)-JH I. This observation indicates that the receptor for JH I must be highly enantioselective.

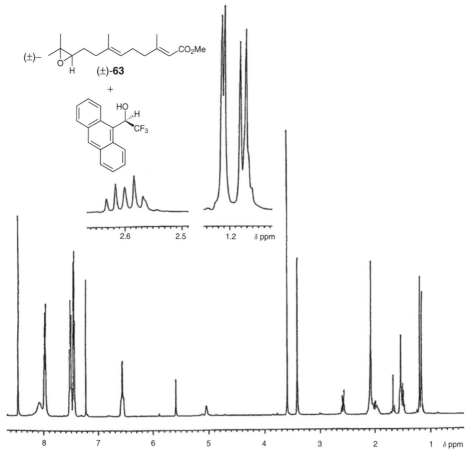

Figure 3.8 *Estimation of the enantiomeric purities of the enantiomers of JH III by using a chiral solvating reagent in ¹H NMR measurements (400 MHz, CDCl₃). Modified by permission of Shokabo Publishing Co., Ltd*

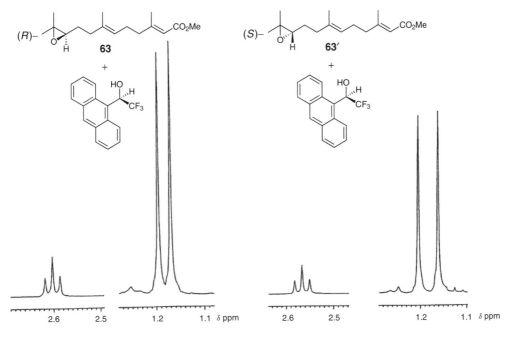

Figure 3.8   *(continued)*

Our synthesis of extremely pure enantiomers of JH I thus enabled the truly enantioselective nature of the perception of JH I to be revealed.

3.1.4.6   Synthesis of (+)-JH I by asymmetric dihydroxylation

Although the use of yeasts as biocatalysts was quite effective in preparing extremely pure enantiomers of JHs, their synthetic routes were lengthy. Indeed, in the case of (+)-JH I (**61**), its overall yield was only 0.34% (21 steps) by the biocatalytic method.[28] We therefore examined the application of Sharpless asymmetric dihydroxylation for the synthesis of (+)-JH I and (+)-JH II.

In 1988, Sharpless invented his asymmetric dihydroxylation (AD), a catalytic process to convert alkenes to optically active 1,2-diols with known absolute configuration.[38] As shown in Figure 3.10, a commercially available reagent AD-mix-$\alpha^{\circledR}$ (Aldrich) gives α-1,2-diol from an alkene, while AD-mix-$\beta^{\circledR}$ affords β-1,2-diol. This reaction was used by Crispino and Sharpless to synthesize (*R*)-(+)-JH III (92% ee).[39] We also employed this AD reaction to synthesize (+)-JH I (**61**) and (+)-JH II (**62**).[40] Figure 3.11 summarizes the synthesis of (+)-JH I by means of an AD reaction.

Cyclopropyl methyl ketone was converted to **A**, which was oxidized to a mixture of **B** and **C**. After acetylation of **C** to diacetate **D**, it was methylated to give **E**. Alcohol **E** gave acetylene **F**, which was subjected to Negishi's carboalumination reaction to give **G** after quenching with formaldehyde. In a similar manner as already shown in Figure 3.7, **G** was converted to **I** via **H**. Asymmetric dihydroxylation of **I** with AD-mix-$\alpha^{\circledR}$ gave **J**, whose enantiomeric purity was 95% ee as judged by HPLC analysis of the corresponding (*R*)-MTPA ester. Finally, dihydroxy ester **J** yielded (+)-JH I (**61**) via mesylate **K**. The overall yield of this synthesis was 1.0% (18 steps). Although this is a remarkable improvement, there is still room for further improvement in future to obtain pure (+)-JH I more efficiently.

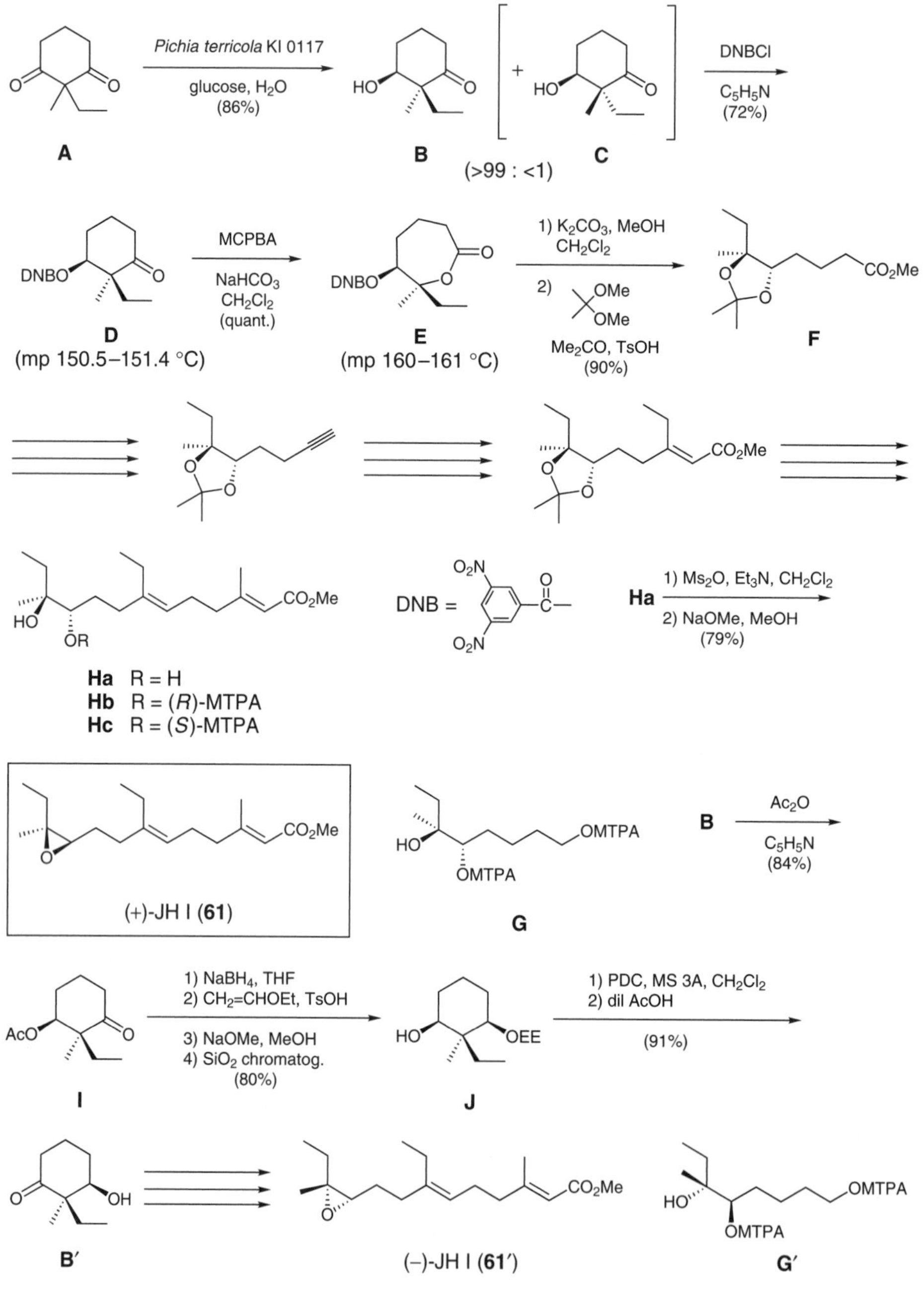

Figure 3.9 *Synthesis of the enantiomers of JH I. Modified by permission of Shokabo Publishing Co., Ltd*

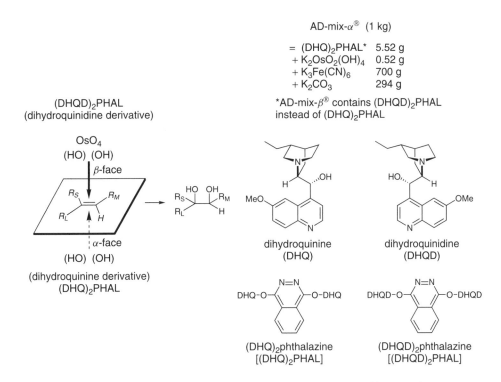

Figure 3.10   *Asymmetric dihydroxylation (AD) as invented by Sharpless*

3.2   Insect antifeedants

Some of the metabolites of higher plants inhibit the feeding activity of insects. Poisons and antifeedants are the defense substances of higher plants against insects. Insect antifeedants have been studied as a part of our efforts to protect our crops against pest insects. A triterpene named azadirachtin (Figure 3.12) is practically used in several countries as a pesticide of natural origin.

Azadirachtin was isolated in 1968 by Morgan and coworkers from the seeds of the neem tree, *Azadirachta indica*. I once saw a big neem tree in front of the Department of Chemistry of Delhi University in India. Insects avoid azadirachtin, and their growth can be retarded by it, because it inhibits the feeding. Azadirachtin shows no toxicity against mammals, birds and fishes, while it is active against insects at concentrations of less than 0.1 ppm.[41] It is therefore a good pesticide in so-called "organic agriculture." The structure of azadirachtin, as depicted in Figure 3.12, was clarified in 1987 independently by three groups headed by Ley, Kraus and Nakanishi. Its total synthesis was reported by Ley and coworkers in 2007.[42]

Now let me describe our own synthetic works on insect antifeedants.

3.2.1   Synthesis of polygodial

(−)-Polygodial (**64**, Figure 3.13) was first isolated in 1962 by Ohsuka as a hot-tasting sesquiterpene of *Polygonum hydropiper*, water-pepper in English or yanagi-tade in Japanese. Its structure was clarified in

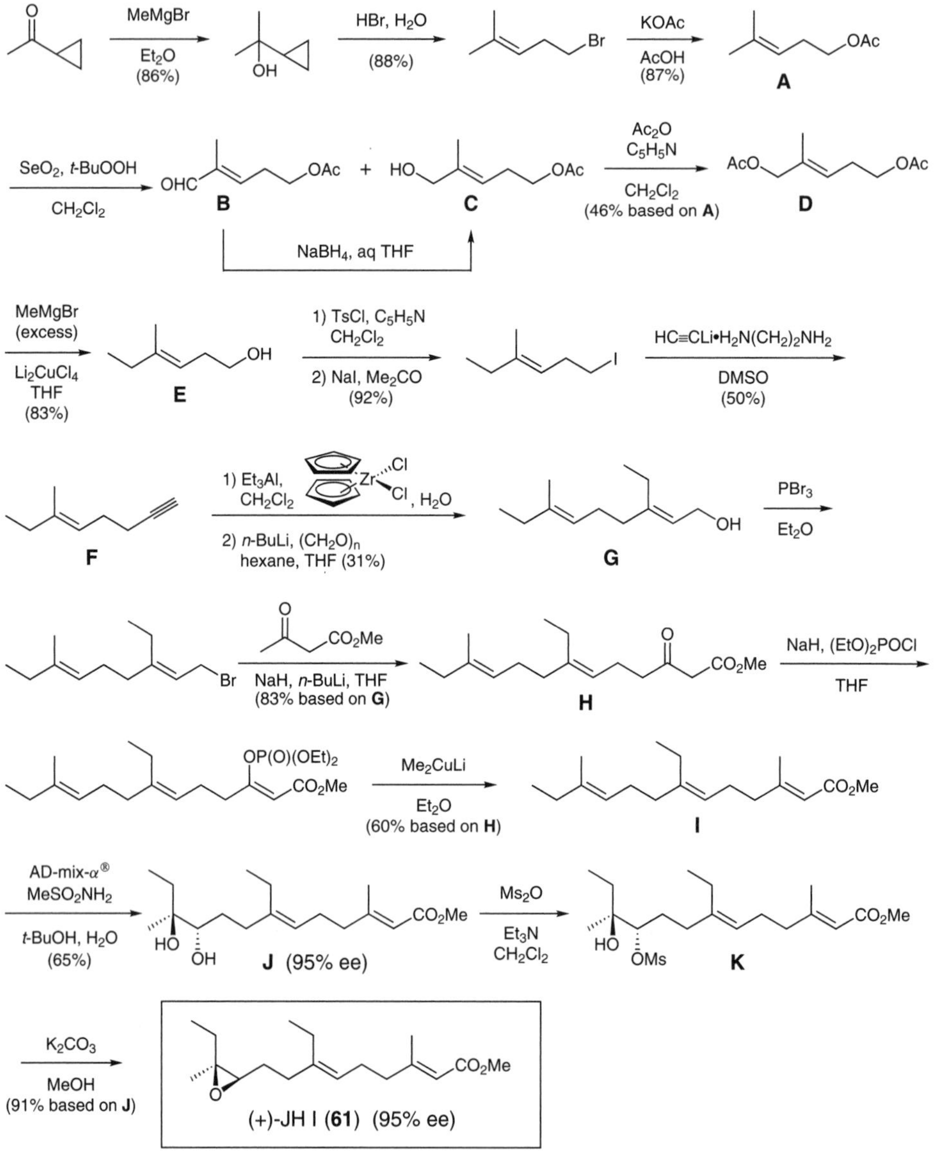

Figure 3.11   *Synthesis of (+)-JH I by AD*

the same year by Barnes and Loder in Australia. In 1976 Kubo, Nakanishi *et al.* discovered the antifeedant activity of **64** against African crop pests such as the army worms *Spodoptera littoralis* and *S. exempta*. Some nudibranch molluscs contain **64** as a chemical defense substance, because it is piscicidal. Kubo suggested a hypothesis that the absolute configuration of the antifeedant appears to govern the hotness of its taste. We became interested in synthesizing the enantiomers of **64** so as to clarify the stereochemistry–bioactivity relationship.

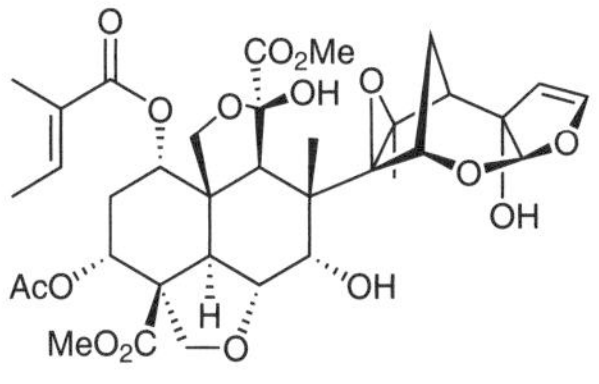

Azadirachtin

Figure 3.12 *Structure of azadirachtin. Reprinted with permission of Shokabo Publishing Co., Ltd*

Figure 3.13 summarizes our synthesis of the enantiomers of polygodial (**64**).[43] The hydroxy ketone **A**, which served as the starting material of our JH III synthesis (Figure 3.7), was converted to diene **B**. Diels–Alder reaction of **B** with dimethyl acetylenedicarboxylate was nonselective, and afforded a 1:1 mixture of **C** and **D**. After chromatographic separation, **D** was converted to (−)-polygodial (**64**), while **C** gave unnatural (+)-polygodial (**64′**). Fortunately, the nonselective Diels–Alder reaction simplified our work to prepare both **64** and **64′**.

When Mr. Watanabe (now Prof. Watanabe at the University of Tokyo) and I tasted the synthetic samples, both (−)-**64** and (+)-**64′** were extremely hot-tasting. Examination of piscicidal activity by Prof. Asakawa and that of antifeedant activity by Prof. Pickett showed no difference between the enantiomers of polygodial. In this particular case, both the enantiomers **64** and **64′** showed the same bioactivity.[44]

3.2.2 Synthesis of warburganal

(−)-Warburganal (**65**, Figure 3.14) was isolated in 1976 as an antifeedant sesquiterpene of African plants, *Warburgia stuhlmannii* and *W. ugandensis*, by Kubo, Nakanishi *et al*. Our 1989 synthesis of (−)-warburganal (**65**) is shown in Figure 3.14.[45]

The hydroxy ketone **A** was converted to bicyclic hydroxy ketone **B**.[46] Treatment of **B** with trifluoromethanesulfonyl chloride and DMAP gave unsaturated ketone **C**.[47] This was hydrogenated to furnish **D**.[45] Conversion of **D** to (−)-warburganal (**65**) was executed as reported by de Groot and coworkers[48] to furnish **65** in 8.1% overall yield based on **A** (20 steps).[45]

3.2.3 Synthesis of 3,4′-dihydroxypropiophenone 3-β-D-glucopyranoside

In 1990, Watanabe at Tokyo University Forest found that the leaves of *Betula platyphylla* var. *japonica* produce (Z)-3-hexen-1-ol in an amount larger than usual, when the tree was attacked by gypsy moths. The resulting higher concentration of (Z)-3-hexen-1-ol even at 5 ppm evoked the increase of the amount of a phenolic constituent in the leaves, and this compound was isolated and identified as 3,4′-dihydroxypropiophenone 3-β-D-glucopyranoside (**66**, Figure 3.15). The attack of the leaves by insects produces (Z)-3-hexen-1-ol as a chemical signal to accelerate the biosynthesis of **66**, and the presence of the excess amount of **66** in the leaves prevents further attack by insects, because it functions as an insect antifeedant.

We planned a synthesis of **66** by an enzymatic method as shown in Figure 3.15, because any chemical synthesis of **66** must involve a cumbersome use of protective groups.[49] The crystalline aglycone **A** possesses both an aliphatic and a phenolic hydroxy groups. Selective glucosidation of the aliphatic hydroxy group of **A** with the glucosyl donor **B** could be achieved by using the lactase from *Kluyveromyces lactis* in a

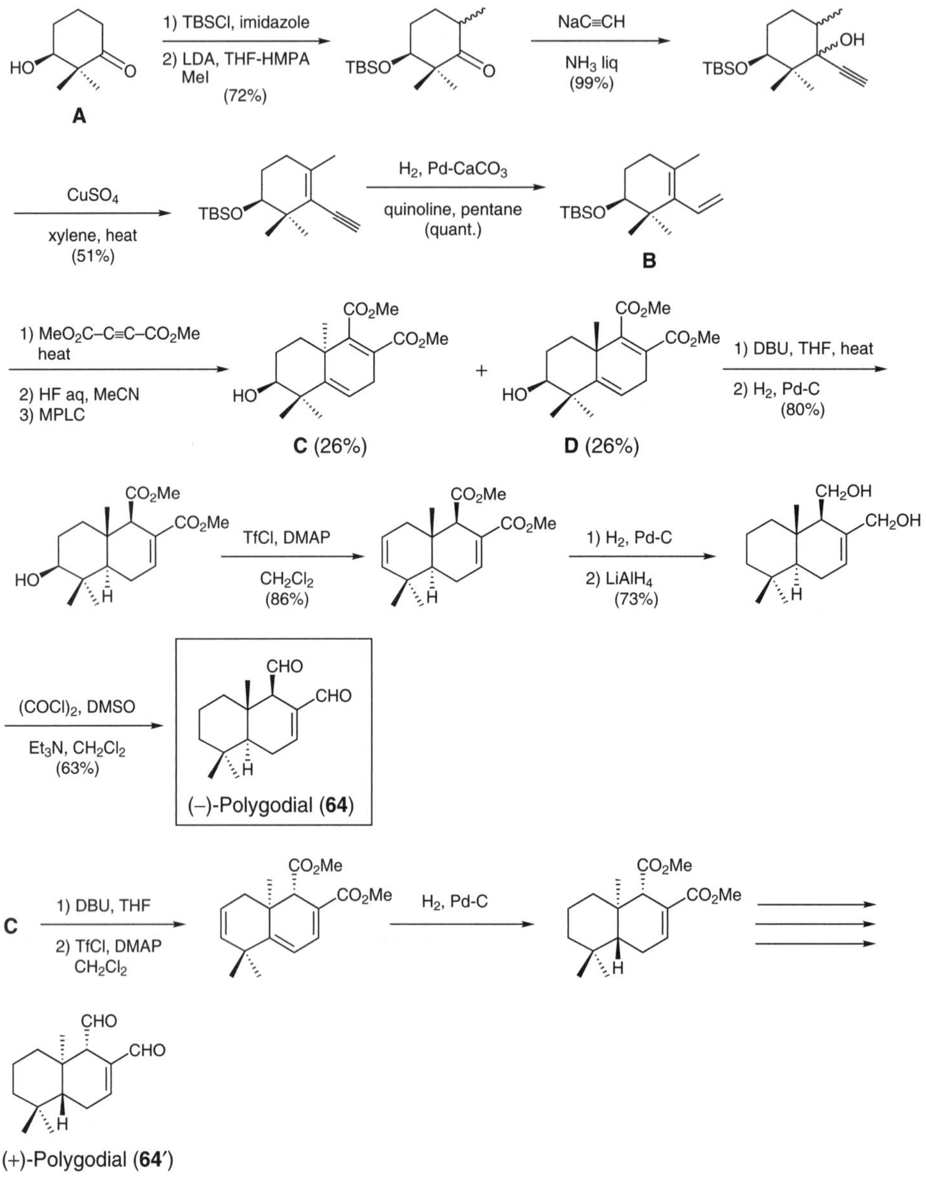

Figure 3.13 *Synthesis of the enantiomers of polygodial. Modified by permission of Shokabo Publishing Co., Ltd*

biphasic system of diisopropyl ether and phosphate buffer (pH 6.5). This enzymatic glucosidation furnished crystalline **66** in 26% yield. Enzymes are useful catalysts for selective reactions. Both **A** and **66** were shown to be weak antifeedants at a dosage of 40 000 ppm (4% aqueous solution). It was also observed that both **A** and **66** repelled the first instar larvae of the gypsy moth at a dosage of $\geq$5000 ppm. The ecological meaning of these modest repellent and antifeedant activities requires further studies.

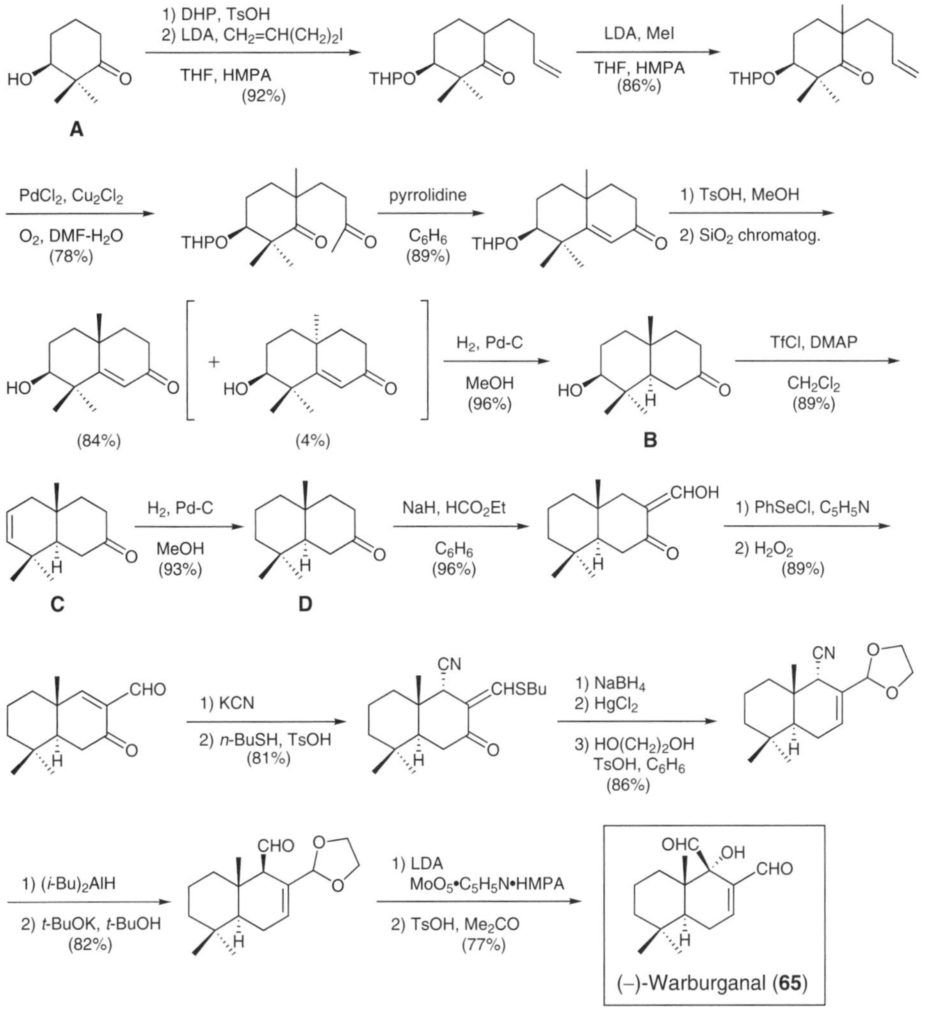

Figure 3.14 Synthesis of warburganal. Modified by permission of Shokabo Publishing Co., Ltd

3.2.4 Synthesis of homogynolide A

(−)-Homogynolide A (**67**, Figure 3.16) is a sesquiterpene lactone isolated from *Homogyne alpina* (Compositae) by Harmatha *et al.* in 1976. It shows antifeedant activity against beetle adults (*Sitophilus granarius, Tribolium confusum*) and larvae (*Trogoderma granarium, Tribolium confusum*), which are pests of stored grains and seeds. Its structure **67** with five stereogenic centers prompted us to synthesize it as shown in Figure 3.16.[50]

Reduction of prochiral 1,3-diketone **A** with baker's yeast gives hydroxy ketone **B** of 99% ee.[10] The ketone **B** was converted to unsaturated ketone **C**, which was treated with (*Z*)-1-propenylmagnesium bromide in the presence of cerium(III) chloride to give a mixture of **D** and **E**. This mixture was directly subjected to oxy-Cope rearrangement to give **F** from **D**. The *exo*-isomer **E** was recovered unchanged.

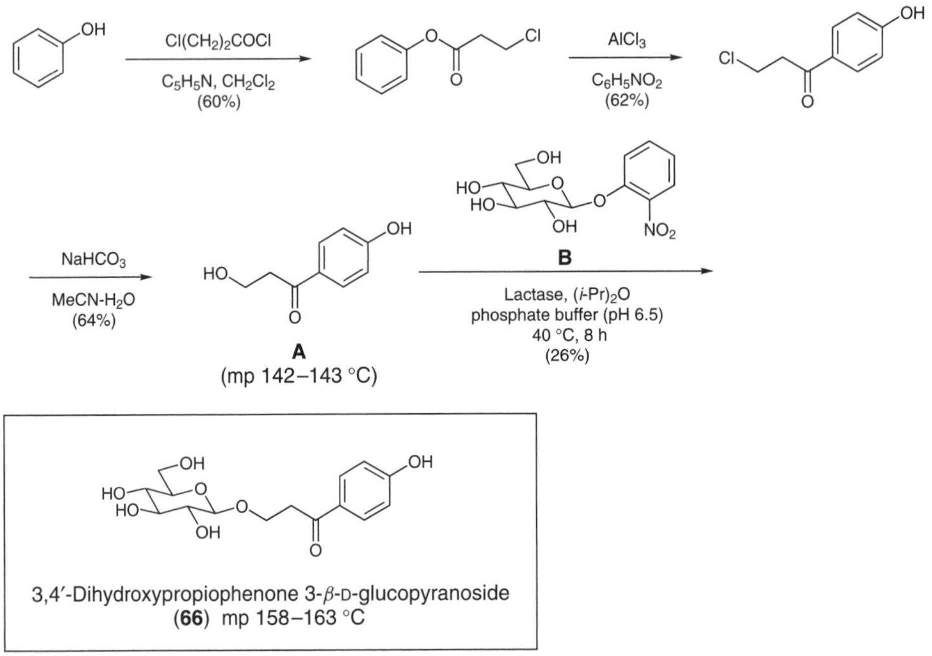

Figure 3.15   *Synthesis of 3,4'-dihydroxypropiophenone 3-β-D-glucopyranoside. Modified by permission of Shokabo Publishing Co., Ltd*

Readers should attempt to rationalize the course of this rearrangement by writing the flow of electrons. The carbonyl group of **F** was protected to give **G**, which was treated with ozone to furnish a keto aldehyde. This gave aldol product **H** upon treatment with base. The double bond of **H** was then hydrogenated to give a saturated ketone **I**. To construct a quaternary carbon at the spiro center, a methoxycarbonyl group was selectively introduced at that position by treating lithium enolate of **I** with methyl cyanoformate via the thermodynamically more stable silyl enol ether derived from **I**. The reaction was stereoselective to give only **J** by the attack of the reagent from the less hindered convex side of **I**. This was converted to (−)-homogynolide A (**67**) in 0.63% overall yield based on **B** (22 steps).[50]

3.3   Insect repellents

Higher plants sometimes produce insect repellents as part of their defense system. Lower animals also produce insect repellents so as to defend themselves from the attack of insects. I will describe two syntheses of ours in this area.

3.3.1   Synthesis of rotundial

In 1995 Watanabe *et al.* at Sumitomo Chemical Co. isolated and identified a degraded monoterpenoid (+)-rotundial (**68**, Figure 3.17) as a new mosquito repellent from *Vitex rotundifolia*, a medicinal plant whose leaves and twigs were used for repelling mosquitoes. They reported that the mosquito-repelling activity

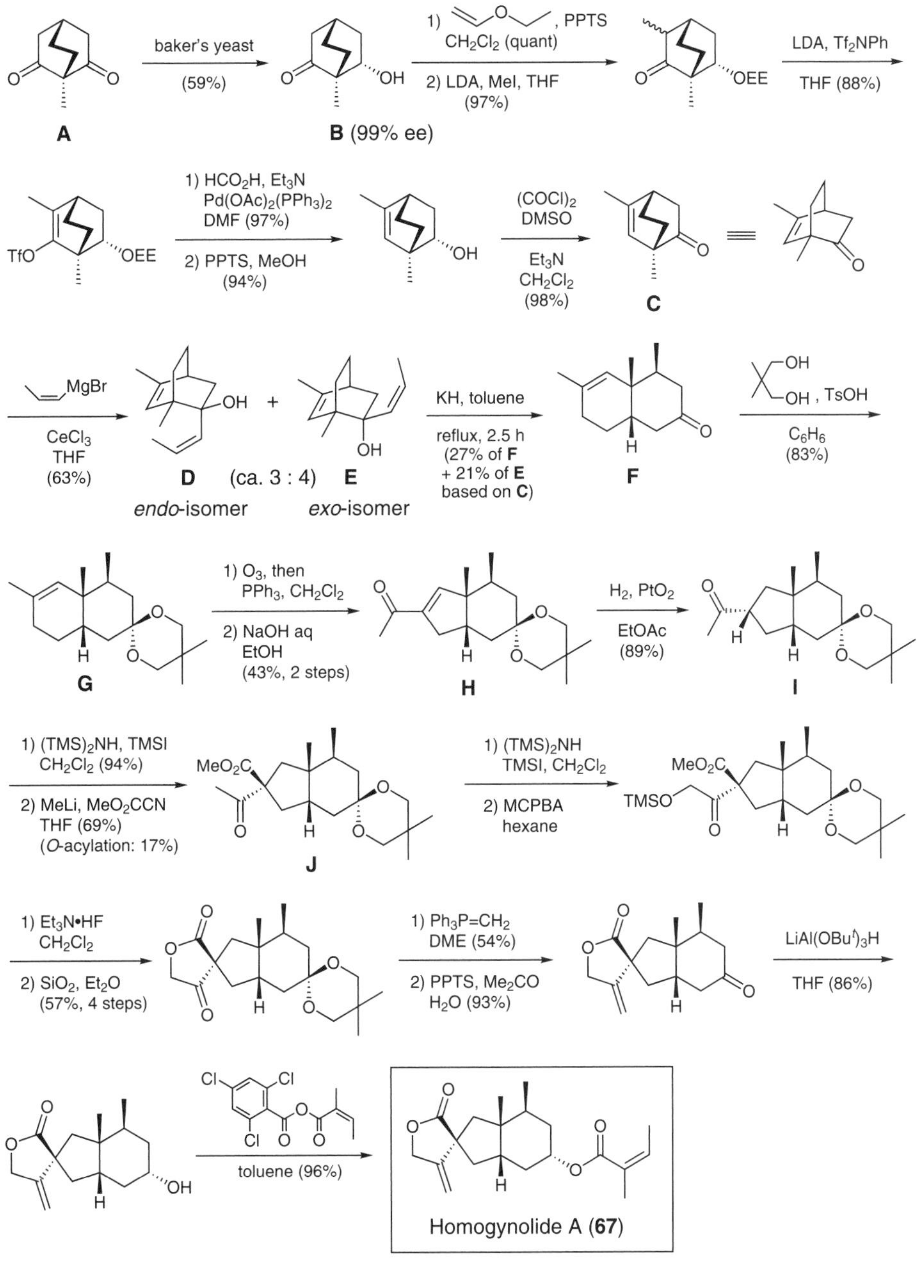

Figure 3.16 *Synthesis of homogynolide A*

of rotundial was superior to that of *N*, *N*-diethyl-*m*-toluamide (Deet®), the active ingredient in almost all the commercial insect-repellent formulations. We synthesized both the enantiomers of **68** so as to determine the absolute configuration of the natural (+)-**68** and also to clarify the stereochemistry–bioactivity relationship.

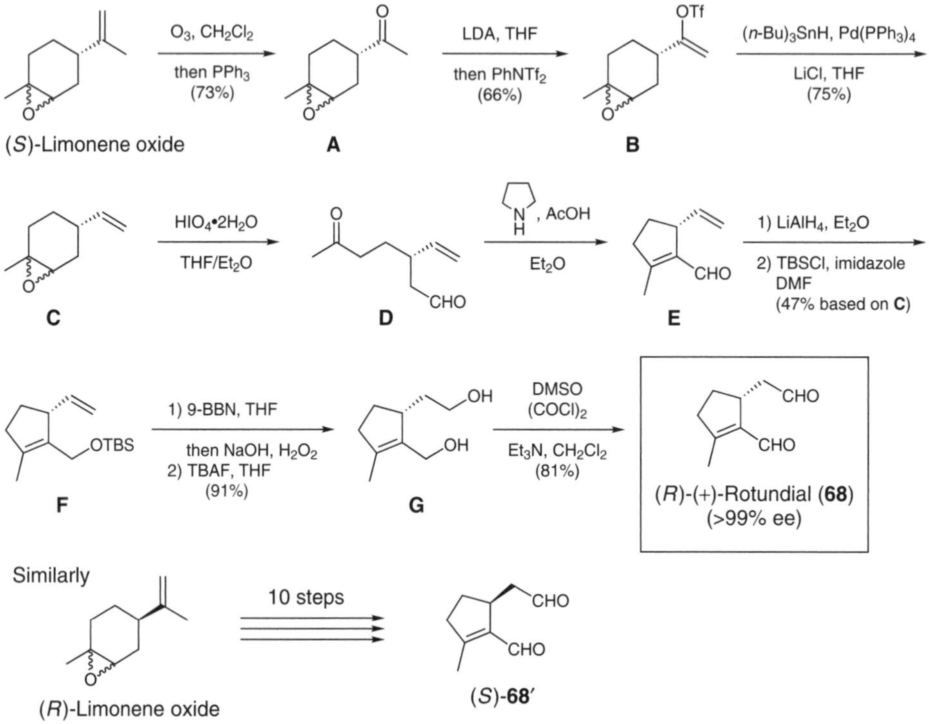

Figure 3.17 *Synthesis of the enantiomers of rotundial*

Figure 3.17 summarizes our synthesis of (*R*)-(+)-rotundial (**68**).[51] Commercially available (*S*)-limonene oxide (99% ee at C-4, a diastereomeric mixture) was chosen as the starting material. Its ozonolysis furnished ketone **A**. The kinetic enolate of **A**, generated with LDA, was treated with *N*-phenyltrifluoromethanesulfonimide (PhNTf₂) to give the corresponding enol triflate **B**. Epoxide **C**, obtained by reduction of **B**, was cleaved with periodic acid dihydrate to give **D**, which was cyclized to **E**. Selective hydroboration-oxidation of diene **F** with 9-BBN and hydrogen peroxide gave diol **G** after deprotection. Swern oxidation of **G** afforded (*R*)-(+)-rotundial (**68**), the naturally occurring repellent. The two enantiomers **68** and **68′** were equally bioactive as the mosquito repellent. Therefore, in this particular case, the absolute configuration of **68** plays no role in its bioactivity.

3.3.2 Synthesis of polyzonimine

In 1975, in the course of their studies on compounds from the defensive glands of a milliped *Polyzonium rosalbum*, Meinwald, Eisner and their respective coworkers isolated and identified (+)-polyzonimine (**69**, Figure 3.18) as a volatile insect repellent, which acts as a topical irritant to predating insects such as ants and cockroaches. Its structure as a monoterpene alkaloid **69** (without assigning absolute configuration) was suggested by the X-ray analysis of a closely related minor component of the secretion, (+)-nitropolyzonamine (**70**). We became interested in establishing the absolute configuration of (+)-polyzoninine.

Our synthesis of (*S*)-(+)-polyzonimine (**69**) is summarized in Figure 3.18.[52,53] Commercially available 2-methylcyclohexanone **A** was converted to aldehyde **B**. This was treated with (*S*)-prolinol methyl ether

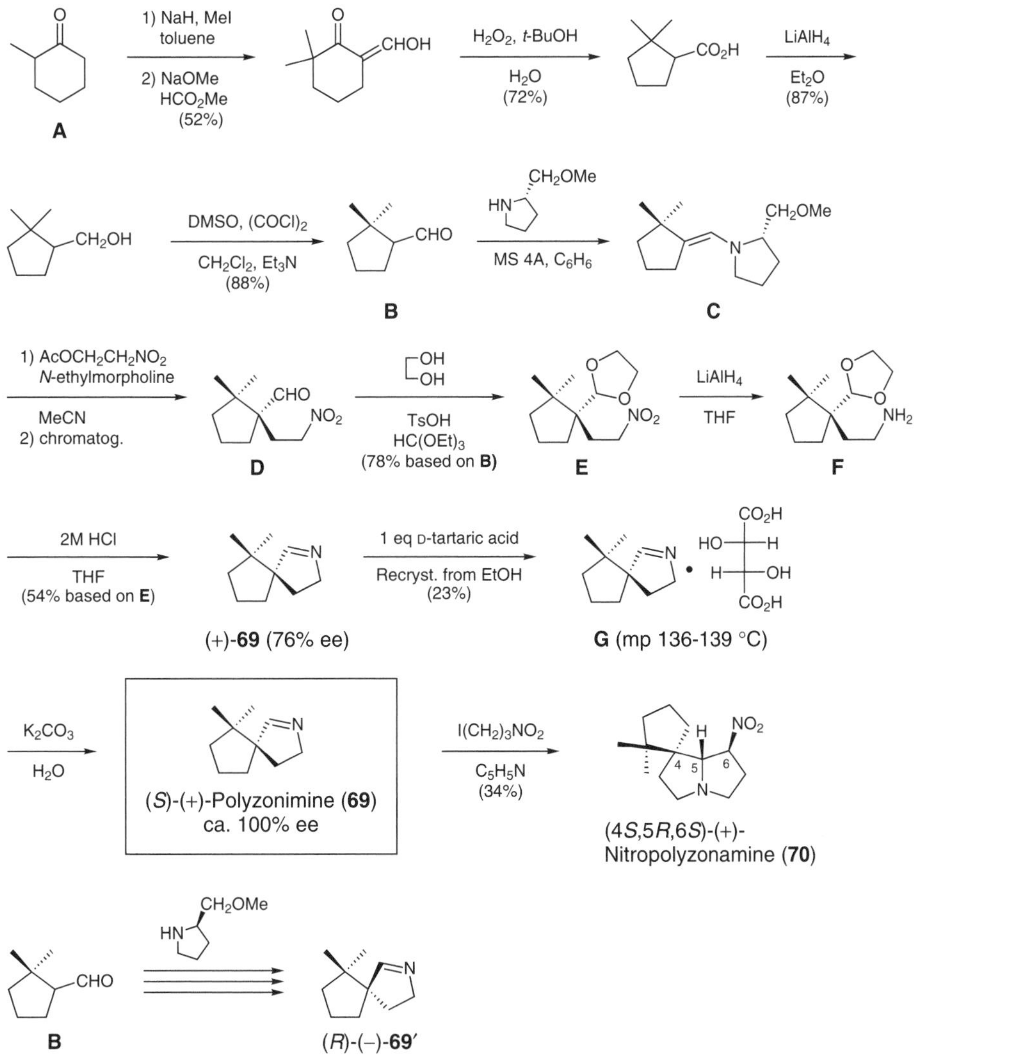

Figure 3.18 *Synthesis of the enantiomers of polyzonimine*

to give enamine **C**. Michael addition of nitroethylene to **C** yielded **D**, which was converted to (+)-**69** of 76% ee via **E** and **F**. Enantiomeric purity of (+)-**69** could be improved by recrystallizing its salt **G** with D-tartaric acid. Three recrystallizations of **G** from ethanol furnished pure **G**, which was treated with base to give (+)-polyzonimine (**69**) of ca. 100% ee as checked by enantioselective GC on a chiral stationary phase. Its absolute configuration was determined as *S* by correlating it with (4*S*,5*R*,6*S*)-(+)-nitropolyzonamine (**70**), whose absolute configuration was known to be (4*S*,5*R*,6*S*) by X-ray analysis. Although malic and mandelic acids were also used for the salt formation, only tartaric acid gave good result. Similarly, (*S*)-(+)-polyzonimine (**69′**) was also synthesized. Both **69** and **69′** showed oviposition deterrant activity against the webbing clothes moth (*Tineola bisselliella*), but did not repel the German cockroach (*Blattella germanica*).

In this chapter we discussed the synthesis of ten insect-related bioregulators. Mimics of juvenile hormones are now used practically as pesticides. Semiochemicals, some of which were discussed here, are interesting compounds for their future utilization as ecofriendly agents for pest control.

References

1. Sláma, K.; Williams, C.M. *Proc. Natl. Acad. Sci. USA* **1965**, *54*, 411–414.
2. Bowers, W.S.; Fales, H.M.; Thompson, M.J.; Uebel, E.C. *Science* **1966**, *154*, 1020–1021.
3. Tutihasi, R.; Hanazawa, T. *J. Chem. Soc. Jpn*. **1940**, *61*, 1045–1047.
4. Momose, T. *Yakugaku Zasshi* (*J. Pharm. Soc. Jpn.*) **1941**, *61*, 288–291.
5. Nakazaki, M.; Isoe, S. *Bull. Chem. Soc. Jpn.* **1963**, *36*, 1198–1204.
6. Mori, K.; Matsui, M. *Tetrahedron Lett*. **1967**, 2515–2518.
7. Mori, K.; Matsui, M. *Tetrahedron* **1968**, *24*, 3127–3138.
8. Nagano, E.; Mori, K. *Biosci. Biotechnol. Biochem*. **1992**, *56*, 1589–1591.
9. Watanabe, H.; Shimizu, H.; Mori, K. *Synthesis* **1994**, 1249–1254.
10. Mori, K.; Nagano, E. *Biocatalysis* **1990**, *3*, 25–36.
11. Röller, H.; Dahm, K.H.; Sweely, C.C.; Trost, B.M. *Angew. Chem. Int. Ed.* **1967**, *6*, 179–180.
12. Mori, K.; Stalla-Bourdillon, B.; Ohki, M.; Matsui, M.; Bowers, W.S. *Tetrahedron* **1969**, *25*, 1667–1677.
13. Mori, K.; Mitsui, T.; Fukami, J.; Ohtaki, T. *Agric. Biol. Chem*. **1971**, *35*, 1116–1127.
14. Akai, H.; Kiguchi, K.; Mori, K. *Appl. Entomol. Zool*. **1971**, *6*, 218–220.
15. Ohtaki, T.; Takeuchi, S.; Mori, K. *Jpn. J. Med. Sci. Biol*. **1971**, *24*, 251–255.
16. Nihmura, M.; Aomori, S.; Mori, K.; Matsui, M. *Agric. Biol. Chem*. **1972**, *36*, 889–892.
17. Mori, K. *Mitt. Schweiz. Entomol. Ges*. **1971**, *44*, 17–35.
18. Mori, K.; Ohki, M.; Sato, A.; Matsui, M. *Tetrahedron* **1972**, *28*, 3739–3745.
19. Mori, K. *Tetrahedron* **1972**, *28*, 3747–3756.
20. Mori, K.; Ohki, M.; Matsui, M. *Tetrahedron* **1974**, *30*, 715–718.
21. Loew, P.; Johnson, W.S. *J. Am. Chem. Soc*. **1971**, *93*, 3765–3766.
22. Prestwich, G.D.; Wawrzeńczyk, C. *Proc. Natl. Acad. Sci. USA* **1985**, *82*, 5290–5294.
23. Imai, K.; Marumo, S.; Mori, K. *J. Am. Chem. Soc*. **1974**, *96*, 5925–5927.
24. Imai, K.; Marumo, S.; Ohtaki, T. *Tetrahedron Lett*. **1976**, 1211–1214.
25. Schooley, D.A.; Bergot, B.J.; Goodman, W.; Gilbert, L.I. *Biochem. Biophys. Res. Commun*. **1978**, *81*, 743–749.
26. Yanai, M.; Sugai, T.; Mori, K. *Agric. Biol. Chem*. **1985**, *49*, 2373–2377.
27. Mori, K.; Mori, H. *Tetrahedron* **1987**, *43*, 4097–4106.
28. Mori, K.; Fujiwhara, M. *Tetrahedron* **1988**, *44*, 343–354.
29. Mori, K.; Fujiwhara, M. *Liebigs Ann. Chem*. **1989**, 41–44.
30. Mori, K.; Fujiwhara, M. *Israel J. Chem*. **1991**, *31*, 223–227.
31. Mori, K.; Fujiwhara, M. *Liebigs Ann. Chem*. **1990**, 369–372.
32. Mori, K. *Synlett* **1995**, 1097–1109.
33. Mori, K. In *Stereoselective Biocatalysis*; Patel, R.N., ed.; Marcel Dekker, New York, 2000; pp. 59–85.
34. Mori, K. In *Biocatalysis in the Pharmaceutical and Biotechnology Industries*, Patel, R.N., ed.; CRC Press; Boca Raton, 2007; pp. 563–589.
35. Mori, K.; Mori, H. *Org. Synth*. **1989**, *68*, 56–63.
36. Kindle, H.; Winistörfer, M.; Lanzrein, B.; Mori, K. *Experientia* **1989**, *45*, 350–360.
37. Sakurai, S.; Ohtaki, T.; Mori, H.; Fujiwhara, M.; Mori, K. *Experientia* **1990**, *46*, 220–221.
38. Kolb, H.C.; VanNieuwenhze, M.S.; Sharpless, K.B. *Chem. Rev*. **1994**, *94*, 2483–2547.
39. Crispino, G.A.; Sharpless, K.B. *Synthesis* **1993**, 777–779.
40. Okochi, T.; Mori, K. *Eur. J. Org. Chem*. **2001**, 2145–2150.
41. Ley, S.V.; Denholm, A.A.; Wood, A. *Nat. Prod. Rep*. **1993**, 109–157.
42. Veitch, G.E.; Beckmann, E.; Burke, B.J.; Boyer, A.; Maslen, S.L.; Ley, S.V. *Angew. Chem. Int. Ed.* **2007**, *46*, 7629–7632.

43. Mori, K.; Watanabe, H. *Tetrahedron* **1986**, *42*, 273–281.
44. Asakawa, Y.; Dawson, G.W.; Griffiths, D.C.; Lallemand, J.-Y; Ley, S.V.; Mori, K.; Mudd, A.; Pezechk-Leclaire, M.; Pickett, J.A.; Watanabe, H.; Woodcock, C.M.; Zhang, Z.-n. *J. Chem. Ecol*. **1988**, *14*, 1845–1855.
45. Mori, K.; Takaishi, H. *Liebigs Ann. Chem*. **1989**, 695–697.
46. Mori, K.; Mori, H.; Yanai, M. *Tetrahedron* **1986**, *42*, 291–294.
47. Mori, K.; Komatsu, M. *Liebigs Ann Chem*. **1988**, 107–119.
48. Jansen, B.J.M.; Sengers, H. H. W. J. M.; Bos, H.J.T; de Groot, Ae. *J. Org. Chem*. **1988**, *53*, 855–859.
49. Mori, K.; Qian, Z.-H.; Watanabe, S. *Liebigs Ann. Chem*. **1992**, 485–587.
50. Mori, K.; Matsushima, Y. *Synthesis* **1995**, 845–850.
51. Takikawa, H.; Yamazaki, Y.; Mori, K. *Eur. J. Org. Chem*. **1998**, 229–232.
52. Mori, K.; Takagi, Y. *Tetrahedron Lett*. **2000**, *41*, 6623–6625.
53. Takagi, Y.; Mori, K. *J. Brazil. Chem. Soc*. **2000**, *11*, 578–583.

4

Synthesis of Pheromones

Pheromone science is one of the new fields of science, whose development in the late 20th century was remarkable. Once chemists knew that the communications among a variety of organisms depend on chemical substances termed pheromones, they isolated, identified and synthesized hundreds of pheromones to use them practically for pest control. In this 21st century, practical application of pheromone science in bioindustries is being actively pursued all over the world. My synthetic works have been focused on this branch of science for almost 40 years. This chapter summarizes my works in pheromone science.

4.1   What are pheromones?

After the discovery of the first insect sex attractant bombykol [(10E,12Z)-10,12-hexadecadien-1-ol] by Butenandt and coworkers,[1] the term "pheromone" was defined by Karlson and Lüscher in 1959.[2] The term was derived from the Greek *pherein* (carry or transfer) and *horman* (stir up or excite). Pheromones are substances that are secreted to the outside by an individual and received by a second individual of the same species, in which they release a specific reaction, for example, a definite behavior or a developmental process.

Pheromones are classified into two categories, releaser pheromones and primer pheromones. Releaser (or signaller) pheromones cause changes of behaviors in the receivers, while primer pheromones cause physiological impacts on the receivers. Releaser pheromones can be further classified as sex pheromones, aggregation pheromones, trail pheromones, etc., according to the type of behavioral change.

Since less than milligram amounts of pheromones are available from organisms by isolation, their synthesis is of the utmost importance for establishing their structures and also for studying their practical applications in pest control. The targets of my pheromone synthesis have been chiral molecules. Accordingly, this chapter treats the subjects of enantioselective synthesis as applied in pheromone science. Several books and reviews are available on pheromone synthesis.[3-7] The significance of chirality in pheromone science has been discussed historically[8] and comprehensively.[9]

4.2   Methods for enantioselective synthesis

Since Biot's discovery in 1815 of the optical rotatory powers of solutions of natural products such as sucrose, many people have studied the relationship between chemical structure and optical rotatory power.

Chemical Synthesis of Hormones, Pheromones and Other Bioregulators   Kenji Mori
© 2010 John Wiley & Sons, Ltd

Pasteur's success in 1848 of the first enantiomer separation (optical resolution) of racemic acid as ammonium sodium (±)-tartrate tetrahydrate together with McKenzie's success in 1904 of the first asymmetric synthesis prompted many chemists to synthesize optically active compounds without recourse to "the vital force" of organisms, although by employing the capacity of a special species of organism, *Homo sapiens*, to discriminate left from right. Pasteur remarked in 1883 that "The universe is dissymmetric." Since then chemists' efforts have been focused on the control of asymmetry in this world of chiral and nonracemic materials.

There are three methods available for the enantioselective synthesis of pheromones: (1) derivation from enantiopure natural products, (2) enantiomer separation (optical resolution), and (3) chemical or biochemical asymmetric synthesis. Practitioners of enantioselective synthesis must be familiar with the analytical methods for the determination of enantiomeric purity of an optically active compound. These basic methods will be explained briefly in this section, and discussed in depth through examples in the later sections of this chapter.

4.2.1   Derivation from enantiopure natural products

Emil Fischer's basic strategy employed in his monumental work in 1891 on the determination of the configuration of D-glucose was to correlate the known sugars by degradation reactions and synthetic transformations. Several terpenes, sugars, α-amino acids and hydroxy acids are abundantly available as enantiopure materials. It is possible to synthesize pheromones from these enantiopure natural products, often called chiral pools, by carefully avoiding racemization in the course of their conversion to the target pheromones. In 1973 when I started my pheromone synthesis, this approach was attractive, since there was no practical chemical reactions for efficient asymmetric synthesis. Another reason why I adopted this Fischer-like method was the fact that I am a scientific "great-grandson" to Emil Fischer, because my thesis adviser Professor M. Matsui was a former student of Professor U. Suzuki, who worked as a research fellow (1903–1905) of Fischer's in Berlin. Emil Fischer's personality, achievements and scientific progeny was thoroughly described by Lichtenthaler.[10]

There are two drawbacks in employing the chiral pool approach. First, a natural product usually exists as a single enantiomer, and therefore only a single enantiomer of a pheromone can be prepared easily. Secondly, a natural product possesses a definite structure, which is often quite different from that of a pheromone. Accordingly, the synthesis may become lengthy to remove the undesired structural features in the starting natural product.

Figure 4.1 shows the structures of pheromones, which have been synthesized by us starting from (+)-tartaric acid and (+)-citronellic acid or (+)-citronellal. Malic acid, glutamic acid, serine, glyceraldehydes, limonene, carvone and others were employed as our starting materials in pheromone syntheses.[7]

4.2.2   Enantiomer separation (optical resolution)

Numerous racemates have been separated into their enantiomers since the first success in 1848 by Pasteur. The traditional method is to derivatize a racemate into a mixture of two diastereoisomers by means of so-called resolving agents, and then to separate the diastereoisomers by recrystallization or chromatography. Recent development of chiral stationary phases for chromatographic separation of the enantiomers made it possible to separate them even without derivatization. Another method of choice is to use enzymes for enantiomer separation. Examples in this chapter will illustrate the use of these methods. After enantiomer separation, the absolute configuration and the enantiomeric purity of the resulting enantiomers must be determined.

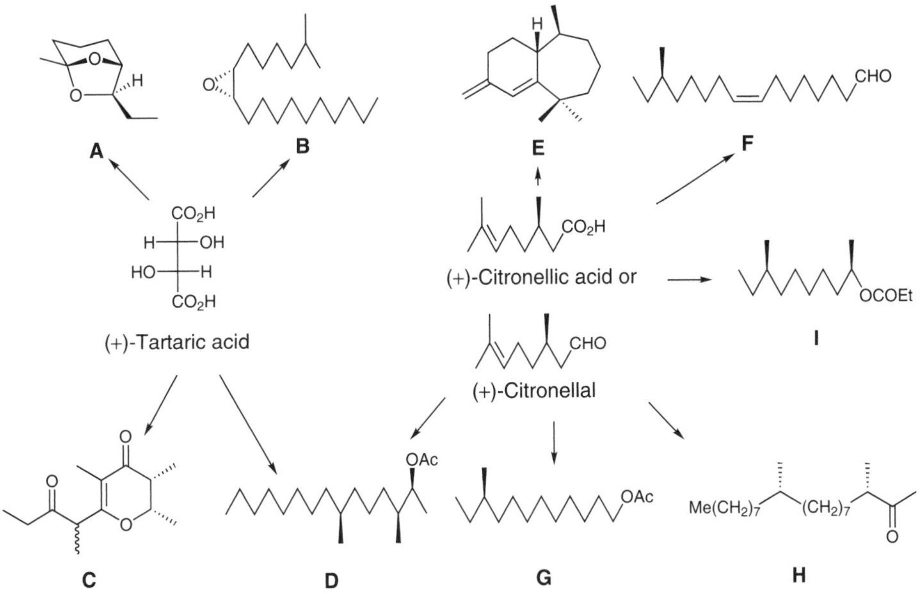

Figure 4.1 *Pheromones synthesized from tartaric acid, citronellic acid, and citronellal. Modified by permission of Shokabo Publishing Co., Ltd*

4.2.3   Asymmetric synthesis

Since the advent of asymmetric epoxidation by Sharpless and Katsuki in 1980, numerous asymmetric reactions have been invented, and some of them are selective and efficient enough to be employed in pheromone synthesis.[6,11] Examples will be given later in this chapter.

4.2.4   Determination of enantiomeric purity

Enantiomeric purity of an enantiomeric mixture is defined as below.

$$\text{Enantiomeric purity} = [(A_+ - A_-)/(A_+ + A_-)] \times 100 \ (\% \ ee)$$

$$\text{Where } A_+ = \% \text{ mole fraction of the dextrorotatory enantiomer}$$

$$A_- = \% \text{ mole fraction of the levorotatory enantiomer}$$

$$ee = \text{enantiomeric excess}$$

Accordingly, estimation of A_+ and A_- of a sample by an appropriate physical method such as GC, HPLC and NMR enables the calculation of its enantiomeric purity.[12]

Direct determination of the enantiomeric purity of an enantiomeric mixture is usually achieved by GC or HPLC on a chiral stationary phase derived from chiral materials such as cellulose and cyclodextrin. NMR analysis in the presence of a chiral shift reagent is also applicable, although with less accuracy than the chromatographic methods.

If the above methods are unsuccessful, an enantiomeric mixture must be derivatized with an enantio-pure derivatizing agent to give a diastereomeric mixture. The diastereomeric ratio can then be estimated by GC, HPLC or NMR analysis. The most popular derivatizing agent is Mosher's acid (α-methoxy-α-trifluoromethylphenylacetic acid), whose pure enantiomers are commercially available.[13,14] Examples of the determination of the enantiomeric purities will frequently appear in this book.

4.3   Why is it meaningful to synthesize enantiopure pheromones?

Although the first insect pheromone bombykol was an achiral molecule, a number of chiral pheromones were isolated and identified in the late 1960s. Such new discoveries of chiral pheromones necessitated the determination of their absolute configurations and enantiomeric purities. All of the chiral pheromones, however, were volatile liquids and could not be analysed by X-rays. In addition, they were scarce materials obtained in milligram or less quantities after laborious isolation works. Their limited availability precluded their degradation reactions to give compounds with known absolute configuration. In short, the conventional methods for determination of their absolute configuration could not be useful in the case of pheromones.

Accordingly, enantioselective synthesis of a pheromone with known absolute configuration was the only realistic method to determine its absolute configuration and supply a sample in an amount sufficient for its biological evaluation. Of course, it was also possible to compare the chiroptical properties of the synthetic pheromone with those of the naturally occurring ones. These biological and physical comparisons of the natural pheromone with those of its synthetic enantiomers were the only way to determine the absolute configuration of a pheromone.

Another important problem in 1973 when I started my pheromone synthesis was the relationship between the absolute configuration of a pheromone and its bioactivity. At that time there was no experimental result about the relationship, just because no one synthesized both the pure enantiomers of a pheromone. I really wanted to know the relationship, because among other bioactive compounds, there were several known cases, as shown in Figure 4.2. Only a single enantiomer of glutamic acid and also that of estrone are bioactive, while both the enantiomers of carvone as well as those of camphor were odoriferous.

It was clear in 1973 that the stereochemistry–bioactivity relationships among pheromones would be elucidated only after successful synthesis of both the pure enantiomers of pheromones. If a synthetic

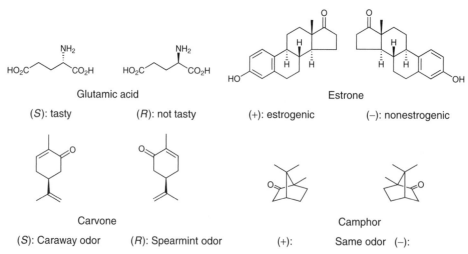

Figure 4.2   *Relationships between absolute configuration and bioactivity. Reprinted with permission of Shokabo Publishing Co., Ltd*

pheromone of 70% ee was used for bioassay, then the contaminating 15% of the wrong enantiomer would totally confuse the bioassay, especially when the major enantiomer (85%) was nonbioactive. The synthetic pheromone would show bioactivity due to the contaminating minor isomer (15%), because a pheromone is extremely bioactive even at a very low concentration. I was attracted by the challenge to prepare extremely pure enantiomers. This was not at all easy in 1973.

In summary, pheromone synthesis is a meaningful endeavor to verify the proposed structure, and also to supply a sufficient amount of a sample to biologists. I will now give a number of examples of pheromone synthesis.

4.3.1 Determination of absolute configuration (1). trogodermal

Trogoderma species of insects are notorious pests of stored products. They use (*R,Z*)-trogodermal (**72**) and/or its (*E*)-isomer (**73**) as their female-produced sex pheromone (Figure 4.3). At the early stage of the pheromone isolation in 1969, alcohol **71** was erroneously identified as the pheromone candidate. The absolute configuration of levorotatory **71** of insect origin was determined in 1973 as *R* by synthesizing dextrorotatory (*S*)-**71**′ from the commercially available (*S*)-2-methyl-1-butanol (**A**, Figure 4.3).[15,16] The isolated amount of **71** was very small. But fortunately **71** could be shown to be dextrorotatory, although it was impossible to determine the magnitude of the rotation. The dextrorotation of (*S*)-**71**′ allowed the absolute configuration of the natural and levorotatory **71** to be deduced as *R*. (*S*)-2-Methyl-1-butanol (**A**) was known as the optically active amyl alcohol obtained as a byproduct of saké (rice wine) fermentation, and its *S*-configuration had been established in connection with that of natural isoleucine, as determined by X-ray analysis.

Although the synthetic route leading to **71** as shown in Figure 4.3 was simple and straightforward, this work in 1973 was the first example of the determination of the absolute configuration of a chiral pheromone. Since it was later shown that the genuine pheromone was not **71** but trogodermals **72** and **73**, we synthesized both their (*R*)- and (*S*)-isomers.[17,18] From the result it became clear that only (*R*)-**72** and (*R*)-**73** show pheromone activity, while (*S*)-isomers are biologically inactive.[18–20]

As shown in Figure 4.3, the specific rotation of (*S*)-**71**′ was reported as +5.31 in chloroform. Recently, in 2008, there was a criticism against this value claiming that the value +5.31 was too large and could not be so, because no strongly absorbing chromophore was present in (*S*)-**71**′. I therefore synthesized again (*R*)-**71** and (*S*)-**71**′, together with their derivatives, as summarized in Figure 4.4.[21]

The key reaction in the new synthesis of **71** was olefin cross-metathesis between (*R*)-**B** and **C** to give acetate (*R*)-**D**. The alcohol obtained by saponification of (*R*)-**D** showed $[\alpha]_D^{21}$ −5.98, while its (*S*)-isomer showed $[\alpha]_D^{21}$ +5.89. Trogodermal enantiomers as well as 3-methylhexadecane enantiomers showed $[\alpha]_D$ values around 6. Accordingly, the criticism against our experimental data turned out to be without experimental support. It is dangerous to criticize others without solid experimental or theoretical support.

4.3.2 Determination of absolute configuration (2). hemiacetal pheromone of *Biprorulus bibax*

The spined citrus bug (*Biprorulus bibax*) is a pest of citrus in southern Australia. The major component (**74**, Figure 4.5) of the male-produced pheromone of *B. bibax* was isolated and identified by the joint work of James in Australia and Oliver *et al.* in the USA. Oliver found by chiral GC analysis that the insect produces a single enantiomer of **74**. In 1992 I became interested in clarifying the absolute configuration of this pheromone, because Dr. Oliver requested me to do so.

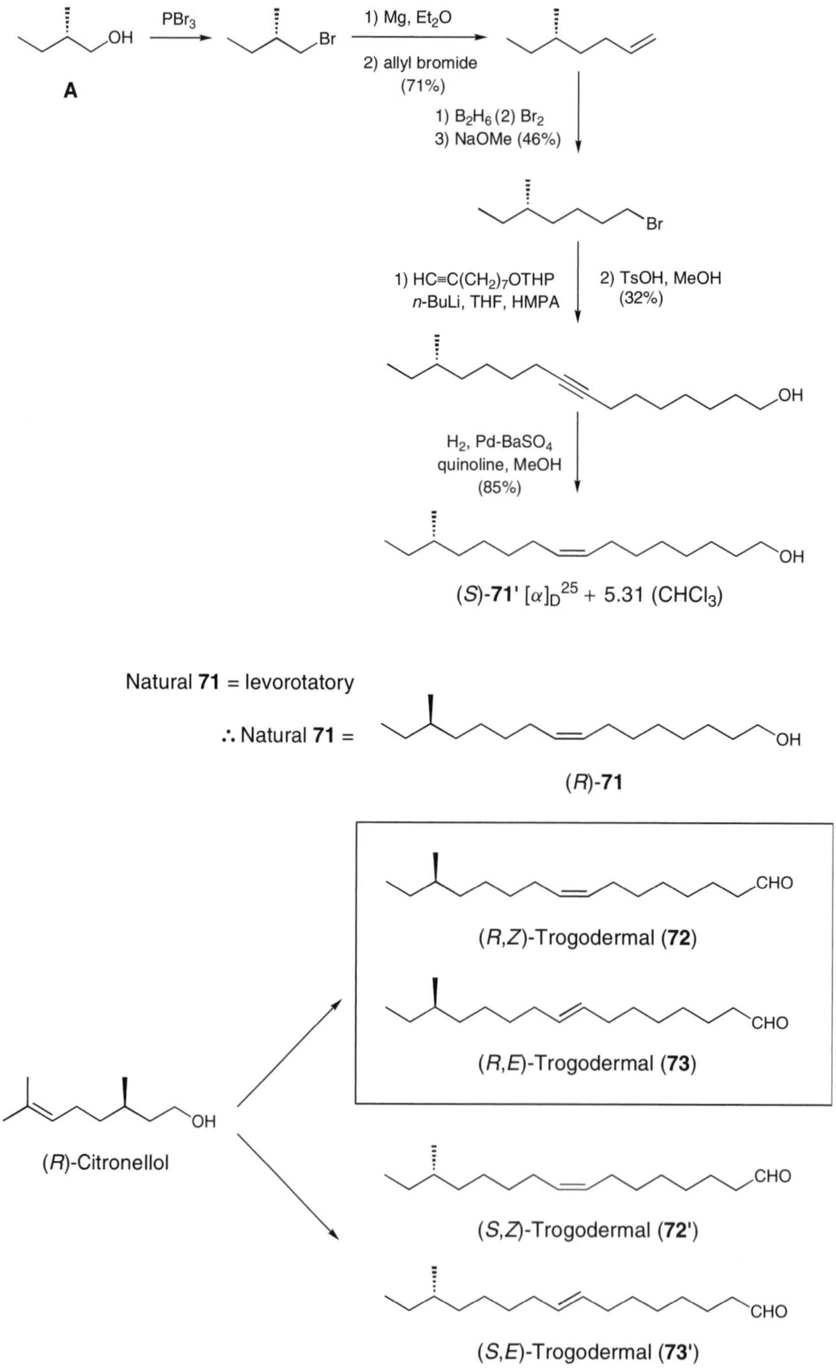

Figure 4.3 *Determination of the absolute configuration of trogodermal. Modified by permission of Shokabo Publishing Co., Ltd*

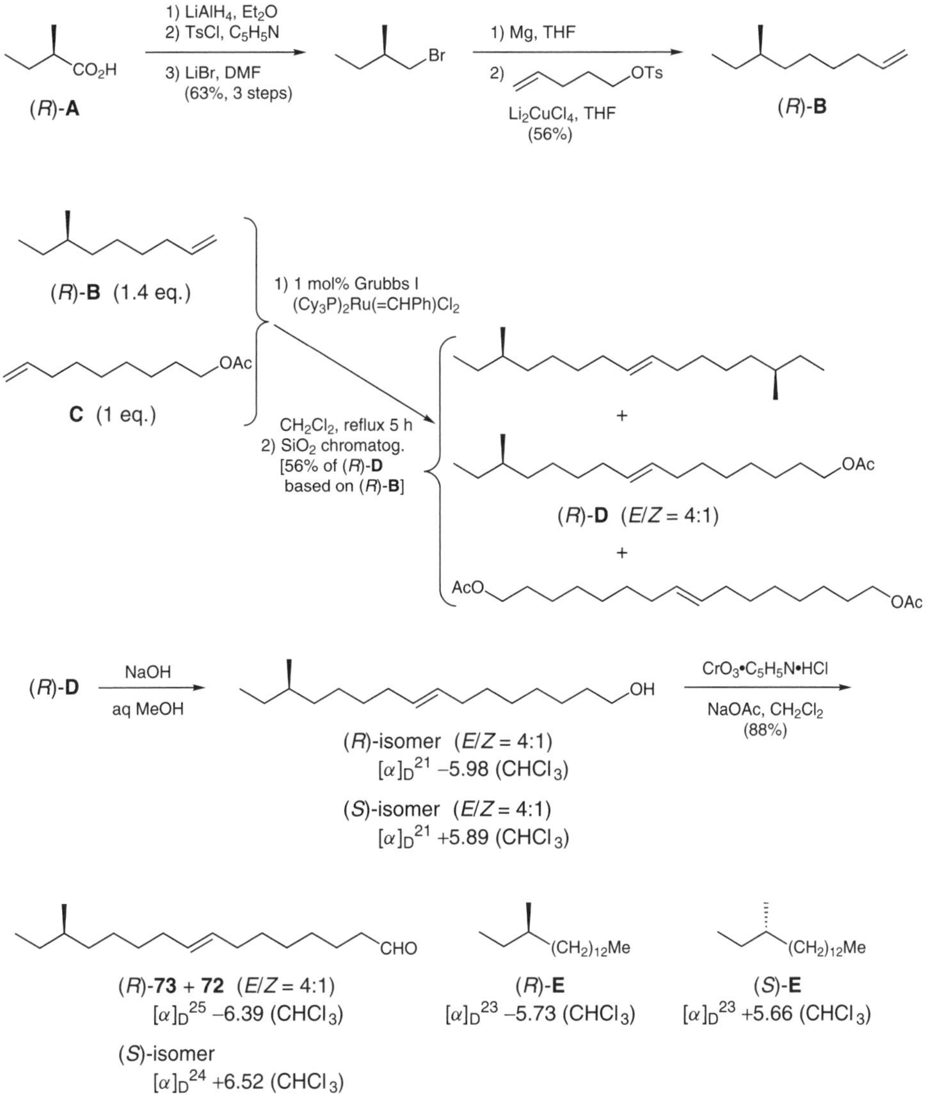

Figure 4.4 *New synthesis of trogodermal and related compounds*

(a) Synthesis of (±)-**74** by employing Diels–Alder reaction

As shown in Figure 4.5, it is well known that *cis*-**A** is the Diels–Alder adduct of 1,3-butadiene and maleic anhydride. Conversion of **A** to *meso*-**B** followed by its oxidation gave (±)-**C**. This was reduced to furnish (±)-**74**.[22] This synthesis, however, was not efficient enough to give a sufficient amount of (±)-**74**. Its overall yield was only 4.6% based on **A**.

(b) Synthesis of (±)-**74** by employing Claisen rearrangement

A more efficient synthesis of (±)-**74** was achieved by employing Ireland's ester enolate Claisen rearrangement, as shown in Figure 4.6.[23] Ester **A** was converted to its lithio-enolate in THF/HMPA, which

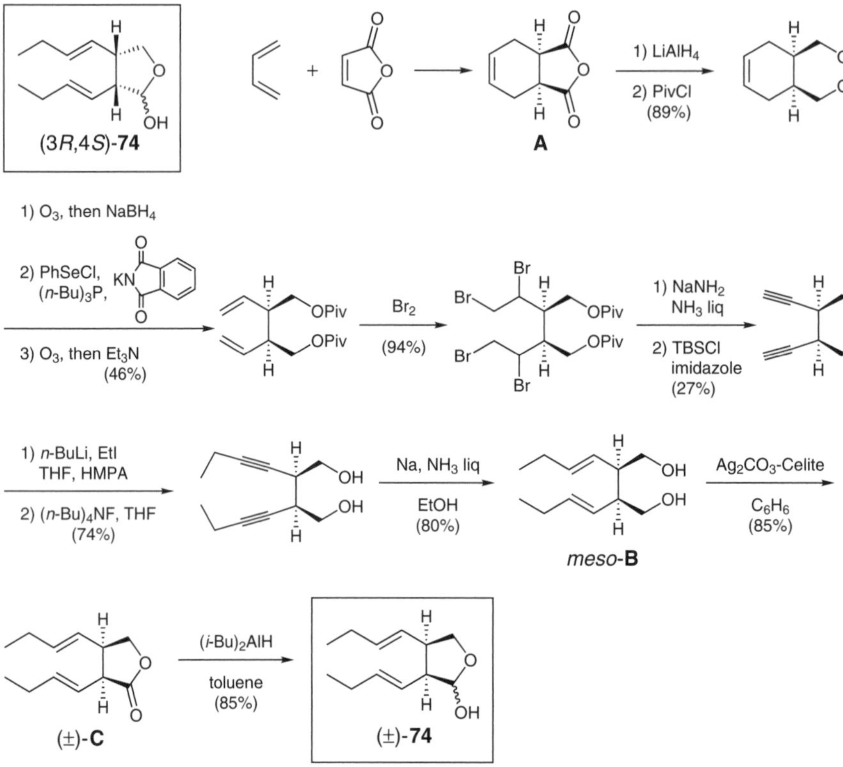

Figure 4.5 *Synthesis of the pheromone of Biprorulus bibax employing Diels–Alder reaction. Modified by permission of Shokabo Publishing Co., Ltd*

was trapped with trimethylsilyl chloride to give **B**. After its Claisen rearrangement, **C** was obtained as the product, because all the three substituents of the cyclohexane-chair-like transition state **B** adopted the more stable equatorial orientation. Accordingly, reduction of **C** gave *meso*-**D** as the major product. The overall yield of (±)-**74** based on propargyl alcohol THP ether was 34% (9 steps). A mixture of (±)-**74** and the minor components (linalool, nerolidol and farnesol) of the pheromone was a potent attractant of *B. bibax*.

(c) Determination of the absolute configuration of the naturally occurring **74**

For the synthesis of optically active **74**, we used lipase to achieve asymmetric hydrolysis of *meso*-diacetate **A** (Figure 4.7) to optically active monoacetate **B**. The stereochemical outcome of this asymmetric hydrolysis **A**→**B** was known to be as depicted through many examples. Acetylation of *meso*-diol **C** afforded *meso*-diacetate **D**, which was treated with lipase AK provided by Amano Enzyme, Inc. The product was (5*R*,6*S*)-**E**, whose enantiomeric purity was 97% *ee* as analysed by HPLC of the corresponding 3,5-dinitrobenzoate on Chiralcel-OJ®. Oxidation of **E** furnished **F**, which was converted to lactone (−)-**G**.

Unambiguous determination of the absolute configuration of (−)-**G** as 3*S*,4*R* was achieved by the correlation experiments depicted in Figure 4.8.[23] The strategy was to prepare (*S*)-alcohol **A**, convert it to (*S*)-**D** and then subject it to the Claisen rearrangement to give (2*S*,3*R*)-**F** via **E**.

Accordingly, (±)-alcohol **A** was oxidized to give **B**. Asymmetric reduction of **B** with Corey's CBS reagent gave (*S*)-**A** (79% ee). The *S*-configuration of **A** was confirmed by its conversion to the known lactone (*S*)-**C**. Then, the ester (*S*)-**D** prepared from (*S*)-**A** was subjected to the Claisen rearrangement to

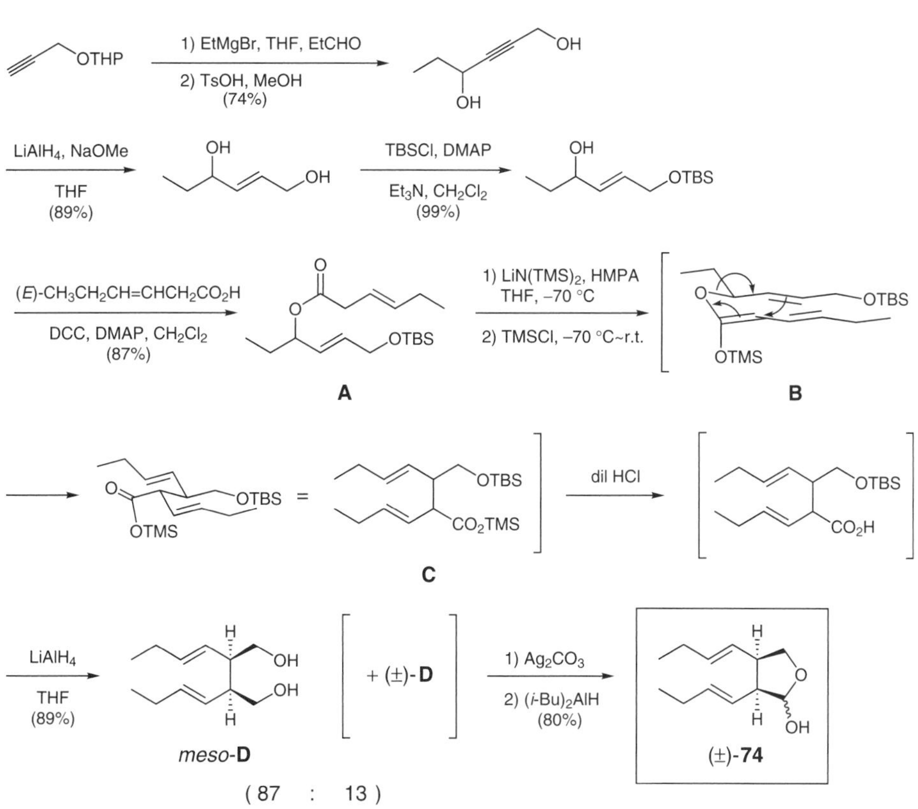

Figure 4.6 *Synthesis of the pheromone of Biprorulus bibax employing Claisen rearrangement. Modified by permission of Shokabo Publishing Co., Ltd*

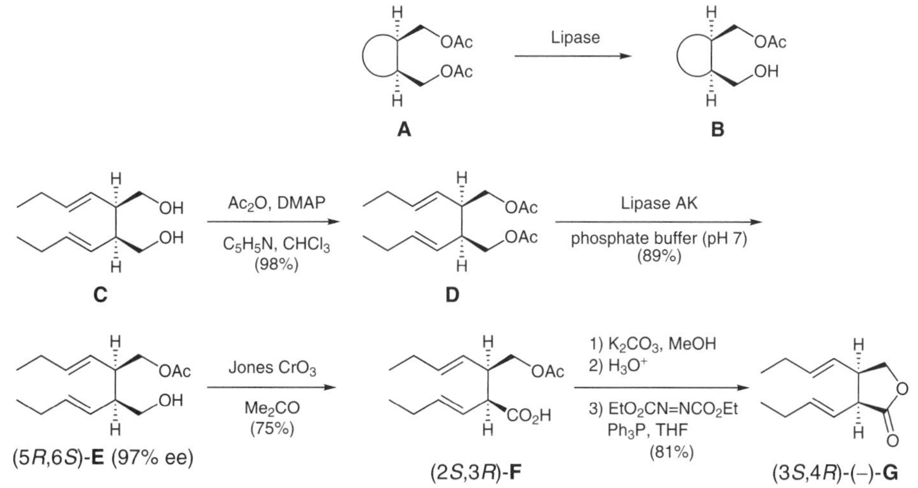

Figure 4.7 *Synthesis of optically active lactone (−)-**G** by asymmetric hydrolysis with lipase. Reprinted with permission of Shokabo Publishing Co., Ltd*

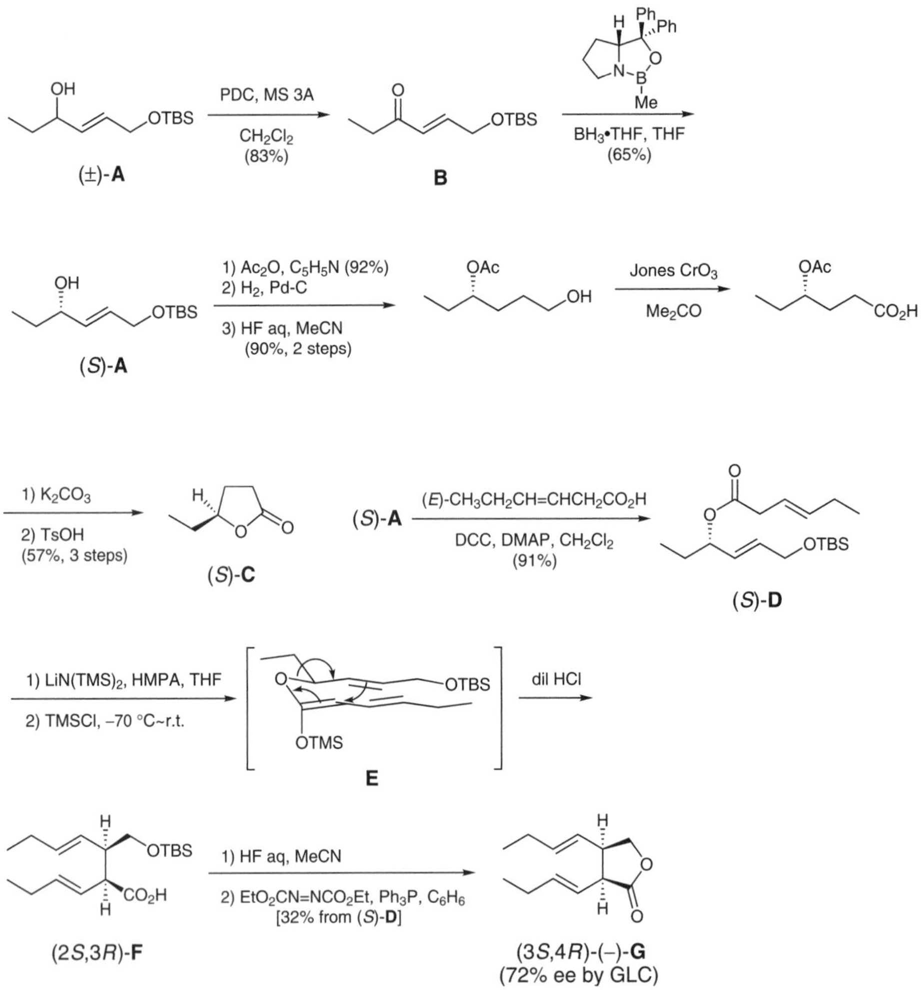

Figure 4.8   *Synthesis of (−)-**G** with (3S,4R)-configuration. Reprinted with permission of Shokabo Publishing Co., Ltd*

give (2S,3R)-**F** via **E**. Conversion of **F** into the lactone afforded (3S,4R)-(−)-**G** of 72% ee. It was therefore concluded that lactone (−)-**G** was with (3S,4R)-configuration.

(d) Synthesis of both the enantiomers of **74**

The next task was to synthesize the pure enantiomers of **74**. Lipase AK served as the pivotal biocatalyst to achieve the goal as shown in Figure 4.9.[23] We know that lipase-catalysed esterification/hydrolysis is a reversible process. Since the hydrolysis of diacetate **D** (Figure 4.7) yielded (5R,6S)-monoacetate in the presence of lipase AK, the antipodal (5S,6R)-monoacetate **B** (Figure 4.9) would be obtained by asymmetric acetylation of *meso*-diol **A** (Figure 4.9) with vinyl acetate and lipase AK. The monoacetate (5S,6R)-**B** (88% ee) was oxidized to give acid (2R,3S)-**C**, which was purified by recrystallization of its salt with (R)-1-(1-naphthyl)ethylamine to give pure (+)-**D**. Lactonization of (+)-**D** gave (3R,4S)-(+)-**E**, whose enantiomeric

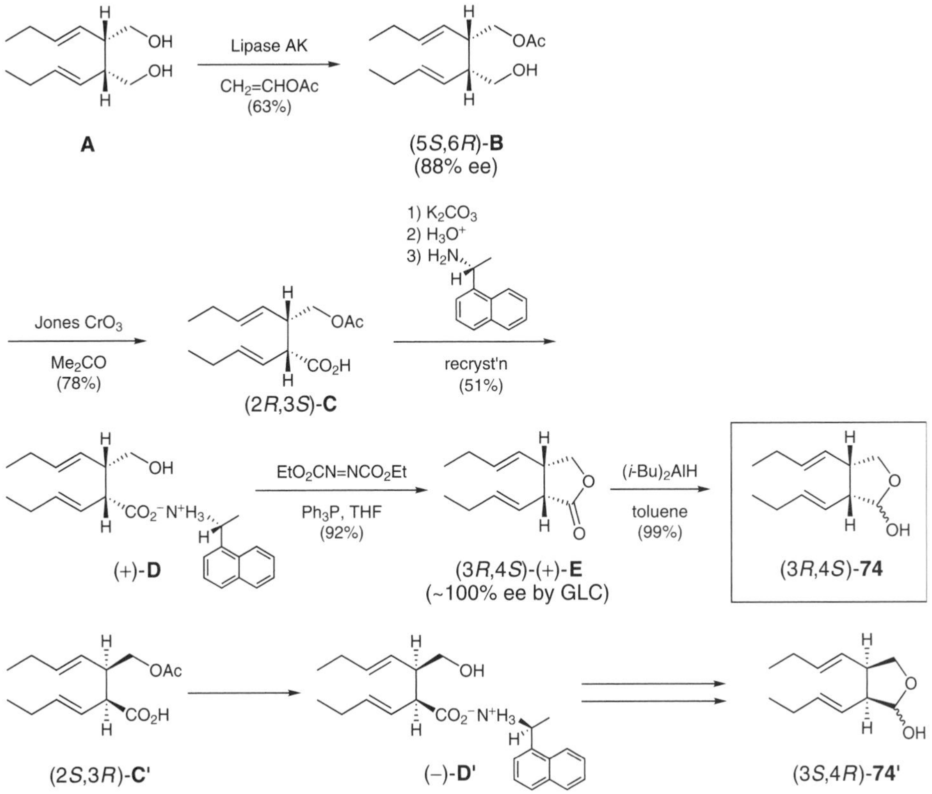

Figure 4.9 *Synthesis of the enantiomers of pheromone of Biprorulus bibax. Modified by permission of Shokabo Publishing Co., Ltd*

purity was nearly 100% ee. Reduction of (3*R*,4*S*)-**E** with diisobutylaluminum hydride furnished (3*R*,4*S*)-**74**. The antipodal acid (2*S*,3*R*)-**C′** was also purified by recrystallization of (−)-**D′**. The purified salt **D′** was converted to pure (3*S*,4*R*)-**74′**.

Bioassays of **74** and **74′** in Australia showed them to be equally active as the pheromone. However, *B. bibax* produces only (3*R*,4*S*)-**74** as revealed by GC analysis.[24] Thus, spined citrus bugs in Australia do not discriminate between the enantiomers of their aggregation pheromone. This is similar to the fact that even a perfumer cannot discriminate between camphor enantiomers.

(e) Efficient synthesis of (±)-**74**

The fact that (±)-**74** was as active as the natural (3*R*,4*S*)-**74** prompted us to prepare (±)-**74** efficiently, as shown in Figure 4.10.[25] In a large-scale synthesis, use of HMPA as the solvent at the stage of the Claisen rearrangement must be avoided due to its toxicity. To circumvent this problem, we chose the (*Z*)-isomer (±)-**C** as the substrate for the Claisen rearrangement. As shown in Figure 4.10, the Claisen rearrangement of (±)-**C** in the absence of HMPA gave (±)-**D** as the major product. The acid (±)-**D** was converted to (±)-**74** via (±)-**E** and (±)-**G**. Unfortunately, the practical application of (±)-**74** to control the population of *B.bibax.* could not be pursued any further, because Dr. James left Australia to live in the USA.

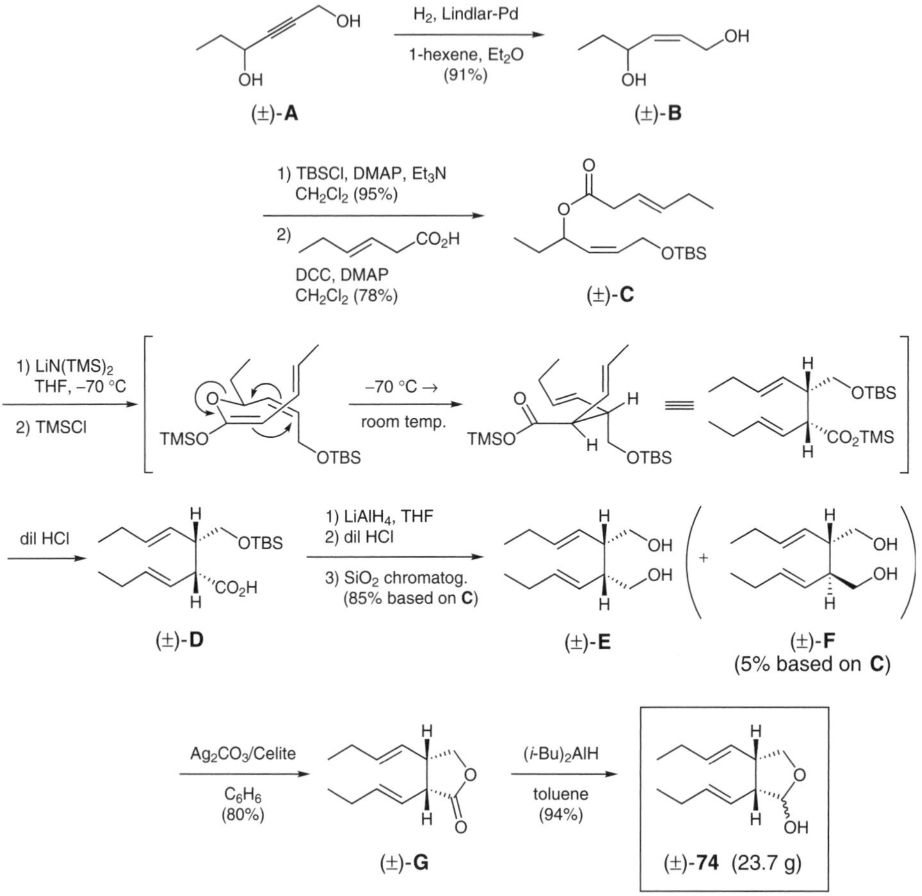

Figure 4.10   *Efficient synthesis of the racemic pheromone of Biprorulus bibax*

4.3.3   Determination of absolute configuration (3). sesquiterpene pheromone of *Eysarcoris lewisi*

Pecky rice, or rice damaged by pest insects, is a serious economic problem in Japan, because consumers do not accept such damaged rice. A stink bug *Eysarcoris lewisi* is known as one of the major species of rice bugs that cause pecky rice in northern Japan. *E. lewisi* usually lives in meadows and fields, and comes to rice paddy fields where it attacks rice plants at the time of their grain formation. Its emergence can hardly be surveyed by conventional means, especially because it cannot be attracted by light traps. The possibility of using its pheromone for monitoring its population was therefore examined by Takita and coworkers. In 2005 Takita proposed the structure of the male-produced aggregation pheromone of *E. lewisi* as (*E*)-2-methyl-6-(4'-methylenebicyclo[3.1.0]hexyl)hept-2-en-1-ol (**A** in Figure 4.11).

(a) Elucidation of the correct gross structure

As I synthesized (±)-sabina ketone in 1976 by intramolecular carbene addition reaction, as shown in Figure 4.11, the similarity of **A** with sabina ketone led me to synthesize the proposed structure **A** of

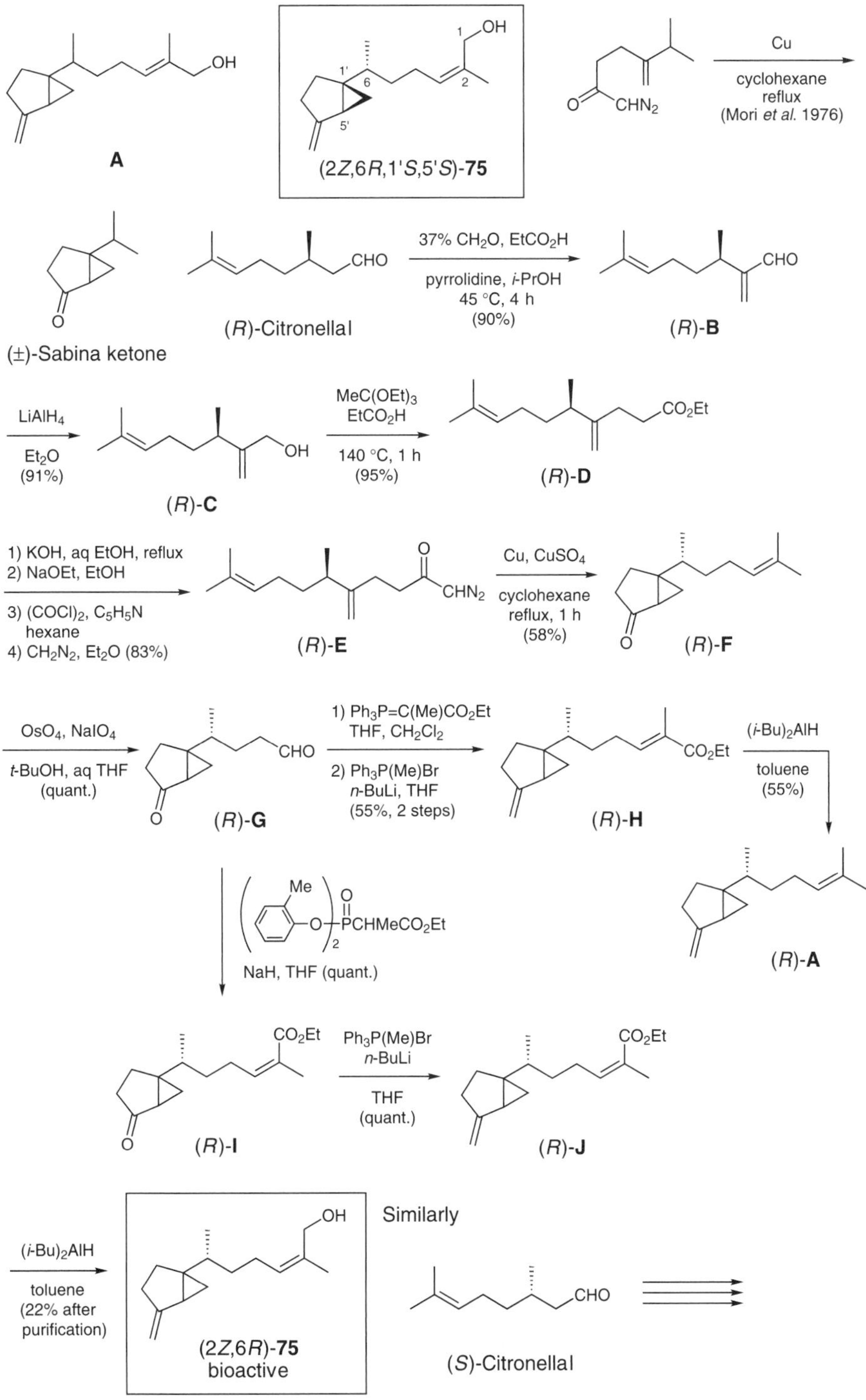

Figure 4.11   *Elucidation of the correct gross structure of the pheromone of Eysarcoris lewisi by synthesis*

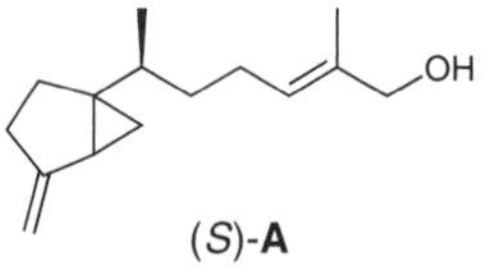

Figure 4.11   *(continued)*

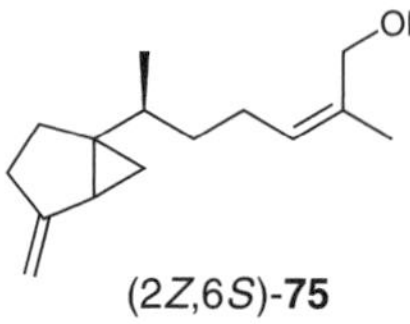

the pheromone of *E. lewisi*. Our synthesis finally determined the correct structure of the pheromone as (2*Z*,6*R*,1′*S*,5′*S*)-**75**. I will now describe the process leading to that conclusion.

The synthesis of (*R*)-**A** started from (*R*)-citronellal, which gave (*R*)-**B** by aldol reaction followed by deydration.[26] The aldehyde (*R*)-**B** was reduced with lithium aluminum hydride to give (*R*)-**C**. Orthoester Claisen rearrangement was employed for the two-carbon elongation of the carbon chain of (*R*)-**C** to give (*R*)-**D**. This was converted to diazoketone (*R*)-**E**. Ketocarbene generated from (*R*)-**E** under copper catalysis gave (*R*)-**F**, which was oxidized to furnish aldehyde (*R*)-**G**. Chain-elongation of (*R*)-**G** with a stabilized ylide gave a keto ester, whose methylenation afforded (*R*)-**H**. Finally, reduction of (*R*)-**H** with diisobutylaluminum hydride yielded (*R*)-**A**.

The synthetic (*R*)-**A**, however, showed ^{1}H- and ^{13}C-NMR data different from those of the natural pheromone. Accordingly, its (*Z*)-isomer, (2*Z*,6*R*)-**I**, was synthesized from (*R*)-**G** by employing Ando's Z-selective olefination reagent. Methylenation of **I** to give **J** was followed by its reduction to give (2*Z*,6*R*)-**75**, whose NMR data around the olefinic double bond were in good accord with those of the natural pheromone. Bioassay revealed (2*Z*,6*R*)-**75** as pheromonally active, while (*R*)-**A**, (*S*)-**A** and (2*Z*,6*S*)-**75** were all inactive.[26]

(b) Determination of the absolute configuration

The next stage was to determine the absolute configuration at the positions of ring fusion (C1′ and C5′) of the natural pheromone.[27] For that purpose, the diastereomeric mixture of ketone (*R*)-**A** (Figure 4.12) was reduced with L-Selectride® to give a mixture of two racemic and diastereomeric alcohols **B** and **B′**. Asymmetric acetylation of this mixture with vinyl acetate in the presence of lipase PS-D (Amano Enzyme) was followed by chromatographic purification to give the recovered alcohol (6*R*,1′*S*,4′*S*,5′*R*)-**B** and the acetate (6*R*,1′*R*,4′*R*,5′*S*)-**C**. The alcohol **B** was oxidized with tetra(*n*-propyl)ammonium perruthenate and *N*-methylmorpholine *N*-oxide to give (6*R*,1′*S*,5′*R*)-**D**. The acetate **C** was also converted to the ketone (6*R*,1′*R*,5′*S*)-**D′**. The circular dichroism (CD) spectra of these two diastereomeric ketones **D** and **D′** were then compared with that of the known (1*R*,5*S*)-(−)-sabina ketone. The ketone (6*R*,1′*S*,5′*R*)-**D**, which showed the CD spectrum antipodal to that of (1*R*,5*S*)-sabina ketone, yielded the bioactive pheromone with its ^{1}H- and ^{13}C-NMR spectra identical to those of the natural pheromone. The pheromone of *E. lewisi* was therefore shown to be (2*Z*,6*R*,1′*S*,5′*S*)-**75**. The other stereoisomer, (2*Z*,6*R*,1′*R*,5′*R*)-**75′**, was biologically inactive.

(c) Asymmetric synthesis of (2*Z*,6*R*,1′*S*,5′*S*)-**75**

The final task was to carry out an enantioselective synthesis of the natural pheromone (2*Z*,6*R*,1′*S*,5′*S*)-**75**. Figure 4.13 summarizes our synthesis of the pheromone.[28] Hodgson's intramolecular cyclopropanation procedure (**D**→**E**) could be employed successfully.[29] The synthetic pheromone amounted to about 400 mg, which was enough to carry out further biological studies. The acetate of (2*Z*,6*R*,1′*S*,5′*S*)-**75** was shown to be identical with the acetate isolated in1982 from an African plant *Haplocarpha scaposa* by Bohlmann and

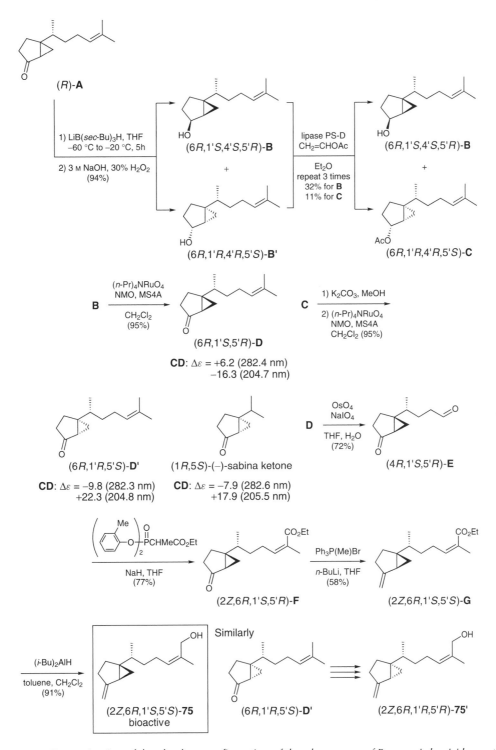

Figure 4.12 *Determination of the absolute configuration of the pheromone of Eysarcoris lewisi by synthesis*

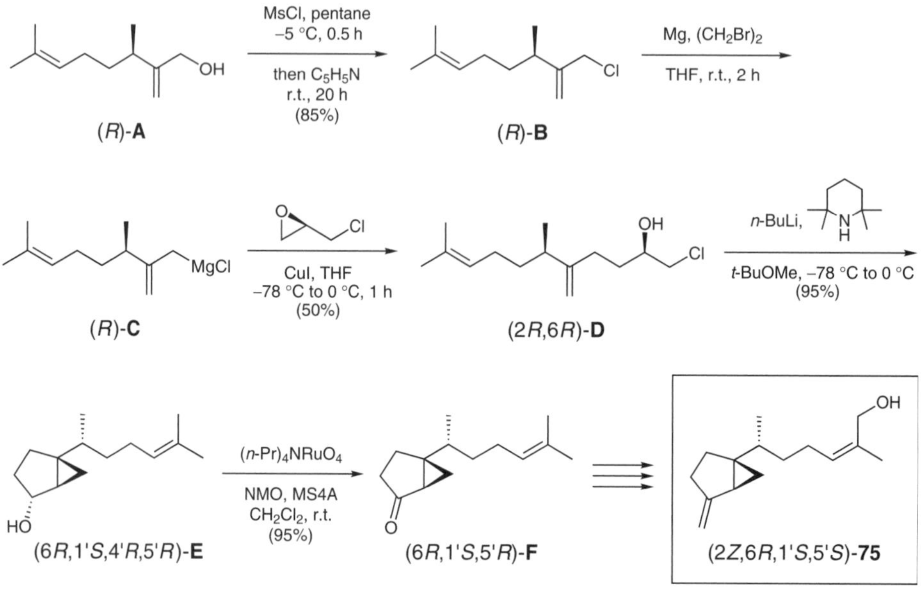

Figure 4.13 *Asymmetric synthesis of the pheromone of Eysarcoris lewisi*

Wallmeyer. It sometimes happens that the same compound is produced by different organisms in different parts of the world.

4.3.4 Clarification of the relationship between absolute configuration and pheromone activity—*exo*-brevicomin

The second pheromone study of mine started in the summer of 1973 to synthesize both the enantiomers of *exo*-brevicomin (**76**, Figure 4.14).[30] It was isolated as the female-produced aggregation pheromone of the western pine beetle (*Dendroctonus brevicomis*), the pest against ponderosa pine in the Pacific coast of the USA. Silverstein *et al.* isolated about 2 mg of the pheromone from 1.6 kg of frass produced by unmated female beetles, and its structure was determined as *exo*-7-ethyl-5-methyl-6,8-dioxabicyclo[3.2.1]octane (**76**).[31] Synthetic (±)-**76** elicited a response from *D. brevicomis*.[31] Nothing was known, however, about its absolute configuration. Indeed in their 1968 paper, Silverstein *et al.* reported that a 0.05% hexane solution of the pheromone showed no optical rotation between 350 and 250 nm.[31] This description puzzled me, considering the highly dissymmetric structure **76**. I thought it might mean that the natural **76** was a racemate. Otherwise, the pheromone should show optical rotation. I therefore decided to synthesize the enantiomers of **76**, as shown in Figure 4.14.[30]

The enantiomers of tartaric acid were chosen as the starting material, because tartaric acid possesses two stereogenic centers corresponding to C-1 and C-7 of **76**. In the period between 1968 and 1977, I taught a course in organic stereochemistry as a young associate professor. This obliged me to teach once a year about tartaric acid and Louis Pasteur. I therefore wanted to use tartaric acid in my own research program. I have the feeling that research and teaching are inseparable.

D-(−)-Tartaric acid, the unnatural enantiomer, was converted to the known ditosylate **A**. Two-carbon elongation of **A** with sodium cyanide was problematic. When the required amount (2 moles) of sodium cyanide was added in one portion to a DMSO solution of **A**, crystalline (2*E*,4*E*)-2,4-hexadienedinitrile

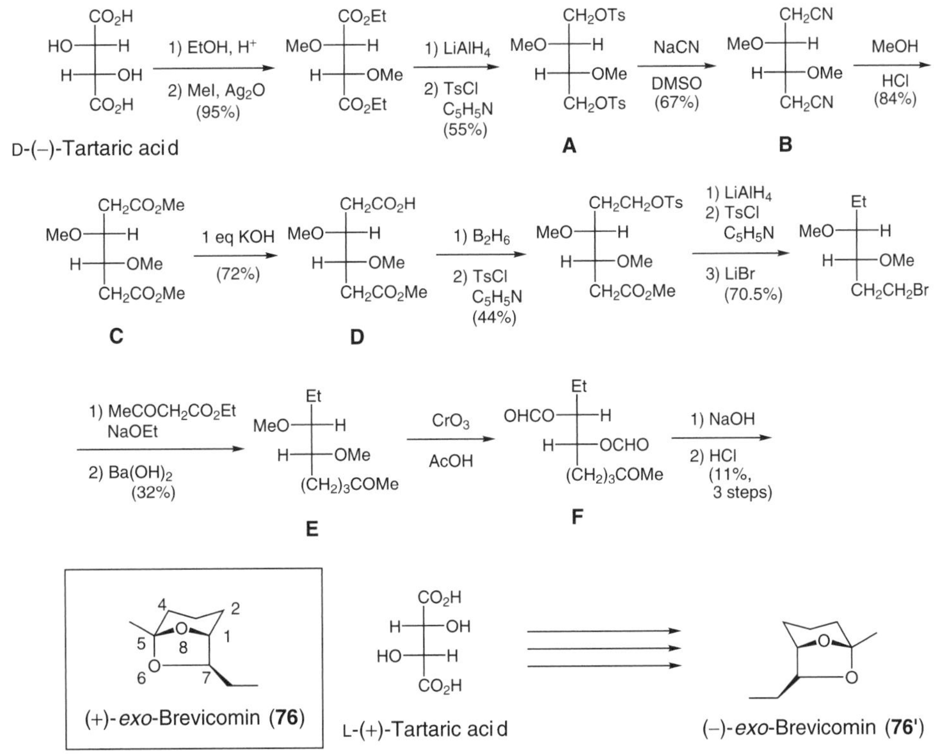

CO₂H ... D-(–)-Tartaric acid

1) EtOH, H⁺ 2) MeI, Ag₂O (95%) → CO₂Et (MeO/OMe)

1) LiAlH₄ 2) TsCl C₅H₅N (55%) → **A** (CH₂OTs, MeO/OMe, CH₂OTs)

NaCN DMSO (67%) → **B** (CH₂CN, MeO/OMe, CH₂CN)

MeOH HCl (84%) →

C (CH₂CO₂Me, MeO/OMe, CH₂CO₂Me)

1 eq KOH (72%) → **D** (CH₂CO₂H, MeO/OMe, CH₂CO₂Me)

1) B₂H₆ 2) TsCl C₅H₅N (44%) → (CH₂CH₂OTs, MeO/OMe, CH₂CO₂Me)

1) LiAlH₄ 2) TsCl C₅H₅N 3) LiBr (70.5%) → (Et, MeO/OMe, CH₂CH₂Br)

1) MeCOCH₂CO₂Et NaOEt 2) Ba(OH)₂ (32%) → **E** (Et, MeO/OMe, (CH₂)₃COMe)

CrO₃ AcOH → **F** (Et, OHCO/OCHO, (CH₂)₃COMe)

1) NaOH 2) HCl (11%, 3 steps) →

(+)-*exo*-Brevicomin (**76**)

L-(+)-Tartaric acid (CO₂H, H/OH, HO/H, CO₂H)

⟹ (–)-*exo*-Brevicomin (**76'**)

Figure 4.14 *Synthesis of the enantiomers of exo-brevicomin. Modified by permission of Shokabo Publishing Co., Ltd*

[a dinitrile corresponding to (2*E*,4*E*)-muconic acid] was obtained as the product. Extremely slow addition of sodium cyanide to the reaction mixture in the course of 3 days could avoid the basification of the mixture to give the desired **B** as crystals. Under basic conditions 2 moles of methanol would be eliminated to give the undesired 2,4-hexadienedinitrile.

Half-saponification of **C** with 1 equivalent of potassium hydroxide gave a single product **D** due to the C_2 symmetry of **C**. Removal of the two methyl protective groups of **E** could be executed to give **F** in a miserable yield by chromic-acid oxidation. The resulting formate **F** gave (+)-**76** after basic hydrolysis and acetalization with acid. The product (+)-**76** was extremely volatile, and its chromatographic purification caused a big loss in the yield. Even the measurement of the IR spectrum of **76** was not easy, because the film of **76** evaporated very quickly prior to the completion of the measurement. Similarly, (−)-**76'** was synthesized by starting from L-(+)-tartaric acid. The specific rotation of (+)-**76** was $[\alpha]_D^{20}$ +84.1 (*c* 2.2, Et₂O), which was large enough to measure. The reason was unclear why Silverstein *et al.* reported **76** to show no optical rotation. Perhaps the pheromone had been lost by evaporation before their $[\alpha]_D$ measurement.

The bioassay of (1*R*,5*S*,7*R*)-(+)-*exo*-brevicomin (**76**) and its (−)-enantiomer (**76'**) was carried out by Professor David L. Wood and coworkers at the University of California, Berkeley. When I visited his laboratory in June 1974, I was deeply impressed by the fact that only (+)-**76** was pheromonally active to attract the beetles. A large-scale field test also indicated that (+)-**76** was the bioactive isomer, while (−)-**76'** was neither active nor inhibitory.[32] Thus, only a single enantiomer of the pheromone was found to be highly bioactive. It became clear that bioactivity depends on the chirality of a pheromone. This result was in accord with the generally accepted belief in 1974 that a single enantiomer is biologically important. At that time, however, I did not extraporate and generalize the above result to regard all the pheromones to show the same stereochemistry–pheromone activity relationship as that observed for *exo*-brevicomin. I therefore continued my pheromone synthesis to provide sufficient amount of samples for bioassay. The results of my endeavor will appear later in this chapter (see Section 4.6).

4.3.5 Clarification of structure (1). lineatin

The available amount of a natural pheromone is so small that it is often difficult to propose even a gross structure. Our synthetic works on lineatin (**77**, Figure 4.15) started in 1979 to confirm its proposed structure. In 1977 Silverstein and coworkers isolated lineatin as a female-produced attractant of the striped ambrosia beetle (*Trypodendron lineatum*), an important pest to forests in both Europe and North America by boring tunnels into the sapwood of a number of coniferous species. The females initiate the attack and produce frass containing the pheromone attractive to both sexes. At the initial stage of structure elucidation, two possible structures **77** and **78** were given to lineatin without assigning its absolute configuration.

The cage-like intramolecular acetal structures of **77** and **78** attracted my attention, and we synthesized both (±)-**77** and (±)-**78**, the former of which showed the spectral data identical to those of the natural lineatin.[33,34] Photocycloaddition of vinyl acetate to 3-methyl-2-cyclopentenone (**A**) gave **B**, from which were prepared two isomeric ketones **C** and **D**. These were separable by chromatography, and their structures could be determined by ^{1}H-NMR analysis. The ketone (±)-**C** was further converted to (±)-**77**, while (±)-**D**

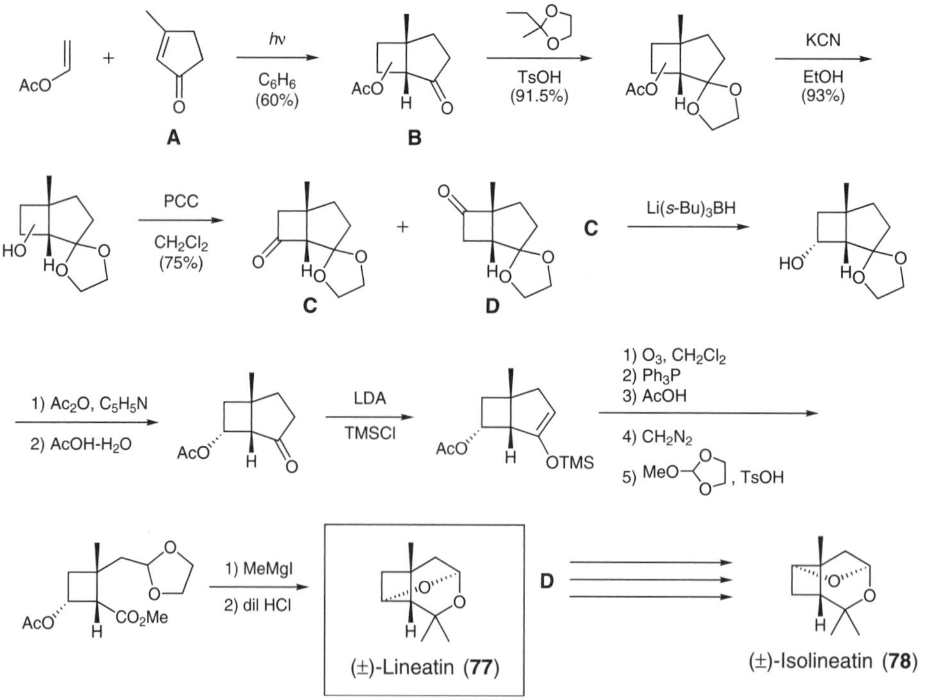

Figure 4.15 *Synthesis of (±)-lineatin and (±)-isolineatin. Modified by permission of Shokabo Publishing Co., Ltd*

afforded (±)-**78**, Spectral analysis of (±)-**77** proved it to be the racemate of lineatin. The isomer (±)-**78** was named isolineatin.

Subsequently, the absolute configuration of the natural lineatin was determined as (1*R*,4*S*,5*R*,7*R*)-**77** by our second synthesis, as shown in Figure 4.16.[35] The first step was the cycloaddition of dichloroketene to isoprene to construct a cyclobutane ring. The symmetrical cyclobutanone **A** was then converted to (±)-bicyclic lactone **B**. Enantiomer separation (optical resolution) of (±)-**B** was executed as follows. Reaction of (±)-**B** with the resolving agent **C** (derived from chrysanthemic acid) yielded a diastereomeric mixture

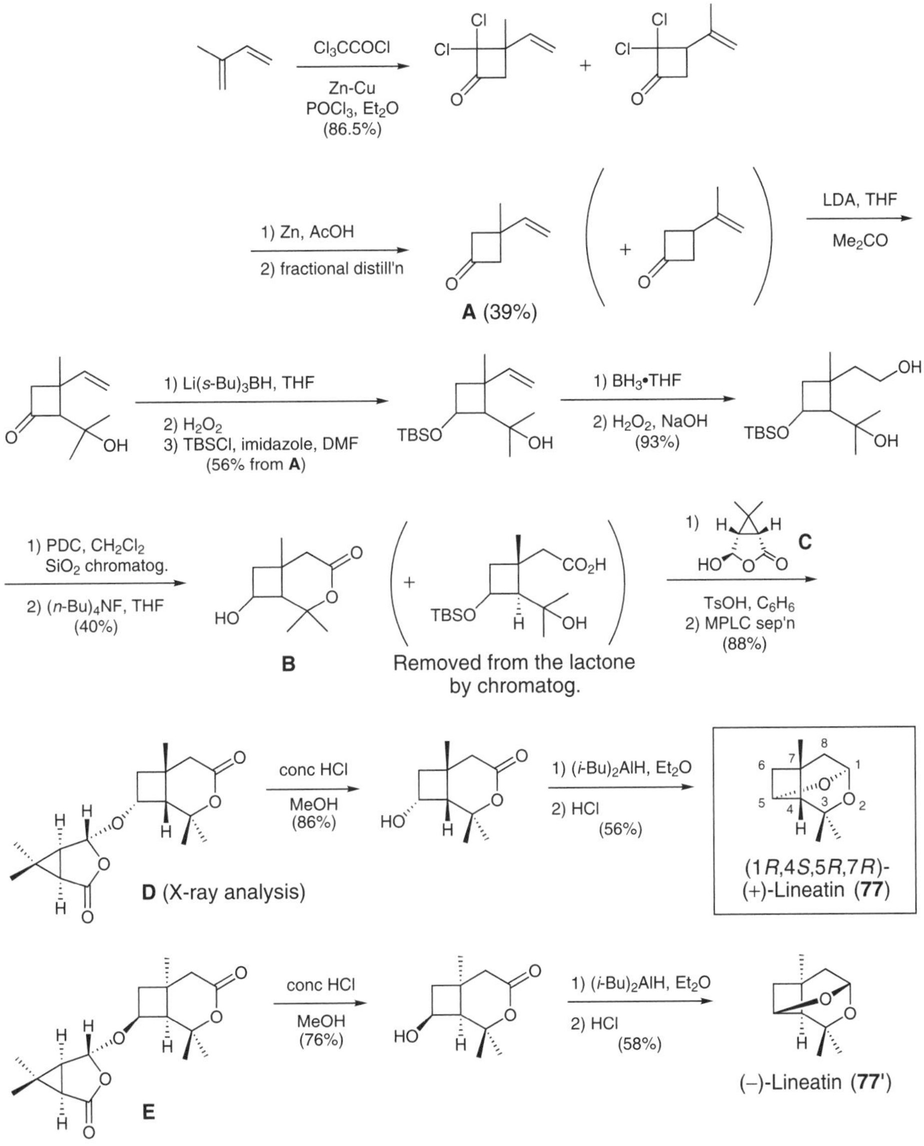

Figure 4.16   *Synthesis of the enantiomers of lineatin. Modified by permission of Shokabo Publishing Co., Ltd*

of **D** and **E**. These two were separated by chromatography to give **D** and **E** as crystals. X-ray analysis of **D** solved its absolute configuration as depicted, because the absolute configuration of **C** was known. The diastereomer **D** was converted to (1R,4S,5R,7R)-(+)-lineatin (**77**), while **E** furnished (−)-**77'**.

(+)-Lineatin (**77**) was pheromonally active, while (−)-**77'** was inactive. By using synthetic enantiomers of lineatin as the reference samples, the natural pheromone of *Trypodendron lineatum, T. signatum*, and *T. domesticum* was shown to be pure (+)-**77**.[36]

4.3.6 Clarification of structure (2). American cockroach pheromone

The female-produced sex pheromone of the American cockroach (*Periplaneta americana*) consists of the four components, periplanones-A, B, C and D, periplanone-B being the major component (Figure 4.17).[37] Periplanones-A and B were isolated in 1974 by Persoons in the Netherlands. The structure of periplanone-B was proposed by him in 1976, confirmed by its synthesis, and its absolute configuration was determined, as depicted in Figure 4.17. We achieved two different enantioselective syntheses of the naturally occurring (−)-periplanone-B.[38,39]

Structure elucidation of periplanone-A, one of the minor components, was difficult, because it was obtained in an extremely small amount. Indeed, only 20 μg of periplanone-**A** was obtained from 75 000 virgin females of *P. americana*. In 1978 Persoons *et al.* proposed **A** in Figure 4.18 as the structure of periplanone-A, and reported it to be highly unstable. Its half-life was said to be two weeks, giving a stable rearrangement product **B**. Subsequently in 1987, Macdonald in the USA proposed the stereochemistry of **B** as depicted. The structure of periplanone-A itself, however, was not definitely elucidated despite a number of attempts to clarify it.[40]

In 1986 Hauptmann in Germany attempted the reisolation of the American cockroach pheromone, and found a potent pheromonally active compound with spectral properties different from those reported for periplanone-A of Persoons. Hauptmann clarified the structure of his compound as **79**, and named it periplanone-A.

In October 1987, there was a pheromone conference in Angers, France, where I was able to listen to the discussion between Drs. Hauptmann and Persoons. I noticed what Hauptmann said to Persoons, "You executed the final purification of your periplanone-A by gas chromatography, but I did it by HPLC." It occurred to me immediately to regard Persoons' periplanone-A as the product of thermal rearrangement of Hauptmann's periplanone-A (**79**). During the conference I recognized that many people were eager to know the true structure of Persoons' periplanone-A. I therefore decided to solve the problem.

So as to check the credibility of Persoons' work, we first attempted the synthesis of (±)-**B** in Figure 4.18 starting from 1,3-butadiene and **C**.[41,42] Persoons reported **B** to be a stable compound. Our synthetic (±)-**B** was stable, and exhibited the ¹H-NMR spectrum identical with that reported for **B**. Persoons' work was therefore confirmed to be correct as far as the structure of the stable rearrangement product **B** was concerned.

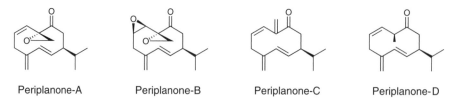

| Periplanone-A | Periplanone-B | Periplanone-C | Periplanone-D |

Figure 4.17 *Structures of the components of the sex pheromone of the American cockroach. Reprinted with permission of Shokabo Publishing Co., Ltd*

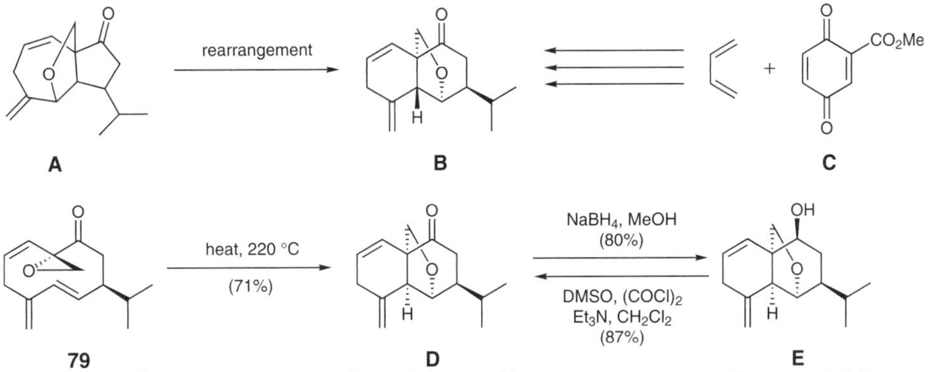

Figure 4.18   *Periplanone-A and isoperiplanone-A. Reprinted with permission of Shokabo Publishing Co., Ltd*

The next task was to synthesize the enantiomers of Hauptmann's periplanone-A. Especially for the naturally occurring (−)-enantiomer of **79**, I asked my capable coworker Dr. Kuwahara to synthesize it in over 100 mg quantities so that we would be able to carry out its thermal decomposition.[43] Our synthesis of the enantiomer of Hauptmann's periplanone-A is summarized in Figure 4.19.[43,44] The starting material for (−)-**79** was (*R*)-3-cyclohexene-1-carboxylic acid (**A**). Aldol condensation of **B** with **C** afforded **D**, which was converted to a separable mixture of **E** and **F**. Oxy-Cope rearrangement of **E** afforded **G**, which was converted to the levorotatory enantiomer of Hauptmann's periplanone-A (**79**). The spectral properties of (−)-**79** were identical with those reported for Hauptmann's periplanone-A. Our synthetic (−)-**79** was pheromonally active even at a dosage of 10^{-5} μg. (+)-Periplanone-A (**79′**) was synthesized from **F**, and it was almost biologically inactive [0.1% as active as (−)-**79**].

We secured about 250 mg of the crystalline and naturally occurring enantiomer (−)-**79**, and examined its pyrolysis. GC-MS analysis of (−)-**79** at the column temperature of 180 °C gave a product with a mass spectrum identical to that reported for Persoons' periplanone-A. Having been encouraged by the preliminary GC-MS experiment, about 80 mg of (−)-**79** was subjected to thermal decomposition at 220 °C on a 3% OV-17 column, which had been employed by Persoons for his purification experiment. After TLC purification of the thermolysis product, we obtained an oil in 71% yield based on (−)-**79**, whose IR and ¹H-NMR spectra were identical with those reported for Persoons' periplanone-A. It was therefore the pyrolysis product of (−)-**79**, although its structure was still unknown.

Since X-ray analysis was believed to be the best method to solve the problem, we attempted to derivatize the oily pyrolysis product to a crystalline compound. Very fortunately, reduction of the oil with sodium borohydride gave a crystalline alcohol, whose structure was solved by X-ray analysis as **E** (Figure 4.18). Swern oxidation of **E** regenerated the product of thermal decomposition of (−)-**79**. The structure of the pyrolysis product must therefore be ketone **D**. The oily **D** was stable at room temperature, and did not give **B**. The ketone **D** regenerated from recrystallized **E** was pheromonally inactive, although the crude **D** obtained by TLC purification was pheromonally active due to the trace amount of contaminating (−)-**79**.[44] The bioactivity observed for Persoons' periplanone-A might have been due to the contaminating (−)-**79**.

The biologically inactive ketone **D** is now called isoperiplanone-A.[37] Use of preparative GC by Persoons in the final purification of periplanone-A (−)-**79** caused its thermal decomposition to **D**, and confused his

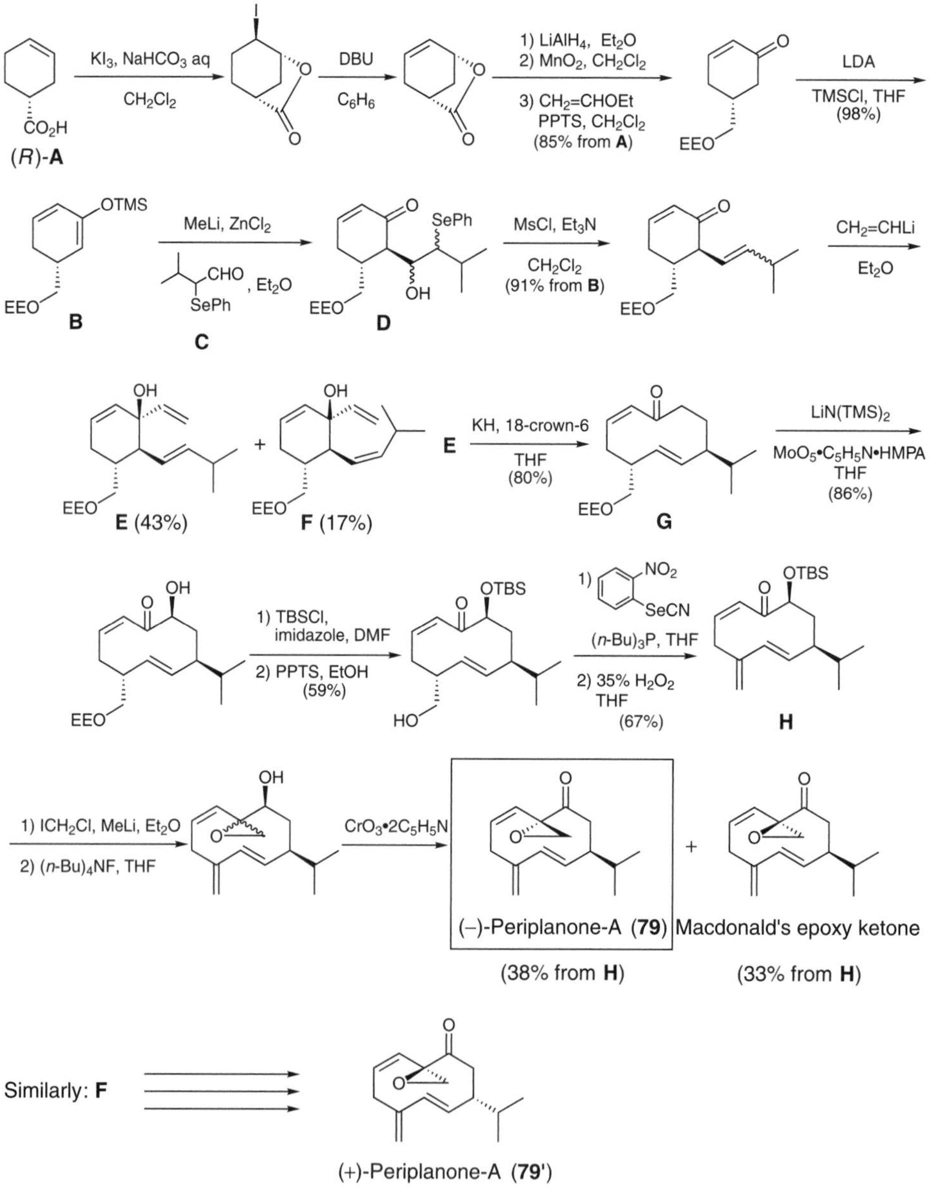

Figure 4.19 *Synthesis of the enantiomers of periplanone-A. Modified by permission of Shokabo Publishing Co., Ltd*

structural studies. This is a good lesson for natural products chemists to avoid excessive heating in the course of purification. The chaotic confusion could be clarified finally by our synthetic and degradative studies on (−)-**79**. Our success in preparing a sufficient amount of (−)-**79** was the key to solve this unfortunate confusion in structural studies. Persoons cannot be blamed, because in early 1970s when he engaged in the structural studies, a good preparative HPLC apparatus was not available.

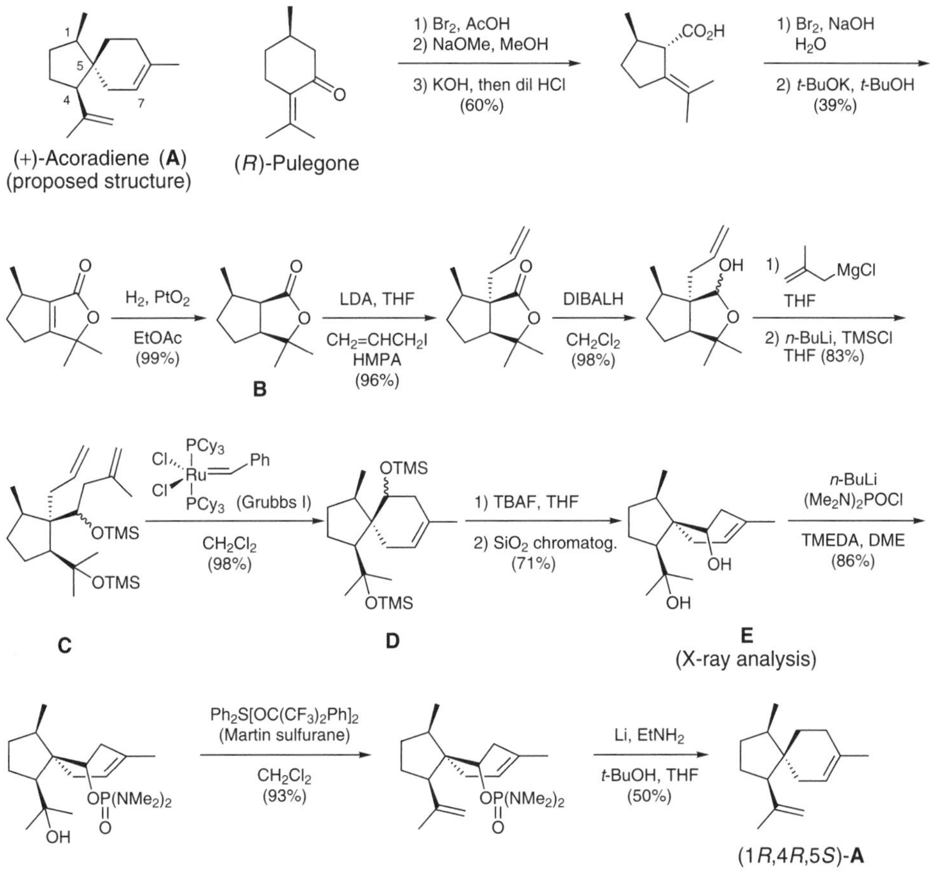

Figure 4.20 Synthesis of (1R,4R,5S)-acoradiene

4.3.7 Clarification of structure (3). acoradiene

In 1998 Tebayashi *et al.* in Japan isolated the major component of the male-produced aggregation pheromone of the broad-horned flour beetle (*Gnatocerus cornutus*). The proposed structure, (+)-acoradiene (**A** in Figure 4.20), was unique as the only pheromone with a spirosesquiterpene structure.

We therefore synthesized **A** in 2001.[45] The known lactone **B** was prepared from (*R*)-pulegone as shown in Figure 4.20. The lactone **B** was converted to diene **C**. Ring-closing olefin metathesis of **C** gave **D**, which afforded crystalline diol **E** after deprotection. The structure of **E** was confirmed by X-ray analysis. Finally, **E** was converted to (1R,4R,5S)-acoradiene (**A**). Its ^{1}H- and ^{13}C-NMR spectra were not identical with those of the natural pheromone. Therefore, the natural pheromone was not **A**.

Because the spectral differences were definite but rather small, we hypothesized that the natural pheromone must be one of the possible stereoisomers of **A**. So as to verify this hypothesis, we synthesized a stereoisomeric mixture of all the possible stereoisomers of **A** as shown in Figure 4.21.[46] (*R*)-Pulegone was converted to a mixture of the four stereoisomers of acoradiene. Fortunately, the mixture was separable by preparative GC, and a fraction with a retention time of 54.3 min showed a ^{1}H-NMR spectrum identical with that of the natural pheromone. Accordingly, the pheromone must be a stereoisomer of acoradiene.

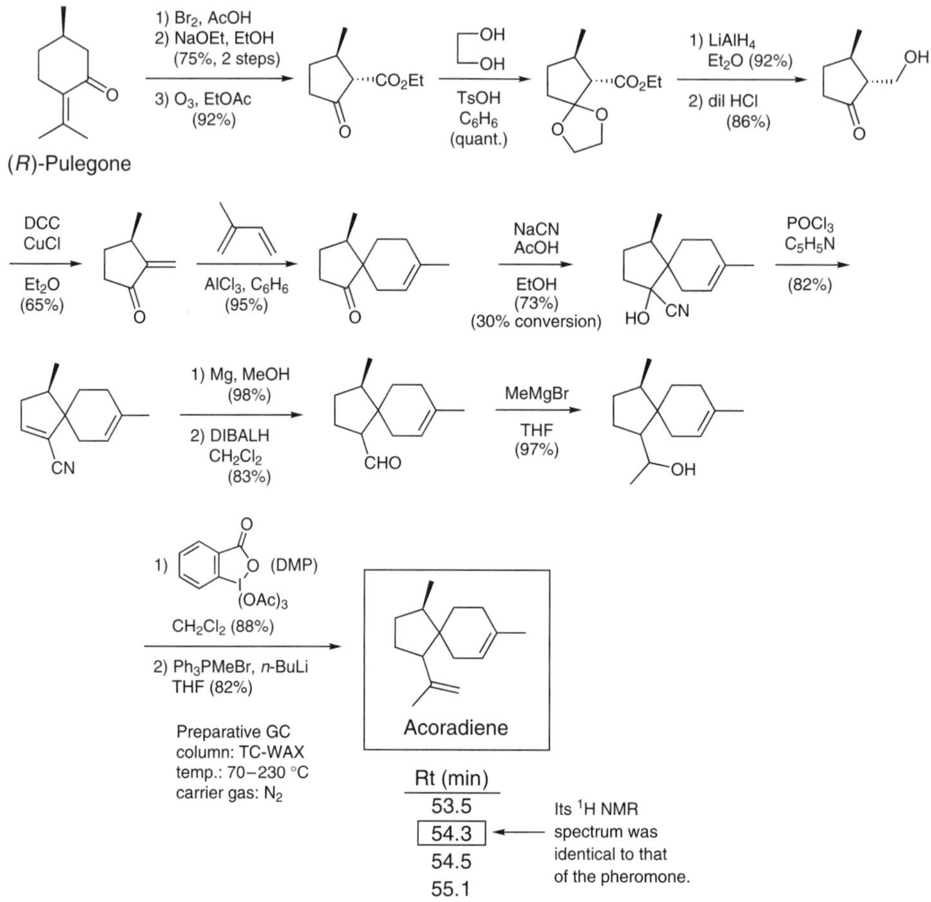

Figure 4.21   *Synthesis of a stereoisomeric mixture of acoradiene*

A subsequent literature survey concerning the NMR and chiroptical data of the stereoisomers of aco-radiene led us to assume the pheromone as (1*S*,4*R*,5*R*)-acoradiene (**80**). At this stage, on May 4, 2002, Dr. D. Joulain of Robertet S. A., France, kindly informed me the same thought. We therefore synthesized (1*S*,4*R*,5*R*)-**80**, as shown in Figure 4.22.[46] (*S*)-Pulegone was employed as the starting material, and the ^{1}H- and ^{13}C-NMR spectra of the resulting (1*S*,4*R*,5*R*)-**80** coincided with those of the natural pheromone. The optical rotation of (1*S*,4*R*,5*R*)-**80** was $[\alpha]_D^{24}$ +38.2 (hexane), which was in good accord with the value ($[\alpha]_D$ +37.1) published for the natural pheromone.

In this case, the misassigned stereochemistry of the pheromone of *G. cornutus* caused a considerable problem for us, and the correct structure could be assigned finally by synthesis.

4.3.8   Clarification of structure (4). himachalene-type pheromone

In 2001, Bartelt *et al.* in the USA isolated and identified four himachalene-type sesquiterpenes (**A–D**, Figure 4.23) as the male-produced aggregation pheromone of the flea beetle (*Aphthona flava*). In order

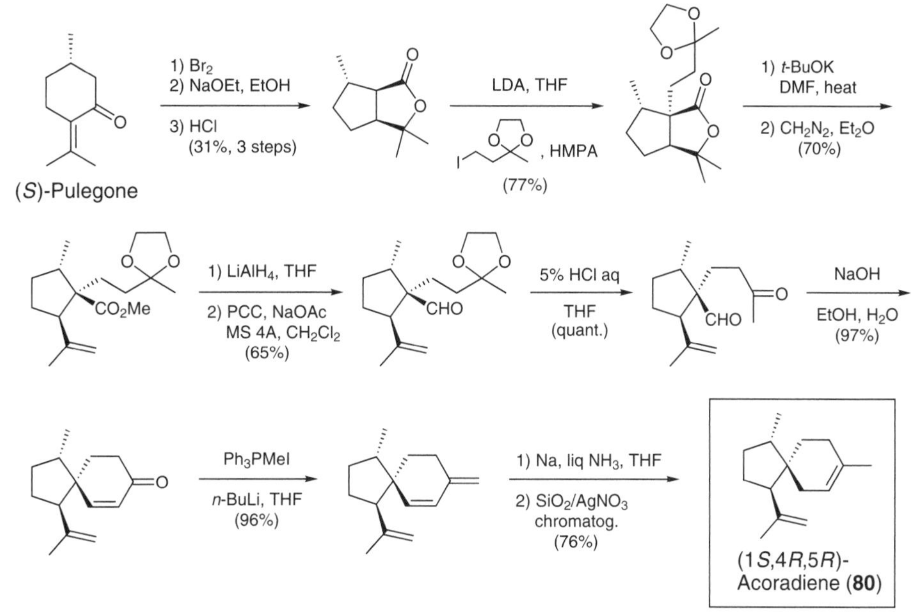

Figure 4.22 *Synthesis of (1S,4R,5R)-acoradiene*

to confirm the proposed stereostructures **A**–**D** (**81′**–**84′**), we carried out their synthesis starting from (*S*)-citronellal, as summarized in Figure 4.23.[47]

(*S*)-Diester **E** prepared from (*S*)-citronellal was cyclized under the Dieckmann conditions to give (*S*)-**F**. Its hydrolysis and decarboxylation were followed by methylation to furnish (*S*)-**G**, whose Robinson annulation yielded (−)-**A** (= **81′**). The structure of the crystalline ketone (−)-**A** was solved by X-ray analysis, and then (−)-**A** was further converted to (−)-**B**, (−)-**C** and (−)-**D**. Although **C** (= **83′**) and **D** (= **84′**) possess the absolute configuration given to the dextrorotatory natural pheromone components, our synthetic **C** and **D** were levorotatory in hexane. In addition, none of them showed pheromone activity.[48]

We then synthesized (+)-**81**, (+)-**82**, (+)-**83** and (+)-**84** starting from (*R*)-citronellal (Figure 4.24).[47] They were all pheromonally active.[48] We were confident about our stereochemical assignment, because the stereochemistry of (1*R*,2*S*)-(−)-**A** (**81′**) was unambiguously determined by X-ray analysis.

Why was our stereochemical conclusion opposite to that of Bartelt's? To answer this question, I synthesized (*R*)-*ar*-himachalene (**84**) by employing Evans asymmetric alkylation as the key step (Figure 4.25).[49] The steric course of the Evans asymmetric process (**A**→**B**) is well established and unambiguous. When the specific rotation of my synthetic (*R*)-*ar*-himachalene (**84**) was measured, I immediately found the reason why Bartelt made a mistake. The hydrocarbon **84** was levorotatory in chloroform, but dextrorotatory in hexane. Although Sukh Dev used chloroform in his rotation measurement, Bartelt did not care about the solvent and used hexane, without knowing the fact that a different solvent may change the sign of rotation.

A similar phenomenon was reported by us in 1976.[50] (1*S*,4*S*,5*S*)-*cis*-Verbenol (Figure 4.25) is a pheromone component of the bark beetle, *Ips paraconfusus*. At that time there was a confusion concerning the name of that pheromone component. Some people called (1*S*,4*S*,5*S*)-*cis*-verbenol as (+)-*cis*-verbenol, while others called it (−)-*cis*-verbenol. Our study of this compound revealed it to be levorotatory in chloroform, while dextrorotatory in methanol and acetone.

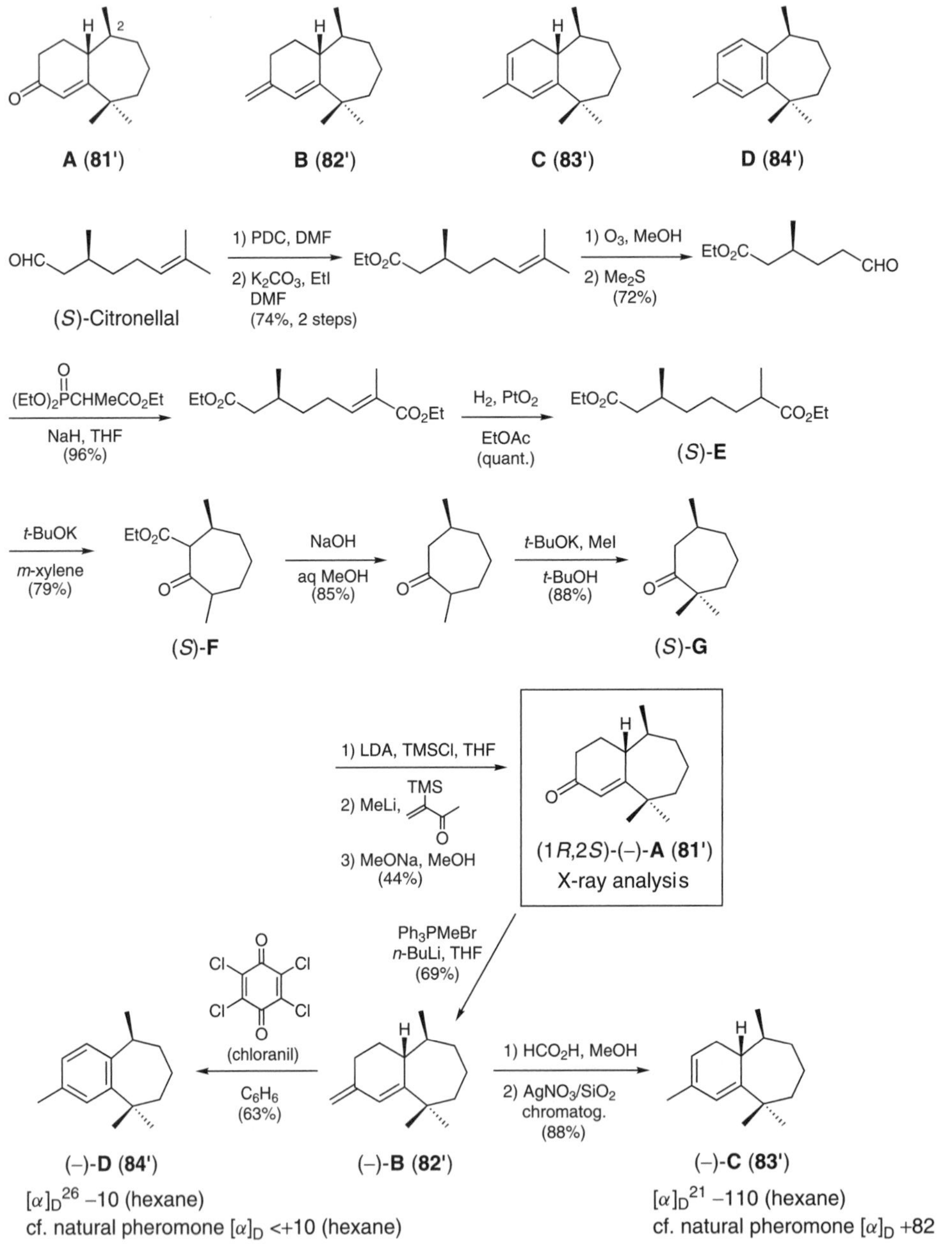

(−)-**A**, (−)-**B**, (−)-**C**, and (−)-**D** were pheromonally inactive.

Figure 4.23 *Synthesis of the unnatural (−)-enantiomers of the pheromone components of Aphthona flava*

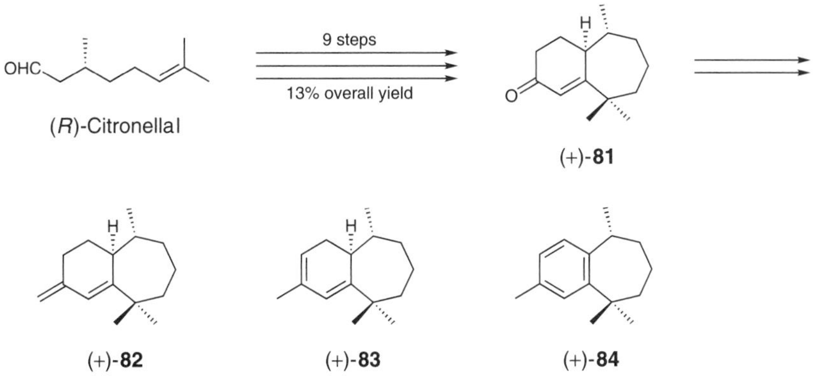

(R)-Citronellal

9 steps

13% overall yield

(+)-**81**

(+)-**82**   (+)-**83**   (+)-**84**

Figure 4.24 Synthesis of the natural (+)-enantiomers of the pheromone components of Aphthona flava

The important lesson is the fact that the sign of optical rotation may change by using a different solvent. We should be careful about this fact.

4.3.9 Preparation of a pure sample for bioassay (1). disparlure

Gypsy moth (*Lymantria dispar*) is an important forest pest originally in Eurasia and recently in North America. This moth was brought into the USA in 1809 from France, and became a big problem for US Forest Service due to the damage caused by its larva. The female-produced sex pheromone of *L. dispar* was studied by Haller and Acree *et al.* at the US Department of Agriculture since 1925. In 1960 Jacobson *et al.* at USDA reported the isolation of about 20 mg of *L. dispar* pheromone from half a million virgin females, named it gyptol, and gave **A** (Figure 4.26) as its structure. Subsequently, Jacobson *et al.* reported the synthesis of ($\pm$)-**A**, (+)-**A** and (−)-**A**. They were all claimed to be pheromonally active at the dosage of 10^{-7} μg to attract male gypsy moths.

In 1967, however, Eiter at Bayer A.G. in Germany synthesized ($\pm$)-gyptol (**A**), and found it to be pheromonally inactive. In 1970 Beroza *et al.* at USDA reinvestigated the gypsy moth pheromone, and isolated the genuine pheromone from 78 000 virgin females. The pheromone was named disparlure, and its structure was shown to be **85**. The structure **85** is totally different from that of gyptol (**A**). The reason why Jacobson proposed **A** as the structure of the gypsy moth pheromone remains a mystery. As I personally met and talked with both Jacobson and Eiter, this discrepancy still puzzles me.

Since 1970 many syntheses of enantiomerically pure disparlure (**85**) have been reported, based on various different strategies. In 1976 we published our synthesis of the enantiomers of disparlure (**85** and **85′**) by employing naturally occurring (+)-tartaric acid as the starting material. The abundant (+)-tartaric acid possesses C_2-symmetry, and we took advantage of its symmetrical nature to prepare both **85** and **85′** as shown in Figure 4.27.[51,52] The synthesis was executed in a fairly large scale at the request of US Department of Agriculture.

L-(+)-Tartaric acid was converted to **A**, the known intermediate used in the synthesis of *exo*-brevicomin (cf. Figure 4.14). Substitution of the tosyloxy group of **A** with an *iso*-amyl group yielded **B**, which was finally converted to (7R,8S)-(+)-disparlure (**85**). When the tosyloxy group of **A** was replaced with an octyl group, **C** was obtained, and it later gave (7S,8R)-(−)-disparlure (**85′**). We synthesized 5.2 g of (+)-**85** and 7.9 g of (−)-**85′** by this synthetic route. (My contract with USDA demanded me to supply each 5 g of the enantiomers of disparlure.) Bioassay of these enantiomers **85** and **85′** revealed (+)-**85** to be the pheromone,

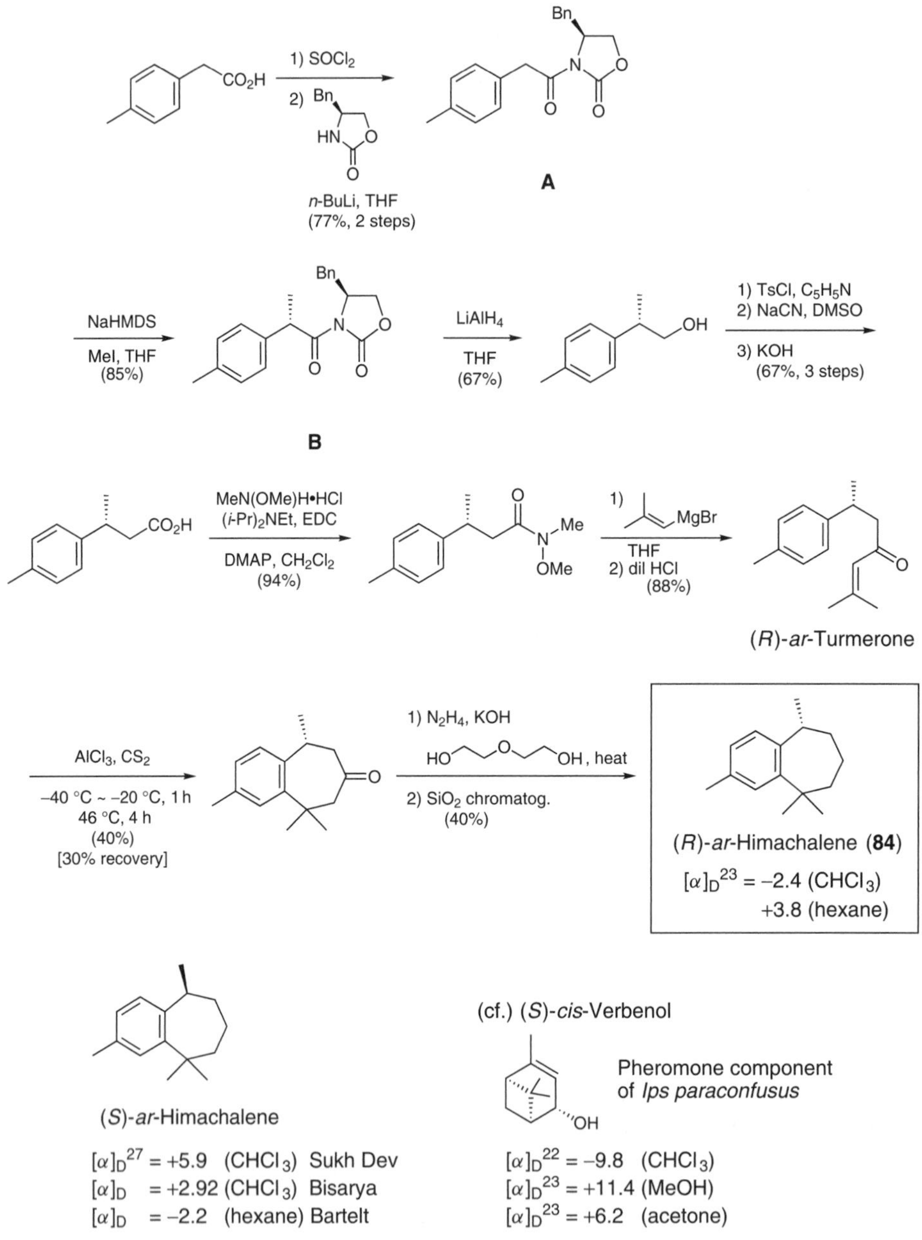

Figure 4.25 *Synthesis of (R)-ar-himachalene*

while (−)-**85′** inhibited the pheromone activity of (+)-**85**.[53,54] A large-scale field test of (+)-**85** in the USA by the US Forest Service confirmed its pheromone activity.

In 1980 my friend Professor Barry Sharpless sent to me two preprints of his papers on the now famous asymmetric epoxidation. He wrote a message at the end of the Katsuki–Sharpless paper: "With best wishes to another lover of epoxides—Barry." Indeed, the Sharpless asymmetric epoxidation dramatically improved

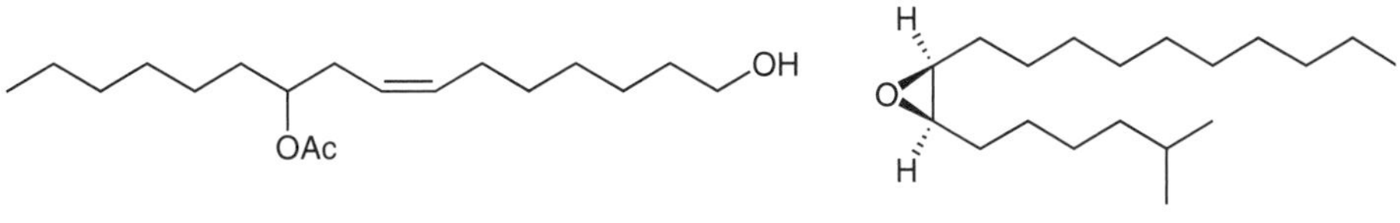

A (Gyptol) **B** [Disparlure (**85**)]

Figure 4.26 *Structure of the gypsy moth pheromone. Modified by permission of Shokabo Publishing Co., Ltd*

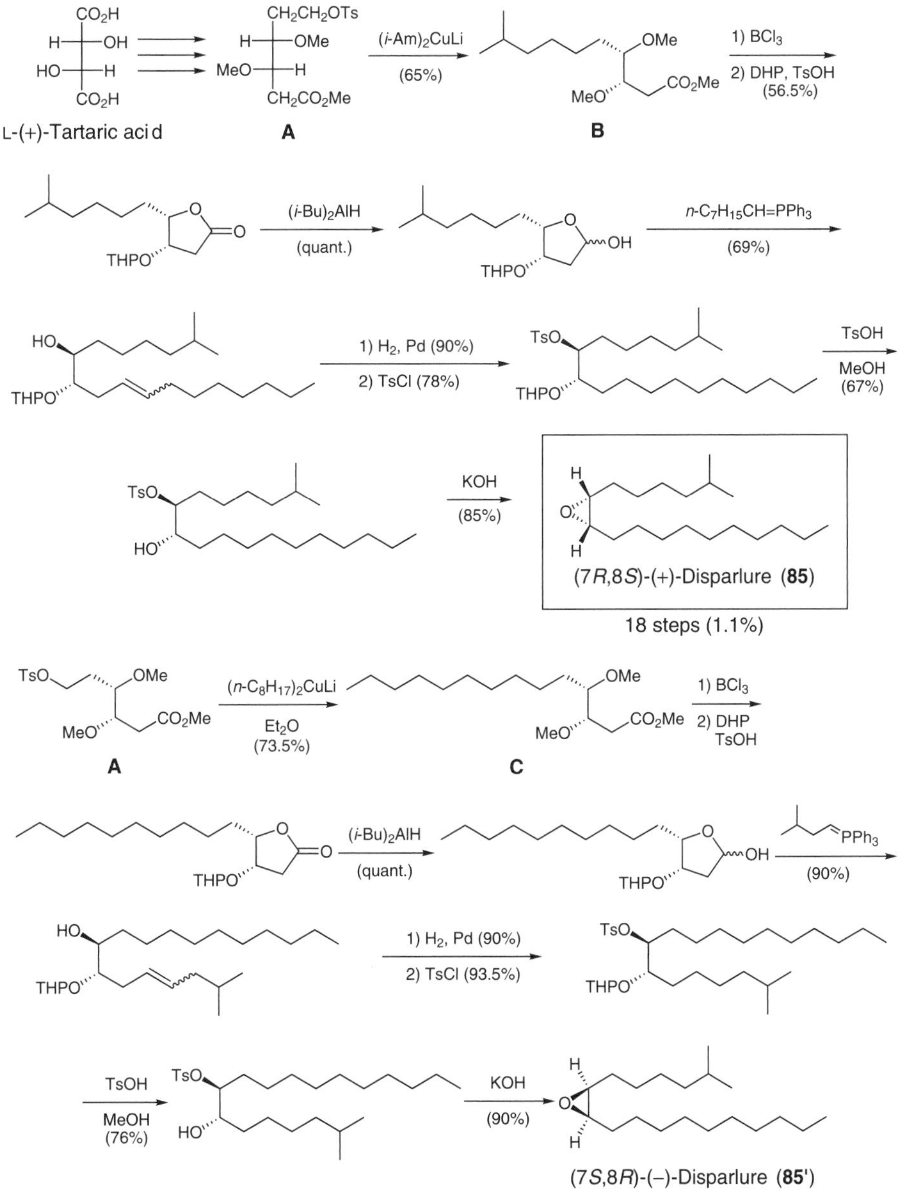

Figure 4.27 *Synthesis of the enantiomers of disparlure from (+)-tartaric acid. Modified by permission of Shokabo Publishing Co., Ltd*

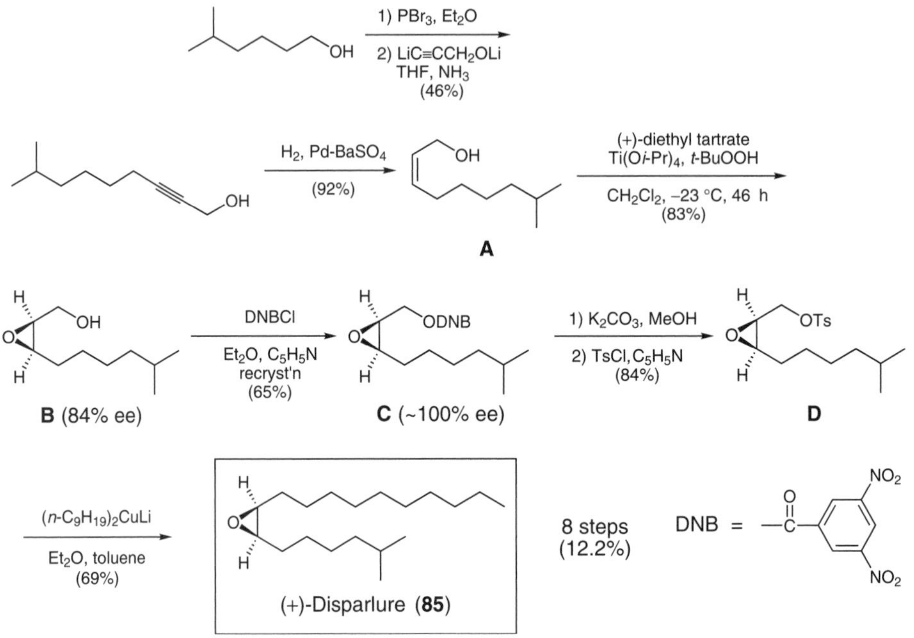

Figure 4.28   *Synthesis of (+)-disparlure employing asymmetric epoxidation. Modified by permission of Shokabo Publishing Co., Ltd*

the efficiency of disparlure synthesis.[55,56] Asymmetric epoxidation of allylic alcohol **A** (Figure 4.28) in the presence of (+)-diethyl tartrate gave epoxy alcohol **B** (84% ee). Purification of **B** was achieved by recrystallization of the corresponding 3,5-dinitrobenzoate **C** to give enantiomerically pure **C**. Treatment of tosylate **D** with lithium di(n-nonyl)cuprate afforded (+)-**85** in 12% overall yield (8 steps), while the overall yield of the synthesis of (+)-**85** from L-tartaric acid was only 1.1% (18 steps).

Subsequently, a number of new epoxide pheromones were identified, and there was a need to develop a general synthetic method for epoxide pheromones. As shown in Figure 4.29, preparation of enantiomerically pure building block **D** facilitated the synthesis of (+)-disparlure **85** and other epoxide pheromones. We prepared **D** by employing lipases.[57,58]

Treatment of *meso*-diacetate **A** with pig-pancreatic lipase (PPL) in a mixture of diisopropyl ether and phosphate buffer (pH 7) gave optically active monoacetate **B** (90% ee). This was purified by recrystallization of the corresponding 3,5-dinitrobenzoate to give pure **C**. The crystalline **C** was treated with potassium carbonate in a methanol/THF mixture to give the desired building block **D**.[57] Conversion of **D** to (+)-disparlure (**85**) proceeded in nearly 44% overall yield (5 steps).

A more convenient preparative method for **D** was devised later, as shown in Figure 4.30 [**D** in Figure 4.29 is (2R,3S)-**A** in Figure 4.30].[58] Asymmetric acetylation of (±)-**A** with vinyl acetate in the presence of lipase PS-C gave acetate (2S,3R)-**B** (98.1% ee) and recovered alcohol (2R,3S)-**A** (99.5% ee) in almost quantitative yields after chromatographic separation. The building block (2R,3S)-**A** was successfully used for the synthesis of leucomalure [(3Z,6S,7R,9S,10R)-**86**], the female sex pheromone of the Satin moth (*Leucoma salicis*).[58,59]

Figure 4.31 illustrates the pheromones synthesized from the epoxy building blocks **A** and **B**.

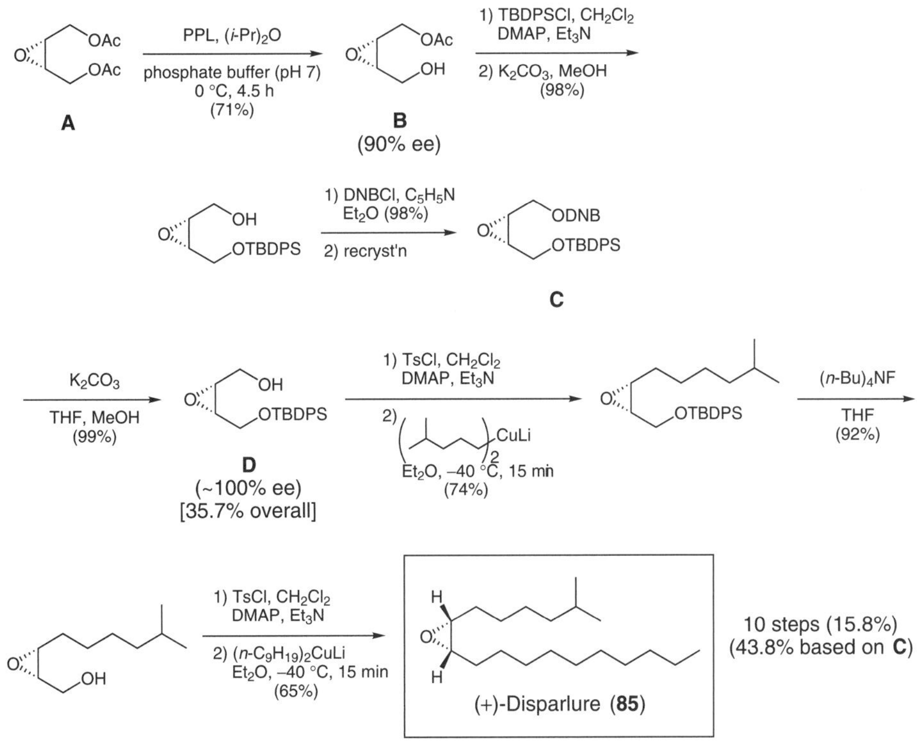

Figure 4.29 *Synthesis of (+)-disparlure employing lipase. Modified by permission of Shokabo Publishing Co., Ltd*

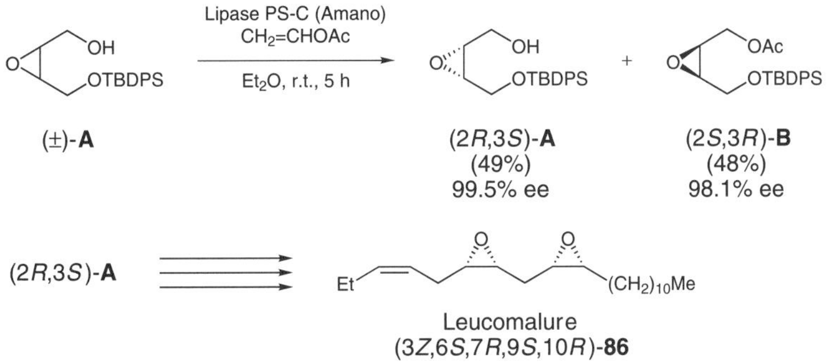

Figure 4.30 *Synthesis of leucomalure employing lipase-catalysed asymmetric acetylation*

4.3.10 Preparation of a pure sample for bioassay (2). japonilure

The Japanese beetle (*Popillia japonica*) was brought into the USA from Japan in the early 20th century, and became a notorious pest there to cause damages to lawn grasses and ornamental trees in gardens and golf courses. Its female-produced sex pheromone was isolated and identified by Tumlinson *et al.* at U.S. Department of Agriculture as (4R,5Z)-5-tetradecen-4-olide (**87**, Figure 4.32). It is also called japonilure.

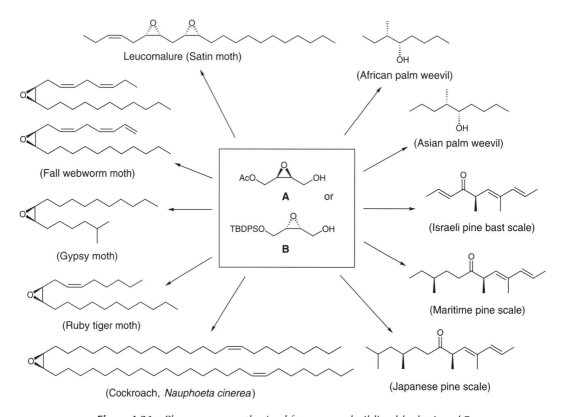

Figure 4.31 *Pheromones synthesized from epoxy building blocks **A** and **B***

Figure 4.32 *Synthesis of (R)-japonilure by Tumlinson et al. Modified by permission of Shokabo Publishing Co., Ltd*

Because synthetic (±)-**87** was pheromonally inactive, Tumlinson *et al.* carefully studied in 1977 the relationship between the enantiomeric purity of **87** and its pheromone activity.[60] They synthesized the enantiomers of **87**, starting from the enantiomers of glutamic acid. The bioactive enantiomer is (*R*)-(−)-**87**, and (*S*)-(+)-**87′** severely inhibits the action of (*R*)-**87**. Accordingly, (*R*)-**87** of 99% ee is only two-thirds as active as pure (*R*)-**87**; that of 90% ee is one-third as active, that of 80% ee is one-fifth as active as pure (*R*)-**87**. Both (*R*)-**87** of 60% ee and (±)-**87** were inactive. These results illustrate convincingly the importance of enantiomeric composition in chemical communication. Later in 1996, Leal found that the sex pheromone of the female scarab beetle (*Anomala osakana*) is (*S*)-**87**, while (*R*)-**87** interrupts the attraction caused by (*S*)-**87**.[61] Thus, chirality accounts for species discrimination.

Tumlinson and coworkers synthesized (*R*)-japonilure (**87**) by the Wittig reaction between **A** and **B** as shown in Figure 4.32.[60] The aldehyde **A** was prepared from D-glutamic acid. In the course of the Wittig

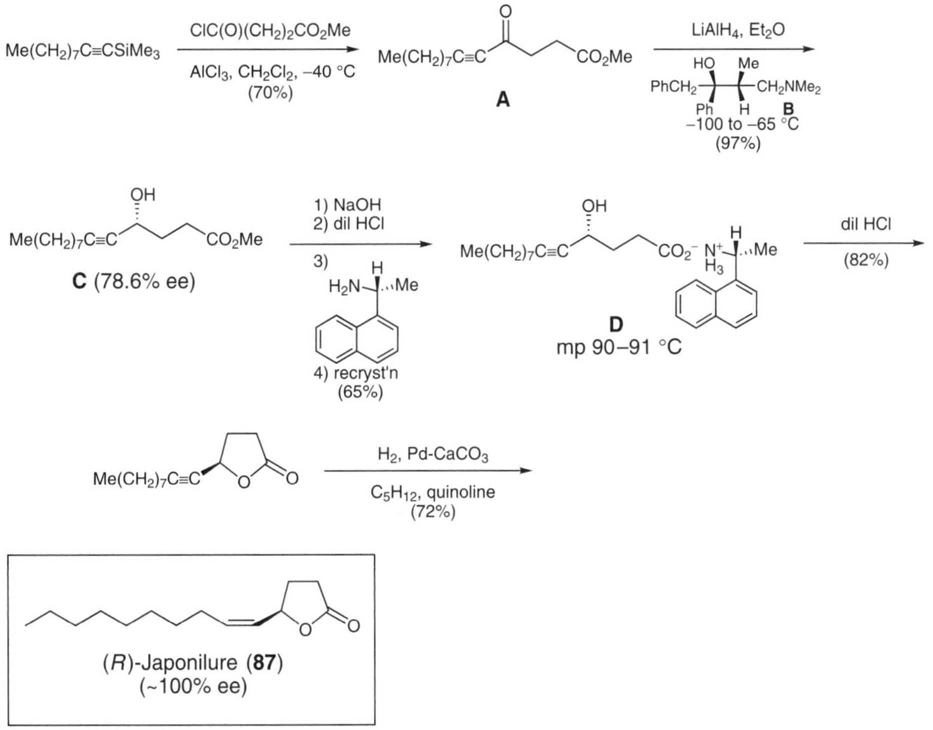

Figure 4.33   *Asymmetric synthesis of (R)-japonilure. Modified by permission of Shokabo Publishing Co., Ltd*

reaction, however, the partial racemization of **A** sometimes occurred due to the slightly basic conditions of the reaction. This resulted in the decrease in the enantiomeric purity of the resulting (*R*)-**87** to give impure and less-bioactive pheromone samples.

In 1983 we developed a synthesis of (*R*)-**87** without recourse to the Wittig reaction.[62] Figure 4.33 summarizes our synthesis. Asymmetric reduction of keto ester **A** with lithium aluminum hydride in the presence of Darvon alcohol **B** furnished hydroxy ester **C** of moderate enantiomeric purity (78.6% ee). Recrystallization of **D** was the successful way to enrich its enantiomeric purity. Accordingly, the hydroxy ester **C** was hydrolysed, and the resulting acid was treated with (*R*)-(+)-1-(1-naphthyl)ethylamine to give **D**, which was recrystallized from acetonitrile to furnish the pure salt **D**. Acidification of **D** was followed by the Lindlar hydrogenation to give enantiomerically pure (*R*)-**87**. This process was once used for the commercial production of (*R*)-**87**.

4.3.11   Preparation of a pure sample for bioassay (3). pheromone of the palaearctic bee, *Andrena wilkella*

Male palaearctic bees (*Andrena wilkella*) in Sweden produce 2,8-dimethyl-1,7-dioxaspiro[5.5]undecane (**88**, Figure 4.34) as their pheromone. Its (2*S*,6*R*,8*S*)-isomer is pheromonally active, while the opposite enantiomer (2*R*,6*S*,8*R*)-**88** is inactive.[63] There are six stereoisomers of 2,8-dimethyl-1,7-dioxaspiro[5.5]undecane (**88**), three as shown in Figure 4.34 and their opposite enantiomers. Due to the oxygen-anomeric effect, (2*S*,6*R*,8*S*)-**88** and its opposite enantiomer are the stable isomers.

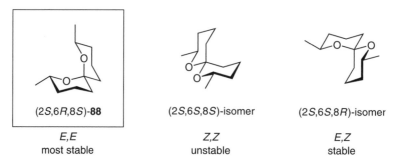

(2S,6R,8S)-**88**	(2S,6S,8S)-isomer	(2S,6S,8R)-isomer
E,E	*Z,Z*	*E,Z*
most stable	unstable	stable

Figure 4.34   *Pheromone of Andrena wilkella and its stereoisomers. Reprinted with permission of Shokabo Publishing Co., Ltd*

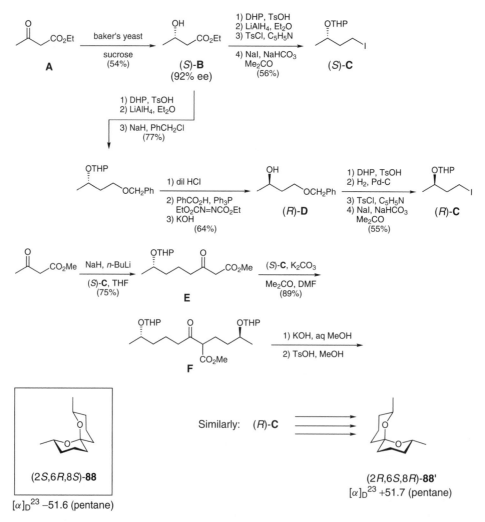

Figure 4.35   *Synthesis of the spiroacetal pheromone of Andrena wilkella (1). Modified by permission of Shokabo Publishing Co., Ltd*

In 1981 we started our works to employ ethyl (*S*)-3-hydroxybutanoate (**B**, Figure 4.35) as a chiral and nonracemic building block in pheromone synthesis. Reduction of ethyl acetoacetate (**A**) with fermenting baker's yeast gives (*S*)-**B**. Synthesis of (2*S*,6*R*,8*S*)-**88** was executed as a part of this project (Figure 4.35).[64,65]

The hydroxy ester (*S*)-**B** (92% ee) was converted to iodide (*S*)-**C** (4 steps). Inversion of the configuration of (*S*)-**B** to (*R*)-**D** was possible by means of Mitsunobu inversion, which was further converted to (*R*)-**C**. Accordingly, both the enantiomers of **C** were prepared from the single enantiomer (*S*)-**B**. Alkylation of the dianion of methyl acetoacetate with (*S*)-**C** gave **E**, whose further alkylation with (*S*)-**C** afforded **F**. Successive treatments of **F** with base followed by acid effected hydrolysis, decarboxylation and acetalization to give (2*S*,6*R*,8*S*)-**88** as a volatile oil. Similarly, (*R*)-**C** afforded (2*R*,6*S*,8*R*)-**88′**. Both **88** and **88′** were purified by chromatography and distillation, and they showed $[\alpha]_D^{23}$ −51.6 (pentane) and +51.7 (pentane), respectively.

In 1984 Isaksson *et al.* in Sweden separated the enantiomers **88** and **88′** by chromatography over a triacetylcellulose column. Their **88** and **88′** were of ≥98% ee, and showed $[\alpha]_D^{24}$ −44.3 ± 0.7 (pentane) and +44.6 ± 0.7 (pentane), respectively. They questioned the higher rotation values of our **88** and **88′**, since we employed ethyl (*S*)-3-hydroxybutanoate of only 92% ee as our starting material. They said "The reason for this is not clear to us."[66] Apparently, they thought that our starting material with 92% ee should have generated **88** and **88′** of 92% ee.

We published a paper in 1986, explaining the reason why the enhancement of the enantiomeric purity of **88** took place by removing an unwanted diastereomer **D** (Figure 4.36) through chromatographic separation.[67] In 1986 we had a method to prepare (*R*)-**A** of 100% ee by ethanolysis of poly β-hydroxybutyrate (PHB), a biopolymer produced by a micro-organism *Zoogloea ramigera*.[68] There was no doubt that **88′** of 100% ee would be prepared from (*R*)-**A** of 100% ee. The specific rotation of pure (2*R*,6*S*,8*R*)-**88′** prepared by us was $[\alpha]_D^{21}$ +57.6 (pentane).[67]

When the synthesis was repeated by starting from (*S*)-**A** of 85% ee, the resulting (2*S*,6*R*,8*S*)-**88** exhibited the rotation value of −56.0 (pentane). Simple comparison of these two values indicated that the enantiomeric purity of the resulting **88** was (56.0/57.6) × 100 = 97.2% ee, which was definitely larger than that (85% ee) of the starting (*S*)-**A**.

The reason for this enhancement of enantiomeric purity is explained in the lower half of Figure 4.36. It can be seen that the calculated value of the enantiomeric purity of **88**, which is prepared from (*S*)-**A** of 85% ee, is 98.7% ee after separating the undesired diastereomer. In a synthesis like this to remove the undesired diastereomer, the enantiomeric purity of the desired diastereomer can be enriched.

Isaksson *et al.* reported the specific rotation of (2*R*,6*S*,8*R*)-**88′** as $[\alpha]_D^{24}$ +44.6 (pentane), while our pure (2*R*,6*S*,8*R*)-**88′** showed $[\alpha]_D^{21}$ +57.6 (pentane). Isaksson's value was too small. The reason for this was not clear to me. Perhaps they did not purify their samples by final distillation. Without distillation an oily sample can be contaminated with the solvents, which lower the $[\alpha]_D$ value of the sample.

We carried out another synthesis of (2*R*,6*S*,8*R*)-**88′**, as shown in Figure 4.37, by employing (*S*)-malic acid as the starting material.[67] In this synthesis, the enantiomeric purity of the intermediate **C** was rigorously proved to be 100% ee as checked by the ^{1}H-NMR analysis of its α-methoxy-α-trifluoromethylphenylacetate (MTPA ester). Subsequently, **C** was converted to (2*R*,6*S*,8*R*)-**88′**, whose rotation value was $[\alpha]_D^{23}$ +60.1 (pentane). As a byproduct of this synthetic study, (2*R*,6*R*,8*S*)-2,8-dimethyl-1,7-dioxaspiro[5.5]undecane (**F′**) was synthesized from **D**, which was obtained by isomerization of **E**. This isomerization was made possible by the fact that the crystalline 3,5-dinitrobenzoate of **D** precipitated, while that of **E** remained oily.

In 1990 Kitching and coworkers synthesized (2*S*,6*R*,8*S*)-**88** of 98% ee (analysed by enantioselective GC on a chiral stationary phase), starting from ethyl (*S*)-lactate. Its specific rotation was reported to be $[\alpha]_D^{23}$ −58.7 (pentane).[69] This was in good agreement with our value. Judgment by a third party is decisive in scientific disputes, too.

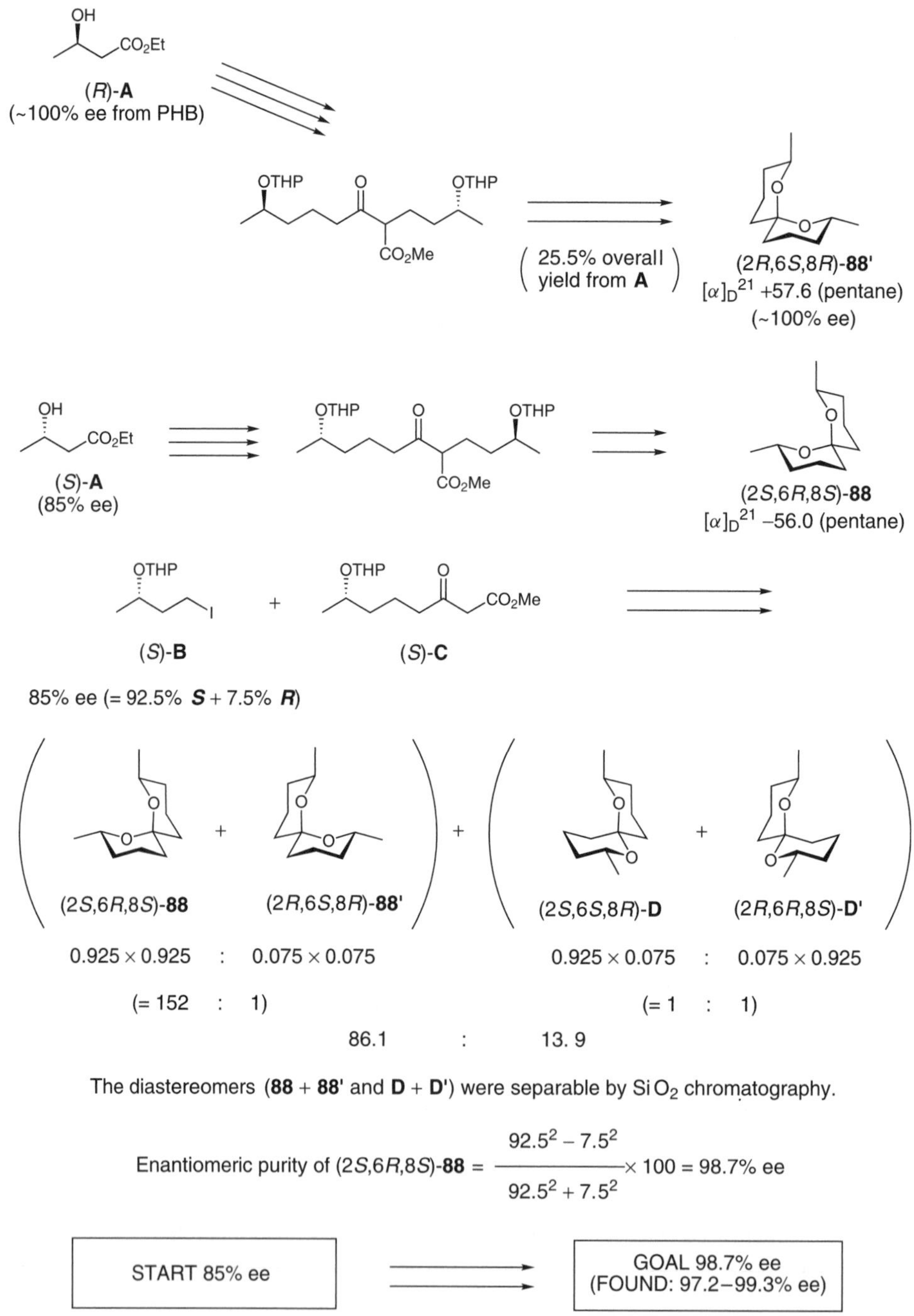

Figure 4.36 *Synthesis of the spiroacetal pheromone of Andrena wilkella (2). Modified by permission of Shokabo Publishing Co., Ltd*

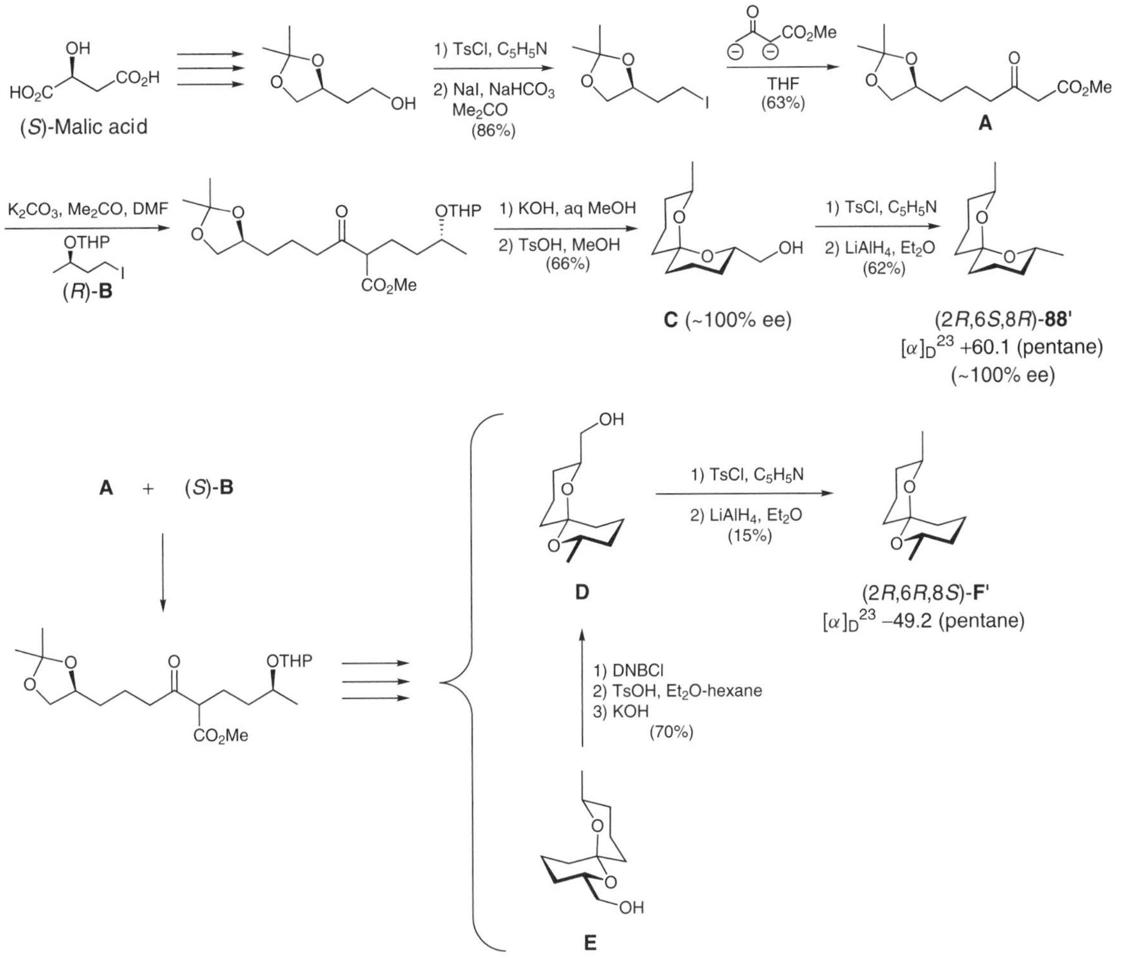

Figure 4.37 *Synthesis of the spiroacetal pheromone of Andrena wilkella (3). Modified by permission of Shokabo Publishing Co., Ltd*

4.4   Chiral pheromones whose single enantiomers show bioactivity

Generally, a single enantiomer of chiral and biofunctional molecule exhibits bioactivity. As shown in Figure 4.2, in the cases of glutamic acid and estrone, one of their enantiomers is bioactive, while the other shows no bioactivity at all. There are many pheromones whose single enantiomer is bioactive. In this section, I will discuss the syntheses of some of them.

4.4.1   Dihydroactinidiolide, a pheromone component of the red imported fire ant

The red imported fire ant (*Solenopsis invicta*) is a widely distributed pest in the southern USA. There exists a queen-recognition pheromone of *S. invicta*, which attracts worker ants, and causes them to bring inanimated objects treated with queen extracts into their nests as if they were real queens. One of the

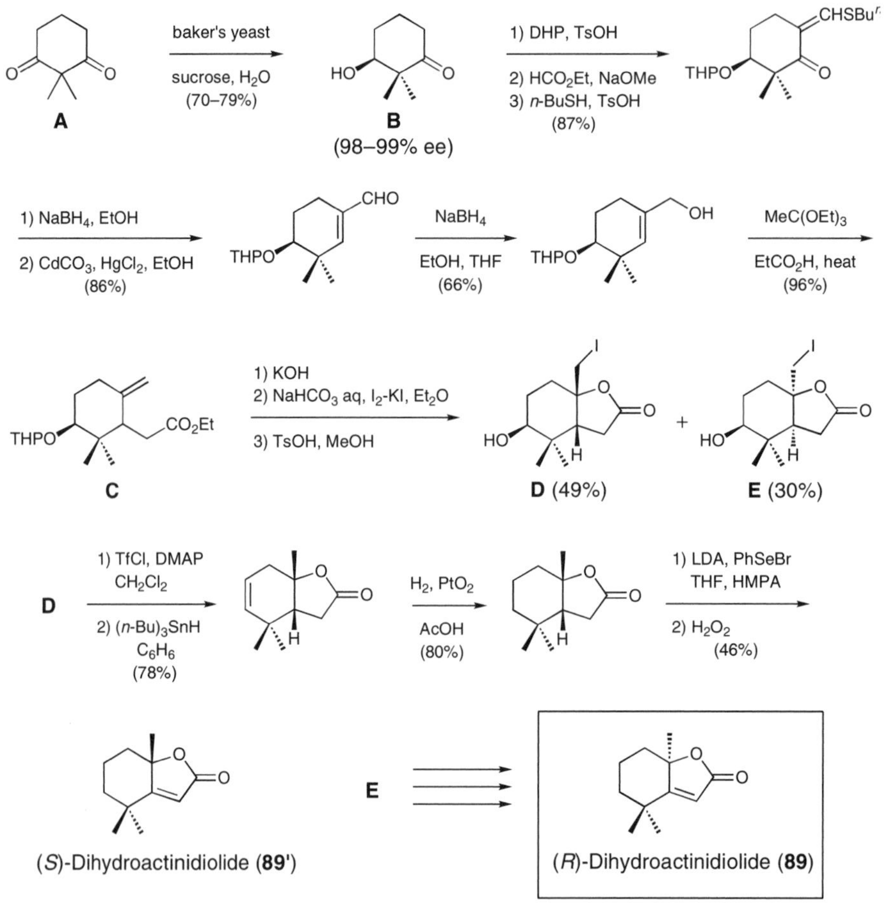

Figure 4.38 *Synthesis of the enantiomers of dihydroactinidiolide. Modified by permission of Shokabo Publishing Co., Ltd*

pheromone components was identified in 1983 by Tumlinson *et al.* as dihydroactinidiolide (**89**, Figure 4.38). Dihydroactinidiolide had been identified by Sakan *et al.* in Osaka in 1967 from the leaves of *Actinidia polygama*.

In 1986, we synthesized both the enantiomers **89** and **89′** of dihydroactinidiolide so as to determine the absolute configuration of the naturally occurring pheromone component (Figure 4.38).[70] The starting material was (*S*)-3-hydroxy-2,2-dimethylcyclohexanone (**B**), which was obtained by reduction of diketone **A** by fermenting baker's yeast and used for the synthesis of glycinoeclepin A (**50**), *O*-methyl pisiferic acid (**57**), juvenile hormone III (**63**), polygodial (**64**), and warburganal (**65**).

The hydroxy ketone **B** was converted to **C**, whose hydrolysis and iodolactonization followed by removal of the THP protective group yielded a separable mixture of two iodolactones **D** and **E**. The former (**D**) furnished (*S*)-dihydroactinidiolide (**89′**), while the latter (**E**) afforded (*R*)-**89**. Bioassay of our synthetic **89** and **89′** by Dr. Tumlinson proved (*R*)-**89** as the bioactive pheromone component, and (*S*)-**89′** was biologically inactive.

A more efficient synthesis of (*S*)-**89′** was achieved in 1993 as shown in Figure 4.39.[71] Asymmetric hydrolysis of acetate (±)-**A** with pig-liver esterase (PLE) gave (*R*)-alcohol **B** and the recovered acetate

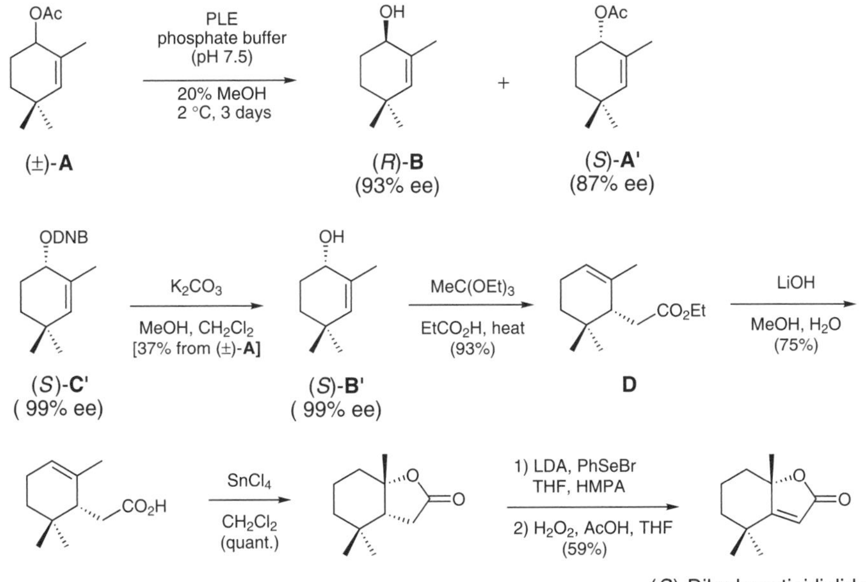

Figure 4.39 *Synthesis of (S)-dihydroactinidiolide. Modified by permission of Shokabo Publishing Co., Ltd*

(*S*)-**A′**. They were converted to the corresponding 3,5-dinitrobenzoates **C** and **C′**, whose enantiopure forms could be obtained by recrystallization.[72] The pure (*S*)-**C′** regenerated (*S*)-**B′** after methanolysis. Orthoester Claisen rearrangement of (*S*)-**B′** with triethyl orthoacetate in the presence of propionic acid gave **D**, which was converted to (*S*)-dihydroactinidiolide (**89′**).

4.4.2 Lardolure, the aggregation pheromone of the acarid mite

Lardolure (**90**, Figure 4.41) was isolated in 1982 by Y. Kuwahara *et al.* as the aggregation pheromone of the acarid mite (*Lardoglyphus konoi*), a primary pest for stored products with high protein content such as dried meat and fish meal. Its structure was proposed as 1,3,5,7-tetramethyldecyl formate (**90**), and confirmed by Kuwahara's synthesis of a racemic and diastereomeric mixture of **90**. In 1986 we clarified the stereochemistry of lardolure, as shown in Figure 4.40.[73]

Two useful hints were available about the stereochemistry of the natural **90**. First, the absolute configuration at C-1 of lardolure was assigned as *R* by the ORD (optical rotatory dispersion) comparison of the natural **90** with (*S*)-1-methylheptyl formate. Secondly, Y. Kuwahara observed that his synthetic mixture consisting of all the eight diastereomers of **90** gave seven peaks when analysed by capillary GC, and the peak showing the shortest retention time coincided with that of the natural **90**.

Our strategy was based on the following consideration. If a mixture consisting of 1,3-*syn*-diastereomers of **90** can be prepared stereoselectively and the natural **90** coincides in GC retention time with one of them, 1,3-*syn*-relationship can be assigned to the natural **90**. The same consideration is applicable to both C-3,5- and C-5,7- relationships, and therefore it is possible to correlate the configuration of the four methyl substituents of **90**. In order to prepare 1,3-*syn*-dimethyl building blocks required for the synthesis of 1,3-*syn*-**90**, 3,5-*syn*-**90** and 5,7-*syn*-**90**, catalytic hydrogenation of methyl-substituted phenols was chosen, considering its *cis*-selectivity and experimental simplicity.

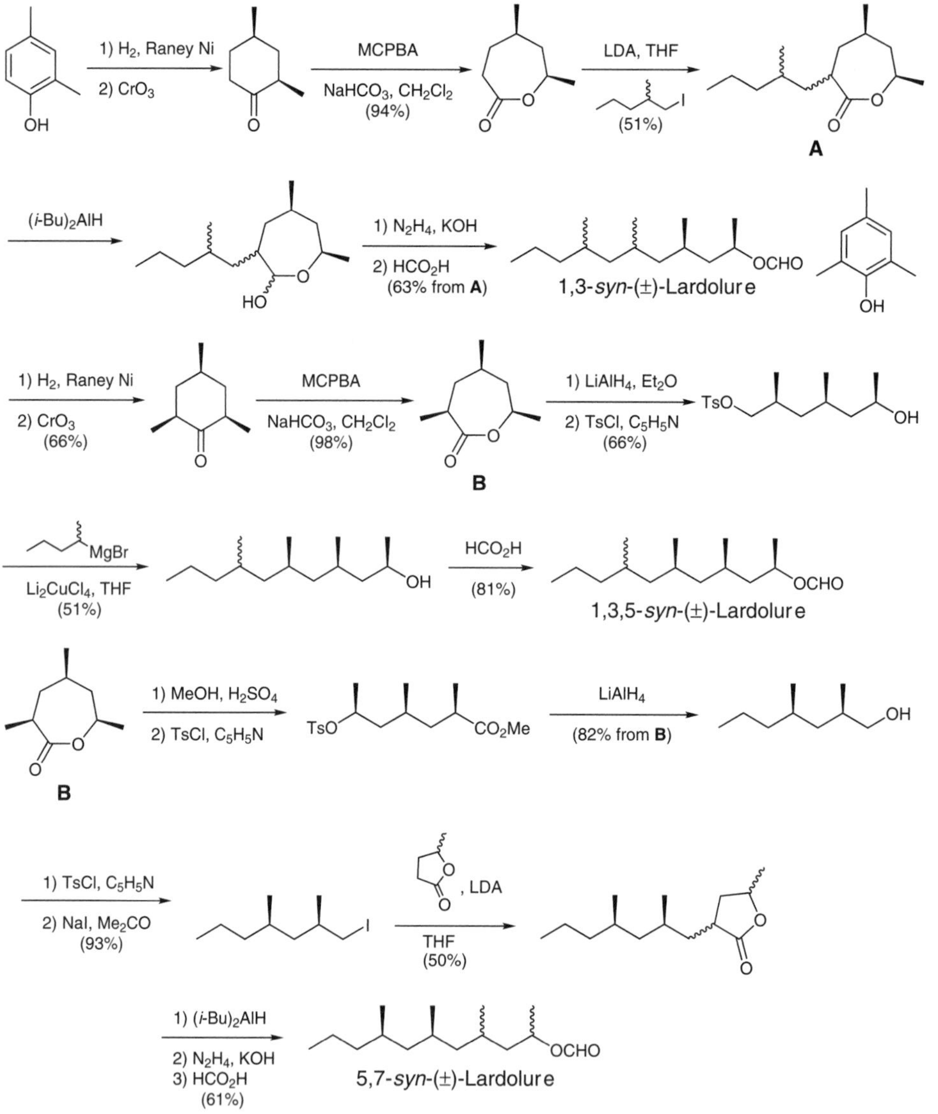

Figure 4.40 *Elucidation of the relative configuration of lardolure by synthesis. Reprinted with permission of Shokabo Publishing Co., Ltd*

Figure 4.40 summarizes our synthesis of 1,3-*syn*-(±)-lardolure, 1,3,5-*syn*-(±)-lardolure and 5,7-*syn*-(±)-lardolure.[73] All of these three synthetic samples contained a component with the same GC retention time as that of the natural **90**. The relative configuration of the natural **90** was therefore proposed as all-*syn*.

We then synthesized the two enantiomers **90** and **90′** of lardolure as shown in Figure 4.41.[74] The first task was the enantiomer separation (optical resolution) of (±)-lactone **A**, which was an intermediate in the synthesis of 1,3,5-*syn*-(±)-lardolure. Treatment of (±)-**A** with (S)-prolinol furnished **B**, whose 3,5-dinitrobenzoates **C** and **D** could be separated by silica-gel chromatography. The absolute configuration of **D** could be determined as depicted by its conversion to the known lactone (−)-**F** via (−)-**E**. The

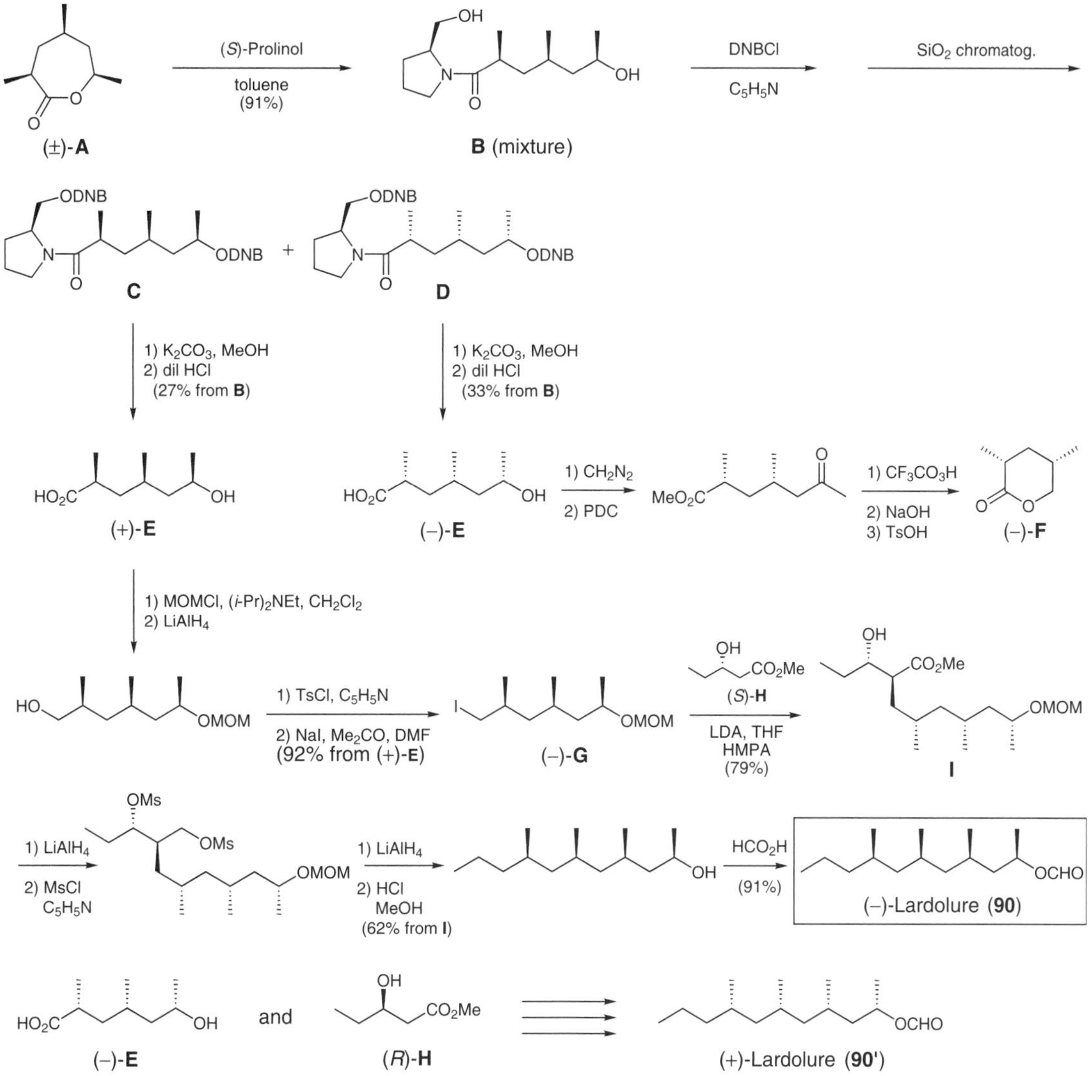

Figure 4.41 *Synthesis of enantiomers of lardolure. Modified by permission of Shokabo Publishing Co., Ltd*

key intermediate (−)-**G** for lardolure synthesis was then prepared from (+)-**E**. Alkylation of the dianion derived from (*S*)-**H**[75] with (−)-**G** afforded **I** generated by *anti*-selective alkylation of the dianion. This hydroxy ester **I** was converted to (1*R*,3*R*,5*R*,7*R*)-(−)-lardolure (**90**). Similarly, (−)-**E** and (*R*)-**H** furnished (1*S*,3*S*,5*S*,7*S*)-(+)-lardolure (**90′**).

Our synthetic (1*R*,3*R*,5*R*,7*R*)-lardolure (**90**) showed the ORD curve identical to that of the natural pheromone,[74] and attracted mites such as *Lardoglyphus konoi*, *Carpoglyphus lactis*, *Aleuroglyphus ovatus* and *Tyrophagus putrescentiae*. The enantiomer (1*S*,3*S*,5*S*,7*S*)-**90′** inhibited the action of **90** on *L. konoi*, but neither attractive nor inhibitory against the action of **90** on *C. lactis*.

In this work, we showed the usefulness of *cis*-selective hydrogenation of aromatic rings in pheromone synthesis, and also illustrated the usefulness of GC comparison in stereochemical investigation.

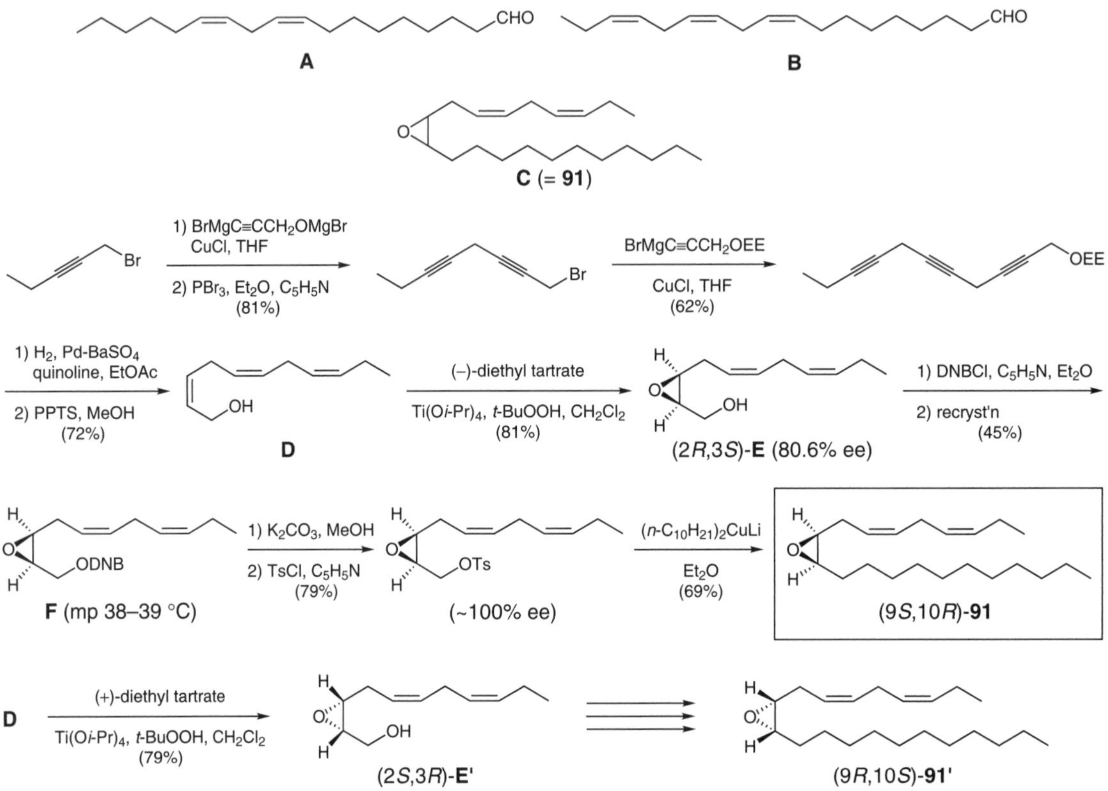

Figure 4.42 *Synthesis of the pheromone components of Hyphantria cunea (1). Modified by permission of Shokabo Publishing Co., Ltd*

4.4.3 Pheromone of the fall webworm moth

The fall webworm moth (*Hyphantria cunea*) is a troublesome pest in Europe and Japan, where its larva attacks fruit trees and ornamental trees such as grape, peach, pear, apple, cherry, poplar and platan. Its female-produced pheromone was first studied by Roelofs and coworkers, who identified three pheromone components **A**, **B** and **C** (Figure 4.42) in 1982.

In order to determine the absolute configuration of **C**, we synthesized (3Z,6Z,9S,10R)-9,10-epoxy-3,6-henicosadiene (**91**) and its enantiomer **91'** employing the Sharpless asymmetric epoxidation as the key step.[55,56] Asymmetric epoxidation of **D** afforded (2R,3S)-**E** of 80.6% ee. This was converted to the corresponding 3,5-dinitrobenzoate **F** and recrystallized to give enantiomerically pure **F**. Further synthetic transformation converted **F** to (9S,10R)-**91**. Bioassay of (9S,10R)-**91** proved it to be pheromonally active, while the enantiomer (9R,10S)-**91'** was inactive. A blend of **A**, **B** and **C** (= **91**), however, was pheromonally inactive when tested against *H. cunea*. Two additional components **92** and **93** were necessary for the pheromone action. In 1987 Dr. H. Arn in Switzerland asked me to synthesize these two compounds. We did this, and in 1989 Tóth, Arn and their coworkers published the identification of **92** and **93**.

Our synthesis of **92** and **93** is shown in Figure 4.43.[76,77] Asymmetric epoxidation of **B** gave epoxide **C** (88% ee), which was purified by recrystallization of the corresponding 3,5-dinitrobenzoate **D**. The pure **D** gave (9S,10R)-**92** via **E**. Similarly, (9S,10R)-**93** was synthesized from **E**. Their enantiomers **92'** and

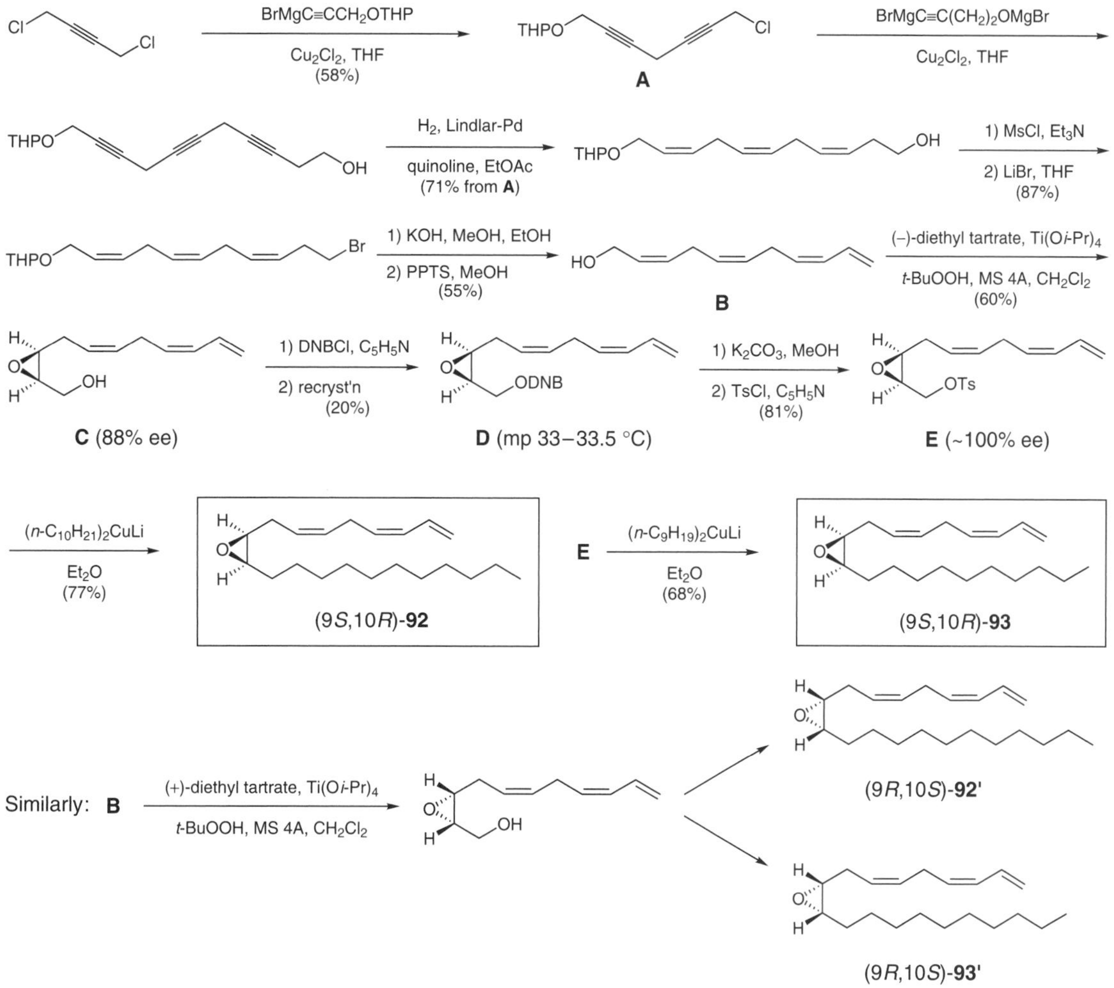

Figure 4.43 *Synthesis of the pheromone components of Hyphantria cunea (2). Modified by permission of Shokabo Publishing Co., Ltd*

93' were also synthesized by starting from **B**. Bioassay revealed (9S,10R)-**92** and (9S,10R)-**93** to be the bioactive enantiomers.

The synthetic pheromone lure containing **A**, **B**, **91**, **92** and **93** was bioactive in both Europe and Japan. Subsequently, in 1993 Senda *et al.* found that a blend of **91**, **B** and **92** could attract *H. cunea*, and this blend was developed as a commercial lure for monitoring the population of *H. cunea*.

In the above-described synthesis, the overall yield of **91** and **92** were 7% (10 steps)[56] and 1.5% (12 steps), respectively.[77] These low overall yields of **91** and **92** led us to develop their new synthesis, as shown in Figure 4.44.[78] Before planning the synthesis, I was informed by Dr. Senda that **91** and **92** with 84–87% ee were as effective attractants as those with >99% ee. The new synthesis was therefore planned to prepare efficiently **91** and **92** with 84–87% ee. Asymmetric acetylation of **A** with vinyl acetate in the presence of lipase PS-C afforded the acetate (2S,3R)-**C** with 84–87% ee. Triflate **E** prepared from **C** was coupled

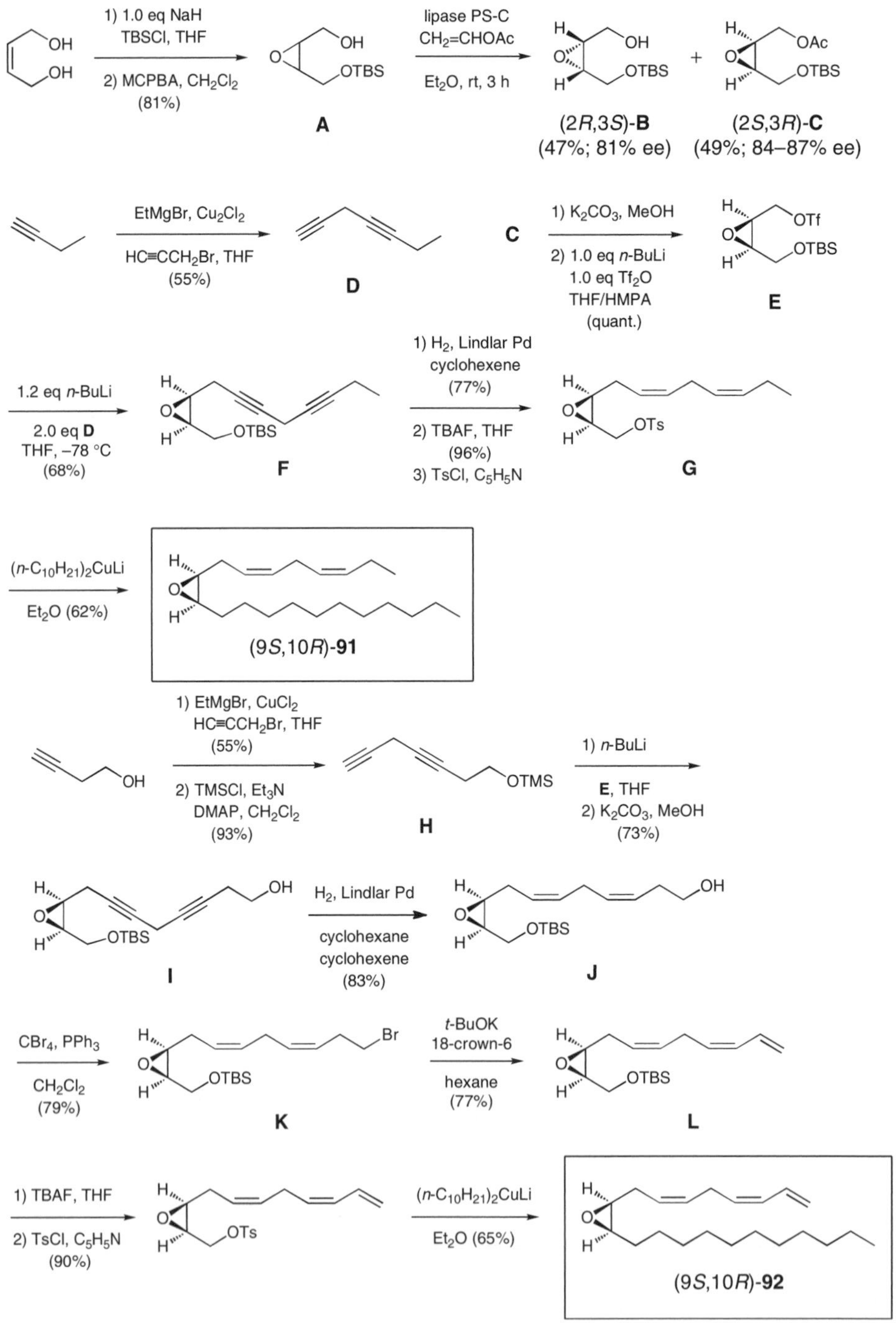

Figure 4.44 *Lipase-assisted synthesis of the pheromone components of Hyphantria cunea*

with diyne **D** to give **F**, which furnished tosylate **G**. Treatment of **G** with lithium di(*n*-decyl)cuprate gave (9*S*,10*R*)-**91**. Its overall yield was 12% (10 steps).

For the synthesis of (9*S*,10*R*)-**92**, its side-chain part **H** was prepared first. It was then coupled with **E** to give **I**. Subsequently, **I** was converted to (9*S*,10*R*)-**92**. Its overall yield was 8.6% (12 steps). This new synthesis will be useful in securing practical amounts of **91** and **92**.

4.4.4 Posticlure, the female sex pheromone of *Orgyia postica*

In 2001 Wakamura *et al.* in Japan isolated, identified and synthesized (6*Z*,9*Z*,11*S*,12*S*)-11,12-epoxyhenicosadiene (**94**, Figure 4.45. trivial name: posticlure) as the female sex pheromone of the tussock moth *Orgyia postica*, a pest on mango and litchi in Okinawa. Their synthesis relied upon Sharpless asymmetric epoxidation, and afforded **94** of 59% ee, which was purified by preparative HPLC on a chiral stationary phase to give pure **94**.

Our synthesis as shown in Figure 4.45, gave pure **94** (99.9% ee) in 25% overall yield (6 steps).[79] The key step was the asymmetric dihydroxylation of **A** to give **B**. The intermediates **B** and **C** were crystalline, and therefore could be purified by recrystallization.

4.4.5 Faranal, the trail pheromone of the pharaoh's ant

In 1977, Ritter and coworkers in the Netherlands isolated and identified faranal (**95**, Figure 4.46), the trail-following pheromone of the workers of the Pharaoh's ant (*Monomorium pharaonis*). The detection threshold of **95** is about 1 pg/cm of a trail. This remarkable bioactivity of **95** attracted the attention of

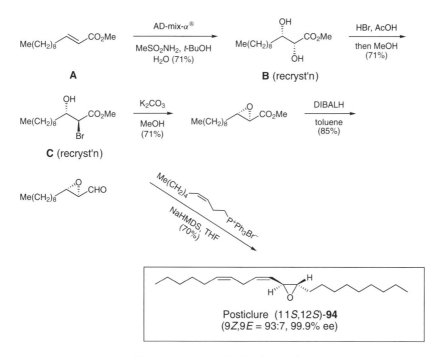

Figure 4.45   *Synthesis of posticlure*

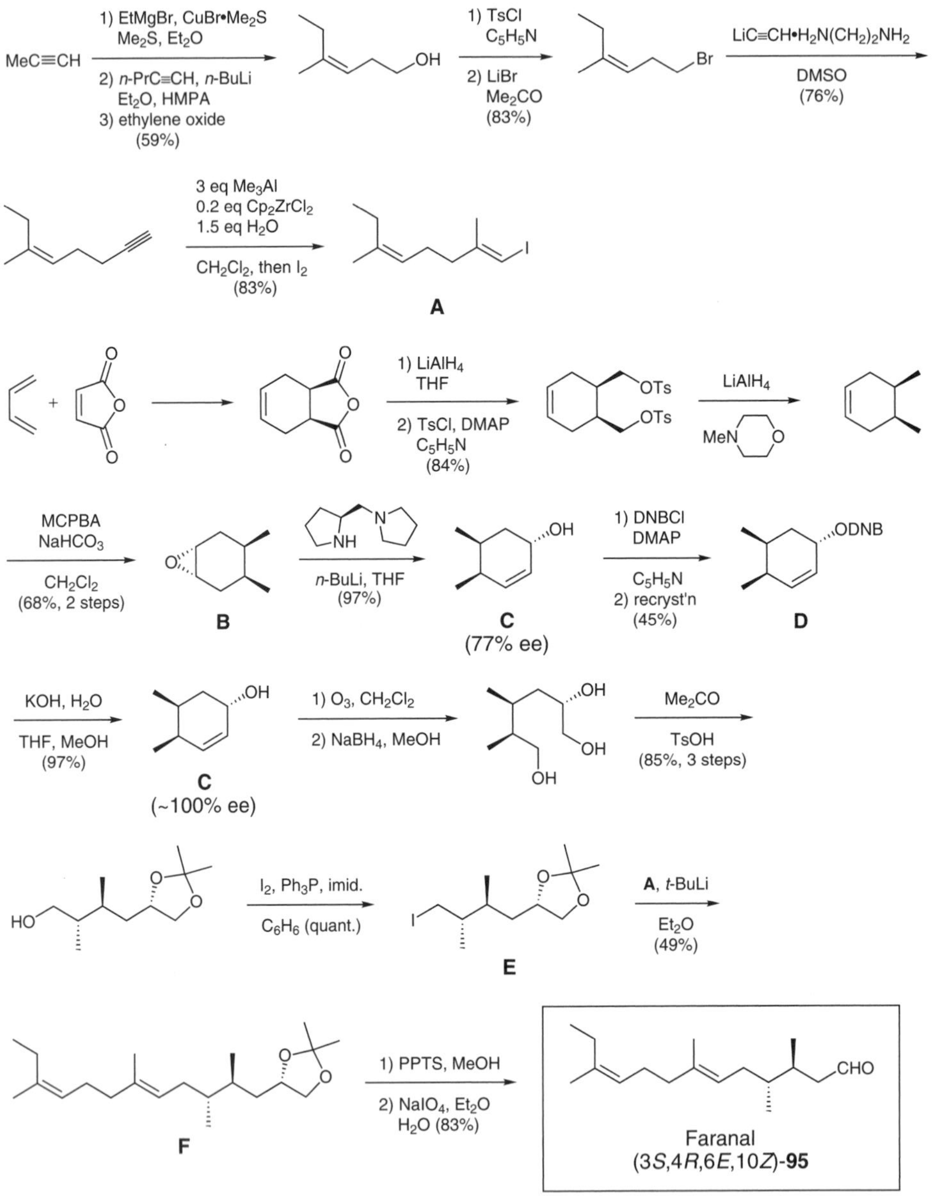

Figure 4.46   *Synthesis of (+)-faranal*

synthetic chemists, and several different syntheses of **95** have been reported to date. The first synthesis of the enantiomers of **95** was achieved in 1980 by Ogura and coworkers by employing farnesyl pyrophosphate synthase for construction of the stereogenic center at C-4 of **95**. In 1981, we reported our own synthesis of (+)- and (−)-**95**, and only (3*S*,4*R*,6*E*,10*Z*)-(+)-**95** was shown to be pheromonally active.

Figure 4.46 summarizes our second synthesis of (+)-**95**.[80] The key step was the asymmetric cleavage of *meso*-epoxide **B** with a chiral lithium amide to give allylic alcohol **C** (77% ee), which was purified as its crystalline 3,5-dinitrobenzoate **D**. Building block **A**, representing the left half of **95**, was prepared from the known 6-methyl-5-octen-1-yne employing Negishi's zirconocene-mediated carboalumination reaction, and was coupled with iodide **E** derived from **C** to furnish **F**. The acetonide **F** gave faranal (**95**). Desymmetrization of a *meso*-compound like **B** by either chemical or enzymatic asymmetric reaction is a useful strategy in enantioselective synthesis.

4.4.6 (1*S*,3*S*,7*R*)-3-Methyl-α-himachalene, the male sex pheromone of the sandfly from Jacobina, Brazil

The sandfly *Lutzomyia longipalpis* is the vector of the protozoan parasite *Leishmania chagasi*, the causative agent of visceral leishmaniasis in South and Central America. Population control of *L. longipalpis* is therefore of urgent importance to prevent the disease. In 1994, Hamilton and coworkers isolated the male-produced pheromone of *L. longipalpis* from Jacobina, Brazil, and proposed its structure as 3-methyl-α-himachalene (**96**, Figure 4.47) with unknown stereochemistry. We first synthesized (1*R**,3*R**,7*S**)-(±)-**96**.[81] Enantiomer separation (optical resolution) of a synthetic intermediate enabled us to prepare both the enantiomers of **96**, and their bioassay and GC comparisons with the natural pheromone showed the latter to be (1*S*,3*S*,7*R*)-**96**.

We then achieved the enantioselective synthesis of (1*S*,3*S*,7*R*)-**96** as shown in Figure 4.47,[82] Evans' chiral auxiliary was attached to acid **A**, giving **B**. Methylation of **B** and subsequent hydrolysis of the product afforded **C**. Acid **C** was converted to **D**. Then, intramolecular Diels–Alder reaction of **D** furnished **E**. Methylenation of **E** with Tebbe reagent yielded the desired (1*S*,3*S*,7*R*)-**96**.[82] It was shown definitely that only (1*S*,3*S*,7*R*)-**96** is bioactive, while other isomers are inactive.[83] It must be added that α-himachalene obtained from Himalayan deodar *Cedrus deodara* possesses the opposite 1*R*,7*S* configuration. Insects and plants sometimes produce similar compounds with different absolute configuration.

4.4.7 (*S*)-9-Methylgermacrene-B, the male sex pheromone of the sandfly from Lapinha, Brazil

In 1996 Hamilton *et al.* proposed 9-methylgermacrene-B (**97**, Figure 4.48) as the structure of the male-produced sex pheromone of *Lutzomyia longipalpis* from Lapinha, Brazil. We synthesized (±)-**97** in 1999,[84] and then both the enantiomers of **97** in 2000, as shown in Figure 4.48.[85] The key step was the cyclization of **A** to give crystalline **B** after deprotection. The intramolecular cyclization reaction itself proceeded in 47–53% yield. The final isopropylidenation could not be achieved by the conventional Wittig reaction. The successful protocol was the use of samarium and chromium.

The natural pheromone was shown to be (*S*)-**97** by GC comparison on a chiral stationary phase.[86] While (*S*)-**97** was highly bioactive, (*R*)-**97**′ was also bioactive to some extent, but did not interfere with the activity of (*S*)-**97**.[86]

4.4.8 (1*S*,5*R*)-Frontalin, the bark beetle pheromone

(1*S*,5*R*)-(−)-Frontalin (**98**, Figure 4.49) is the active component of the aggregation pheromone of the southern pine beetle (*Dendroctonus frontalis*), the western pine beetle (*Dendroctonus brevicomis*) and the Douglas-fir beetle (*Dendroctonus pseudotsugae*). Mori's 1975 synthesis of the enantiomers of frontalin via enantiomer separation (optical resolution) of an intermediate[87] enabled their bioassay, and only (1*S*,5*R*)-**98** was bioactive as the pheromone component of *D. brevicomis*.[32] A recent study on female *D. frontalis* revealed its (1*S*,5*R*)-**98** to be of about 91% ee.[88]

In 1996, Seybold in the USA requested me to prepare 10 g of (1*S*,5*R*)-**98** of >76% ee so as to use it for a field test to attract the Jeffrey pine beetle (*Dendroctonus jeffreyi*), which is an aggressive pest of

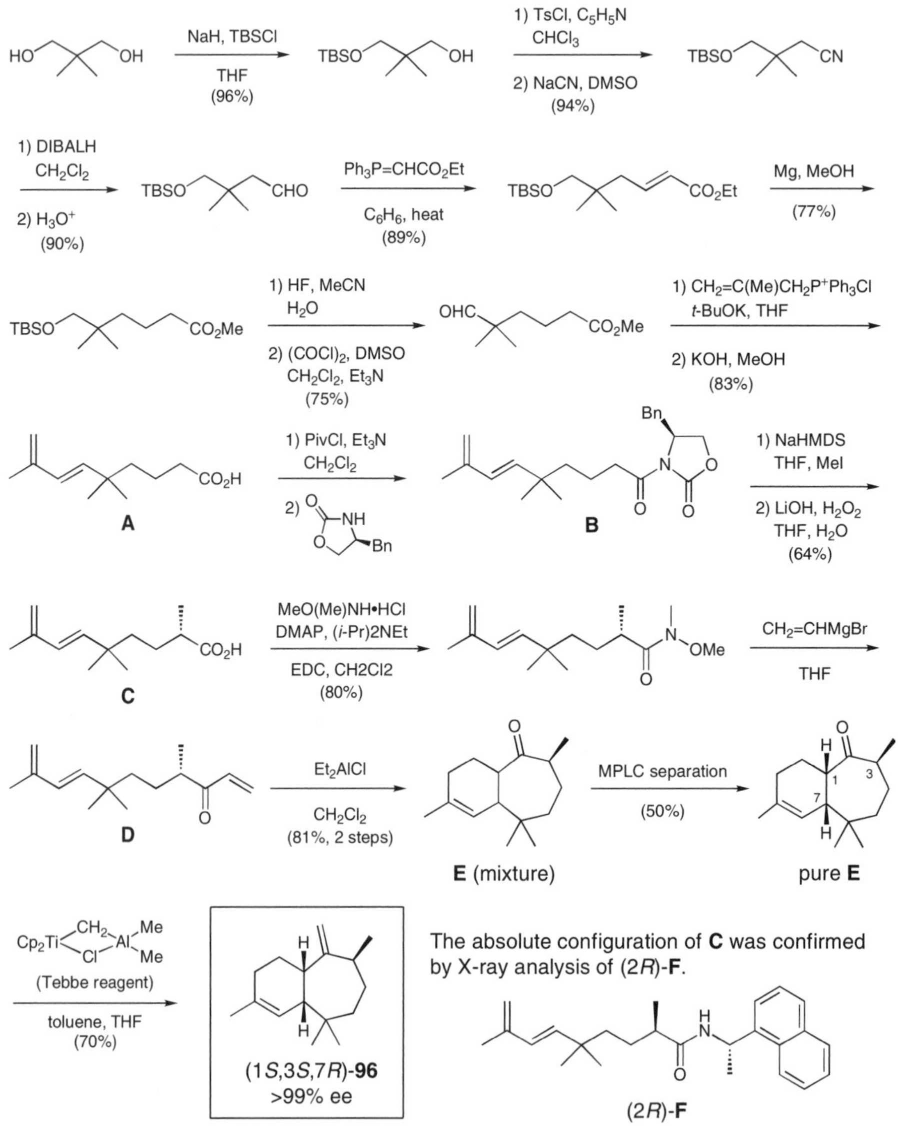

Figure 4.47 *Synthesis of (1S,3S,7R)-3-methyl-α-himachalene*

Jeffrey pine in America. Figure 4.49 summarizes our frontalin synthesis, which successfully provided over 10 g of (1S,5R)-**98**.[89] The first and important step in our synthesis was the asymmetric reduction of β-keto ester **A** with baker's yeast to give hydroxy ester **B** of 97.7% ee. Methylation of the dianion derived from **B** gave **C** diastereoselectively, which was oxidized to give **D**, Baeyer–Villiger oxidation of **D** afforded **E** with retention of configuration. Subsequent conversion of **E** to **98** proceeded smoothly to give 10 g of the sample in an overall yield of 7.8% based on **A** (10 steps).

Now, let me tell you a story about Asian elephants (*Elephas maximus*) in connection with frontalin. In 1996, Rasmussen and coworkers found that female Asian elephants use a 97:3 mixture of (*Z*)- and

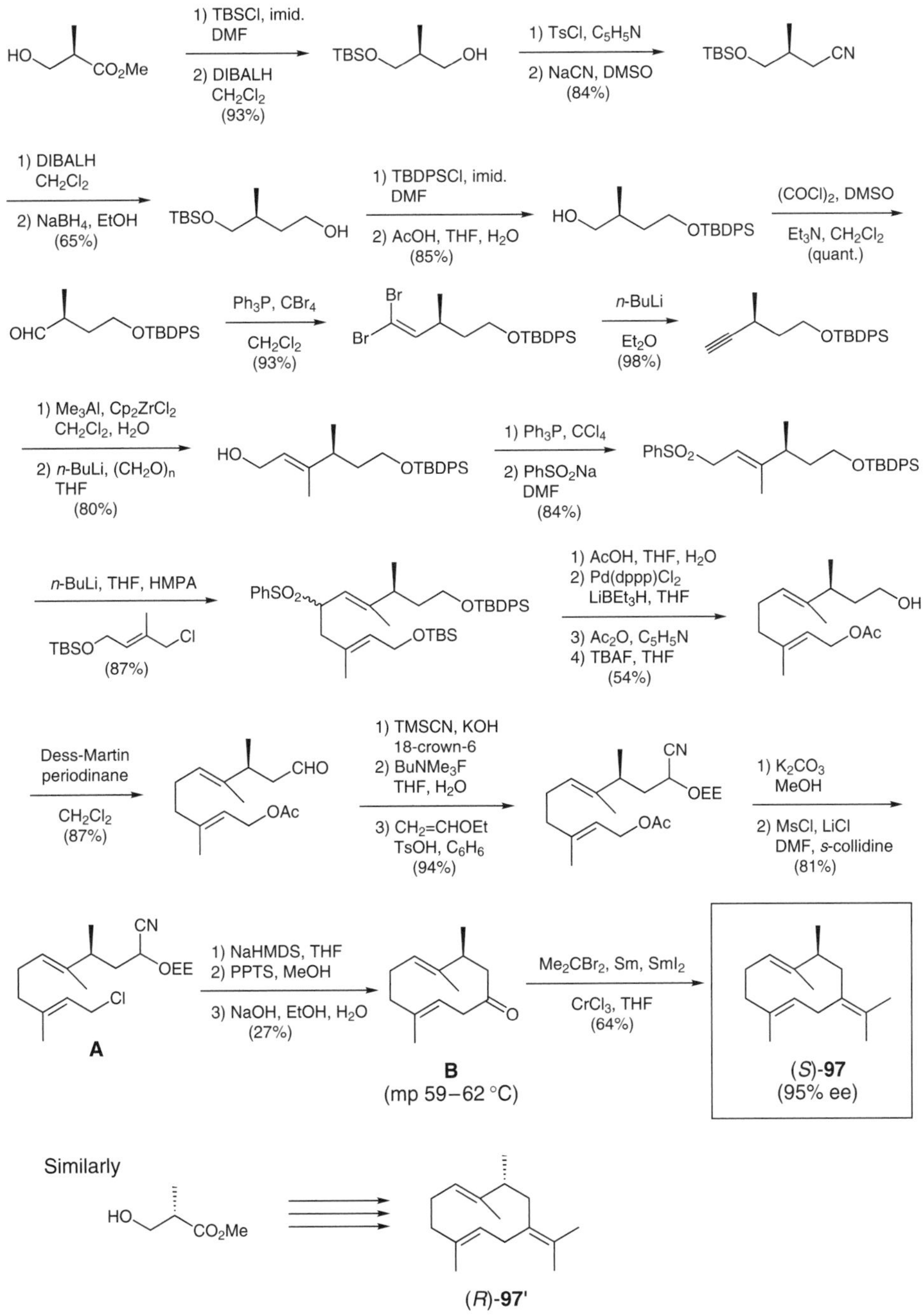

Figure 4.48 *Synthesis of the enantiomers of 9-methylgermacrene-B*

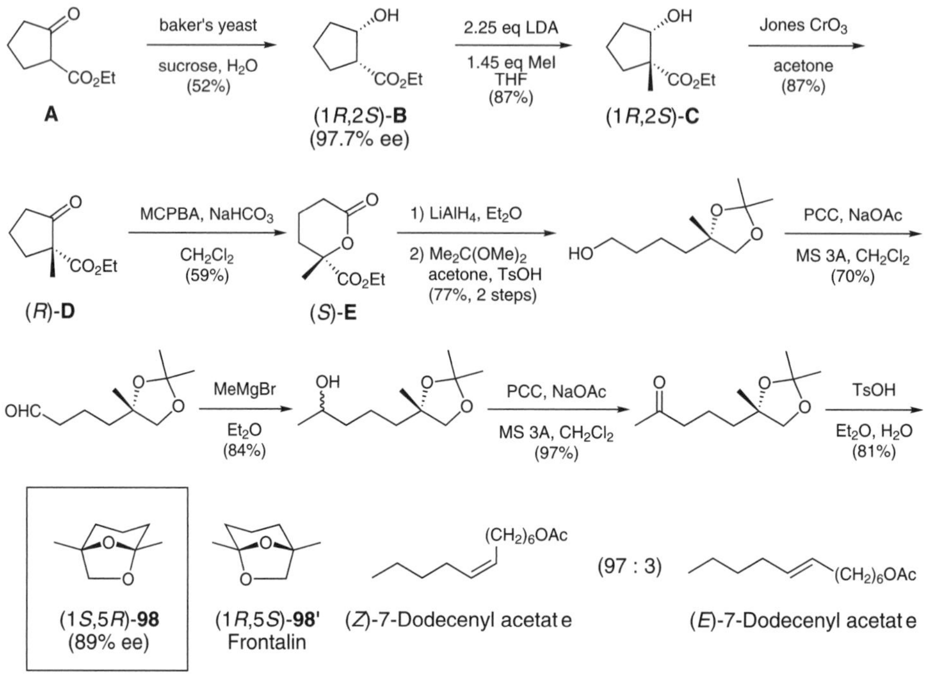

Figure 4.49   *Synthesis of (1S,5R)-frontalin*

(*E*)-7-dodecenyl acetates (Figure 4.49) as their sex pheromone to show their readiness to mate. The same acetates are employed by the females of 126 species of insects including the moth *Trichoplusia ni*.[90]

According to Rasmussen's subsequent study, male Asian elephants release frontalin from the temporal gland on the face during musth, which is an annual period of sexual activity and aggression.[91] The ratio of frontalin enantiomers enables other elephants to distinguish both the maturity of male elephants in musth and the phase of musth. In young males, significantly more (1*R*,5*S*)-(+)-**98′** than (1*S*,5*R*)-(−)-**98** is released. As the elephant matured, the ratio becomes almost equal to emit (±)-frontalin. Musth periods get longer as males age. Secretions containing high concentration of frontalin at racemic ratios attracted follicular phase females, whereas the secretions repulsed males as well as luteal phase and pregnant females. The importance of the enantiomeric composition of frontalin in the behavior of Asian elephants could be noticed only after the development of enantioselective GC on a chiral stationary phase. It must be added that bark beetles employ (1*S*,5*R*)-(−)-frontalin as their pheromone component.

4.4.9   (1*R*,5*S*,7*R*)-3,4-Dehydro-*exo*-brevicomin and (*S*)-2-*sec*-butyl-4,5-dihyrothiazole, the pheromone components of the male mouse

In 1984, 3,4-dehydro-*exo*-brevicomin (**99**, Figure 4.50) and 2-*sec*-butyl-4,5-dihydrothiazole (**100**) were isolated by Novotny and coworkers from the urine of the male house mouse (*Mus musculus*) as pheromone components.[92] They act, synergistically, in promoting intermale aggression, sex attraction, and estrus synchronization in female mice. It was later found that the natural components are (1*R*,5*S*,7*R*)-**99** and (*S*)-**100**. Additional achiral compounds **101** and **102** were also identified as possible pheromone components.

Our synthesis of the enantiomers of **99** in 1986 started from the enantiomers of tartaric acid.[93] Our second synthesis of (1*R*,5*S*,7*R*)-**99** in 1999 as shown in Figure 4.50 employed asymmetric dihydroxylation

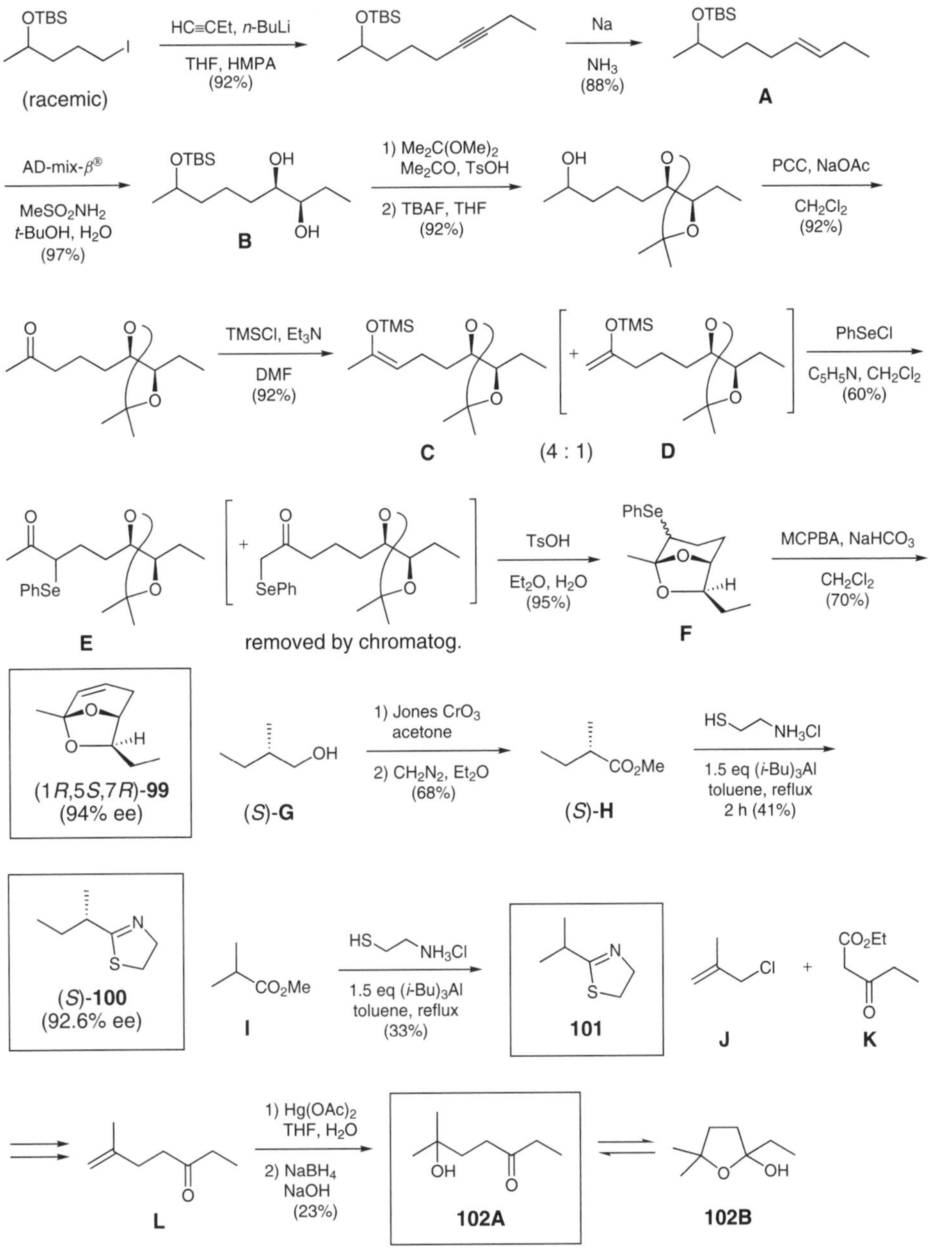

Figure 4.50   *Synthesis of (1R,5S,7R)-3,4-dehydro-exo-brevicomin and (S)-2-sec-butyl-4,5-dihydrothiazole, the mouse pheromone*

of **A** to give **B** as the key step.[94] TMS enol ether **C** was then treated with phenylselenenyl chloride to give **E**, whose acid treatment furnished **F**. Finally, oxidation of **F** with MCPBA afforded (1R,5S,7R)-**99** in 27% overall yield based on the racemic starting material (10 steps).

Figure 4.50 also shows the synthesis of (S)-**100**.[94] Commercially available (S)-2-methyl-1-butanol (**G**) was oxidized and methylated to give (S)-**H**. This was treated with a complex generated by the addition of triisobutylaluminum to a suspension of cysteamine hydochloride in toluene to give (*S*)-**100** of 92.6% ee. This dihydrothiazole (*S*)-**100** was unstable and racemized within 5 days at room temperature. Fortunately, its 5% solution in hexane could be kept intact at room temperature. The decrease in the ee of (*S*)-**100** as 5% hexane solution was only 0.6% after 9 days. Optically active compounds are sometimes unstable, and subject to racemization even under mild conditions. In the same manner methyl isobutyrate (**I**) was converted to isopropyl-4,5-dihydrothiazole (**101**).[95] Hydroxy ketone **102A**, which is in equilibrium with cyclic hemiacetal **102B**, was synthesized from methallyl chloride (**J**) and ethyl 3-oxopentanoate (**K**) by oxymercuration-reduction of **L**.[95]

In many wild animals, older males are often preferred by females, because they carry "good" genes that account for their viability. In the case of the house mouse (*M. musculus*), higher levels of (1*R*,5*S*,7*R*)-**99**, (*S*)-**100** and **101** were detected in the urine of aged male mice than in that of normal adult males, while a lower level of **102** was observed.[96] When (1*R*,5*S*,7*R*)-**99**, (*S*)-**100** and **101** were added to the urine of normal adult males, their urine showed an enhanced attractiveness against female mice. Addition of **102** had no effect at all. Accordingly, it is established in the case of the house mouse that pheromones control the mate-selection process.[96] The search to clarify the roles of pheromones in higher animals including humans will continue to be an interesting area with potential impacts on perfume industries.

4.5   Chiral pheromones whose stereochemistry–bioactivity relationships are diverse and complicated

The relationships between stereochemistry and bioactivity among pheromones are diverse and far from straightforward. Organisms utilize chirality to enrich and diversify their communication systems.[8,9] It must be emphasized that such diversity could be found only through experiments by using pure pheromone enantiomers of synthetic origin. In this section, several examples will be given to illustrate the diverse stereochemistry–bioactivity relationships.

4.5.1   Sulcatol, the pheromone of *Gnathotrichus sulcatus*

In 1974, Silverstein and coworkers isolated and identified sulcatol (**103**, Figure 4.51) as the male-produced aggregation pheromone of *Gnathotrichus sulcatus*, an economically important ambrosia beetle in the Pacific coast of North America. They showed the natural pheromone to be a 35:65 mixture of (*R*)-**103'** and (*S*)-**103** by ^{1}H-NMR analysis of its Mosher ester (α-methoxy-α-trifluoromethylphenylacetate). The reason why the beetle produces a mixture of enantiomers was unclear at the time of its discovery.

The enantiomers of sulcatol were synthesized in 1975 by myself, as shown in Figure 4.51.[97] The starting (*R*)-glutamic acid was treated with nitrous acid to give lactone **A** with retention of configuration via double inversion. The lactone **A** was converted to crystalline **B**, which could be purified by recrystallization. The tosylate **B** finally yielded (*S*)-sulcatol (**103**). Similarly, (*S*)-glutamic acid furnished (*R*)-**103'**. I imagined that either **103** or **103'** must be bioactive. The bioassay results, by Professor John Borden in Canada, however, was different: neither **103** nor **103'** was bioactive. The maximum response of *G. sulcatus* was to racemic (50:50) mixture of the enantiomers, and the response to (±)-**103** was significantly greater than that to a 35:65 mixture. It thus became clear that the beetles must produce a mixture of enantiomers of **103** if they are to communicate with each other.[98] This discovery in 1976 was the first example of a synergistic response based on enantiomers. It must be added that in a closely related species, *Gnathotrichus retusus*, the insect produces and goes to (*S*)-sulcatol (**103**). Therefore, in the case of sulcatol, both the enantiomers are necessary in one species (*G. sulcatus*), but one enantiomer is active in a closely related species (*G. retusus*).

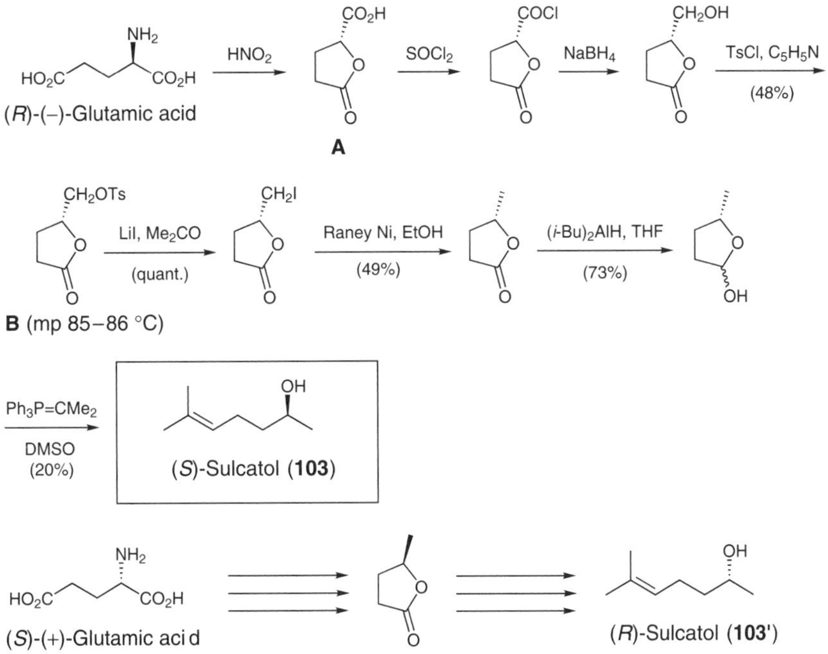

Figure 4.51 *Synthesis of the enantiomers of sulcatol (1). Modified by permission of Shokabo Publishing Co., Ltd*

Later in 1981, I synthesized (*S*)-sulcatol (**103**) from ethyl (*S*)-3-hydroxybutanoate (87% ee) obtained by reduction of ethyl acetoacetate with fermenting baker's yeast (*Saccharomyces cerevisiae*).[99] In 1987 we found that reduction of ethyl acetoacetate with fermenting *Saccaromyces bailii* KI 0116 (Kirin Brewery Co.) gives (*S*)-**A** (Figure 4.52) of 96% ee.[68,100] Its 3,5-dinitrobenzoate could be purified further by recrystallization, and gave pure (*S*)-**A** (100% ee) after removal of 3,5-dinitrobenzoic acid. The pure (*S*)-**A** furnished pure (*S*)-sulcatol (**103**). As to pure (*R*)-sulcatol (**103′**), it could be synthesized from pure (*R*)-**A** prepared by ethanolysis of a biopolymer, poly-β-hydroxybutyrate (PHB).[100] PHB was isolated from a micro-organism *Zoogloea ramigera* at that time.[68] It is now commercially available. These biocatalytic syntheses of the enantiomers of sulcatol were more convenient than my 1975 synthesis from glutamic acid.

4.5.2   Sex pheromone components of female German cockroach

From the cuticular wax of sexually matured virgin females of the German cockroach (*Blattella germanica*), Nishida *et al.* isolated 3,11-dimethyl-2-nonacosanone (**104**, Figure 4.53) in 1974 as the major component of the female-produced sex pheromone, which, upon contact with male antennae, elicited wing-raising and direction-turning response from the male adults at the first stage of their sequential courtship behavior. Subsequent studies by Nishida *et al.* proved the presence of two additional components, 29-hydroxy-3,11-dimethyl-2-nonacosanone (**105**) and 3,11-dimethyl-29-oxo-2-nonacosanone (**106**). The amounts of these compounds as isolated by them was 239 mg of **104** and 1.7 mg of **105** from 224 000 virgin females, and 20 μg of **106** from 2000 virgin females. The biological activities of these pheromone components as expressed by the concentration for 50% biological response were 3.7×10^{-6} M for **104**, 3.9×10^{-7} M for

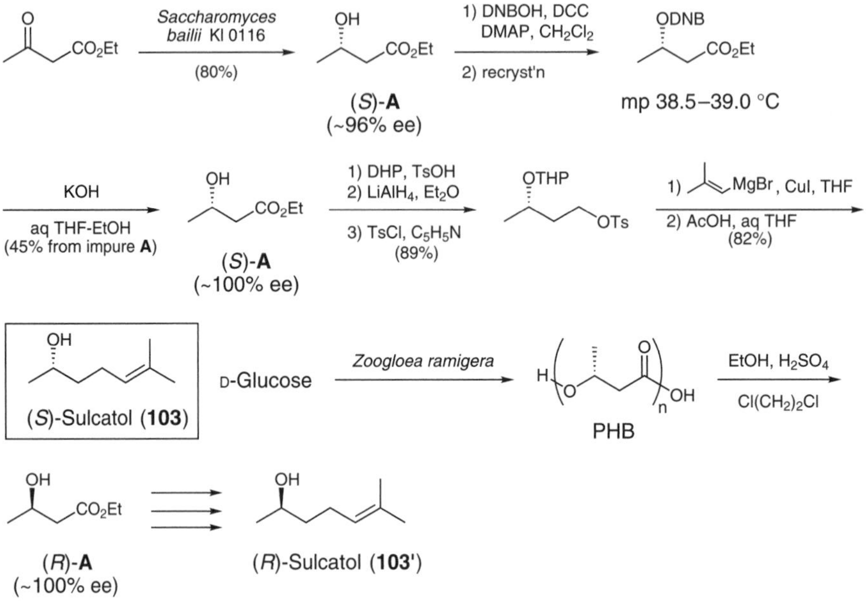

Figure 4.52 *Synthesis of the enantiomers of sulcatol (2). Modified by permission of Shokabo Publishing Co., Ltd*

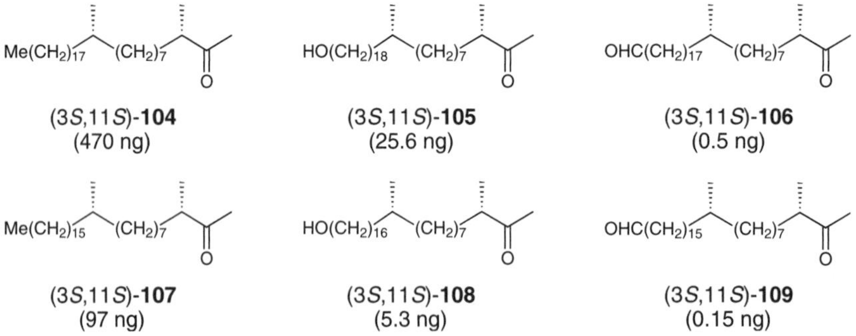

Figure 4.53 *Structures of the components of the sex pheromone of the German cockroach*

105 and 1.9×10^{-6} M for **106**. In other words, **105** was about 10 times more active than **104**, while **106** was about twice as active as **104**. Synthesis of all the four stereoisomers of **104** as well as those of **105** by us in 1981 allowed us to establish the absolute configuration of **104** and **105** as 3*S*,11*S*.[101,102]

Other minor components of the sex pheromone of *B. germanica* were subsequently isolated by Schal *et al.* in the USA as **107**, **108** and **109**. The amounts of these components in the cuticular surface of an adult female *B. germanica* were estimated on the basis of GC analysis of the cuticular hydrocarbons, and shown in Figure 4.53.

Let me first explain the manner by which we determined the absolute configuration of **104** as 3*S*,11*S*.[101,102] At the time of the structure determination of **104**, Nishida *et al.* proposed the 3*S*-configuration on the basis of its ORD (optical rotatory dispersion) spectrum coupled with NMR studies employing a chiral shift reagent. No information was available to assign the absolute configuration at C-11,

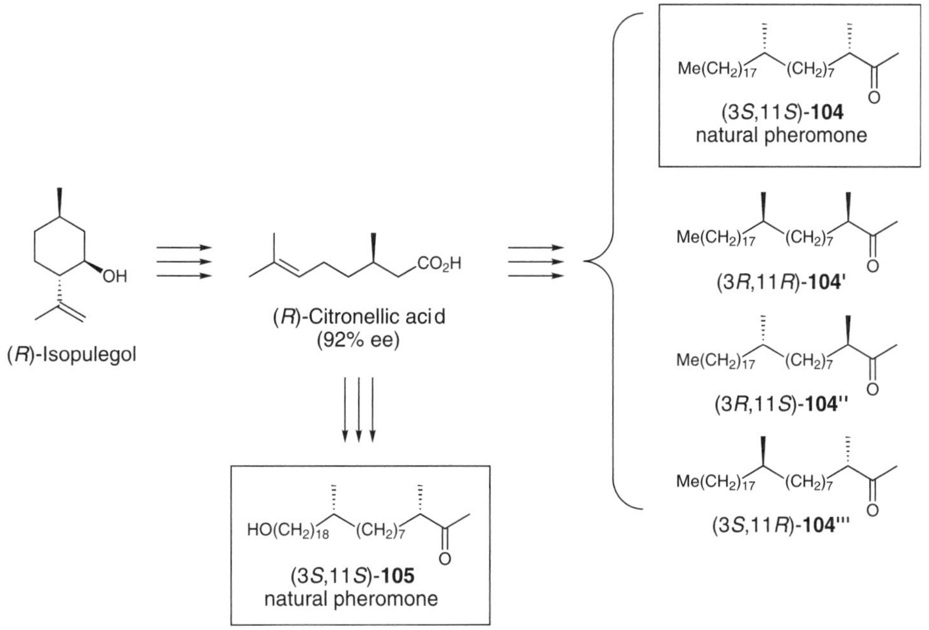

Figure 4.54 *Summary of the 1978 synthesis of the pheromone components of the German cockroach*

because the stereocenter at C-11 was separated from the C-3 stereocenter by seven methylene groups. As shown in Figure 4.54, we synthesized all of the four stereoisomers of **104** by starting from naturally occurring (*R*)-isopulegol. In 1978 (*R*)-citronellal of 97% ee was not available yet as a commercial product of Takasago International Corporation. Because the two stereocenter of **104** are separated by seven methylene groups, its four stereoisomers show identical ^{1}H- and ^{13}C-NMR spectra. Their chromatographic behaviors are also without difference. However, their IR spectra measured as nujol mulls (that is, as solid states not as solutions) show subtle differences.

Their optical rotations and melting points (Table 4.1) were very important in assigning the absolute configuration of the natural pheromone. The natural ketone **104** was dextrorotatory in hexane, and (3*S*,11*S*)-**104** as well as (3*S*,11*R*)-**104** showed positive rotations, while (3*R*,11*R*)- and (3*R*,11*S*)-**104** were levorotatory.

Table 4.1 *Specific rotations, IR and mps of the natural and synthetic steroisomers of **104** and their mixture mp with natural **104***

Sample	$[\alpha]_D$ (in hexane)	IR (nujol)	mp (°C)	Mixture mp with the natural **104**
natural **104**	+5.1 (*c* 3.54)		45–46	————
(3*S*,11*S*)-**104**	+5.98 (*c* 0.9)	—— same	44–44.5	no depression
(3*R*,11*R*)-**104'**	−5.63 (*c* 4.0)	> different	44.5–45	
(3*R*,11*S*)-**104''**	−5.68 (*c* 4.0)	—— same	39–39.5	depression
(3*S*,11*R*)-**104'''**	+5.73 (*c* 2.04)		38–38.5	

The natural **104** must therefore be either (3*S*,11*S*)- or (3*S*,11*R*)-**104**. All the stereoisomers of **104** showed IR spectra (as chloroform solutions) identical to each other. When their IR spectra were measured as nujol mulls, the stereoisomeric and crystalline ketones showed small differences in the spectra due to the differences in their crystal structures. Thus, the natural **104** showed a IR spectrum identical to those of (3*S*,11*S*)- and (3*R*,11*R*)-**104**, but different from those of other two isomers. The natural **104** seemed to be (3*S*,11*S*)-**104** at this stage.

To support this conclusion, melting points (mps) of the four stereoisomers of **104** were measured, and the mixture mp determination of the four isomers with the natural **104** were executed. As can be seen from Table 4.1, (3*S*,11*S*)- and (3*R*,11*R*)-**104** showed the same mp as that of the natural **104**. Mixture mp determinations revealed (3*S*,11*S*)-**104** to be the natural pheromone, because it showed no mp depression. The classical method of mixture mp test was still useful in establishing the identity of two like samples. Similarly, the absolute configuration of the natural **105** could be established as 3*S*,11*S*, too.[102] In analogy with **104** and **105**, all the remaining pheromone components **106–109** are assumed to possess 3*S*,11*S* configuration due to the same biosynthetic origin.

I will now outline the details of our second synthesis of **104** published in 1990.[103] The purpose of our 1990 synthesis was to provide extremely pure products. In our 1978 synthesis of **104** all the stereoisomers were pheromonally active. In that synthesis, however, we employed (*R*)-citronellic acid of 92% ee as our starting material (cf. Figure 4.54). In our 1990 synthesis, we chose enantiomerically pure (*R*)-citronellol and ethyl (*R*)-3-hydroxybutanoate as our starting materials to ensure the preparation of the stereoisomers of **104** in 100% enantiomerically pure forms. Bioassay of the products then unambiguously would verify the previous conclusion that all the stereoisomers are bioactive.

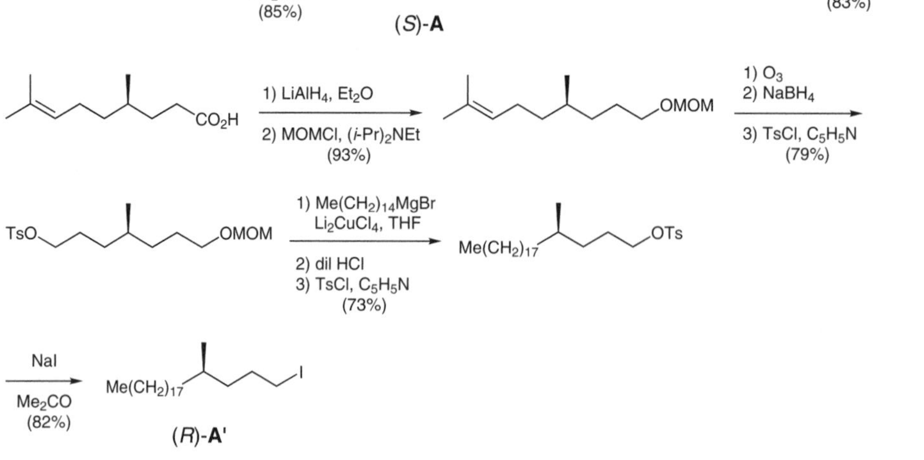

Figure 4.55 *Synthesis of 3,11-dimethyl-2-nonacosanone (1). Reprinted with permission of Shokabo Publishing Co., Ltd*

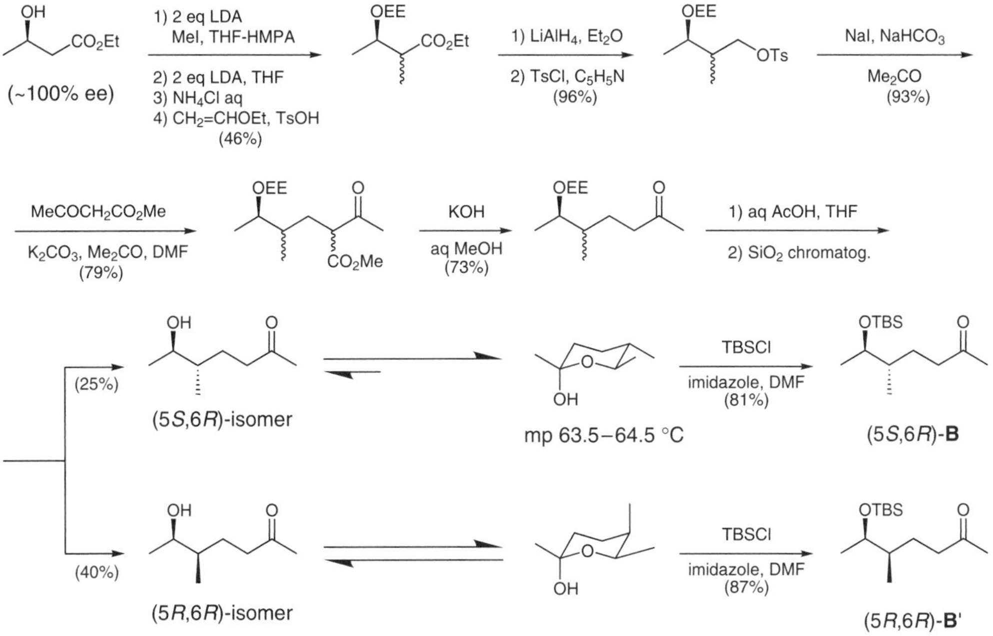

Figure 4.56 *Synthesis of 3,11-dimethyl-2-nonacosanone (2). Reprinted with permission of Shokabo Publishing Co., Ltd*

Figure 4.55 shows the synthesis of alkyl-chain part (*S*)-**A** and (*R*)-**A′** of **104**. Since enantiomerically pure (*R*)-citronellol was the starting material, both **A** and **A′** must be enantiomerically pure.

Conversion of enantiomerically pure ethyl (*R*)-3-hydroxybutanoate to (5*S*,6*R*)-**B** and (5*R*,6*R*)-**B′**, the other building blocks, is summarized in Figure 4.56. The crucial step was the chromatographic separation of a mixture of (5*S*,6*R*)- and (5*R*,6*R*)-6-hydroxy-5-methyl-2-heptanone. This could be achieved, because the former was less polar due to the facile hemiacetal formation. The hemiacetal was crystalline.

Finally, the building blocks **A**, **A′** and **B**, **B′** were coupled, and the coupling products were further processed to give the four stereoisomers of **104** with >99% de and 100% ee (Figure 4.57). They showed slightly higher mps than those of our products prepared in 1978.

Bioassay of our pure stereoisomers of **104** was carried out by Schal and coworkers in the USA, and indeed all of them were pheromonally active. More remarkably, the natural pheromone (3*S*,11*S*)-**104** was the least effective of the four isomers at eliciting courtship responses in males.[104] The cockroach produces the least active (3*S*,11*S*)-**104** due to the stereochemical restriction in the course of its biosynthesis. Nature does not always provide the best thing.

In 2007 Professor Coby Schal in the USA requested me to synthesize (3*S*,11*S*)-**109** and (3*S*,11*S*)-**108** to confirm the structures proposed by him for the minor components of *B. germanica* pheromone. Figure 4.58 shows my synthetic plan for them. Since the enantiomers of citronellal (97% ee) are available now from Takasago International Corporation, they can serve as the starting materials. The plan enabled the synthesis of all of the six pheromone components of *B. germanica* only by changing the alkyl group R of the building block **A**.[105]

Preparation of the building block **C** (Figure 4.58) is detailed in Figure 4.59 together with the synthesis of intermediates for the preparation of another building block. Alkylation of acetylene **F** (Figure 4.60) with iodide **D** or **E** in Figure 4.59 would give the left part of the target molecules.

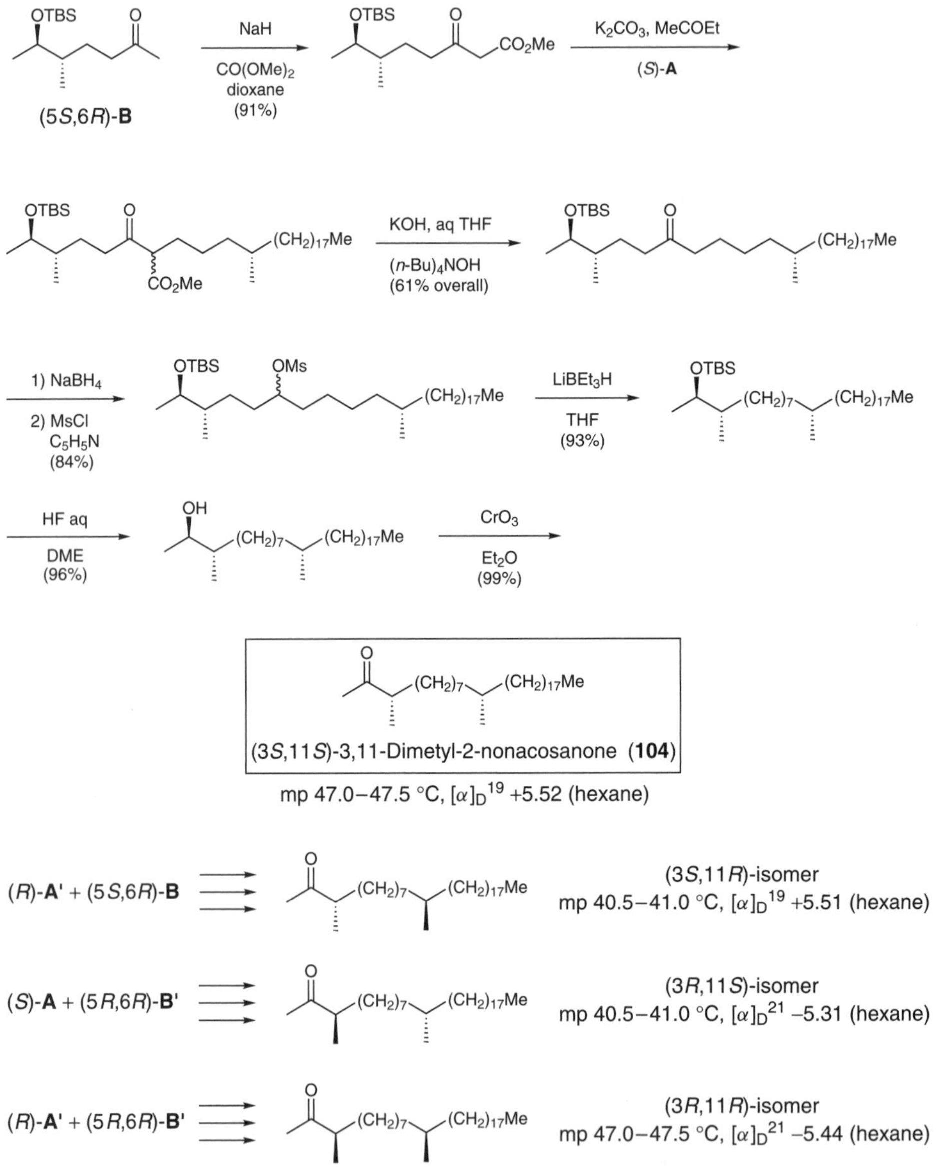

Figure 4.57 *Synthesis of 3,11-dimethyl-2-nonacosanone (3). Modified by permission of Shokabo Publishing Co., Ltd*

Coupling of the building blocks and further transformation to **104–109** are shown in Figure 4.60. Palladium-catalysed Wacker oxidation of **G** yielded **104**, **107** and two other ketones. Final oxidation of **105** and **108** to **106** and **109**, respectively, was achieved with Dess–Martin periodinane.[105] Thus, all the six components **104–109** of the female-produced contact sex pheromone of the German cockroach were synthesized from the enantiomers of citronellal.

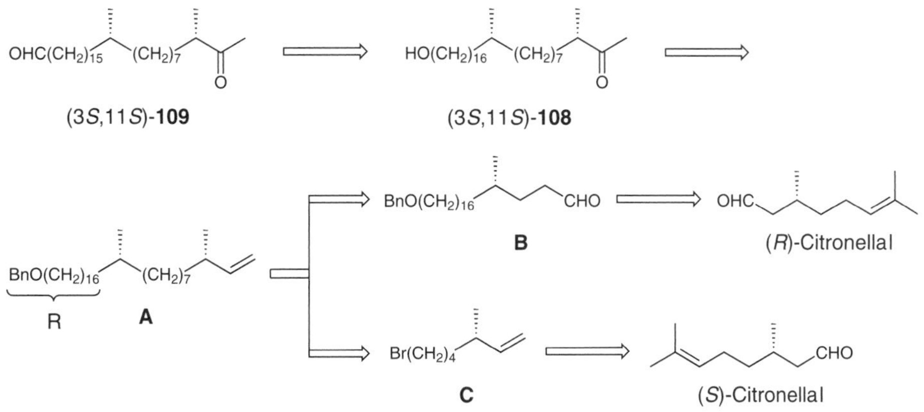

R = n-C$_{18}$H$_{37}$, n-C$_{16}$H$_{33}$, BnO(CH$_2$)$_{18}$, BnO(CH$_2$)$_{16}$

Figure 4.58 *Synthetic plan for the minor components 108 and 109 of the sex pheromone of the German cockroach*

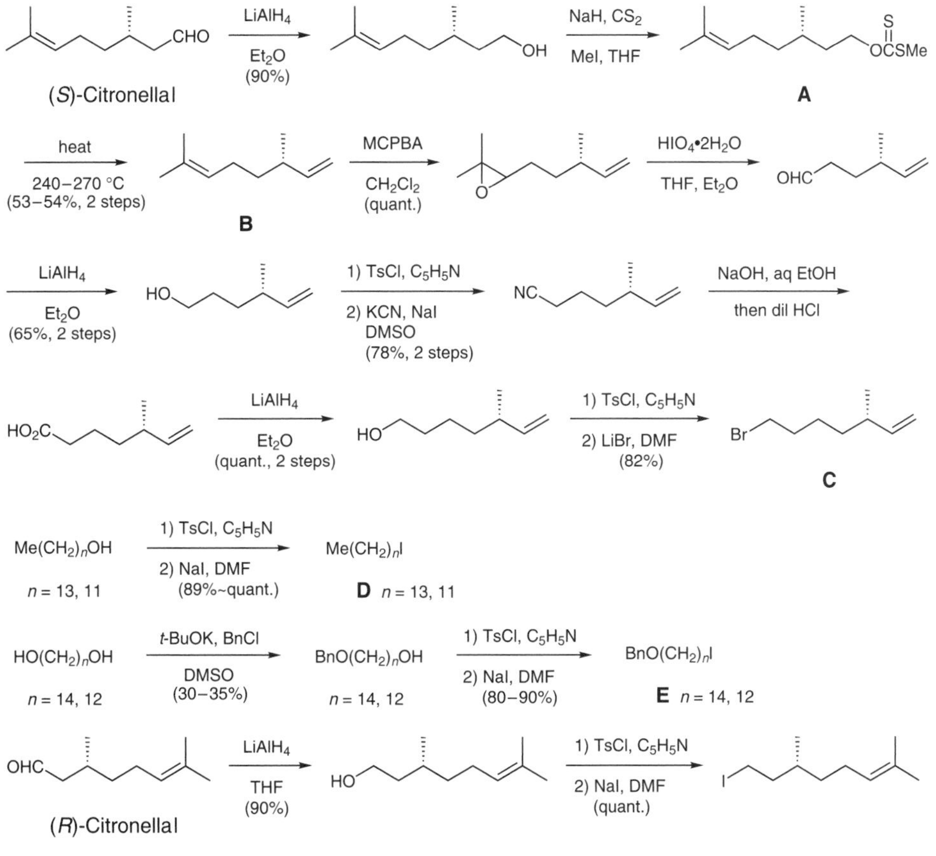

Figure 4.59 *Synthesis of all six components of the sex pheromone of German cockroach (1)*

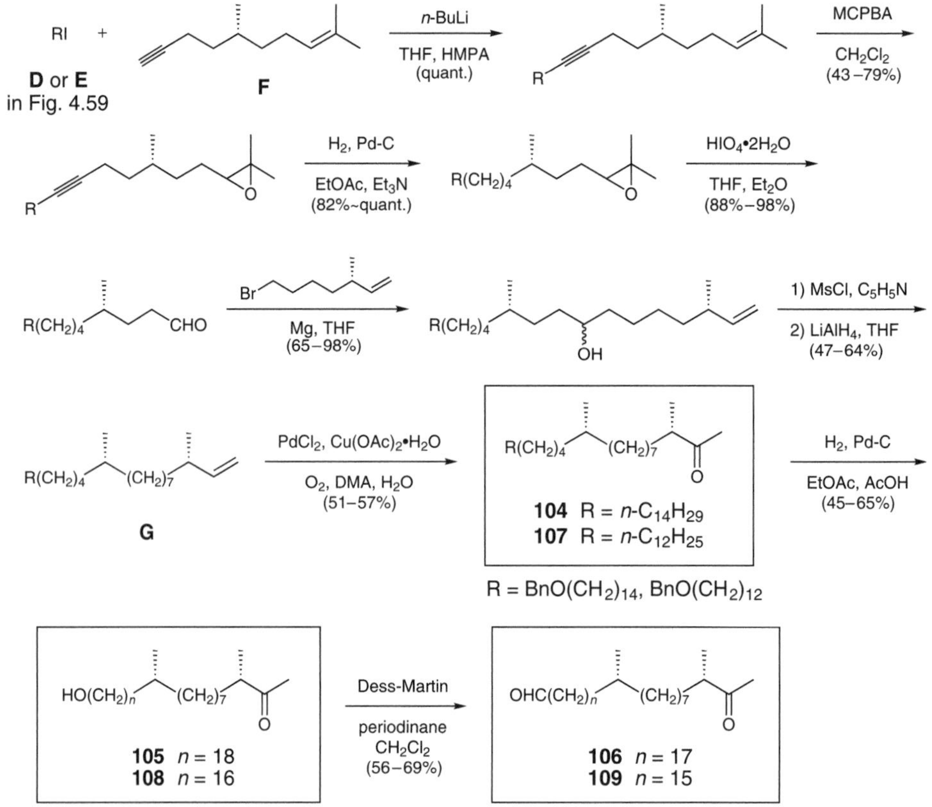

Figure 4.60   *Synthesis of all six components of the sex pheromone of German cockroach (2)*

Bioassay of these compounds indicated that both the hydroxy ketones **105** and **108** are about ten-fold more active than the respective parent ketones **104** and **107** of the same chain-length.[106] Each of the six pheromone components can independently elicit the complete repertoire of sex response with no synergism among others.[106]

4.5.3   Stigmolone, the pheromone of a myxobacterium *Stigmatella aurantiaca*

Myxobacteria are unique prokaryotes that undergo multicellular development including swarming and aggregation of their cells and formation of fruiting bodies. In 1998 Plaga and coworkers in Germany reported the isolation and identification of stigmolone (**110**, Figure 4.61), a myxobacterial pheromone of *Stigmatella aurantiaca* to induce the formation of fruiting bodies.

In the same year we achieved the synthesis of the enantiomers of **110**, as shown in Figure 4.61.[107] (*S*)-Citronellal was converted to olefinic ester **A**, which was epoxidized to give **B**. Treatment of **B** with isopropylmagnesium chloride in the presence of copper(I) bromide furnished lactone **C**, which afforded diol **D** by treatment with methylmagnesium bromide. Finally, Dess–Martin oxidation of **D** yielded (*S*)-stigmolone (**110**), which was in equilibrium with (2*R*,3*S*)-**E**. Similarly, (*R*)-citronellal afforded (*R*)-**110′**.

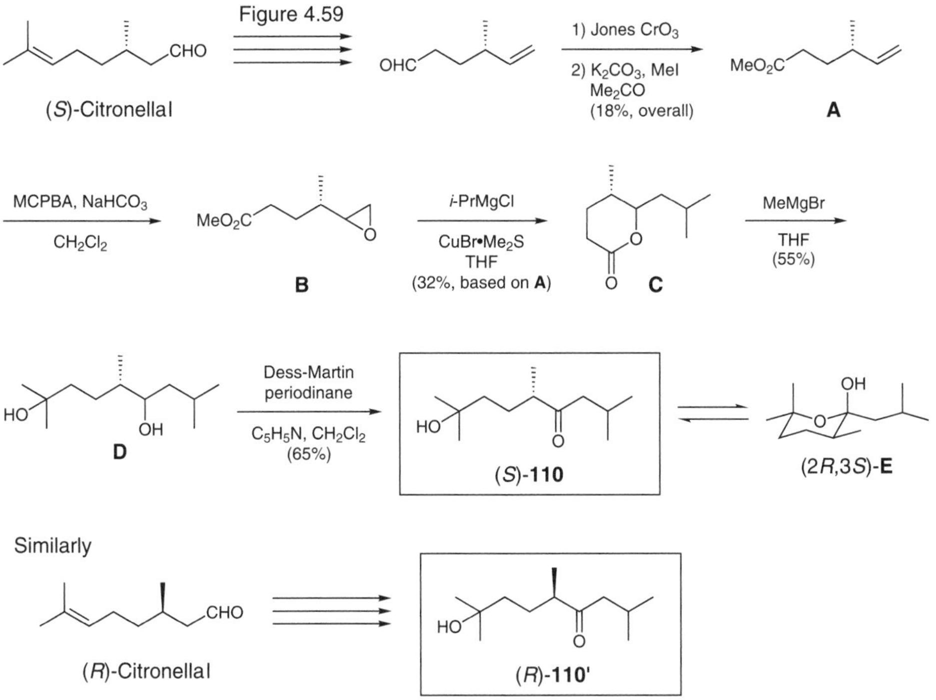

Figure 4.61   *Synthesis of the enantiomers of stigmolone*

Both (*R*)- and (*S*)-stigmolone induced the formation of fruiting bodies of *S. aurantiaca* at a concentration of 0.4–1.0 nM. A racemic mixture of **110** and **110′** was also active. Subsequently, the natural stigmolone was shown to be an enantiomeric mixture.[108]

Since racemic stigmolone was bioactive, we carried out its synthesis, as shown in Figure 4.62.[109] The overall yield of (±)-**110** was 48% based on methyl isobutyl ketone (4 steps). The most efficient synthesis of (±)-**110**, however, is that by Epstein and Kulinkovich.[110] Their synthesis proceeded in 67% yield in 2 steps from ethyl isovalerate as shown in the lower part of Figure 4.62. Kulinkovich employed his own reaction to prepare (±)-**110** efficiently.

4.5.4   Ipsenol and ipsdienol, pheromones of *Ips* bark beetles

Ipsenol (**111**, Figure 4.63), ipsdienol (**112**) and *cis*-verbenol (**A**) were isolated and identified in 1966 by Silverstein *et al.* as the components of the male-produced aggregation pheromone of a bark beetle, California five-spined ips (*Ips paraconfusus*).

I determined in 1975 the absolute configuration of ipsenol as (*S*)-**111** through a synthesis of its enantiomers, as shown in Figure 4.64. (*S*)-Leucine was converted to (*S*)-epoxide **A**, which finally gave (*S*)-ipsenol (**111**).[111,112] A much simpler and more efficient synthesis of ipsenol was subsequently developed as shown in Figure 4.65.[113] Treatment of the epoxide (*S*)-**A** with a Grignard reagent prepared from chloroprene gave (*S*)-ipsenol (**111**) in 50% yield.

Domon, Mori

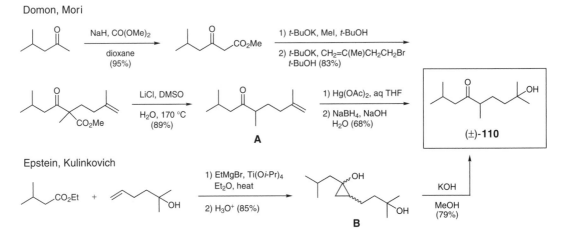

Figure 4.62 *Synthesis of (±)-stigmolone*

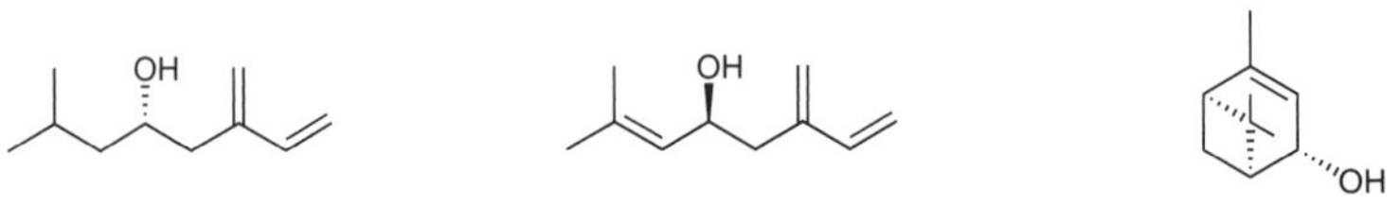

(S)-Ipsenol (**111**)   (S)-Ipsdienol (**112**)   (S)-*cis*-Verbenol (**A**)

Figure 4.63 *Components of the aggregation pheromone of Ips paraconfusus. Modified by permission of Shokabo Publishing Co., Ltd*

Professor Jean-Pierre Vité's bioassay revealed (S)-**111** to be bioactive also as the aggregation pheromone of *Ips grandicollis*, while (R)-**111'** was neither bioactive nor inhibitory.[114]

In the case of ipsdienol (**112**), the stereochemistry–bioactivity relationship is much more complicated and interesting. My first synthesis of (R)-(−)-ipsdienol (**112'**) in 1976 from (R)-glyceraldehyde allowed the assignment of S-configuration to (+)-ipsdienol isolated by Silverstein.[115] Our second synthesis to provide both the enantiomers of ipsdienol employed the enantiomers of malic acid as starting materials.[113] That synthesis, however, was lengthy and yielded the final products of 90% ee due to the partial racemization in the course of the synthesis.

Our third synthesis of the enantiomers of ipsdienol started from the enantiomers of serine as shown in Figure 4.66.[116] (S)-Serine was converted to epoxide A, which was treated with a Grignard reagent prepared from chloroprene to give hydroxy ester B. Subsequently, B afforded (R)-ipsdienol (**112'**, ≥96% ee). Similarly, (R)-serine furnished (S)-**112**.

Later biological studies on the enantiomers of ipsdienol revealed the following interesting facts. *Ips paraconfusus* employs (S)-(+)-**112** as the pheromone, while *Ips calligraphus* and *Ips avulsus* use (R)-(−)-**112'**. *Ips pini* in New York employs a mixture of (R)-**112'** and (S)-**112** = 32–56:68–44, while that in California uses a mixture of (R)-**112'** and (S)-**112** = 89–98: 11–2. Thus, different enantiomers are employed by different species or subspecies.

4.5.5 Serricornin, the cigarette beetle pheromone

Serricornin (**113**, Figure 4.67) was isolated and identified in 1979 by Chuman *et al.* in Japan as the female-produced sex pheromone of the cigarette beetle (*Lasioderma serricorne*), which is a serious pest

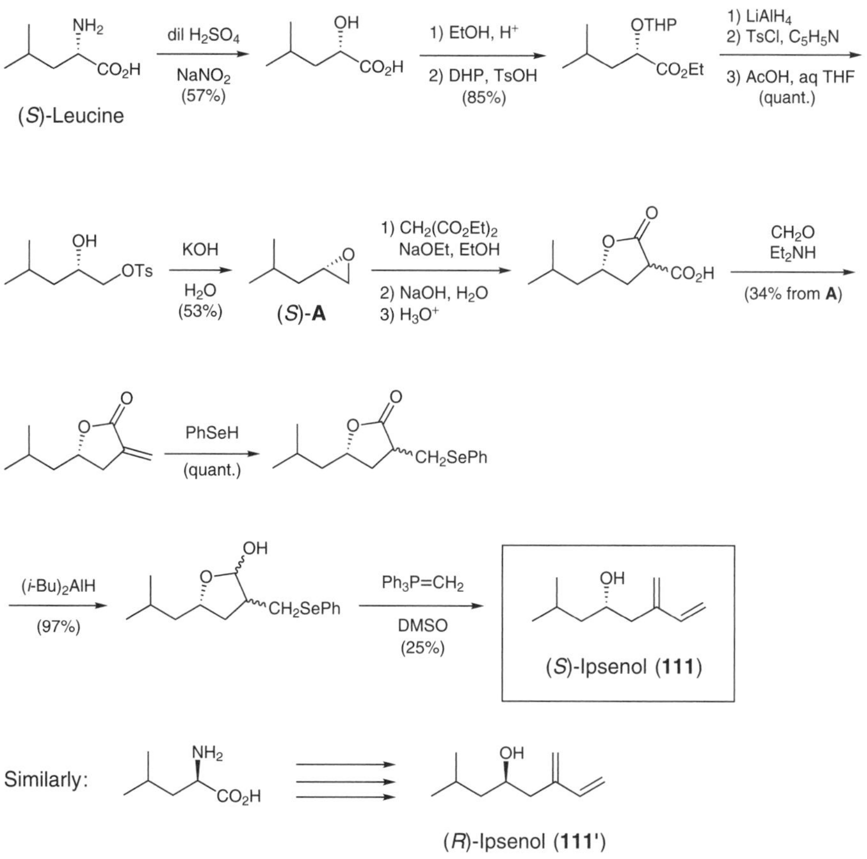

Figure 4.64 *Synthesis of the enantiomers of ipsenol (1). Modified by permission of Shokabo Publishing Co., Ltd*

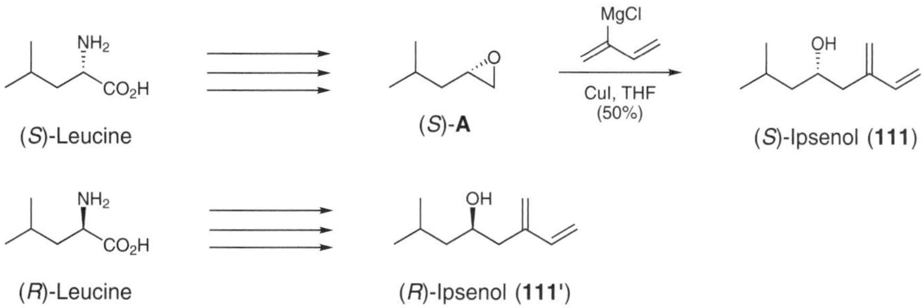

Figure 4.65 *Synthesis of the enantiomers of ipsenol (2). Modified by permission of Shokabo Publishing Co., Ltd*

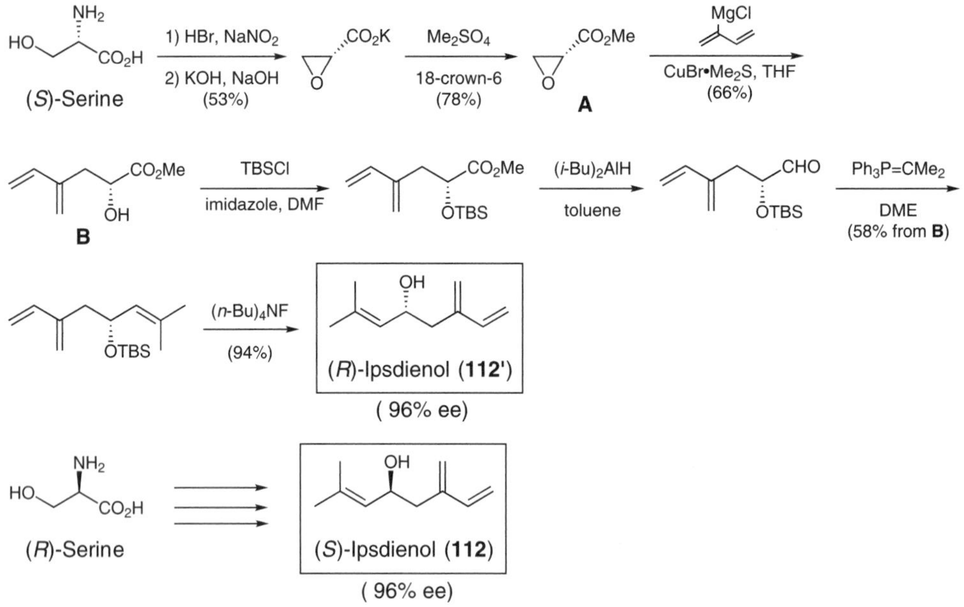

Figure 4.66 *Synthesis of the enantiomers of ipsdienol. Modified by permission of Shokabo Publishing Co., Ltd*

of cured tobacco leaves and dried foods. Joint work of Chuman *et al.* and ours enabled us to propose (4S,6S,7S)-**113** as the stereostructure of serricornin.

In 1981, we synthesized the enantiomers of serricornin as shown in Figure 4.67.[117,118] The starting material for (4S,6S,7S)-**113** was (2R,3R)-β-methylaspartic acid, which was converted to iodide **A**. Alkylation of the SAMP hydrazone of diethyl ketone with **A** was followed by regeneration of the carbonyl and hydroxy groups to give (4S,6S,7S)-serricornin (**113**). Similarly, (2S,3S)-β-methylaspartic acid furnished (4R,6R,7R)-**113′**. Natural serricornin was shown to be (4S,6S,7S)-**113** by comparing the physical properties including $[\alpha]_D$ values of serricornin acetate with those of the acetate of our synthetic (4S,6S,7S)-**113**.

In 1984, we had a chance to prepare a racemic and diastereomeric mixture of all the possible stereoisomers of serricornin. TLC analysis of the mixture, as shown in the lower part of Figure 4.68, revealed an interesting fact that the mixture was readily separable into three fractions. Each fraction was acetylated and analysed by GC. The least polar fraction was (4S*,6R*,7S*)-**113**, and the most polar fraction was (4R*,6S*,7S*)-**113**. The broad spot between the least and the most polar fractions contained (4S*,6S*,7S*)-**113** and (4R*,6R*,7S*)-**113**. As shown in Figure 4.68, the open-chain forms **113a** are in equilibrium with the hemiacetal forms **113b**, the latter being less polar than the former. The stereoisomer with a large R_f value is the one with the tendency to readily cyclize to **113b**.

It is therefore possible to separate serricornin (4S,6S,7S)-**113** from its (4R,6S,7S)-isomer, and we achieved a new serricornin synthesis as summarized in the upper part of Figure 4.68. The starting material was methyl (R)-3-hydroxypentanoate **A**, which was converted to **B**. Mitsunobu inversion of the secondary hydroxy group of **B** with 3,5-dinitrobenzoic acid as a reactant gave **C**. Recrystallization of **C** guaranteed its high chemical and enantiomeric purities. Then, iodide **D** was prepared from pure **C**. Diethyl ketone was alkylated with **D** to give **E** as a stereoisomeric mixture. Deprotection of the TBS group of **E** was followed by chromatographic separation of (4S,6S,7S)-serriconin (**113**) from its (4R,6S,7S)-isomer. The overall yield of the present synthesis was 7.6% based on **A**.[119]

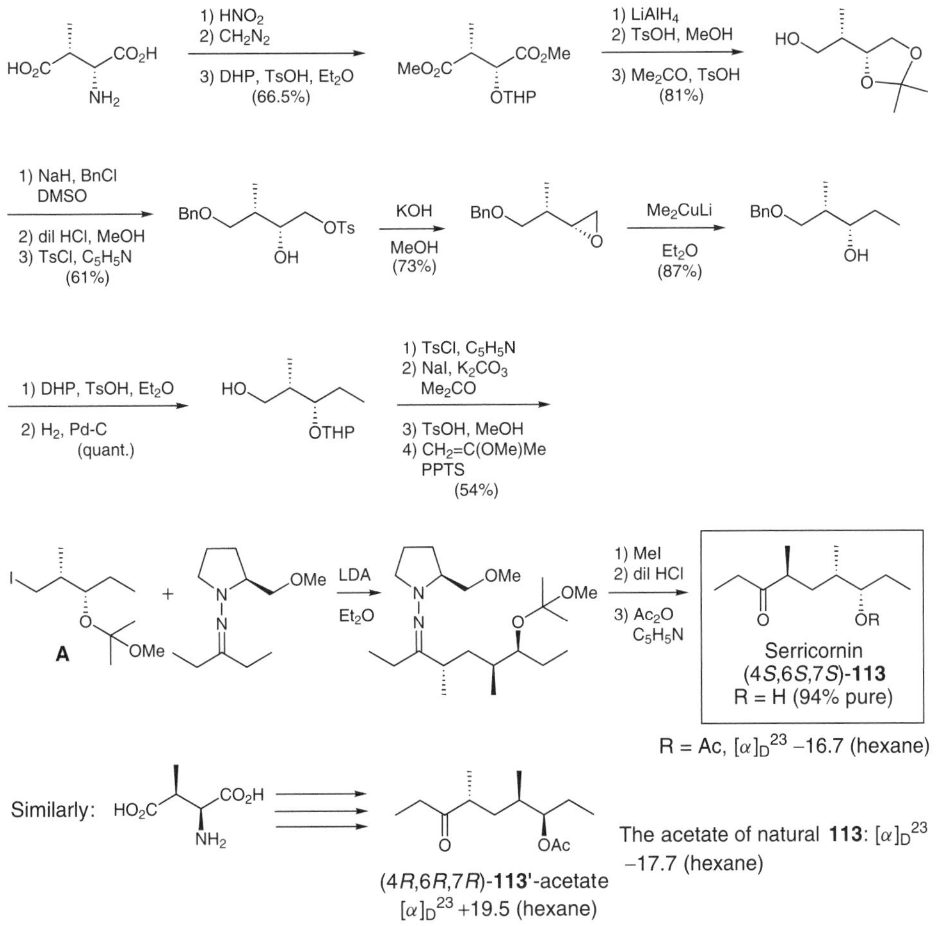

Figure 4.67 *Synthesis of the enantiomers of serricornin. Modified by permission of Shokabo Publishing Co., Ltd*

Biological studies on our synthetic stereoisomers of serricornin were carried out at Japan Tobacco Corporation to reveal interesting results as follows. Pheromone activity of natural serricornin (4*S*,6*S*,7*S*)-**113** cannot be inhibited by its opposite enantiomer (4*R*,6*R*,7*R*)-**113′**. The latter is biologically inactive. However, (4*S*,6*S*,7*R*)-isomer strongly inhibits the pheromone activity of (4*S*,6*S*,7*S*)-**113**. Accordingly, practically useful serricornin must be free from (4*S*,6*S*,7*R*)-isomer. Indeed, a commercial product of serricornin as manufactured by Fuji Flavor Co. contains no (4*S**,6*S**,7*R**)-isomer so that it can exhibit high attractancy against the cigarette beetle. Synthesis and biological evaluation of all the possible stereoisomers are always required to develop an excellent commercial product of chiral pheromones.

4.5.6 Stegobinone, the drugstore beetle pheromone

Drugstore beetle (*Stegobium paniceum*) is a serious pest of a wide variety of commodities and stored products. In 1978 Y. Kuwahara *et al.* proposed the structure of stegobinone, the major and crystalline component of the female-produced sex pheromone of *S. paniceum*, as **114** (Figure 4.69). Nothing was

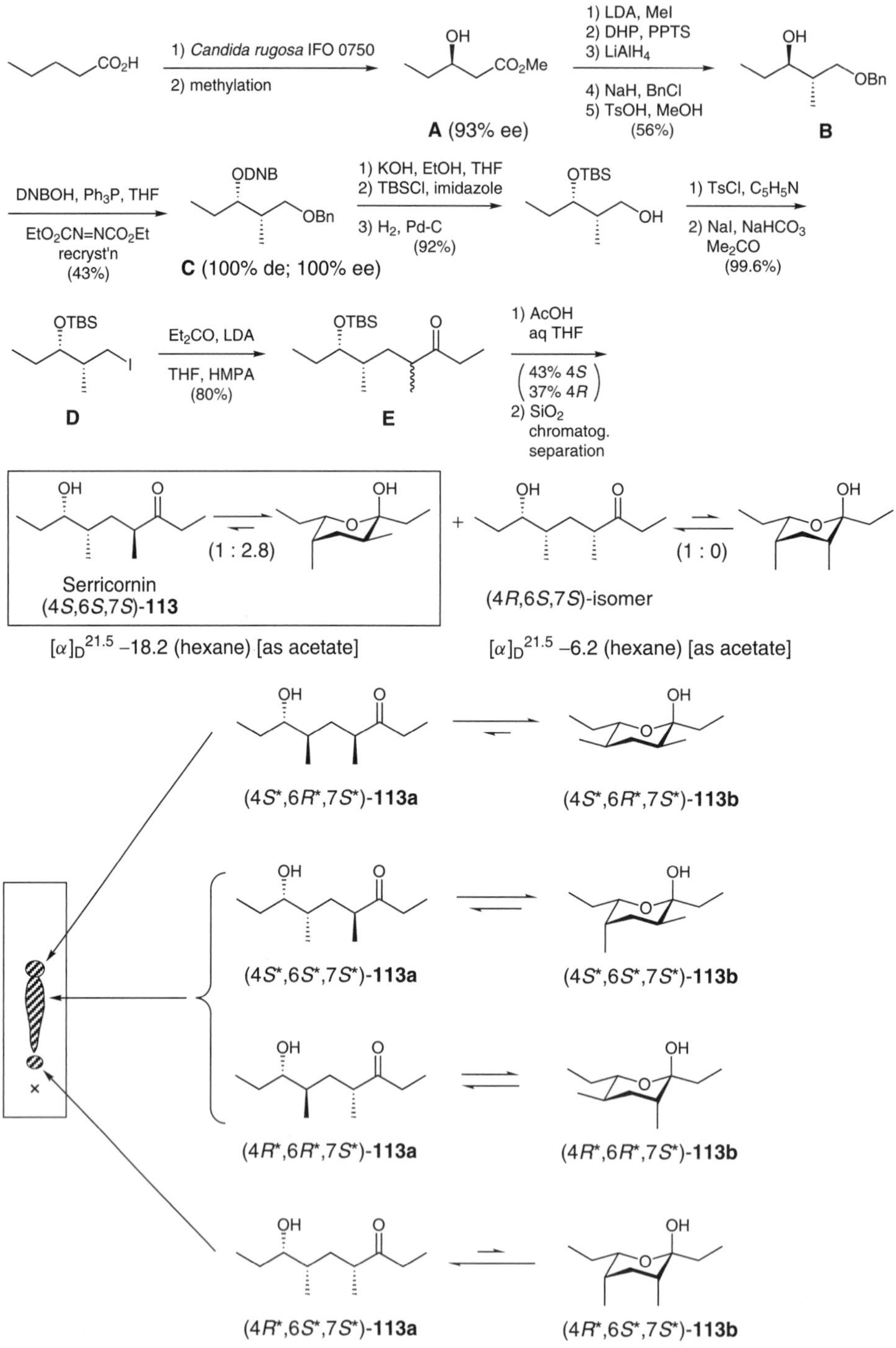

TLC separation of the stereoisomers of serricornin.

Figure 4.68 *Synthesis of (4S,6S,7S)-serricornin. Modified by permission of Shokabo Publishing Co., Ltd*

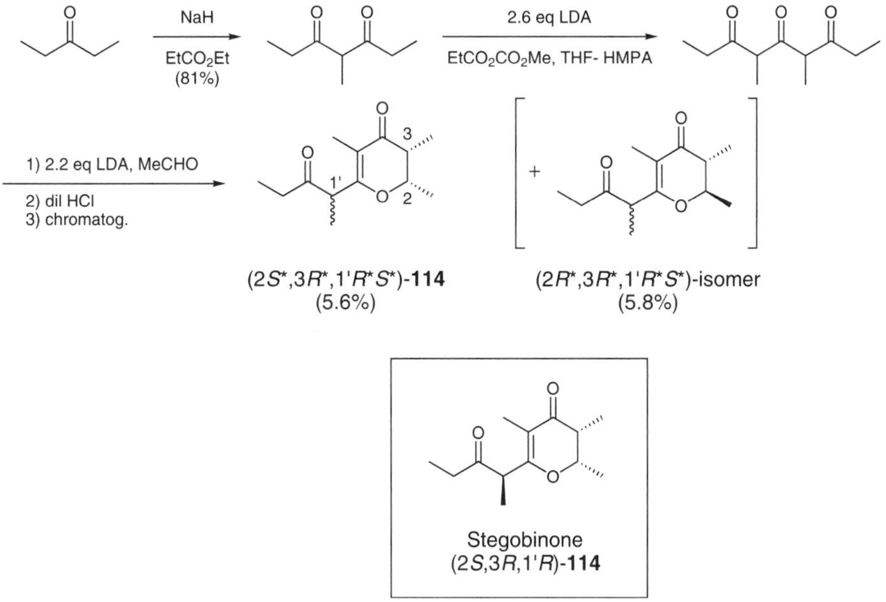

Figure 4.69 *Synthesis of a steroisomeric mixture of stegobinone. Modified by permission of Shokabo Publishing Co., Ltd*

known, however, about its absolute configuration except that the two vicinal methyl groups are with *cis*-relationship.

In 1979, we reported a biomimetic synthesis of (2*S**,3*R**,1'*R***S**)-**114** in a low yield (5.6%) as a racemic and diastereomeric mixture.[120,121] Although the synthetic product was pheromonally active, its activity was shown at a dosage of about 10^{-4} µg, while the natural stegobinone was active at 3×10^{-7} µg. This fact indicated that the synthetic (2*S**,3*R**,1'*R***S**)-**114** contained a certain inhibitory stereoisomer. It took about 20 years to solve this problem completely. We published our preliminary results of this synthesis in *Tetrahedron Letters*.[120] I was surprised in the following week to see the next issue of the same journal, which contained a paper by Professor A. Hassner reporting the same biomimetic synthesis of **114**.[122] Quite often the same idea occurs to the minds of different people.

Subsequently, in 1981, we published our second synthesis so as to determine the absolute configurations at C-2 and C-3.[121] As shown in Figure 4.70, the enantiomers of tartaric acid were converted to (2*S*,3*R*,1'*RS*)-**114** and (2*R*,3*S*,1'*RS*)-**114**. The natural stegobinone showed CD (circular dichroism) spectrum similar to that of (2*S*,3*R*,1'*RS*)-**114** at 350 nm region, and therefore the natural pheromone was thought to possess 2*S*,3*R*-stereochemistry. Unfortunately, however, the pheromone activity of (2*S*,3*R*,1'*RS*)-**114** was considerably weaker than that of the natural stegobinone. In the same year of 1981, Hoffmann *et al.* in Germany established the absolute configuration at C-1' of the natural stegobinone as *R* by the X-ray analysis of the crystalline C-1' epimer [(2*S*,3*R*,1'*S*)-epistegobinone] secured by their own synthesis.[123]

Our first synthesis of the natural (2*S*,3*R*,1'*R*)-**114** was published in 1986.[124] As shown in Figure 4.71, the required building block **B** was prepared from ethyl (*R*)-3-hydroxybutanoate of 100% ee via **A**. Another building block **C** was synthesized from methyl (*R*)-3-hydroxy-2-methylpropanoate (97% ee). Esterification of **B** with **C** gave the key intermediate **D**, whose intramolecular cyclization afforded the required pyranone system. The synthetic (2*S*,3*R*,1'*R*)-**114** showed spectral properties identical to those of the natural stegobinone, although it remained as an oil due to the contamination with trace amounts of impurities. Our

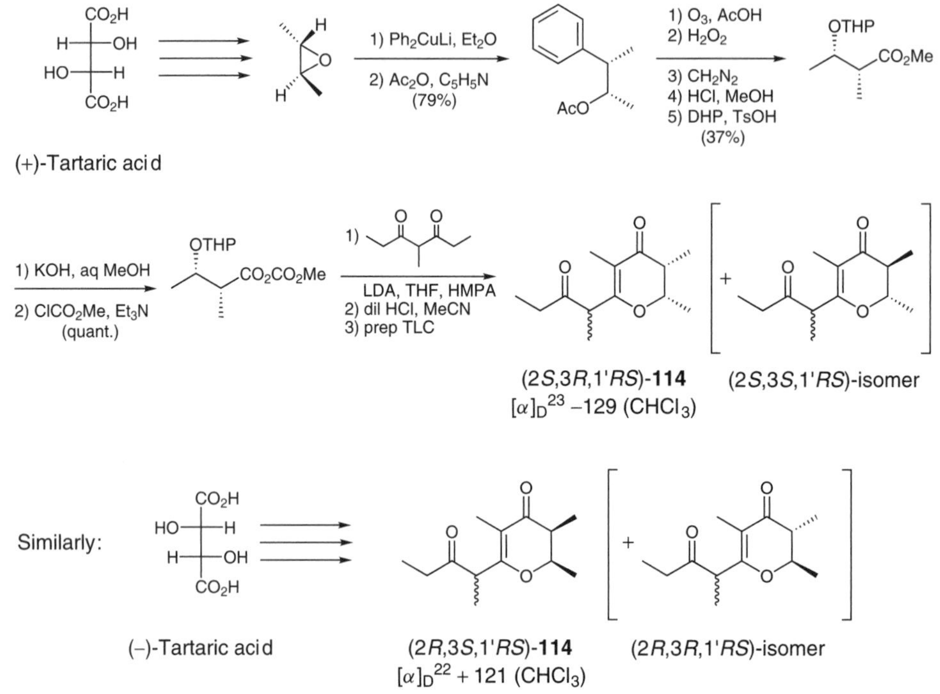

Figure 4.70   *Synthesis of (2S,3R,1'RS)-stegobinone. Modified by permission of Shokabo Publishing Co., Ltd*

synthetic (2S,3R,1'R)-**114** was pheromonally far more potent than (2S,3R,1'RS)-**114**, but less potent than the extract of the female *S. paniceum*.[124] The researchers at Japan Tobacco Corporation later isolated and identified a minor pheromone component stegobiol (see Figure 4.72). They also found that (2S,3R,1'S)-**114** inhibits the action of stegobinone.

Two important works were reported by others in the 1990s. First, Oppolzer in Switzerland found that the intramolecular cyclization as shown in **D**→**114** of Figure 4.71 could best be accomplished with titanium tetrachloride in the presence of ethyldiisopropylamine in dichloromethane at −78 °C.[125] Then, Matteson in the USA synthesized pure and crystalline (2S,3R,1'R)-**114** by employing his organoborane chemistry.[126] This prompted us to attempt our second synthesis of (2S,3R,1'R)-**114** to obtain it as crystals.

Figure 4.72 summarizes our second synthesis of (2S,3R,1'R)-stegobinone (**114**) published in 1998.[127] The building block (4R,5S)-**A** was prepared by an enzymatic process, while another building block (2S,3S)-**C** was synthesized via Sharpless asymmetric epoxidation. Coupling of **A** with **C** gave ester **D**, which was cyclized under Oppolzer's conditions to give crystalline stegobiol (**E**). When **E** was oxidized with Dess–Martin periodinane, tetra(n-propyl)ammonium perruthenate or Jones chromic acid, crystalline stegobinone (**114**) was obtained. Swern oxidation or oxidation with 2,2,6,6-tetramethylpiperidin-1-oxyl (TEMPO) of **E** afforded oily materials.[127] Our synthetic stegobinone was subjected to X-ray crystallographic analysis, and the resulting perspective view of (2S,3R,1'R)-**114** was also published.[127] Immediately after the publication of our work, Professor R.W. Hoffmann in Germany wrote to me: "I was very curious to see what the difference between stegobinone and 1'-epistegobinone really meant in three dimensions." He compared his own X-ray result of 1'-epistegobinone with ours, and found that the configuration at the C-1'

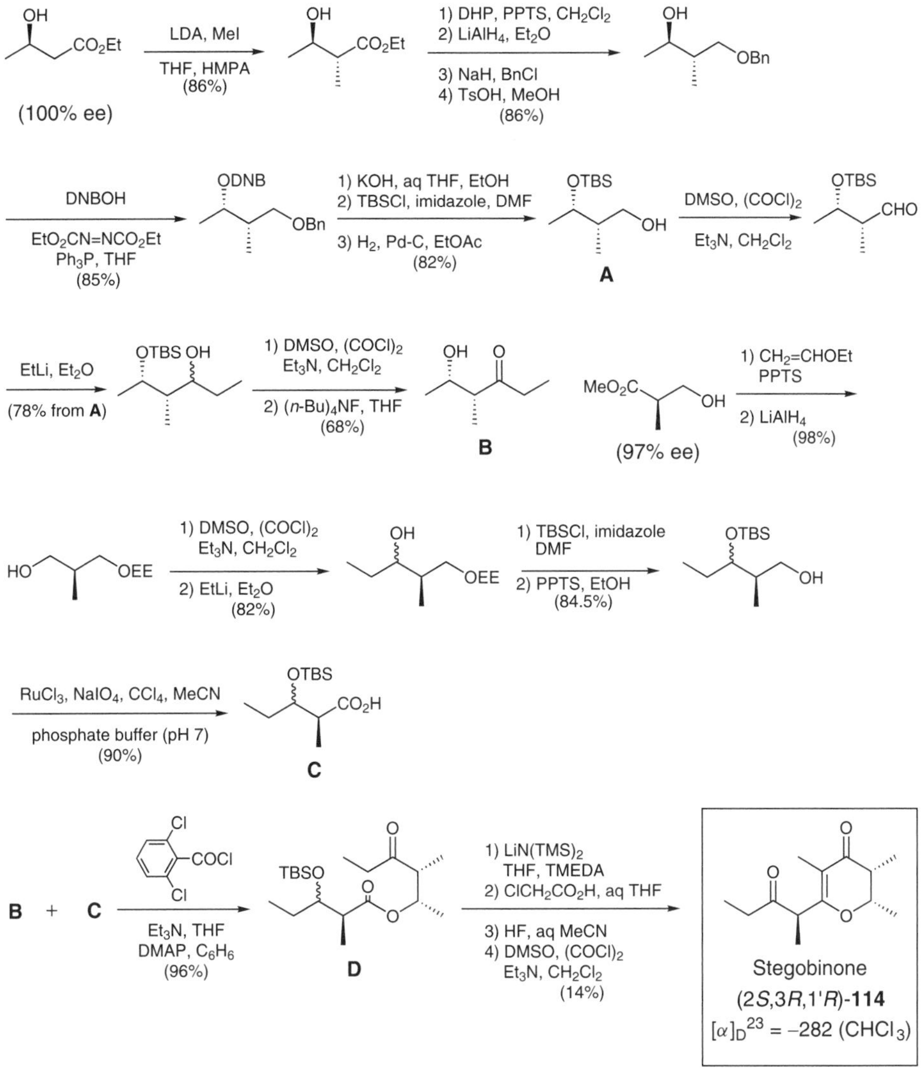

Figure 4.71 Synthesis of (2S,3R,1'R)-stegobinone (1). Modified by permission of Shokabo Publishing Co., Ltd

carbon atom influences the conformation of the dihydropyran ring, influencing the helicity of the half-chair conformation of the ring. Thus, a change in the absolute configuration at C-1' was crucial to generate an inhibitor of the pheromone action.

(2S,3R,1'R)-Stegobinone (**114**) was readily epimerizable, and quickly lost its pheromone activity within a week, although it was stable as crystals. The synthetic (2S,3R,1'R)-stegobinone (**114**) could not be used practically due to this facile racemization at C-1'. Stegobiene (bottom, Figure 4.72) was designed and developed as a practical lure for drugstore beetle by scientists at Fuji Flavor Co. Stegobiene is fairly stable, and does not lose its pheromone activity even after months.

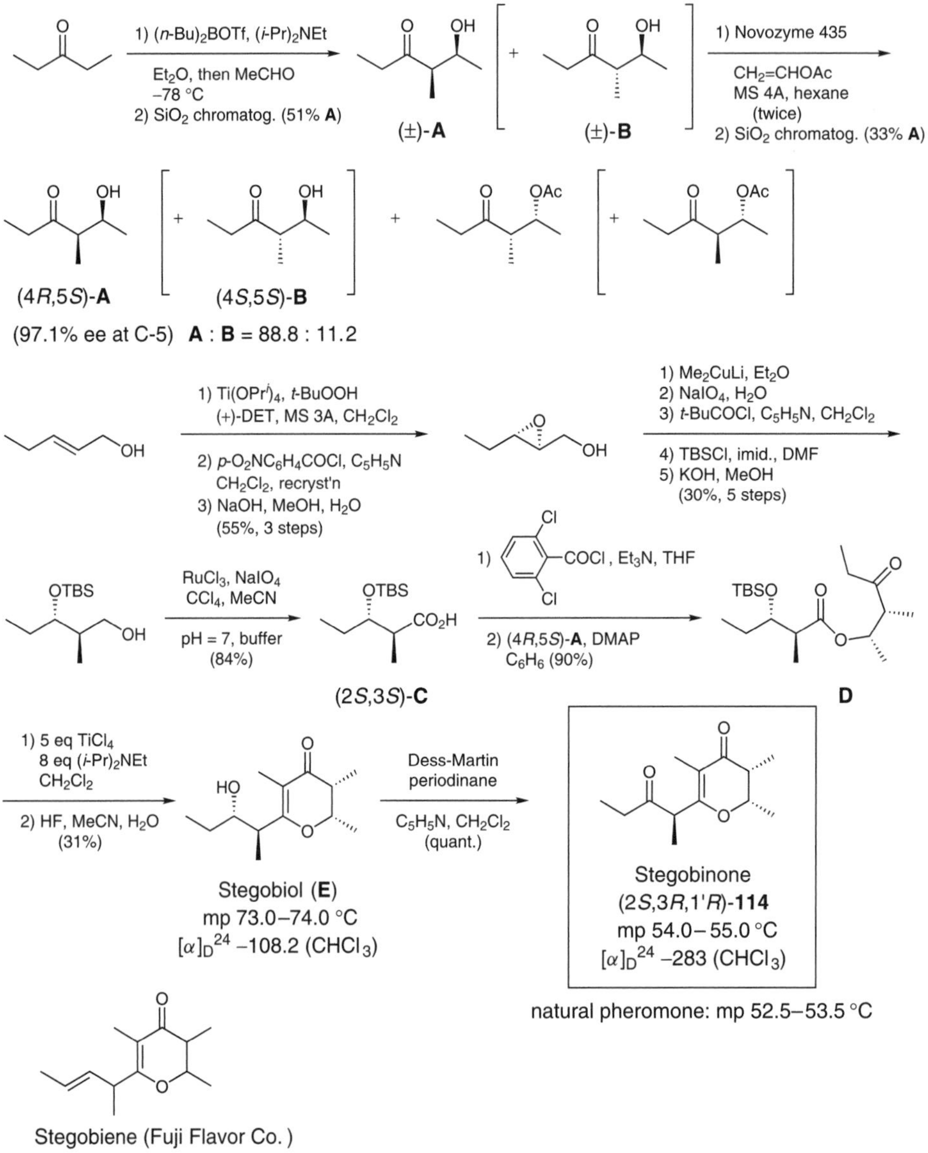

Figure 4.72 *Synthesis of (2S,3R,1′R)-stegobinone (2)*

4.5.7 Supellapyrone, the sex pheromone of the brownbanded cockroach

In 1993 Roelofs and coworkers in the USA isolated and identified supellapyrone (**115**, Figure 4.73) as the female-produced sex pheromone of the brownbanded cockroach (*Supella longipalpa*). Meinwald proposed its stereochemistry as 2R,4R. After preparing (2R*,4R*)-**115**[128] and (2R,4R)-**115**,[129] we synthesized all of the four stereoisomers of **115** as shown in Figure 4.73.[130]

The key-steps of our 2001 synthesis was enzymatic desymmetrization of *meso*-**B**, enzymatic enantiomer separation of (±)-**B**, and Reformatsky-type cyclization of **E** to give **F**. The natural supellapyrone

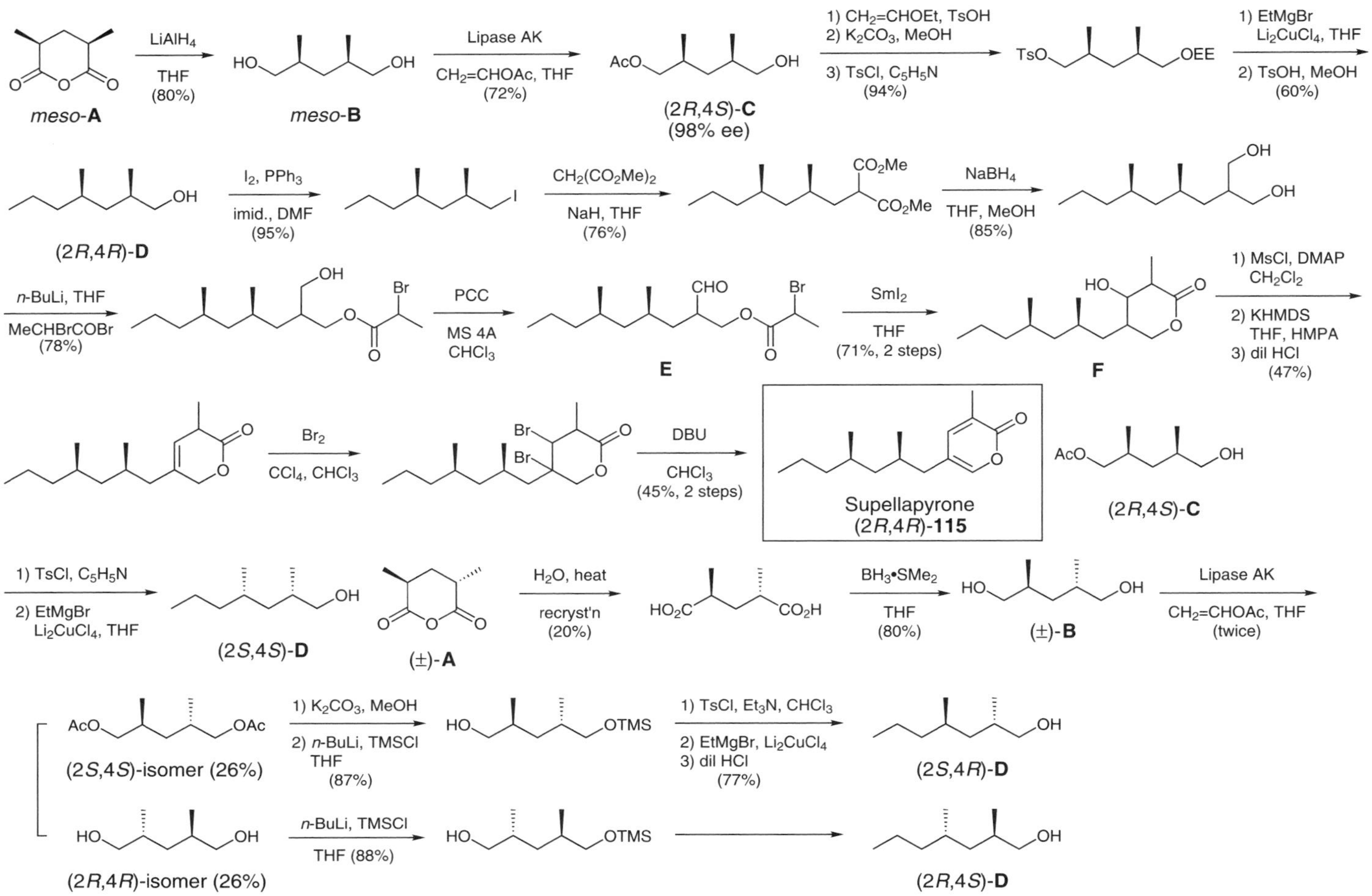

Figure 4.73 *Synthesis of the stereoisomers of supellapyrone*

(2*R*,4*R*)-**115** was synthesized from (2*R*,4*R*)-**D**. Similar conversions of (2*S*,4*S*)-, (2*S*,4*R*)- and (2*R*,4*S*)-**D** to supellapyrone stereoisomers were also achieved.[130]

Behavioral responses of the brownbanded cockroach to the four stereoisomers of supellapyrone (**115**) were studied carefully by Schal and coworkers.[131] In field tests, males are attracted to the natural (2*R*,4*R*)-**115**, but also to the high concentrations of (2*S*,4*R*)-isomer. In an olfactometer in the laboratory, (2*R*,4*R*)-**115** was the most active isomer with just 0.3 pg being sufficient to elicit 50% male response. Males are also attracted to (2*S*,4*R*)- and (2*S*,4*S*)-isomers in the olfactometer, but at much higher dosage (100×) than the natural isomers. At any of the doses tested, (2*R*,4*S*)-**115** did not elicit behavioral responses. In this case of *S. longipalpa*, the stereochemistry–pheromone activity relationships are not simple but complicated.

4.5.8   Olean, the sex pheromone of the olive fruit fly

The olive fruit fly (*Bactrocera oleae*) is the major pest of olive trees in Mediterranean countries such as Greece, Israel, Italy and Spain. In 1980 Baker, Francke and their respective coworkers isolated and identified the female-produced sex pheromone of *B. oleae* as 1,7-dioxaspiro[5.5]undecane (**116**, Figure 4.74), and named it olean.[132] Its racemate (±)-**116** can readily be synthesized as shown in Figure 4.74, and is pheromonally active.[132]

In June 1980, I was talking and tasting good wine at Professor Wittko Francke's home near Hamburg, Germany. We talked on olean, and noticed its axial chirality. We became interested in knowing the effect of axial chirality on its pheromone activity. After my return to Tokyo, we started the synthesis of olean enantiomers, as shown in Figure 4.75.

How can we synthesize the enantiomers of **116** with known axial chirality? In the case of the substituted spiroacetals such as 2,8-dimethyl-1.7-dioxaspiro[5.5]undecane (**88**, Figure 4.34), its stable conformation is that with two equatorial methyl groups [(2*S*,6*R*,8*S*)-**88**]. The absolute configuration at the spiro carbon atom will be fixed automatically owing to the oxygen anomeric effect. It therefore would be possible to synthesize the enantiomers of **116**, if one can tentatively attach substituents on the tetrahydropyranyl rings to control the stereochemistry at the spiro center. The substituents must be removed easily without causing any racemization at the spiro center. As such substituents we chose hydroxy groups, whose reductive removal is a well-established process. Consequently, (4*S*,6*S*,10*S*)-4,10-dihydroxy-1,7-dioxaspiro[5.5]undecane (**B** in Figure 4.75) and its enantiomer **B′** were the key intermediates in our 1984 synthesis of the enantiomers of **116**.[133,134]

Our synthesis started from (*S*)-malic acid, which furnished **A**. Removal of the acetonide and thioacetal groups of **A** afforded (4*S*,6*S*,10*S*)-**B** as crystals, whose structure was solved by X-ray analysis. The two hydroxy groups were equatorially oriented and the *S*-configured spiro center was established. Reductive removal of the two hydroxy groups afforded (*S*)-olean (**116**) as a volatile oil. (*R*)-Olean was also synthesized from (4*S*,6*S*,10*S*)-**B** by the following inversion process at the spiro center. The diol (4*S*,6*S*,10*S*)-**B** was oxidized to give crystalline and sublimable diketone **C**, which was reduced with lithium tri(*sec*-butyl)borohydride (L-selectride®) to give (4*R*,6*S*,10*R*)-**D**. This diol **D** was subjected to X-ray analysis, and

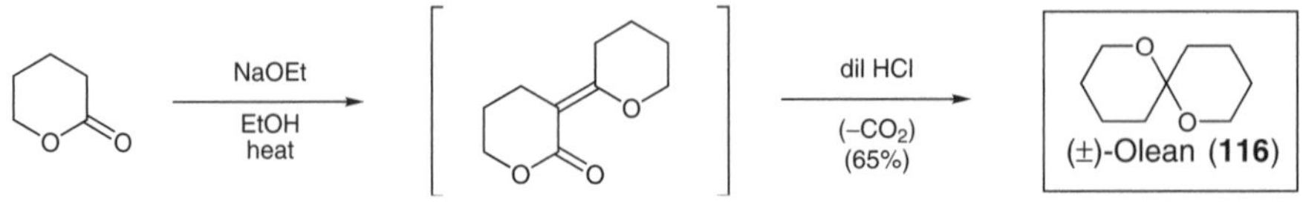

Figure 4.74   *Synthesis of (±)-olean. Modified by permission of Shokabo Publishing Co., Ltd*

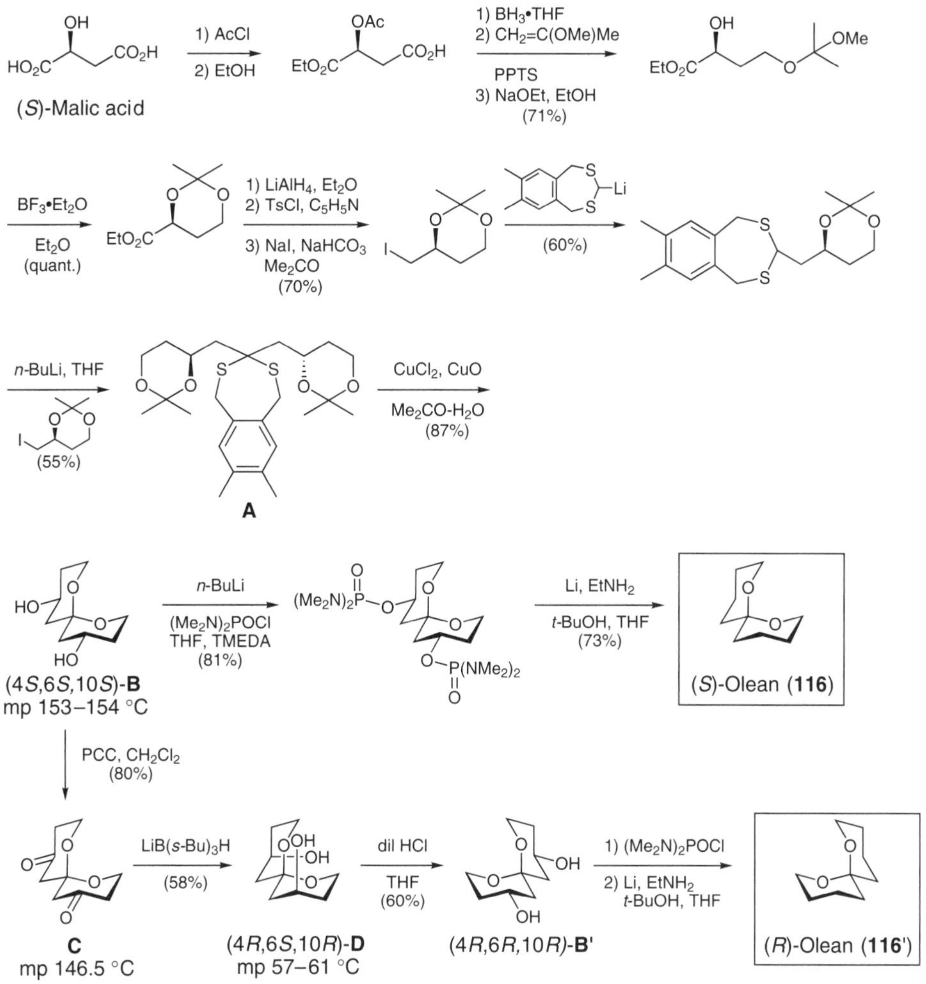

Figure 4.75   *Synthesis of the enantiomes of olean (1). Modified by permission of Shokabo Publishing Co., Ltd*

confirmed to possess the two hydroxy groups in axial orientation. Due to the presence of the two axial substituents, **D** was unstable, and isomerized under acidic conditions to the more stable (4*R*,6*R*,10*R*)-**B'**. This molecular acrobat allowed us to have the opposite enantiomer **B'** of (4*S*,6*S*,10*S*)-**B** by inversion of axial chirality at the spiro center. In other words, the two hydroxy groups served as handles to invert the configuration at the spiro center. Finally, removal of the two hydroxy groups of (4*R*,6*R*,10*R*)-**B'** gave (*R*)-olean (**116'**).

After the above success, we thought that even the presence of a single hydroxy group instead of two might be enough for fixing the configuration at the spiro center, and carried out another synthesis of the enantiomers of olean, as shown in Figure 4.76.[135] (*S*)-Malic acid was converted to (4*S*,6*S*)-**C** via **A**, and removal of the hydroxy group of (4*S*,6*S*)-**C** gave (*S*)-olean (**116**). (*R*)-Olean (**116'**) was also synthesized from (4*S*,6*S*)-**C** via ketone **D**, axial alcohol (4*R*,6*S*)-**B'** and equatorial alcohol (4*R*,6*R*)-**C'**. This second synthesis provided sufficient amounts of olean enantiomers for bioassay.

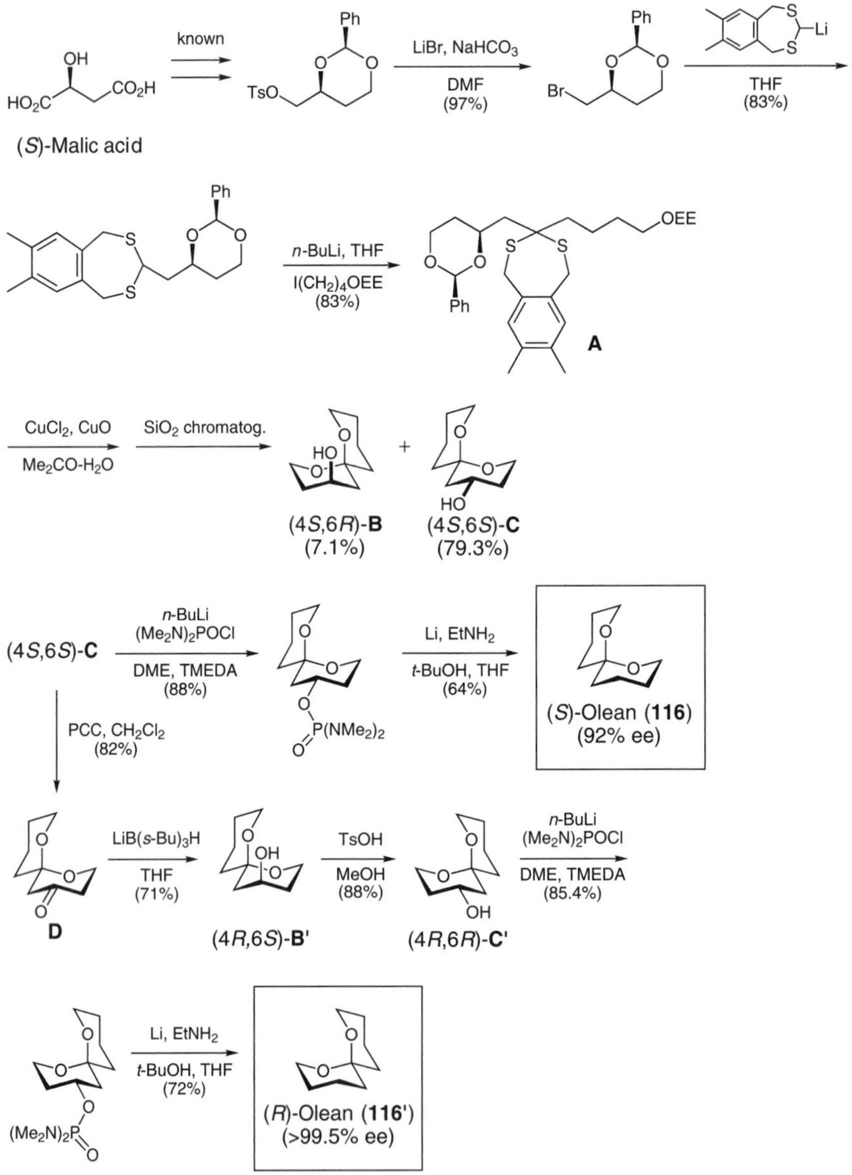

Figure 4.76   *Synthesis of the enantiomes of olean (2). Modified by permission of Shokabo Publishing Co., Ltd*

The enantiomers of olean were bioassayed in Greece by Haniotakis *et al.*[136] Surprisingly, (*R*)-**116′** was active against males, whereas (*S*)-**116** was active against females. GC analysis on a chiral stationary phase of natural olean by Schurig revealed it to be (±)-**116**.[136] Thus, the female-produced sex pheromone activates male olive fruit flies and the female herself. This is a very unusual stereochemistry–pheromone activity relationship, and is beyond our imagination.

4.5.9   13,23-Dimethylpentatriacontane as the sex pheromone of a tsetse fly

Tsetse flies are notorious vectors of African trypanosomiasis, well known as sleeping sickness in humans and as nagana in cattle. A female-produced sex stimulant pheromone of a tsetse fly (*Glossina pallidipes*) was identified by Whitehead *et al.* and also by Carlson *et al.* in 1981–1982 as 13,23-dimethylpentatriacontane (**117**, Figure 4.77). Due to the symmetrical nature of **117**, there are three stereoisomers: (13*R*,23*R*)-, (13*S*,23*S*)- and *meso*-**117**.

In the beginning, I thought either (13*R*,23*R*)- or (13*S*,23*S*)-**117** to be the optically and biologically active isomer, and the three stereoisomers of **117** were synthesized in 1983, as shown in Figure 4.77.[137] The synthesis started from enantiomerically pure (*R*)-citronellic acid, which was converted to the enantiomers of iodide **A**. Alkylation of the dianion of methyl acetoacetate with (*R*)- or (*S*)-**A** gave **B**, which was further alkylated with (*R*)- or (*S*)-**A** to give **C**. Three isomers of **C** were converted to (13*R*,23*R*)-, (13*S*,23*S*)- and *meso*-**117**, respectively. Bioassay of these three samples in Nairobi, Kenya, gave an interesting result. Neither (13*R*,23*R*)- nor (13*S*,23*S*)-**117** were pheromonally active, but optically inactive *meso*-**117** was bioactive as the sex stimulant pheromone of *G. pallidipes*.[138]

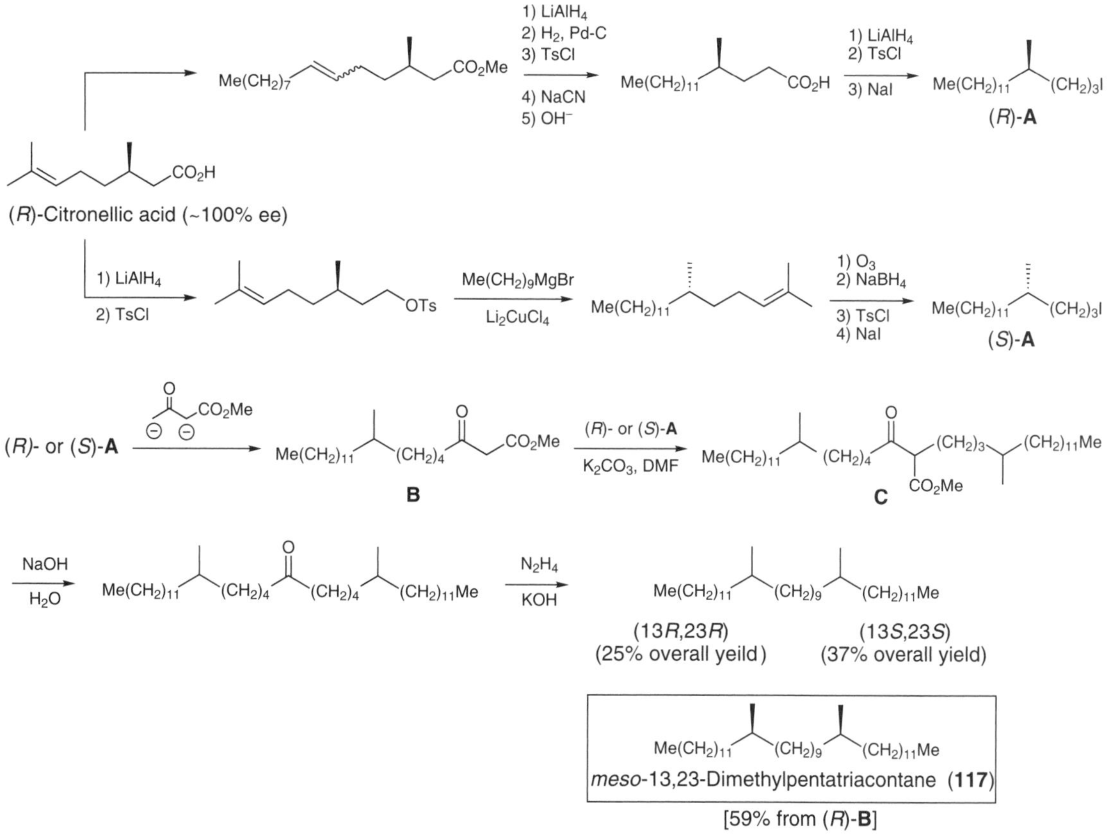

Figure 4.77 *Synthesis of the sex pheromone of a tsetse fly, Glossina pallidipes. Modified by permission of Shokabo Publishing Co., Ltd*

4.6   Significance of chirality in pheromone science

I continued my pheromone synthesis for nearly forty years to provide pure enantiomers of pheromones. As the results, absolute configurations could be assigned to many naturally occurring pheromones, and diverse stereochemistry–pheromone activity relationships could be determined. My comprehensive review on this subject was published in 2007.[9] Here in this section, a brief summary of stereochemistry–bioactivity relationships, classified into ten categories, will be given, as shown in Figure 4.78. It must be emphasized that these ten categories were found only through bioassay by using pure pheromone stereoisomers of synthetic origin.

(1) Only a single enantiomer is bioactive and its opposite enantiomer does not inhibit
the response to the active isomer
This is the most common relationship, and the majority (about 60%) of the chiral pheromones belong to this category. (1*R*,5*S*,7*R*)-*exo*-Brevicomin (**76**), the aggregation pheromone of the western pine beetle, and (3*S*,4*R*)-faranal (**95**), the trail pheromone of the Pharaoh's ant, are the typical members of this group.

(2) Only one enantiomer is bioactive, and its opposite enantiomer inhibits the response
to the pheromone
Response of the gypsy moth to the enantiomers of disparlure (**85**) showed that (7*R*,8*S*)-**85** was bioactive, while (7*S*,8*R*)-**85** was inhibitory. The very strong inhibitory action of the opposite (*S*)-isomer of japonilure (*R*)-**87** is remarkable. In practical application of these pheromones, their pure enantiomers have to be manufactured.

(3) Only one enantiomer is bioactive, and its diastereomer inhibits the response to the pheromone
As already discussed, response to (4*S*,6*S*,7*S*)-serricornin (**113**) can be inhibited by its diastereomer (4*S*,6*S*,7*R*)-**113′**, and that to (2*S*,3*R*,1′*R*)-stegobinone (**114**) by its (2*S*,3*R*,1′*S*)-diastereomer.

(4) The natural pheromone is a single enantiomer, and its opposite enantiomer
or diastereomer is also active
I explained this category in the case of the German cockroach pheromone (**104**), and also discussed the spined citrus bug pheromone [(3*R*,4*S*)-**74**], whose opposite enantiomer was as active as the pheromone itself. Females of the maritime pine scale (*Matsucoccus feytaudi*) use (3*S*,7*R*)-**118** as the sex pheromone. Its (3*R*,7*R*)-isomer also showed bioactivity similar to the natural pheromone, while *M. feytaudi* males responded very weakly to the two other stereoisomers. It therefore seems that only the stereochemistry at C-7 of **118** is important for the expression of pheromone activity.[139]

(5) The natural pheromone is a mixture of enantiomers or diastereomers, and both
the enantiomers or all the diastereomers are separately active
Females of the Douglas-fir beetle (*Dendroctonus pseudotsugae*) produce an average of a 55:45 mixture of (*R*)- and (*S*)-**119**. The combined effect of the enantiomers was additive rather than synergistic, and both enantiomers are required for maximum response.[140]

   The azuki bean beetle (*Callosobruchus chinensis*) uses callosobruchusic acid (**120**) as the pheromone. Although (*R*)-**120** is the major component of the natural pheromone (*R*/*S* = 3.3–3.4:1), (*R*)-**120** is only a half as active as (*S*)-**120**.[141,142]

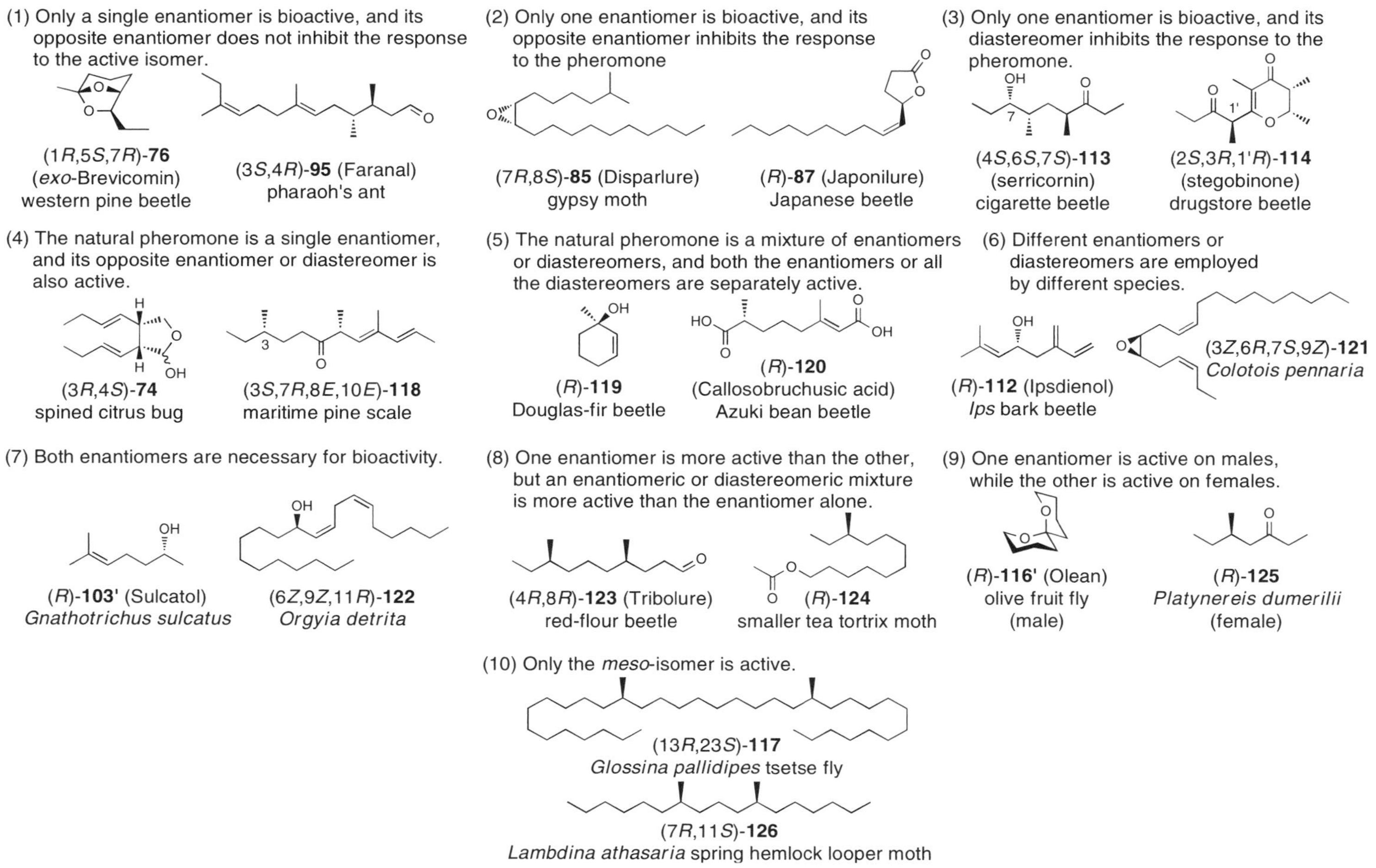

(1) Only a single enantiomer is bioactive, and its opposite enantiomer does not inhibit the response to the active isomer.

(1*R*,5*S*,7*R*)-**76**
(*exo*-Brevicomin)
western pine beetle

(3*S*,4*R*)-**95** (Faranal)
pharaoh's ant

(2) Only one enantiomer is bioactive, and its opposite enantiomer inhibits the response to the pheromone

(7*R*,8*S*)-**85** (Disparlure)
gypsy moth

(*R*)-**87** (Japonilure)
Japanese beetle

(3) Only one enantiomer is bioactive, and its diastereomer inhibits the response to the pheromone.

(4*S*,6*S*,7*S*)-**113**
(serricornin)
cigarette beetle

(2*S*,3*R*,1'*R*)-**114**
(stegobinone)
drugstore beetle

(4) The natural pheromone is a single enantiomer, and its opposite enantiomer or diastereomer is also active.

(3*R*,4*S*)-**74**
spined citrus bug

(3*S*,7*R*,8*E*,10*E*)-**118**
maritime pine scale

(5) The natural pheromone is a mixture of enantiomers or diastereomers, and both the enantiomers or all the diastereomers are separately active.

(*R*)-**119**
Douglas-fir beetle

(*R*)-**120**
(Callosobruchusic acid)
Azuki bean beetle

(6) Different enantiomers or diastereomers are employed by different species.

(*R*)-**112** (Ipsdienol)
Ips bark beetle

(3*Z*,6*R*,7*S*,9*Z*)-**121**
Colotois pennaria

(7) Both enantiomers are necessary for bioactivity.

(*R*)-**103'** (Sulcatol)
Gnathotrichus sulcatus

(6*Z*,9*Z*,11*R*)-**122**
Orgyia detrita

(8) One enantiomer is more active than the other, but an enantiomeric or diastereomeric mixture is more active than the enantiomer alone.

(4*R*,8*R*)-**123** (Tribolure)
red-flour beetle

(*R*)-**124**
smaller tea tortrix moth

(9) One enantiomer is active on males, while the other is active on females.

(*R*)-**116'** (Olean)
olive fruit fly
(male)

(*R*)-**125**
Platynereis dumerilii
(female)

(10) Only the *meso*-isomer is active.

(13*R*,23*S*)-**117**
Glossina pallidipes tsetse fly

(7*R*,11*S*)-**126**
Lambdina athasaria spring hemlock looper moth

Figure 4.78   *Sterochemistry–pheromone activity relationships*

(6) Different enantiomers or diastereomers are employed by different species

I already discussed the case of ipsdienol (**112**). Chirality is important to discriminate between two species of the winter-flying geometrid moths in Middle Europe. Thus, (6*R*,7*S*)-**121** is the pheromone of *Colotois pennaria*, while *Erannis defoliaria* uses (6*S*,7*R*)-**121′** as its pheromone.[143]

(7) Both enantiomers are necessary for bioactivity

The sulcatol (**103**) story was already described. The tussock moth *Orgyia detrita* uses a 1:3.5 mixture of (*R*)-**122** and (*S*)-**122** as its pheromone.[144]

(8) One enantiomer is more active than the other, but an enantiomeric or diastereomeric mixture is more active than the enantiomer alone

Tribolure [(4*R*,8*R*)-**123**] is the male-produced aggregation pheromone of the red-flour beetle (*Tribolium castaneum*). It was found that (4*R*,8*R*)-**123** was as active as the natural pheromone, while a mixture of (4*R*,8*R*)-**123** and its (4*R*,8*S*)-isomer in a ratio of 4:1 was about ten-fold more active than (4*R*,8*R*)-**123** alone.[145]

The smaller tea tortrix moth (*Adoxophyes honmai*) uses (*R*)-**124** as a minor component of its pheromone bouquet, and (*R*)-**124** was slightly more active than (*S*)-**124**. Further field tests suggest that there is an optimum *R/S* ratio of 95:5 for trapping of males.[146]

(9) One enantiomer is active on males, while the other is active on females

The story of olean (**116**) was given already. Another example is 5-methyl-3-heptanone (**125**), which is the pheromone in the coelomic fluid of gravid specimens of Nereid marine polychaetes (*Platynereis dumerilii*). It is responsible for the induction of the nuptial dance behavior prior to the release of gametes in *P. dumerilii*. The female-produced (*S*)-**125** attracts the males, while the male-produced (*R*)-**125** is active on females.[147,148]

(10) Only the *meso*-isomer is active

I already described the story of the tsetse fly pheromone (**117**). The female-produced sex pheromone components of the spring hemlock looper moth (*Lambdina athasaria*) are 7-methylheptadecane and 7,11-dimethylheptadecane (**126**). After the synthesis of all of their stereoisomers, a mixture of (*S*)-7-methylheptadecane and (7*R*,11*S*)-**116** (*meso*-**116**) was found to be pheromonally active.[149]

Extensive joint works by biologists and chemists revealed that diversity is the keyword of pheromone response. I never dreamed of such diversity when I began my pheromone research. At present, this kind of diversity can be clarified only through experiments. It is therefore a prerequisite to study the relationship between stereochemistry (including *cis/trans*-isomerism) and bioactivity, if we want to use a pheromone practically.

Pheromone science is truly interdisciplinary. Without cooperation between biologists and chemists, no good result will emerge. Mutual respect on both sides will make the cooperation fruitful.

References

1. Butenandt, A.; Beckmann, R.; Stamm, D.; Hecker, E. *Z. Naturforsch.* **1959**, *14b*, 283–284.
2. Karlson, P.; Lüscher, M. *Nature* **1959**, *183*, 55–56.
3. Mori, K. In *Recent Developments in the Chemistry of Natural Carbon Compounds*, Bognár, R.; Bruckner, V.; Szantáy, C., eds., Akadénuau Kiadó, Budapest, 1979, Vol. 9, pp. 9–123.

4. Mori, K. In *The Total Synthesis of Natural Products*, ApSimon, J., ed., John Wiley, New York, 1981, Vol. 4, pp. 1–183.

5. Mori, K. In *The Total Synthesis of Natural Products*, ApSimon, J., ed., John Wiley, New York, 1992, Vol. 9, pp. 1–534.

6. Mori, K. *In The Chemistry of Pheromones and Other Semiochemicals I*, Schulz, S., ed., *Topics in Current Chemistry* **2004**, *239*, 1–50, Springer, Berlin.

7. Mori, K. *Tetrahedron* **1989**, *45*, 3233–3298.

8. Mori, K. In *Chirality in Natural and Applied Science*, Lough, W.J.; Wainer, J.W., eds., Blackwell and CRC, Osney Mead and Boca Raton, 2002, pp. 241–259.

9. Mori, K. *Bioorg. Med. Chem.* **2007**, *15*, 7505–7523.

10. Lichtenthaler, F.W. *Eur. J. Org. Chem.* **2002**, 4095–4122.

11. Christmann, M.; Bräse, S. *Asymmetric Synthesis–The Essentials*, 2nd edn, Wiley-VCH, Weinheim, 2008, pp. 355.

12. Mori, K. In *Methods in Chemical Ecology, Chemical Methods*, Millar, J.G., Haynes, K.F., eds., Kluwer, Norwell, 1998, pp. 295–338.

13. Dale, J.A.; Dull, D.L.; Mosher, H.S. *J. Org. Chem.* **1969**, *34*, 2543–2549.

14. Dale, J.A.; Mosher, H.S. *J. Am. Chem. Soc.* **1973**, *95*, 512–519.

15. Mori, K. *Tetrahedron Lett.* **1973**, 3869–3872.

16. Mori, K. *Tetrahedron* **1974**, *30*, 3817–3820.

17. Mori, K.; Suguro, T.; Uchida, M. *Tetrahedron* **1978**, *34*, 3119–3123.

18. Mori, K.; Kuwahara, S.; Levinson, H.Z.; Levinson, A.R. *Tetrahedron* **1982**, *38*, 2291–2297.

19. Silverstein, R.M.; Cassidy, R.F.; Burkholder, W.E.; Shapas, T.J.; Levinson, H.Z.; Levinson, A.R.; Mori, K. *J. Chem. Ecol.* **1980**, *6*, 911–917.

20. Levinson, H.Z.; Levinson, A.R.; Mori, K. *Naturwissenschaften* **1981**, *67*, 480–481.

21. Mori, K. *Tetrahedron* **2009**, *65*, 3900–3909.

22. Mori, K.; Amaike, M.; Oliver, J.E. *Liebigs Ann. Chem.* **1992**, 1185–1190.

23. Mori, K.; Amaike, M.; Watanabe, H. *Liebigs Ann. Chem.* **1993**, 1287–1294.

24. James, D.G.; Mori, K. *J. Chem. Ecol.* **1995**, *21*, 403–406.

25. Amaike, M.; Mori, K. *Liebigs Ann.* **1995**, 1451–1454.

26. Mori, K. *Tetrahedron:Asymmetry* **2007**, *18*, 838–846.

27. Mori, K.; Tashiro, T.; Yoshimura, T.; Takita, M.; Tabata, J.; Hiradate, S.; Sugie, H. *Tetrahedron Lett.* **2008**, *49*, 354–357.

28. Tashiro, T.; Mori, K. *Tetrahedron:Asymmetry* **2008**, *19*, 1215–1223.

29. Hodgson, D.M.; Chung, Y.K.; Nuzzo, I.; Freixas, G.; Kulikiewicz, K.K.; Cleator, E.; Paris, J.-M. *J. Am. Chem. Soc.* **2007**, *129*, 4456–4462.

30. Mori, K. *Tetrahedron* **1974**, *30*, 4223–4227.

31. Silverstein, R.M.; Brownlee, R.G.; Bellas, T.E.; Wood, D.L.; Browne, L.E. *Science* **1968**, *159*, 889–891.

32. Wood, D.L.; Browne, L.E.; Ewing, B.; Lindahl, K.; Bedard, W.D.; Tilden, P.E.; Mori, K.; Pitman, G.B.; Hughes, P.R. *Science* **1976**, *192*, 896–898.

33. Mori, K.; Sasaki, M. *Tetrahedron Lett.* **1979**, 1329–1332.

34. Mori, K.; Sasaki, M. *Tetrahedron* **1980**, *36*, 2197–2208.

35. Mori, K.; Uematsu, T.; Minobe, M.; Yanagi, K. *Tetrahedron* **1983**, *39*, 1735–1743.

36. Schurig, V.; Weber, R.; Klimetzek, D.; Kohnle, U.; Mori, K. *Naturwissenschaften* **1982**, *69*, 602–603.

37. Persoons, C.J.; Ritter, F.J.; Verwiel, P.E J.; Hauptmann, H.; Mori, K. *Tetrahedron Lett.* **1990**, *31*, 1747–1750.

38. Kitahara, T.; Mori, M.; Mori, K. *Tetrahedron* **1987**, *43*, 2689–2699.

39. Kuwahara, S.; Mori, K. *Tetrahedron* **1990**, *46*, 8075–8082.

40. Mori, K.; Kuwahara, S.; Igarashi, Y. *Pure Appl. Chem.* **1990**, *62*, 1307–1310.

41. Mori, K.; Igarashi, Y. *Tetrahedron Lett.* **1989**, *30*, 5145–5148.

42. Mori, K.; Igarashi, Y. *Tetrahedron* **1990**, *46*, 5101–5112.

43. Kuwahara, S.; Mori, K. *Tetrahedron Lett.* **1989**, *30*, 7447–7450.

44. Kuwahara, S.; Mori, K. *Tetrahedron* **1990**, *46*, 8083–8092.
45. Kurosawa, S.; Bando, M.; Mori, K. *Eur. J. Org. Chem.* **2001**, 4395–4399.
46. Tashiro, T.; Kurosawa, S.; Mori, K. *Biosci. Biotechnol. Biochem.* **2004**, *68*, 663–670.
47. Muto, S.; Bando, M.; Mori, K. *Eur. J. Org. Chem.* **2004**, 1946–1952.
48. Tóth, M.; Csonka, E.; Bartelt, R.J.; Cossé, A.A.; Zilkowski, B.W.; Muto, S.; Mori, K. *J. Chem. Ecol.* **2005**, *31*, 2705–2720.
49. Mori, K. *Tetrahedron:Asymmetry* **2005**, *16*, 685–692 (Corrigendum: *Tetrahedron:Asymmetry* **2005**, *16*, 1721).
50. Mori, K.; Mizumachi, N.; Matsui, M. *Agric. Biol. Chem.* **1976**, *40*, 1611–1615.
51. Mori, K.; Takigawa, T.; Matsui, M. *Tetrahedron Lett.* **1976**, 3953–3956.
52. Mori, K.; Takigawa, T.; Matsui, M. *Tetrahedron* **1979**, *35*, 833–837.
53. Vité, J.P.; Klimetzek, D.; Loskant, G.; Hedden, R.; Mori, K. *Naturwissenschaften* **1976**, *63*, 582–583.
54. Miller, J.R.; Mori, K.; Roelofs, W.L. *J. Insect Physiol.* **1977**, *23*, 1447–1453.
55. Mori, K.; Ebata, T. *Tetrahedron Lett.* **1981**, *22*, 4281–4282.
56. Mori, K.; Ebata, T. *Tetrahedron* **1986**, *42*, 3471–3478.
57. Brevet, J.-L.; Mori, K. *Synthesis* **1992**, 1007–1012.
58. Muto, S.; Mori, K. *Eur. J. Org. Chem.* **2003**, 1300–1307.
59. Szöcs, G; Tóth, M.; Mori, K. *Chemoecology* **2005**, *15*, 127–128.
60. Tumlinson, J.H.; Klein, M.G.; Doolittle, R.E.; Ladd, T.L.; Proveaux, A.T. *Science* **1977**, *197*, 789–792.
61. Leal, W.S. *Proc. Natl. Acad. Sci. USA* **1996**, *93*, 12112–12113.
62. Senda, S.; Mori, K. *Agric. Biol. Chem.* **1983**, *47*, 2595–2598.
63. Tengö, J.; Agren, L.; Baur, B.; Isaksson, R.; Lilijefors, T.; Mori, K.; König, W.; Francke, W. *J. Chem. Ecol.* **1990**, *16*, 429–441.
64. Mori, K.; Tanida, K. *Heterocycles* **1981**, *15*, 1171–1174.
65. Mori, K.; Tanida, K. *Tetrahedron* **1981**, *37*, 3221–3225.
66. Isaksson, R.; Lilijefors, T.; Reinholdsson, P. *J. Chem. Soc., Chem. Commun.* **1984**, 137–138.
67. Mori, K.; Watanabe, H. *Tetrahedron* **1986**, *42*, 295–304.
68. Sugai, T.; Fujita, M.; Mori, K. *Nippon Kagaku Kaishi (J. Chem. Soc. Jpn.)* **1983**, 1315–1321.
69. Perkins, M.V.; Kitching, W.; König, W.A.; Drew, R.A I. *J. Chem. Soc., Perkin Trans. 1*, **1990**, 2501–2506.
70. Mori, K.; Nakazono, Y. *Tetrahedron* **1986**, *42*, 283–290.
71. Mori, K.; Khlebnikov, V. *Liebigs Ann. Chem.* **1993**, 77–82.
72. Mori, K.; Puapoomchareon, P. *Liebigs Ann. Chem.* **1991**, 1053–1056.
73. Mori, K.; Kuwahara, S. *Tetrahedron* **1986**, *42*, 5545–5550.
74. Mori, K.; Kuwahara, S. *Tetrahedron* **1986**, *42*, 5539–5544.
75. Mori, K.; Mori, H.; Sugai, T. *Tetrahedron* **1985**, *41*, 919–925.
76. Tóth, M.; Buser, H.R.; Peña, A.; Arn, H.; Mori, K.; Takeuchi, T.; Nikolaeva, L.N.; Kovalev, B.G. *Tetrahedron Lett.* **1989**, *30*, 3405–3408.
77. Mori, K.; Takeuchi, T. *Liebigs Ann. Chem.* **1989**, 453–457.
78. Nakanishi, A.; Mori, K. *Biosci. Biotechnol. Biochem.* **2005**, *69*, 1007–1013.
79. Muto, S.; Mori, K. *Eur. J. Org. Chem.* **2001**, 4635–4638.
80. Mori, K.; Murata, N. *Liebigs Ann.* **1995**, 2089–2092.
81. Sano, S.; Mori, K. *Eur. J. Org. Chem.* **1999**, 1679–1686.
82. Tashiro, T.; Bando, M.; Mori, K. *Synthesis* **2000**, 1852–1862.
83. Spiegel, C.N.; Jeanbourquin, P.; Guerin, P.M.; Hooper, A.M.; Claude, S.; Tabacci, R.; Sano, S.; Mori, K. *J. Insect Physiol.* **2005**, *51*, 1366–1375.
84. Muto, S.; Nishimura, Y.; Mori, K. *Eur. J. Org. Chem.* **1999**, 2159–2165.
85. Kurosawa, S.; Mori, K. *Eur. J. Org. Chem.* **2000**, 955–962.
86. Hamilton, J.G C.; Hooper, A.M.; Ibbotson, H.C.; Kurosawa, S.; Mori, K.; Muto, S.; Pickett, J.A. *Chem. Commun.* **1999**, 2335–2336.
87. Mori, K. *Tetrahedron* **1975**, *31*, 1381–1384.
88. Sullivan, B.T.; Shepherd, W.P.; Pureswaran, D.S.; Tashiro, T.; Mori, K. *J. Chem. Ecol.* **2007**, *33*, 1510–1527.
89. Nishimura, Y.; Mori, K. *Eur. J. Org. Chem.* **1998**, 233–236.
90. Rasmussen, L.E L.; Lee, T.D.; Roelofs, W.L.; Zhang, A.; Daves, Jr., G.D. *Nature* **1996**, *379*, 684.

91. Greenwood, D.R.; Comeskey, D.; Hunt, M.B.; Rasmussen, L.E L. *Nature* **2005**, *438*, 1097–1098.
92. Wiesler, D.P.; Schwende, F.J.; Carmack, M.; Novotny, M. *J. Org. Chem*. **1984**, *49*, 882–884.
93. Mori, K.; Seu, Y.-B. *Tetrahedron* **1986**, *42*, 5901–5904.
94. Tashiro, T.; Mori, K. *Eur. J. Org. Chem*. **1999**, 2167–2173.
95. Tashiro, T.; Osada, K.; Mori, K. *Biosci. Biotechnol. Biochem*. **2008**, *72*, 2398–2402.
96. Osada, K.; Tashiro, T.; Mori, K.; Izumi, H. *Chem. Senses* **2008**, *33*, 815–823.
97. Mori, K. *Tetrahedron* **1975**, *31*, 3011–3012.
98. Borden, J.H.; Chong, L.; McLean, J.A.; Slessor, K.N.; Mori, K. *Science* **1976**, *192*, 894–896.
99. Mori, K. *Tetrahedron* **1981**, *37*, 1341–1342.
100. Mori, K.; Puapoomchareon, P. *Liebigs Ann. Chem*. **1987**, 271–272.
101. Mori, K.; Suguro, T.; Masuda, S. *Tetrahedron Lett*. **1978**, 3447–3450.
102. Mori, K.; Masuda, S.; Suguro, T. *Tetrahedron* **1981**, *37*, 1329–1340.
103. Mori, K.; Takikawa, H. *Tetrahedron* **1990**, *46*, 4473–4486.
104. Eliyahu, D.; Mori, K.; Takikawa, H.; Leal, W.S.; Schal, C. *J. Chem. Ecol*. **2004**, *30*, 1839–1848.
105. Mori, K. *Tetrahedron* **2008**, *64*, 4060–4071.
106. Eliyahu, D.; Nojima, S.; Mori, K.; Schal, C. *J. Chem. Ecol*. **2008**, *34*, 229–237.
107. Mori, K.; Takenaka, M. *Eur. J. Org. Chem*. **1998**, 2181–2184.
108. Morikawa, Y.; Takayama, S.; Fudo, R.; Yamanaka, S.; Mori, K.; Isogai, A. *FEMS Microbiol. Lett*. **1998**, *165*, 29–34.
109. Domon, K.; Mori, K. *Eur. J. Org. Chem*. **1999**, 979–980.
110. Epstein, O.L.; Kulinkovich, O.G. *Tetrahedron Lett*. **2001**, *42*, 3757–3758.
111. Mori, K. *Tetrahedron Lett*. **1975**, 2187–2190.
112. Mori, K. *Tetrahedron* **1976**, *32*, 1101–1106.
113. Mori, K.; Takigawa, T.; Matsuo, T. *Tetrahedron* **1979**, *35*, 933–940.
114. Vité, J.P.; Hedden, R.; Mori, K. *Naturwissenschaften* **1976**, *63*, 43–44.
115. Mori, K. *Tetrahedron Lett*. **1976**, 1609–1612.
116. Mori, K.; Takikawa, H. *Tetrahedron* **1991**, *47*, 2163–2168.
117. Mori, K.; Nomi, H.; Chuman, T.; Kohno, M.; Kato, K.; Noguchi, M. *Tetrahedron Lett*. **1981**, *22*, 1127–1130.
118. Mori, K.; Nomi, H.; Chuman, T.; Kohno, M.; Kato, K.; Noguchi, M. *Tetrahedron* **1982**, *38*, 3705–3711.
119. Mori, K.; Watanabe, H. *Tetrahedron* **1985**, *41*, 3423–3428.
120. Sakakibara, M.; Mori, K. *Tetrahedron Lett*. **1979**, 2401–2402.
121. Mori, K.; Ebata, T.; Sakakibara, M. *Tetrahedron* **1981**, *37*, 709–713.
122. Ansell, J.M.; Hassner, A.; Burkholder, W.E. *Tetrahedron Lett*. **1979**, 2497–2498.
123. Hoffmann, R.W.; Ladner, W.; Steinbach, K.; Massa, W.; Schmidt, R.; Snatzke, G. *Chem. Ber*. **1981**, *114*, 2780–2801.
124. Mori, K; Ebata, T. *Tetrahedron* **1986**, *42*, 4413–4420.
125. Oppolzer, W.; Rodriguez, I. *Helv. Chim. Acta* **1993**, *76*, 1275–1281.
126. Matteson, D.S.; Man, H.-W.; Ho, O.C. *J. Am. Chem. Soc*. **1996**, *118*, 4560–4566.
127. Mori, K.; Sano, S.; Yokoyama, Y.; Bando, M.; Kido, M. *Eur. J. Org. Chem*. **1998**, 1135–1141.
128. Mori, K.; Takeuchi, Y. *Nat. Prod. Lett*. **1995**, *5*, 275–280.
129. Mori, K.; Takeuchi, Y. *Proc. Jpn. Acad. Ser. B* **1994**, *70*, 143–145.
130. Fujita, K.; Mori, K. *Eur. J. Org. Chem*. **2001**, 493–502.
131. Gemeno, C.; Leal, W.S.; Mori, K.; Schal, C. *J. Chem. Ecol*. **2003**, *29*, 1797–1811.
132. Baker, R.; Herbert, R.; Howse, P.E.; Jones, O.T.; Francke, W.; Reith, W. *J. Chem. Soc., Chem. Commun*. **1980**, 52–53.
133. Mori, K.; Uematsu, T.; Watanabe, H.; Yanagi, K.; Minobe, M. *Tetrahedron Lett*. **1984**, *25*, 3875–3878.
134. Mori, K; Uematsu, T.; Yanagi, K.; Minobe, M. *Tetrahedron* **1985**, *41*, 2751–2758.
135. Mori, K.; Watanabe, H.; Yanagi, K.; Minobe, M. *Tetrahedron* **1985**, *41*, 3663–3672.
136. Haniotakis, G.; Francke, W.; Mori, K.; Redlich, H.; Schurig, V. *J. Chem. Ecol*. **1986**, *12*, 1559–1568.
137. Kuwahara, S.; Mori, K. *Agric. Biol. Chem*. **1983**, *47*, 2599–2606.
138. McDowell, P.G.: Hassanali. A.; Dransfield, R. *Physiol. Entomol*. **1985**, *10*, 183–190.
139. Jactel, H.; Manassieu, P.; Letteré, M.; Mori, K.; Einhorn, J. *J. Chem. Ecol*. **1994**, *20*, 2159–2170.

140. Lindgren, B.S.; Gries, G.; Pierce, Jr., H.D.; Mori, K. *J. Chem. Ecol.* **1992**, *18*, 1201–1208.
141. Yajima, A.; Akasaka, K.; Yamamoto, M.; Ohmori, S.; Nukada, T.; Yabuta, G. *J. Chem. Ecol.* **2007**, *33*, 1328–1335.
142. Mori, K.; Ito, T.; Tanaka, K.; Honda, H.; Yamamoto, I. *Tetrahedron* **1983**, *39*, 2303–2306.
143. Szöcs, G.; Tóth, M.; Francke, W.; Schmidt, F.; Philipp, P.; König, W.A.; Mori, K.; Hansson, B.S.; Löfstedt, C. *J. Chem. Ecol.* **1993**, *19*, 2721–2735.
144. Gries, R.; Khaskin, G.; Khaskin, E.; Foltz, J.L.; Schaefer, P.W.; Gries, G. *J. Chem. Ecol.* **2003**, *29*, 2201–2212.
145. Suzuki, T.; Kozaki, J.; Sugawara, R.; Mori, K. *Appl. Entomol. Zool.* **1984**, *19*, 15–20.
146. Tamaki, Y.; Noguchi, H.; Sugie, H.; Kariya, A.; Arai, S.; Ohba, M.; Terada, T.; Suguro, T.; Mori, K. *Jpn. J. Appl. Entomol. Zool.* **1980**, *24*, 221–228.
147. Zeeck, E.; Hardege, J.D.; Willig, A.; Krebber, R.; König, W.A. *Naturwissenschaften* **1992**, *79*, 182–183.
148. Hardege, J.D. *Hydrobiologia* **1999**, *402*, 145–161.
149. Duff, C.M.; Gries, G.; Mori, K.; Shirai, Y.; Seki, M.; Takikawa, H.; Sheng, T.; Slessor, K.N.; Gries, R.; Maier, C.T.; Ferguson, D.C. *J. Chem. Ecol.* **2001**, *27*, 431–442.

5

Synthesis of Biofunctional Molecules of Microbial Origin

Micro-organisms produce hormones and pheromones so as to regulate their own lives. They also generate diverse secondary metabolites such as toxins and antibiotics to control the lives of the organisms. Some of the secondary metabolites possess rather complicated structures, and have attracted the attention of chemists. Medicinally important antibiotics were especially well studied. This chapter describes our synthetic works on biofunctional molecules of microbial origin.

5.1 Microbial hormones

Since ancient times we human beings have utilized micro-organisms for our benefit. In every country we have our own fermentation industries to produce fermented foods such as cheese, soysauce (shoyu), and miso (fermented soybeans). Alcoholic beverages such as beer, saké (rice wine), wine and whiskey are also the products of fermentation industries. Antibiotics are indispensable for our medical care.

Micro-organisms, however, do not exist here on earth just to be utilized by us. They have been here on earth prior to mankind, and have continuously produced biofunctional molecules to develop and regulate their own lives. Scarcity of these highly bioactive molecules makes their isolation and identification difficult, and only after the 1970s did their chemical studies become an important branch of natural products chemistry to clarify the morphogenesis of micro-organisms such as sporulation and fruiting-body formation. The present section describes our synthetic studies on microbial hormones.

5.1.1 A-factor

In autumn of 1980 on my flight to Sapporo, Hokkaido, I learned of a substance called "A-factor", when I was reading a Japanese book "Biseibutsu to Hakkoseisan (Micro-organisms and Fermentation Manufacturing)" written by Samejima and Nara. According to Dr. T. Nara of Kyowa Hakko Co., "A-factor (**127**, Figure 5.1) induces streptomycin production by *Actinomyces streptomycini*. As the structure **127** of A-factor indicates, it cannot be a precursor of streptomycin. It seems to work as a hormone to induce streptomycin biosynthesis." I noticed that the absolute configuration of A-factor was not established yet by synthesis, and wanted to know more about this compound. I therefore asked my friend Professor T. Beppu

---

Chemical Synthesis of Hormones, Pheromones and Other Bioregulators   Kenji Mori
© 2010 John Wiley & Sons, Ltd

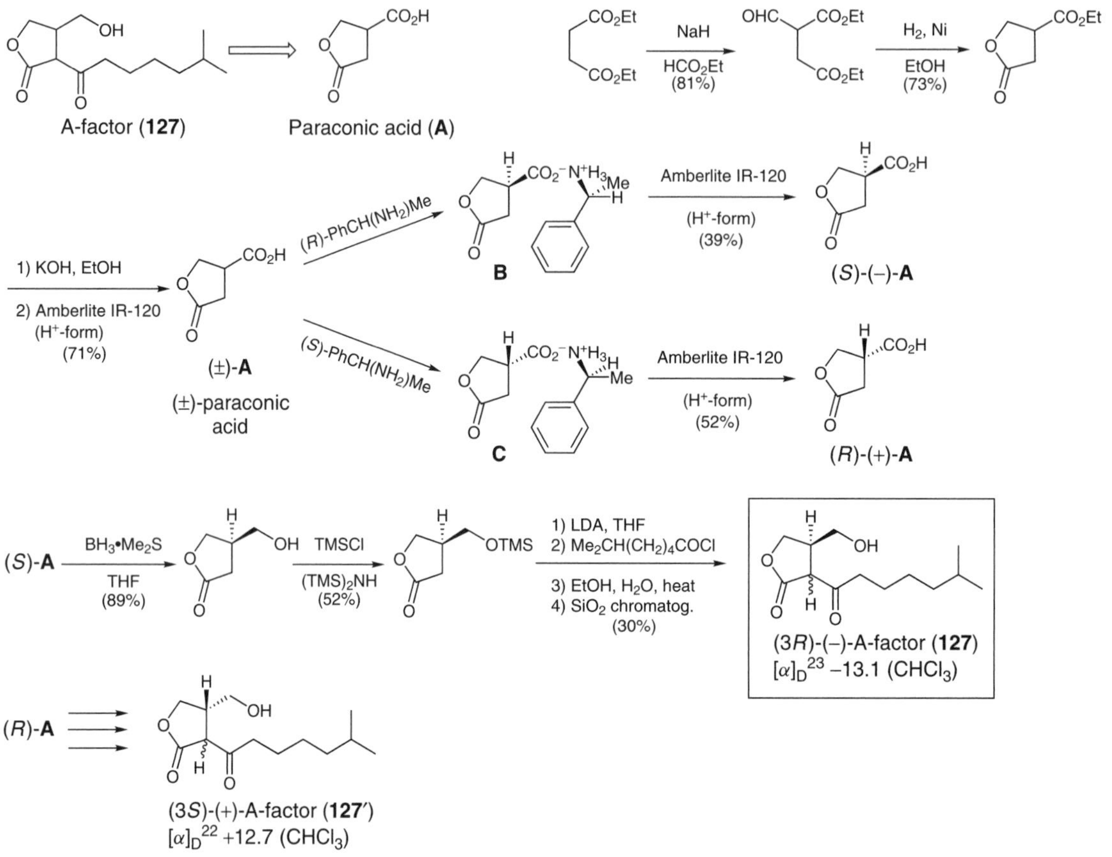

Figure 5.1 *Synthesis of A-factor and its opposite enantiomer. Modified by permission of Shokabo Publishing Co., Ltd*

at the University of Tokyo: "Do you know what A-factor is?" He was surprised, because he was actually studying the biology of A-factor. We agreed to cooperate, and I promised to synthesize A-factor (**127**). I often read books in biology. By doing so I can broaden my perspectives in science.

Khokhlov *et al.* in Moscow discovered A-factor in 1976 as the inducer of the biosynthesis of streptomycin in inactive (= with no ability to generate streptomycin) mutants of *Streptomyces griseus*. Beppu and Hara rediscovered A-factor in 1980 as a compound to restore the biosynthetic ability of a mutant of *S. griseus* inactivated by X-ray irradiation. Khokhlov also found that A-factor induced the formation of spores in asporophological modifications of *S. griseus*. The structure of A-factor was assigned as **127** by chemical and spectroscopic studies, and supported by the synthesis of (±)-**127**. The absolute configuration of the naturally occurring (−)-A-factor was first proposed as $2S, 3S$ by Khokhlov. The keto–enol equilibrium of A-factor (**127**) at C-2, however, made me question that conclusion.

In 1981, I synthesized (−)-A-factor (**127**) as shown in Figure 5.1.[1] The starting material was (−)-paraconic acid (**A**) obtained by enantiomer separation (optical resolution) of (±)-**A**. In the next year in cooperation with Mr. K. Yamane both (−)-**127** and (+)-**127′** were synthesized from the enantiomers of paraconic acid. The identity of (−)-**127** with the natural A-factor was established by their identical spectral properties including the circular dichroism (CD) spectra.[2]

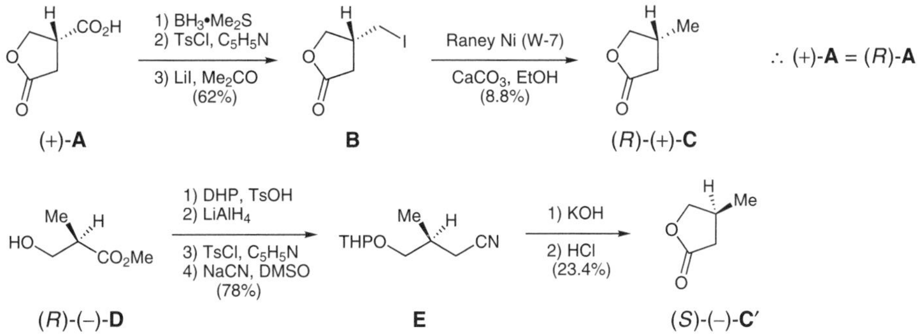

Figure 5.2 *(+)-Paraconic acid possesses R-configuration. Modified by permission of Shokabo Publishing Co., Ltd*

Publication of the above synthetic work made me relax, believing that the A-factor research must have been concluded. In January 1983, I received a letter from Dr. J. Buckingham, the editor of "Dictionary of Organic Compounds" and "Atlas of Stereochemistry." He informed me that there are two papers with different conclusions on the absolute configuration of paraconic acid, and advised me to check our conclusion of the absolute configuration of (−)-A-factor. Indeed, in 1965 Tocanne and Asselineau gave (R)-configuration to (−)-paraconic acid, while in 1968 they revised it to S.

In order to determine the absolute configuration of (+)-paraconic acid beyond doubt, I correlated it to methyl (R)-3-hydroxy-2-methylpropanoate (**D**), as shown in Figure 5.2.[3] Namely, (+)-paraconic acid (**A**) was converted to (+)-3-methyl-4-butanolide (**C**) via **B**. On the other hand, methyl (R)-3-hydroxy-2-methylpropanoate (**D**) was converted to (S)-**C′** via **E**. The (S)-lactone (**C′**) was levorotatory. The absolute configuration of (+)-**C** must therefore be R, and the starting (+)-paraconic acid (**A**) must possess the R-configuration. A number of monographs and reference works had adopted the 1965 conclusion of Tocanne and Asselineau that (+)-paraconic acid must be S-configurated. Earlier in 1981 I believed that description, and reported that the natural (−)-A-factor must be with S-configuration.[1,2] The earlier S-conclusion was revised in 1983 as R.[3] The configuration of A-factor at C-2 is not fixed due to enolization of the ketone carbonyl group. Dr. Buckingham's advice was crucial in revising the (2S,3S)-configuration proposed by Khokhlov to the present (3R)-configuration. We should not rely too much on the secondary literature like reference works.

The hormone activity of our synthetic enantiomers of A-factor was studied by Hara and Beppu by measuring the streptomycin production upon addition of the enantiomers **127** and **127′** to *S. griseus* FT-1 strain, which does not biosynthesize A-factor. The bioactivity of the naturally occurring (3R)-(−)-A-factor (**127**) was 2.6 times higher than that of the unnatural (3S)-(+)-A-factor (**127′**).[4] Later, Miyake of Beppu's group synthesized tritium-labelled A-factor, and identified the A-factor binding protein of *S. griseus*.[5]

Subsequently in 1989, we prepared (S)-(−)-paraconic acid by lipase-catalysed reaction,[6] and synthesized other microbial hormones with structural similarity to A-factor.[7] The biological significance of A-factor in the life cycle of *Streptomyces griseus* was studied in detail by Horinouchi and Beppu.[8]

5.1.2 Sch II and relatives, the fruiting-inducing cerebrosides

Fruiting-body formation in *Basidiomycetes* is indeed a spectacular phenomenon especially to those who love to taste mushrooms. In every country mushrooms are highly appreciated by gourmands. The mechanism of fruiting-body formation, however, is still a mystery in spite of the intensive efforts to clarify it.

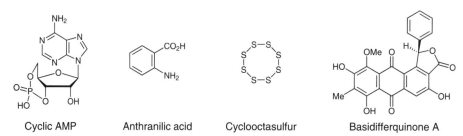

Figure 5.3   *Structures of fruiting-body inducers for mushrooms. Reprinted with permission of Shokabo Publishing Co., Ltd*

Figure 5.3 lists four substances that have been identified as fruiting inducers. In 1973 cyclic AMP was identified by Uno and Ishikawa as a fruiting inducer of a certain mushroom. Then, in 1985, Murao *et al.* isolated anthranilic acid and cyclooctasulfur as the fruiting inducers against *Polyporus (Favolus) arcularius*. An interesting anthraquinone derivative, basidifferquinone A was also isolated from *Streptomyces* sp. by Azuma, Beppu *et al.* as a fruiting-inducer against *Polyporus arcularius*. I will outline later more about basidifferquinone C.

In 1982, Kawai and Ikeda found that the fruiting-body formation of *Schizophyllum commune* (Japanese name: suéhiro také) can be stimulated by some cerebrosides in its mycelia. They then identified one of the active cerebrosides as (4*E*,8*E*,2*S*,3*R*,2'*R*)-*N*-2'-hydroxyhexadecanoyl-1-*O*-β-D-glucopyranosyl-9-methyl-4,8-sphingadienine (**128**, Figure 5.4), which had been isolated previously from a sea anemone (*Metridium senile*) by Karlsson *et al.* Only 0.1 µg of **128** induces the formation of the fruiting body of *S. commune*. A trivial name, Sch II, was given to **128**. We became interested in synthesizing **128** because of its remarkable bioactivity and also to develop synthetic methods for sphingolipids.[9–11]

Cerebrosides are glycosides of ceramides. Accordingly, ceramides must be prepared prior to the synthesis of cerebrosides by glycosidation. Our synthetic plan for **128** is shown in Figure 5.4. Sch II (**128**) can be

Figure 5.4   *Synthesis of Sch II, a fruiting-body inducer for Schizophyllum commune (1). Modified by permission of Shokabo Publishing Co., Ltd*

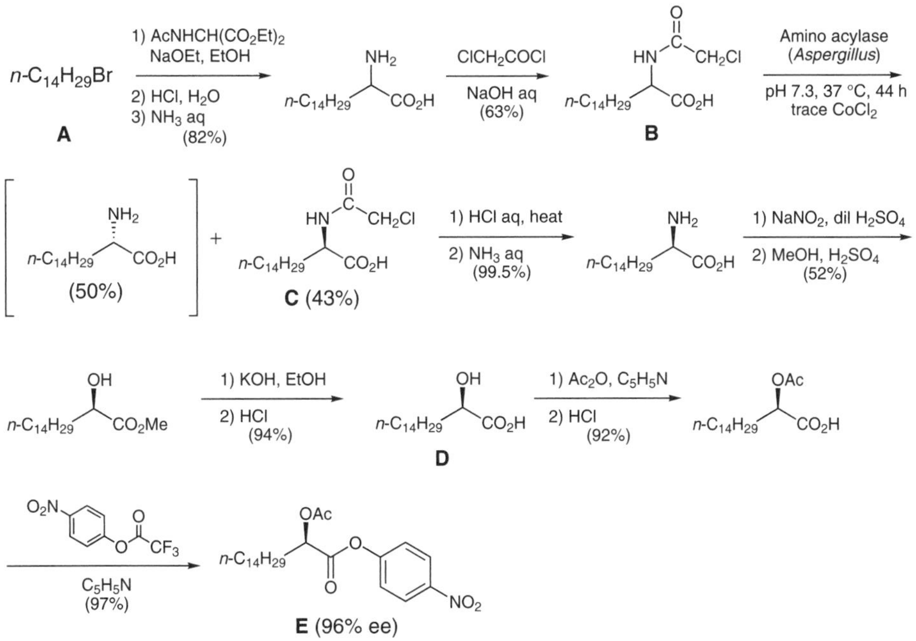

Figure 5.5 *Synthesis of Sch II, a fruiting-body inducer for Schizophyllum commune (2). Reprinted with permission of Shokabo Publishing Co., Ltd*

constructed by connecting α-hydroxy acid **B**, sphingadienine **C** and D-glucose **D**. The acid **B** can be prepared from α-amino acid **E**, and **C** is to be prepared from **F** and an unstable aldehyde **G**, which can be synthesized from L-serine (**H**).

Figure 5.5 shows the synthesis of α-hydroxy acid **D**. 1-Bromotetradecane (**A**) was converted to (±)-2-chloroacetylaminohexadecanoic acid (**B**), which was treated with amino acylase of *Aspergillus* origin to effect asymmetric hydrolysis. This enzyme is known to hydrolyse (*S*)-**B** to give (*S*)-2-aminohexadecanoic acid, while (*R*)-**C** remains intact. Acid hydrolysis of (*R*)-**C** was followed by deamination with nitrous acid to give (*R*)-**D** with retention of configuration. The corresponding α-acetoxy acid was converted to the activated ester **E**, which served as the acyl donor to the sphingadienine part.

The synthesis of the sphingadienine part and the coupling of the three building blocks to give Sch II (**128**) are summarized in Figure 5.6. Cyclopropyl methyl ketone (**A**) was converted to diacetate **B**, which was treated with octylmagnesium bromide under the Schlosser conditions to give **C** after deacetylation. Further chain elongation of **C** gave acetylene **D**, which served as the hydrophobic part of sphingadienine. As to the preparation of the polar and hydrophilic part, its synthesis started from (*S*)-serine, which was converted to the unstable aldehyde **F**. Reaction of **F** with alkenylalane **E** prepared from **D** yielded **G** as crystals after chromatographic purification to remove **H**. Subsequently, **G** was converted to Sch II (**128**) as depicted.[11] Synthetic Sch II (**128**) showed spectral and biological properties identical to those of the natural product. The structure of Sch II was thus confirmed as **128**.[12]

Ceramide (without D-glucose) corresponding to **128** was also bioactive, while the diastereomeric ceramide derived from **H** was only 1/7.5 as active as the ceramide corresponding to **128**. Stereochemistry at C-3 of the sphingadienine part was important for bioactivity.[10]

After completion of the synthesis of Sch II (**128**) in 1985, we continued our studies on the cerebrosides inducing the fruiting-body formation in *Schizophyllum commune*, and the cerebrosides shown in

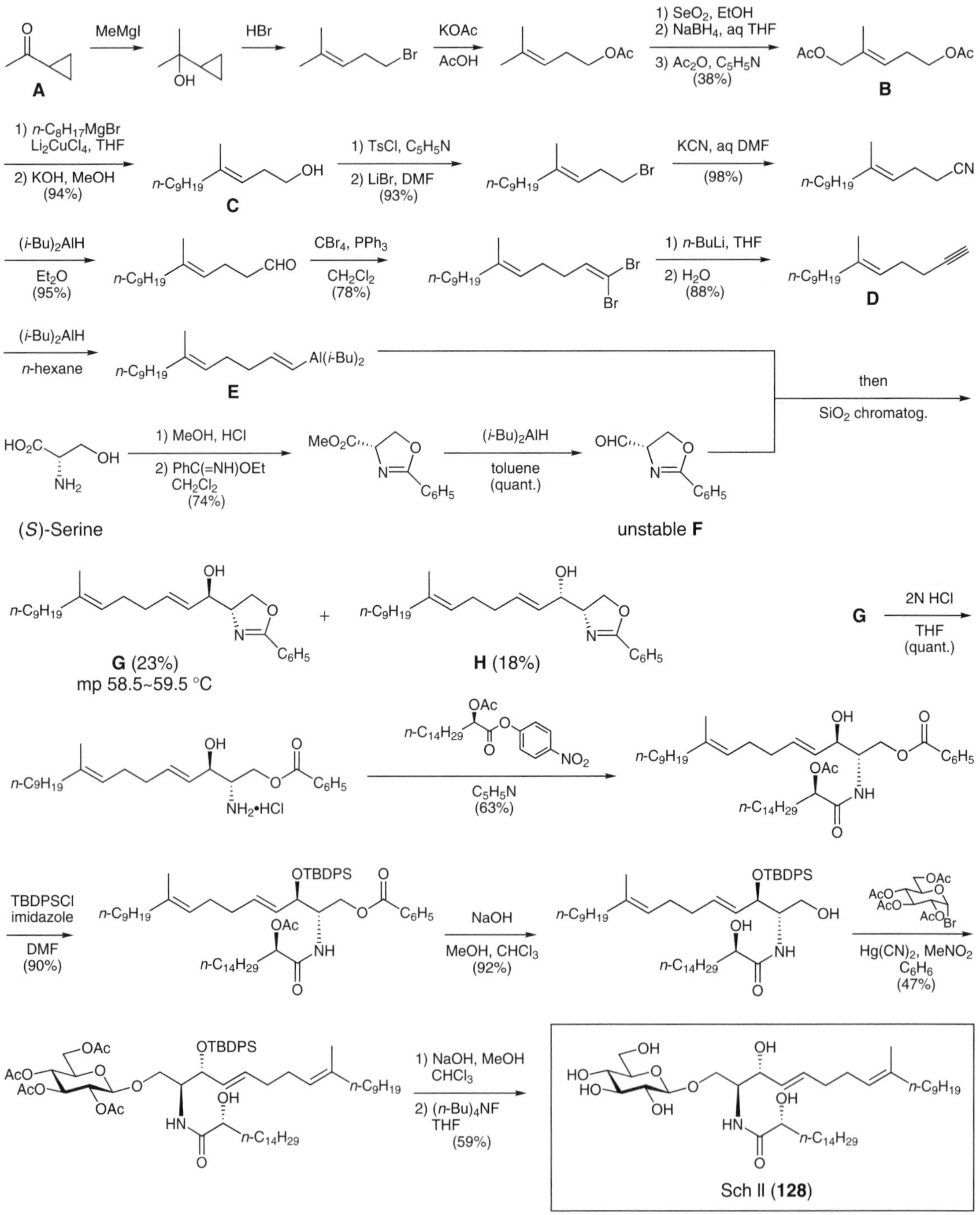

Figure 5.6 *Synthesis of Sch II, a fruiting-body inducer for Schizophyllum commune (3). Modified by permission of Shokabo Publishing Co., Ltd*

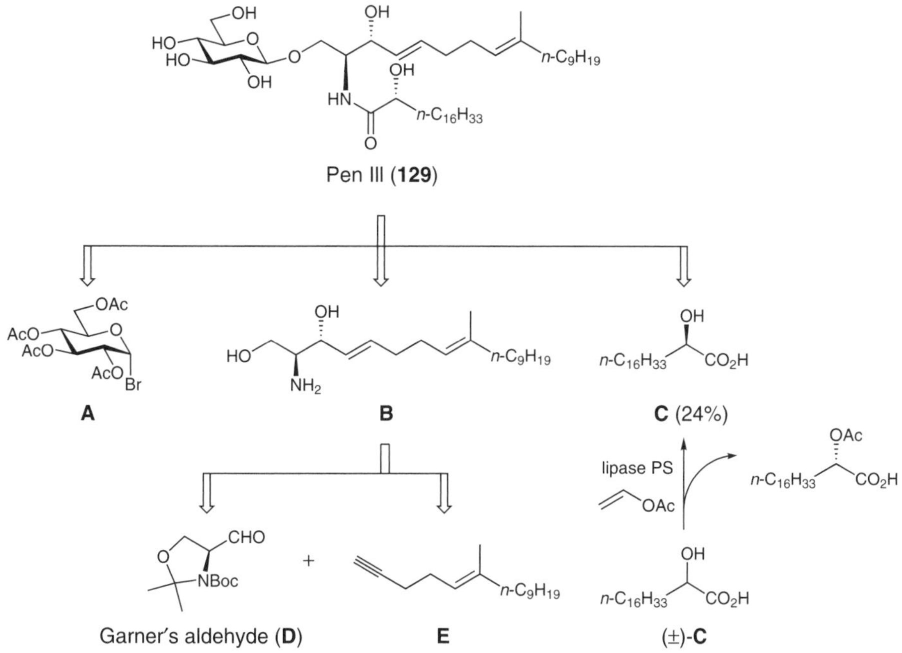

Figure 5.7 *Synthesis of Pen III, a metabolite of Penicillium funiculosum with fruiting-body inducing activity against Schizophyllum commune*

Figures 5.7–5.9 were synthesized. Pen III (**129**) is a metabolite of *Penicillium funiculosum* isolated as a fruiting inducer against *Schizophyllum commune*. Its synthesis was carried out, as shown in Figure 5.7, employing Garner's aldehyde (**D**) as a starting material.[13] The aldehyde **D** is not so unstable, and does not racemize easily. For the preparation of (*R*)-hydroxy acid **C**, (±)-**C** was subjected to asymmetric acetylation with vinyl acetate in the presence of lipase PS.[13]

Pen II (**130**) is another metabolite of *P. fumiculosum*, and also induces fruiting-body formation of *S. commune*. For the synthesis of the sphingadienine part **C** of **130**, a new and simpler route was adopted, as shown in Figure 5.8.[14] Garner's aldehyde **I** was employed in this case, too.

In 1986, Kawai *et al.* isolated Whe II (**131**) from wheat grain, which showed a fruiting-inducing effect on *S. commune*. As shown in Figure 5.9, the sphingenine part **A** of **131** was synthesized by the cleavage of epoxide **C** with Grignard reagent **B**.[15] The epoxide **C** was prepared from D-tartaric acid via epoxy diester **D**.

The reason why these cerebrosides induce fruiting-body formation of *S. commune* is not yet clear.

5.1.3 Basidifferquinone C

In the early 1990s, Azuma, Beppu and their coworkers isolated basidifferquinone C (**132**, Figure 5.10) from *Streptomyces* sp. B-412 as an inducer of fruiting-body formation in a mushroom *Polyporus arcularius*. Its unique structure led us to attempt its synthesis, but we were unable to synthesize it until 2008, when my former student H. Takikawa was successful in synthesizing it, as shown in Figure 5.10.[16]

The synthesis started from commercially available 3,5-dihydroxy-2-naphthoic acid (**A**), which was converted to lactonic naphthoquinone **B**. Diels–Alder reaction of **B** with diene **C** gave a 1:1 mixture of the

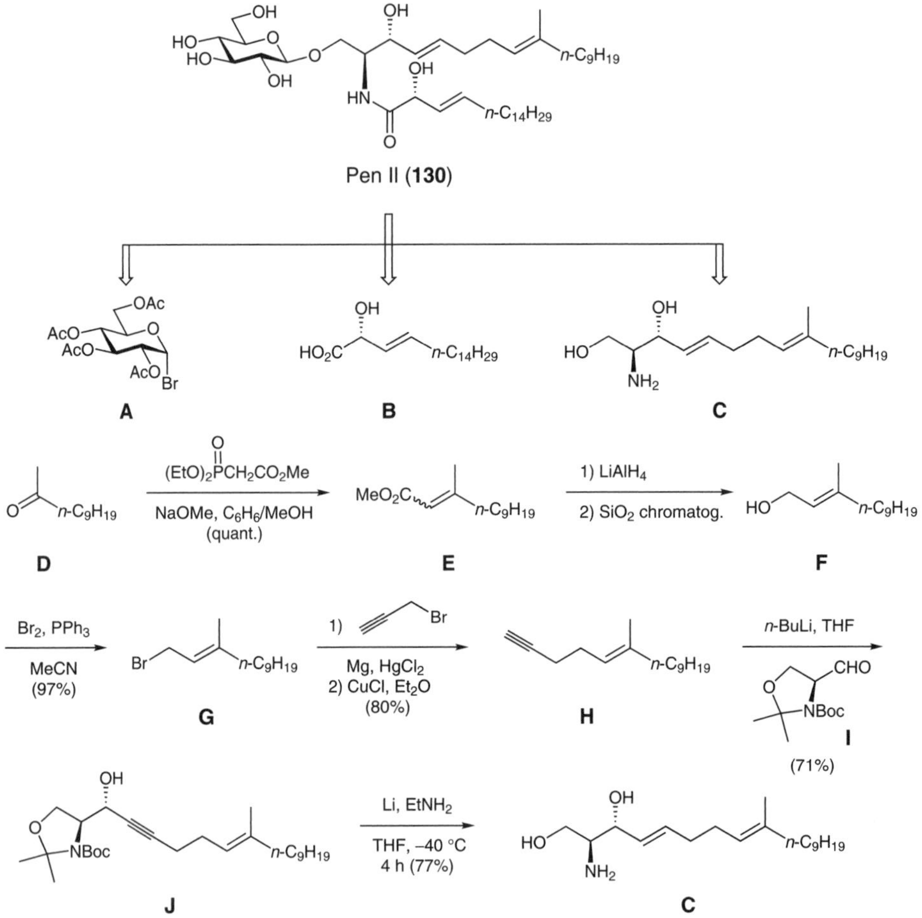

Figure 5.8 *Synthesis of Pen II, a metabolite of Penicillium funiculosum with fruiting-body inducing activity against Schizophyllum commune*

adducts **D** and **E**. Fortunately, these two were separable by preparative TLC, and **D** was treated with acid to remove the MOM protective group, yielding (±)-basidifferquinone C (**132**).[16] In our earlier synthetic efforts, we attempted to construct the lactone part in later stages of the synthesis, and we failed. In any synthesis, the order of the construction of the structural motifs in the target molecule is very important and the key to a successful synthesis.

5.1.4 Sclerosporin

In 1978, Marumo and Katayama isolated sclerosporin (**133**, Figure 5.11) as the major sporogenic substance of a fungus *Sclerotinia fructicola*. It induced the formation of asexual arthrospores in the fungal mycelium at a very low concentration of 1 ng/mL. Initially, they reported the structure of sclerosporin as a *trans*-guaiane-type sesquiterpene carboxylic acid **A**. Our synthetic (±)-**B**, however, showed no sporogenic activity

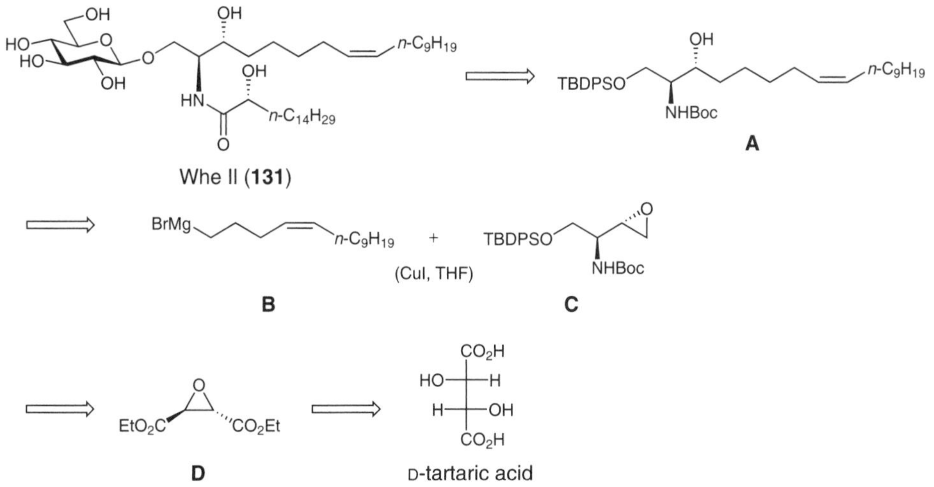

Figure 5.9 *Synthesis of Whe II, a cerebroside isolated from wheat grain with fruiting-body inducing activity against Schizophyllum commune*

at all.[17] We then synthesized guaiane hydrocarbons **C** and **D**, which were the proposed structures of sclerosporene, a hydrocarbon congener of sclerosporin isolated from the culture broth of *S. fructicola*. The mass spectrum of the natural sclerosporene was different from those of **C** and **D**.[18] The structure **A** proposed for sclerosporin was therefore considered to be in error.

The above result of ours made Marumo and Katayama reisolate sclerosporin. They cultured *S. fructicola* by using 5050 Petri dishes to secure 101 L of the culture broth, and obtained 3.8 mg of sclerosporin. Its NMR analysis allowed them to propose **133** as the revised structure.

We synthesized the enantiomers of sclerosporin by starting from carvone employing an intramolecular Diels–Alder reaction (**E**→**F**) as the key step.[19,20] Bioassay proved (+)-**133** to be highly sporogenic, while (−)-**133** was only marginally bioactive. The absolute configuration of sclerosporin was therefore established as depicted in **133**.

5.1.5 Sporogen-AO 1

Aspergillus oryzae is the most important fungus in Japanese fermentation industry, and employed widely for the production of saké (rice wine), shoyu (soy sauce), and miso (fermented soybeans). The fungus is also used for the production of industrial enzymes such as acylase, amylase, and protease. The quality as well as the yield of the fermented products is known to be affected by the extent of sporulation of *A. oryzae*. It is therefore worthwhile to clarify the sporulation phenomenon. In 1984, Marumo and coworkers isolated 3 mg of a sporogenic substance from the culture broth of *A. oryzae*. They named it sporogen-AO 1 and clarified its structure as depicted in **134** (Figure 5.12). They also deduced its absolute configuration by analyzing its CD (circular dichroism) spectrum. Sporogen-AO 1 was identical with 13-deoxyphomenone isolated from *Hansfordia pulvinata*, and induced sporulation of *A. oryzae* even at the dosage of 4.4 μg/disc.

We became interested in synthesizing both the enantiomers of sporogen-AO 1 so as to clarify the stereochemistry–bioactivity relationship and also to provide a sufficient amount of material for its evaluation as a bioregulator useful in fermentation industry. To secure both enantiomers, we adopted the

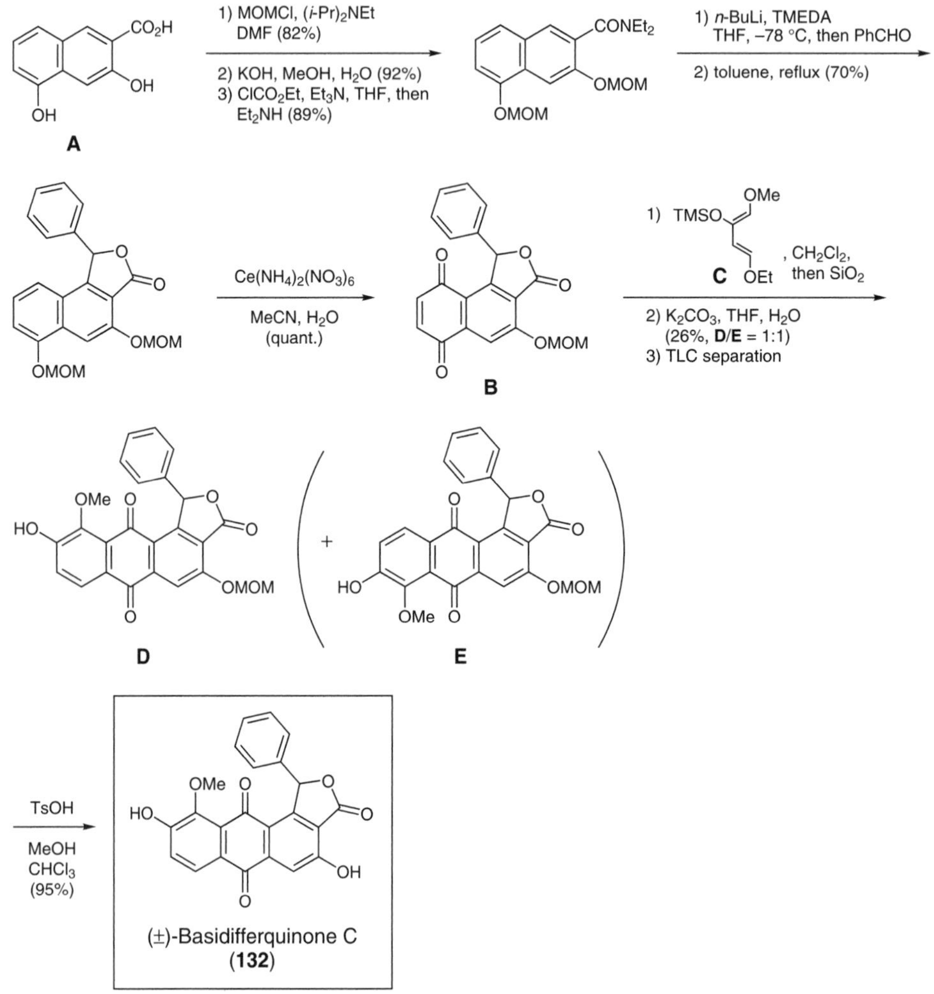

Figure 5.10   *Synthesis of (±)-basidifferquinone C*

enantiomer separation (optical resolution) of an intermediate as our key step, as shown in Figure 5.12.[21] Commercially available 2,3-dimethylhydroquinone (**A**) was converted to lactone (±)-**C** via (±)-**B**. Then, (±)-**C** gave (±)-**D**, which was treated with camphenyl chloride. The resulting mixture of **E** and **F** could be separated by chromatography, and **E** furnished the known (+)-**D**, while **F** gave (−)-**D**. The absolute configuration of (+)-**D** was confirmed by its ORD (optical rotatory dispersion) spectral analysis.

Further conversion of (+)-**D** to sporogen-AO 1 [(+)-**134**] is summarized in Figure 5.13.[21] Similarly, (−)-**D′** gave the unnatural isomer (−)-**134′**. Bioassay of **134** and **134′** on *Asp. oryzae* showed the former to be as sporogenic as the natural sporogen-AO 1, while the latter showed no bioactivity at all. In the same year of 1988, we achieved an enantioselective synthesis of (+)-**134**, which will not be discussed here.[22]

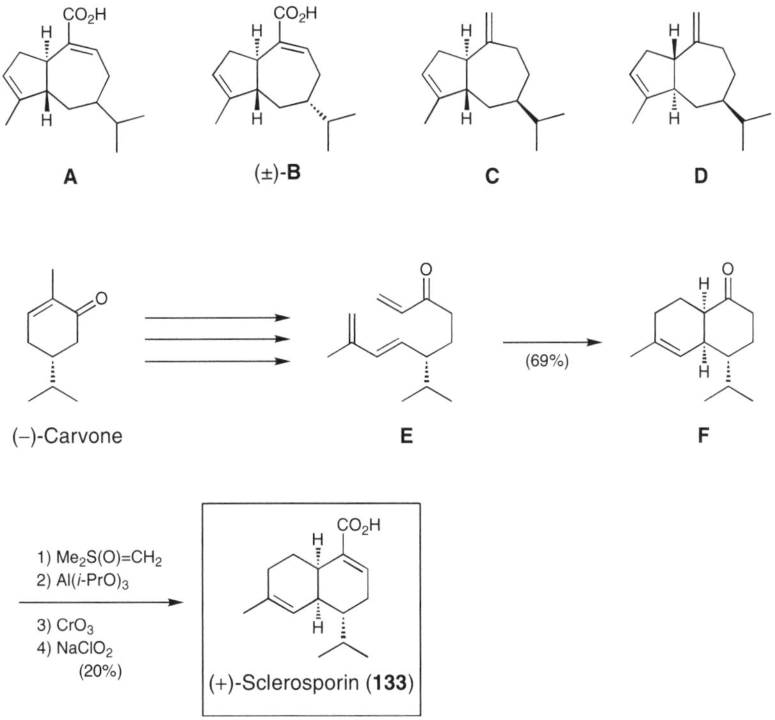

Figure 5.11   *Synthesis of sclerosporin. Modified by permission of Shokabo Publishing Co., Ltd*

5.1.6   Differolide

In 1986, Keller-Schierlein, Zähner and their respective coworkers isolated (±)-differolide (**135**, Figure 5.14) from cultures of an actinomycete, *Streptomyces aurantiogriseus* Tü 3149. This compound (±)-**135**, whose structure was determined by X-ray diffraction studies, was reported to enhance the formation of aerial mycelium and spores of *Streptomyces glaucescens*. Especially noteworthy was the fact that the natural product was racemic. There are some examples of the occurrence of racemic and bioactive natural products such as magnosalicin (**10**) and olean (**116**), but they are rare. Although there was a straightforward Diels–Alder synthesis of (±)-**135** by dimerization of 2-vinyl-2-buten-4-olide, this method would not afford the enantiomers of **135**.

We became interested in synthesizing both the enantiomers of differolide to clarify whether one or both of them are bioactive. Our synthesis is summarized in Figure 5.14 and 5.15.[23] Because both the enantiomers **135** and **135'** were necessary for bioassay, we adopted enantiomer separation (optical resolution) of an intermediate as our key step (Figure 5.14). A crystalline acetal (−)-**B** was obtained from (±)-**A** and (−)-menthol, and analysed by X-ray to reveal its structure as (−)-**B**, basing on the known absolute configuration of (−)-menthol. When (+)-menthol was used for acetal formation, crystalline (+)-**B'** was obtained in a similar manner. We thus secured both (−)-**B** and (+)-**B'** as pure crystals.

Conversion of (−)-**B** to (−)-differolide (**135**) is summarized in Figure 5.15. Acid treatment of (−)-**B** afforded (−)-**A**, which was converted to **C**. Baeyer–Villiger oxidation of **C** gave **D** together with the

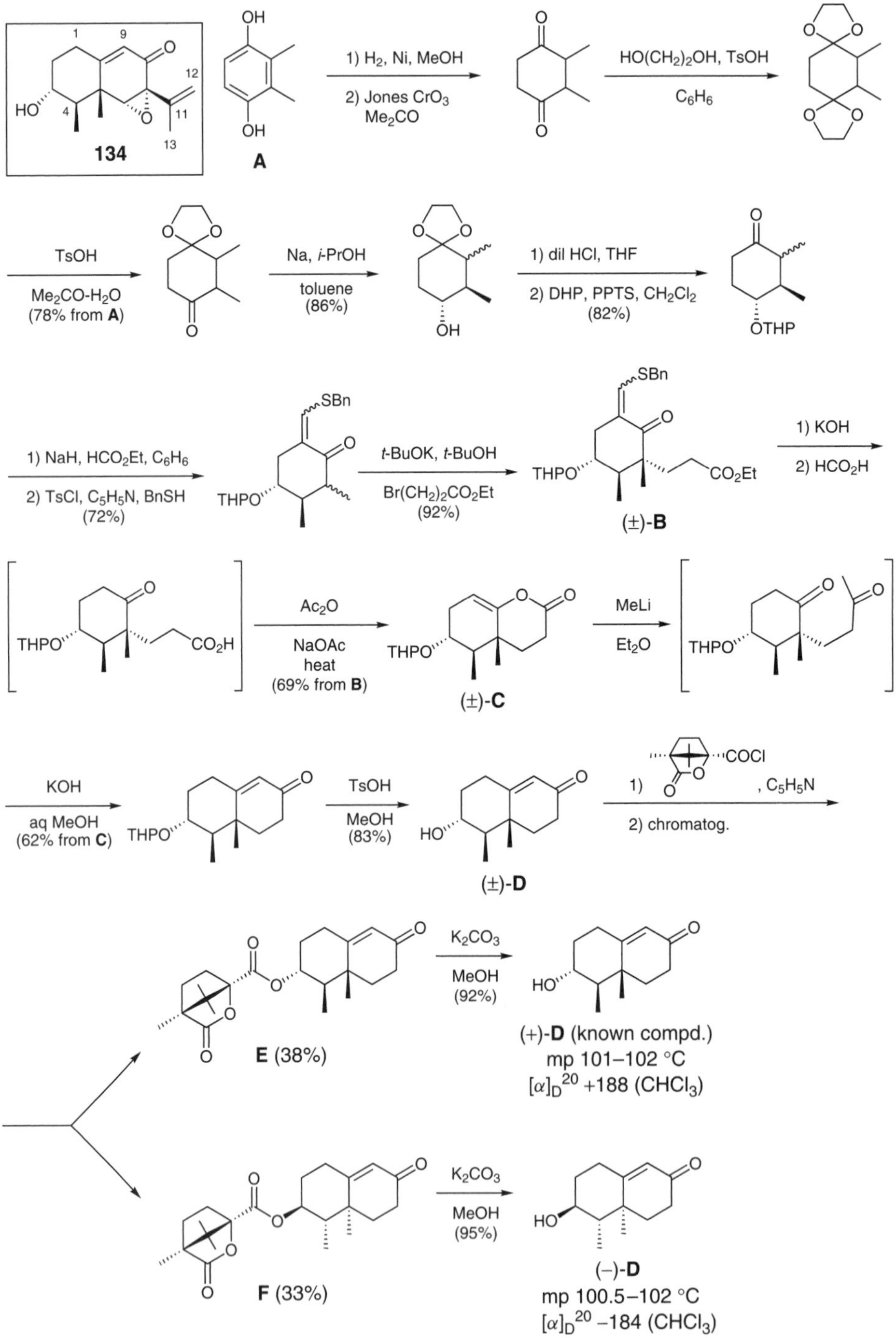

Figure 5.12   *Synthesis of the enantiomers of sporogen-AO 1 (1). Modified by permission of Shokabo Publishing Co., Ltd*

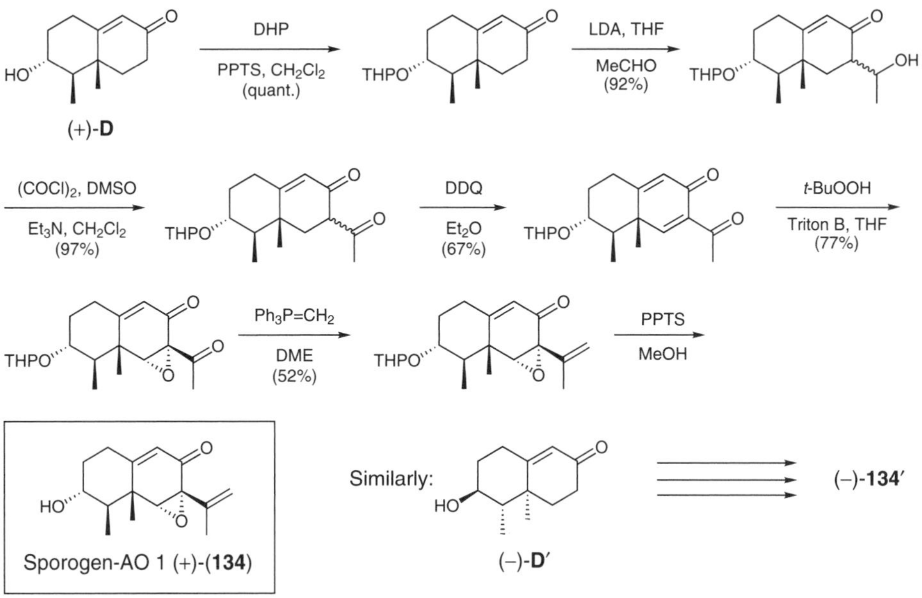

Figure 5.13   *Synthesis of the enantiomers of sporogen-AO 1 (2). Modified by permission of Shokabo Publishing Co., Ltd*

undesired **E**. Separation of these two isomers were possible by chromatography at the stage of (−)-**F**. The ester (−)-**F** gave (−)-differolide (**135**) as crystals. Similarly, (+)-differolide (**135′**) was obtained from (+)-**B′**. The synthetic enantiomers of differolide showed the IR, MS, ^{1}H- and ^{13}C-NMR spectra identical to those reported for the natural differolide.

Prof. Horinouchi at the University of Tokyo bioassayed (−)-**135**, (+)-**135′**, and (±)-**135**, and found them to show no sporogenic activity against *Streptomyces glaucescens* ssp. *glaucescens* Tü 49. Dr. Schülz at Tübingen University also confirmed that (−)-**135**, (+)-**135′** and (±)-**135** exhibited no sporogenic activity. Accordingly, differolide is not a sporogenic substance. The reason remains unclear. For the study of bioactive compounds, we should first provide a reliable and reproducible bioassay system. Otherwise, nonreproducible results can be published even in an esteemed and peer-reviewed journal.

5.2   Antibiotics

There are many different kinds of micro-organisms in the natural environment. Microbiologists have long noticed antagonism to inhibit the growth of other microbes.

It was Fleming in the UK who discovered penicillin in 1928 (Figure 5.16). This antibiotic was the first one to be used by humans as a medicinal against infectious diseases. Fleming found that *Penicillium notatum* produces a substance to inhibit the growth of *Staphyllococcus aureus*, and named it penicillin. Florey and Chain in the UK further studied the phenomenon, and found in 1941 that the crude penicillin powder was extremely useful as a chemotherapeutic agent against infectious diseases such as pneumonia. Almost simultaneously in 1943 Waksman in the USA discovered streptomycin as a metabolite of a streptomycete among soil micro-organisms.

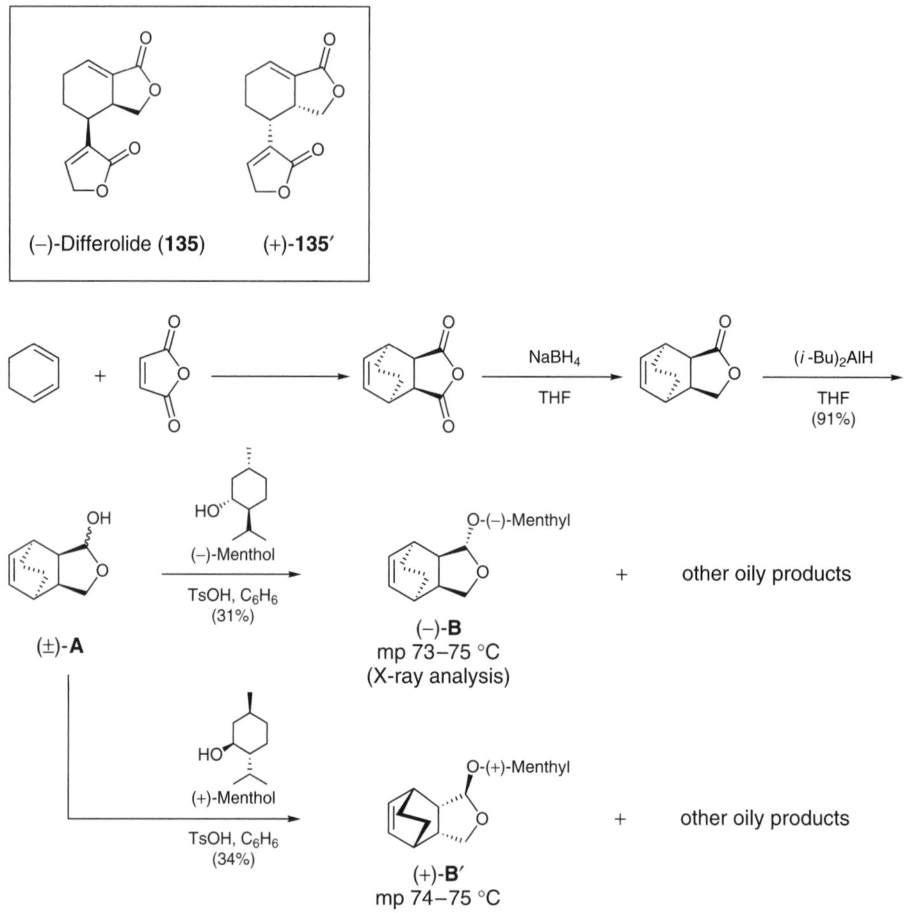

Figure 5.14 *Synthesis of the enantiomers of differolide (1). Modified by permission of Shokabo Publishing Co., Ltd*

Antibiotics research is now very important both scientifically and practically. Some of our syntheses of antibiotics will be treated in this section.

5.2.1 Ascochlorin

(−)-Ascochlorin was isolated in 1968 by G. Tamura *et al.* as an antiviral metabolite of *Ascochyta viciae*, and was shown to be **136** (Figure 5.17) by X-ray analysis. Ando, Tamura and their coworkers also isolated (−)-ascofuranone (**137**) from the mycelia of *A. viciae* as an antitumor compound. Very recently, in 2005, (−)-ascochlorin (**136**) was shown to be a selective inhibitor of breast cancer cells. Colletochlorin B (Figure 5.17) is a related metabolite isolated from phytopathogenic microbe *Colletotrichum nicotianae*.

In the late 1970s to the early 1980s, we were interested in the synthesis of these microbial metabolites with a hexa-substituted benzene ring. The first phase of our work was the synthesis of colletochlorin B to establish a synthetic method for the hexa-substituted benzene ring.[24] The racemates of ascochlorin and asco-furanone were then synthesized,[25] and finally the naturally occurring (−)-ascochlorin and (−)-ascofuranone

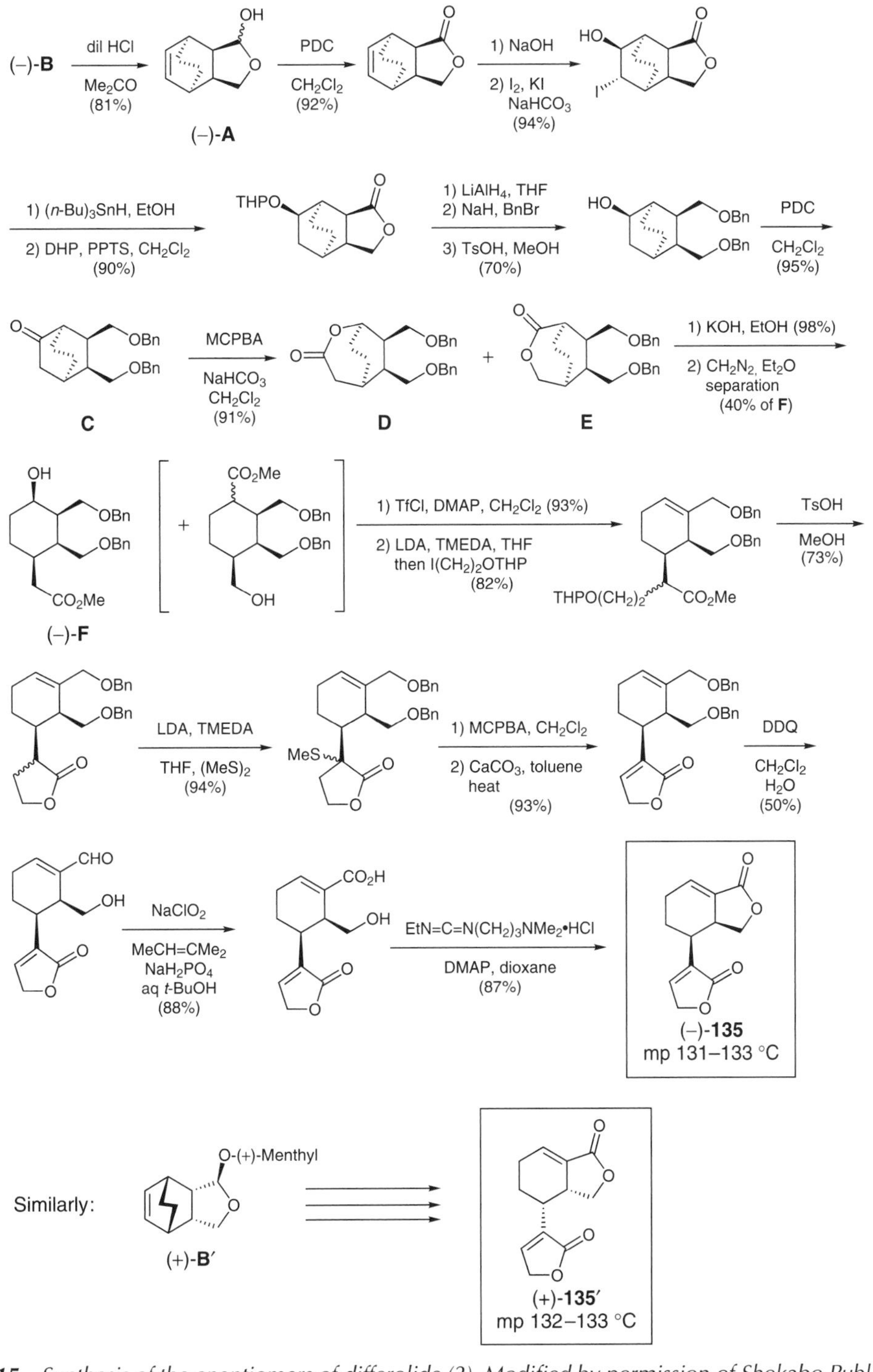

Figure 5.15 *Synthesis of the enantiomers of differolide (2). Modified by permission of Shokabo Publishing Co., Ltd*

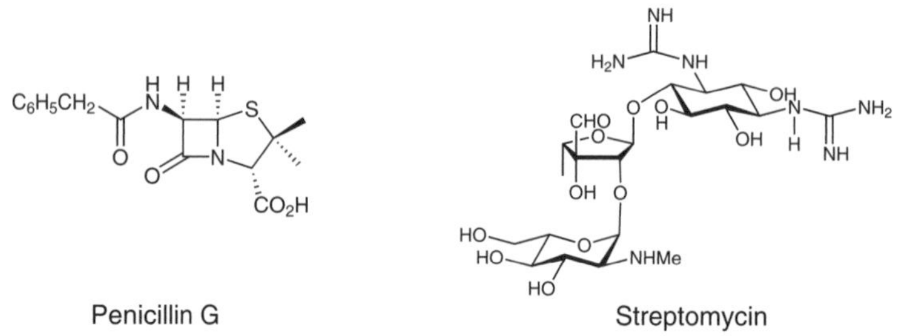

Figure 5.16 *Structures of penicillin G and streptomycin. Reprinted with permission of Shokabo Publishing Co., Ltd*

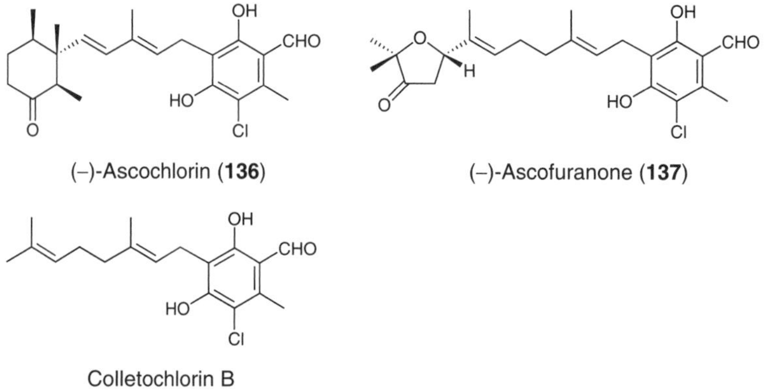

Figure 5.17 *Structures of ascochlorin and its relatives. Modified by permission of Shokabo Publishing Co., Ltd*

were synthesized to conclude the research.[26] The driving force to achieve these syntheses was given by Prof. K. Arima, the great microbiologist (cf. 2.1.2.4). When the structure of (−)-ascochlorin was solved as **136** by X-ray analysis, he showed me the X-ray perspective view of **136**, and said, "Dr. Mori, the structure was solved by X-ray. You may feel difficulty in synthesizing ascochlorin." This comment of Prof. Arima led me to achieve the synthesis of **136**.

Figure 5.18 and 5.19 show our ascochlorin synthesis.[26] (*R*)-Pulegone served as the starting material for the synthesis of the sesquiterpene part (−)-**D** of (−)-ascochlorin. Accordingly, (*R*)-pulegone was converted to (+)-**A** by the known method, and (+)-**A** furnished **C** via (+)-**B**. Fortunately, at the stage of (−)-**D**, it could be separated from its unwanted but crystalline (*Z*)-isomer. The oily (−)-**D** afforded chloride **E**, to which was attached the benzene ring part, as shown in Figure 5.19.

Alkylation of dihydroorcinol dimethyl ether (**F**) with **E** gave **G**. Chlorination of **G** with *N*-chlorosuccinimide provided **H**, which was aromatized by dehydrochlorination to give **I**. Introduction of a formyl equivalent to **I** furnished **J**, whose acid treatment gave (−)-ascochlorin (**136**) as crystals.[26]

5.2.2 Ascofuranone

Synthesis of (−)-ascofuranone was achieved by enantiomer separation (optical resolution) of an intermediate as shown in Figure 5.20 and 5.21.[26] Geraniol was adopted as our starting material.

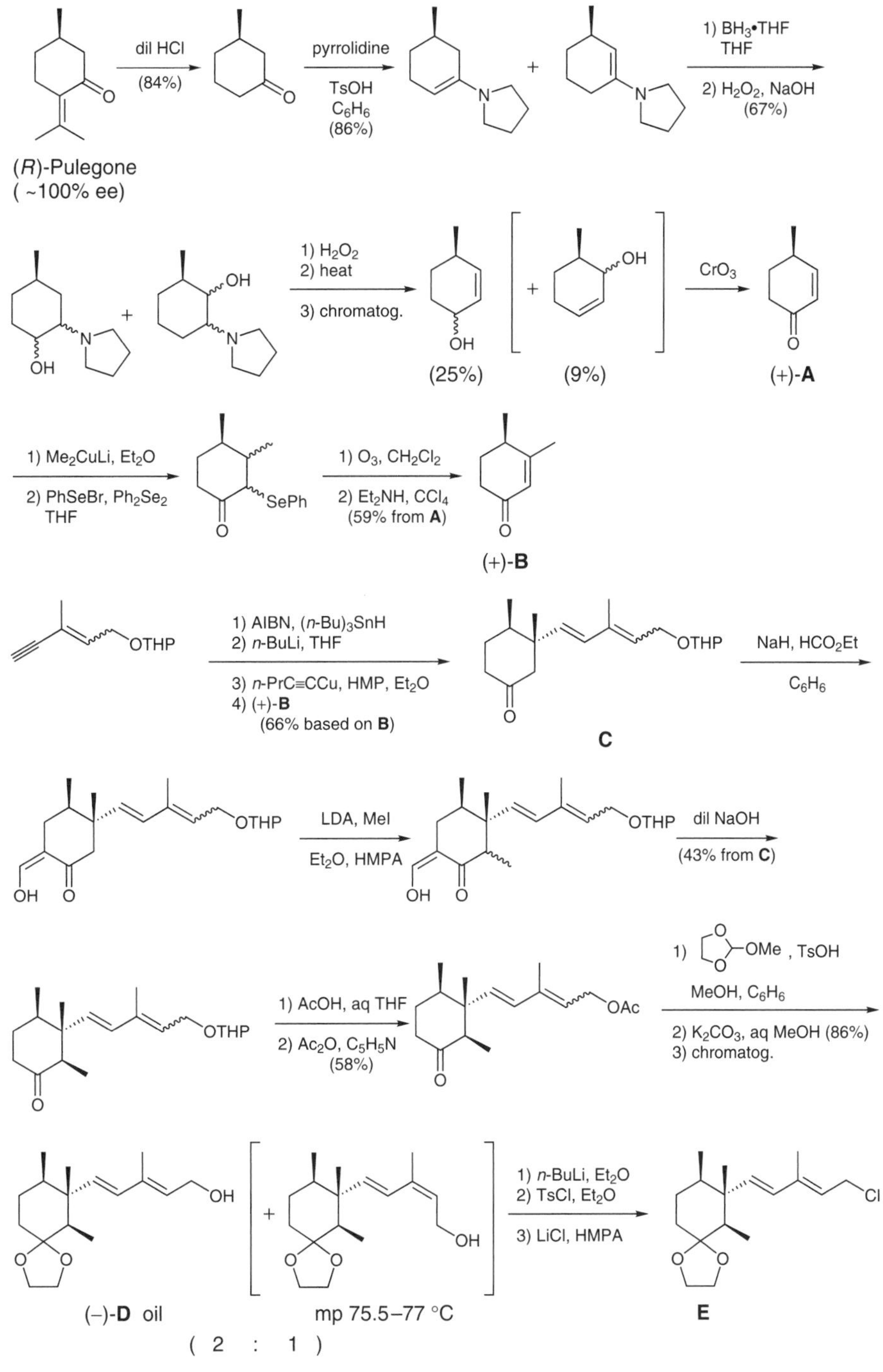

Figure 5.18 *Synthesis of (−)-ascochlorin (1). Reprinted with permission of Shokabo Publishing Co., Ltd*

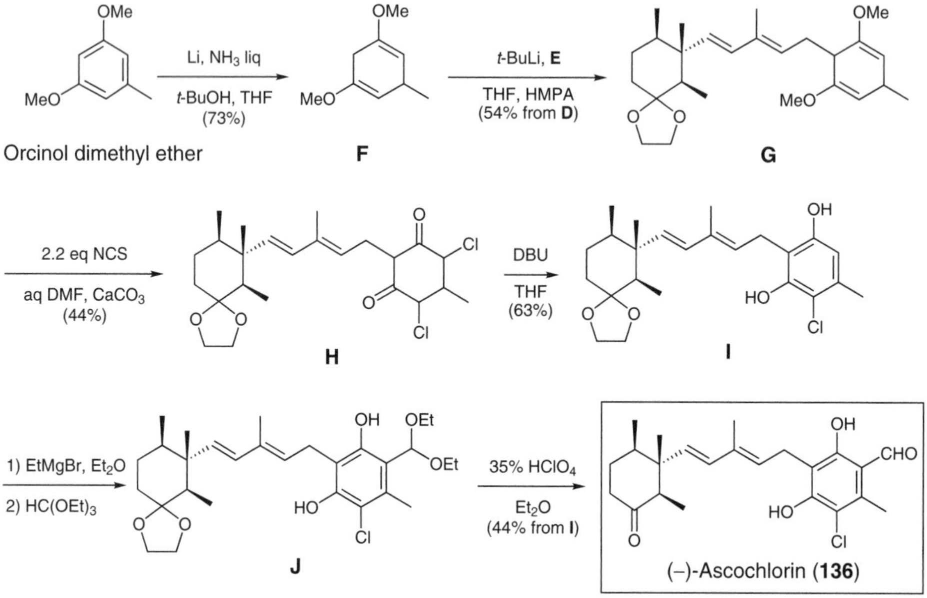

Figure 5.19   *Synthesis of (−)-ascochlorin (2). Modified by permission of Shokabo Publishing Co., Ltd*

Dehydration of keto diol (±)-**A** derived from geraniol furnished furanone (±)-**B**. Its reduction with sodium borohydride yielded alcohol (±)-**C** as the major product. Treatment of (±)-**C** with the isocyanate derived from (*R*)-1-(1-naphthyl)ethylamine gave a mixture of carbamates **E** and **F**. These were separable by medium-pressure liquid chromatography (MPLC), and **E** gave (+)-**C**, while **F** afforded (−)-**C**.

Figure 5.21 summarizes further conversion of (+)-**C** into crystalline (−)-ascofuranone (**137**). (−)-Ascofuranol, the immediate precursor of **137**, was also known as a metabolite of *Ascochyta viciae*. Similarly, (−)-**C** afforded unnatural (+)-ascofuranone.

5.2.3   Trichostatin A

In 1976, Tsuji *et al*. at Shionogi Pharma isolated trichostatin A (**138**, Figure 5.22) as an antifungal antibiotic produced by *Streptomyces hygroscopicus*. Later, in 1985, Yoshida and Beppu at the University of Tokyo rediscovered **138** as a very strong inducer of differentiation of Friend leukemic cells. Simultaneously, a group of researchers at Ajinomoto Co. also identified **138** as a differentiation inducer.

Although (±)-**138** had been synthesized in 1983 by I. Fleming *et al*. in the UK, the absolute configuration of the naturally occurring (+)-**138** was unknown in 1985. We therefore started our work to synthesize both the enantiomers of trichostatin A so as to establish the absolute configuration of (+)-**138** and also to clarify the stereochemistry–bioactivity relationship.

Methyl 3-hydroxy-2-methylpropanoate (**A**) was chosen as the starting material, because both enantiomers were commercially available. Our synthesis of trichostatin A (**138**) from **A** is summarized in Figure 5.22.[27] All the synthetic steps were executed carefully to avoid racemization, and (*R*)-trichostatic acid was obtained as plates, mp 88–89 °C, $[\alpha]_D^{23} = +138$ (MeOH). Trichostatic acid prepared by hydrolysis of trichostatin A was reported to be crystals with mp 138–140 °C, $[\alpha]_D = +3.8$ (MeOH). This large difference in physical

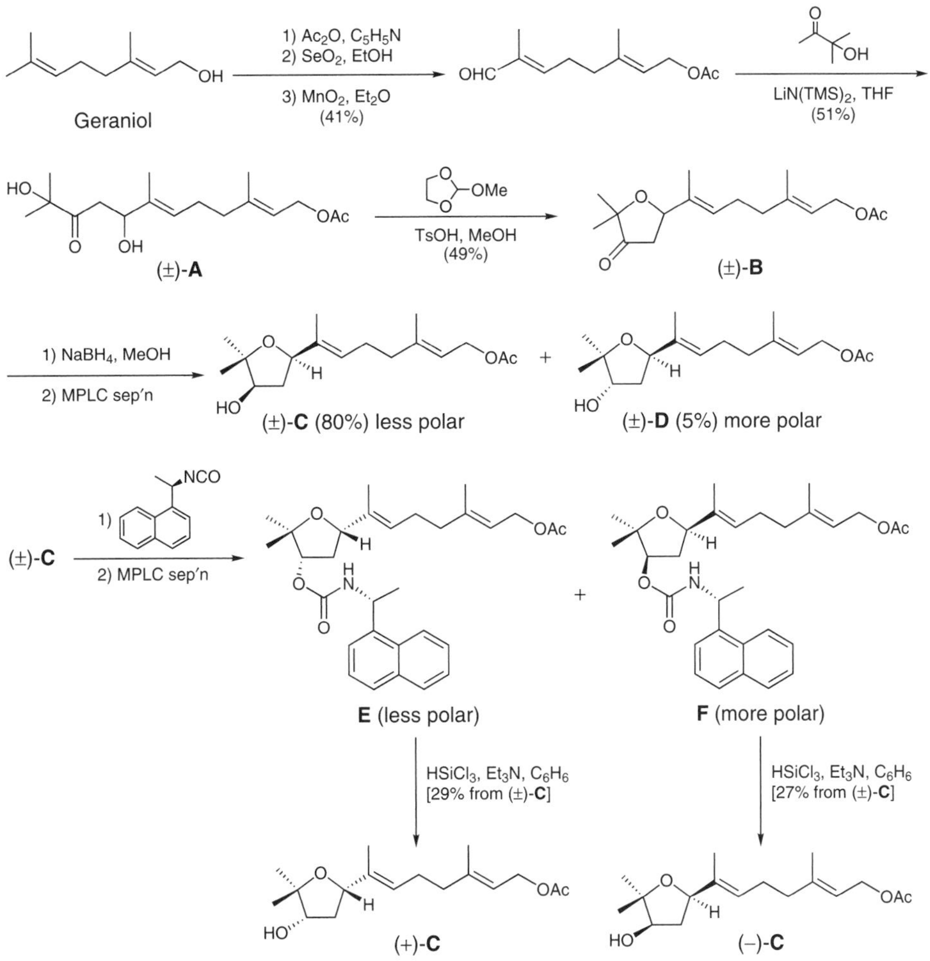

Figure 5.20   *Synthesis of (−)-ascofuranone (1). Reprinted with permission of Shokabo Publishing Co., Ltd*

constants was due to racemization in the course of hydrolysis of (+)-**138**. Conversion of (*R*)-trichostatic acid to (*R*)-trichostatin A (**138**) was realized under mild conditions to give **138** as needles, mp 146–150 °C, $[\alpha]_D^{22} = +96$ (MeOH). The natural product was reported to be with mp 150–151 °C, $[\alpha]_D = +63$ (MeOH). The absolute configuration of the naturally occurring trichostatin A was therefore established as *R*.

Biological evaluation of both (*R*)-**138** and (*S*)-**138′** revealed that only the naturally occurring (*R*)-isomer shows bioactivity as an inducer of differentiation of Friend leukemic cells.[28]

5.2.4   Koninginin A

In 1989, Cutler *et al.* isolated (−)-koninginin (**139**, Figure 5.23) as a metabolite of a soil micro-organism *Trichoderma koningii*, and found it to be a weak inhibitor against the growth of etiolated wheat coleoptiles.

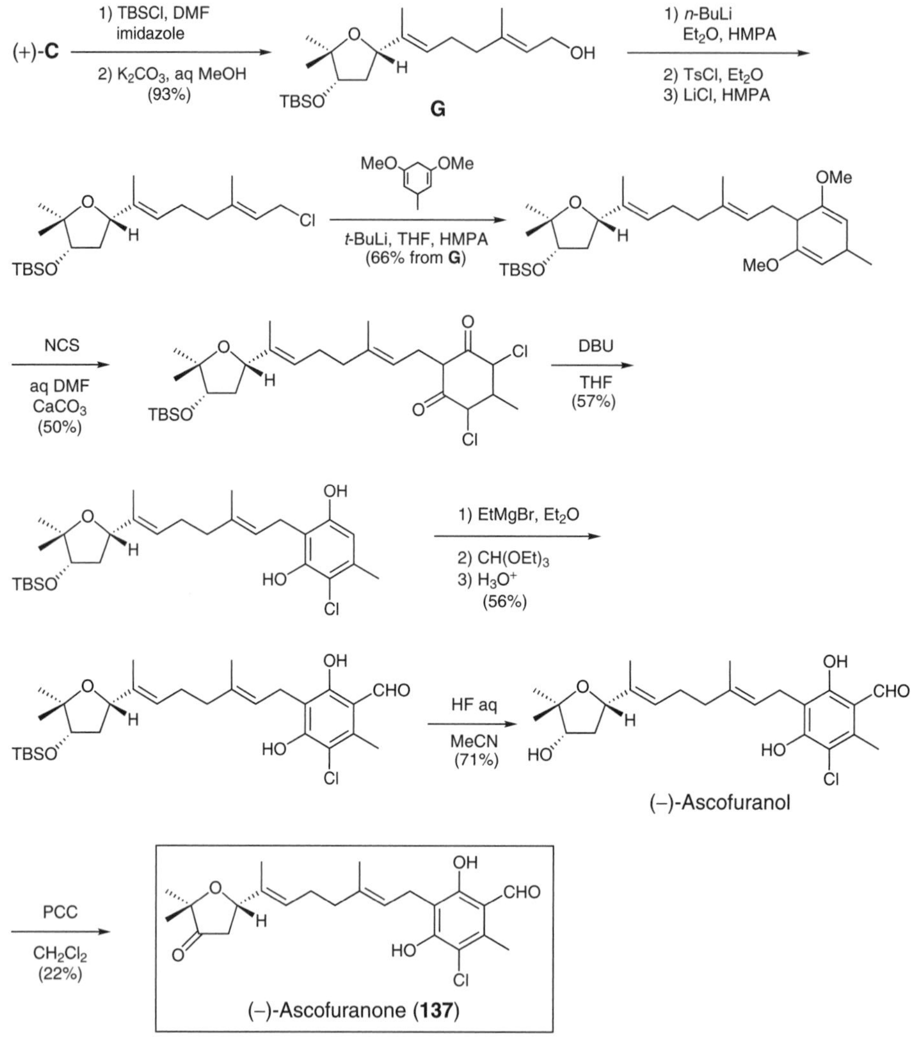

Figure 5.21   *Synthesis of (−)-ascofuranone (2). Modified by permission of Shokabo Publishing Co., Ltd*

They proposed **139** as its structure, although nothing was known about its stereochemistry. Subsequently, in 1991, Ghisalberti and coworkers isolated (−)-koninginin A as a metabolite of *Trichoderma harzianum* with antibiotic activity against the take-all fungus, *Gaeumannomyces graminis* var. *tritici*. By NMR analysis, Ghisalberti deduced the relative stereochemistry of (−)-koninginin A as depicted in **A**, but he was unable to propose its absolute configuration.

The unique structure **A** with six stereogenic centers made it an attractive synthetic target. We therefore carried out a synthesis of racemic koninginin A in 1994,[29] and then that of (−)-koninginin A itself in 1995.[30] The proposed structure **A**, however, was challenged by Xu and Zhu in the same year of 1995,[31] and their new structure **139** was confirmed in 2002 by the X-ray analysis of our synthetic (−)-koninginin A.[32] Here, in this section I will outline the story about structure revision.

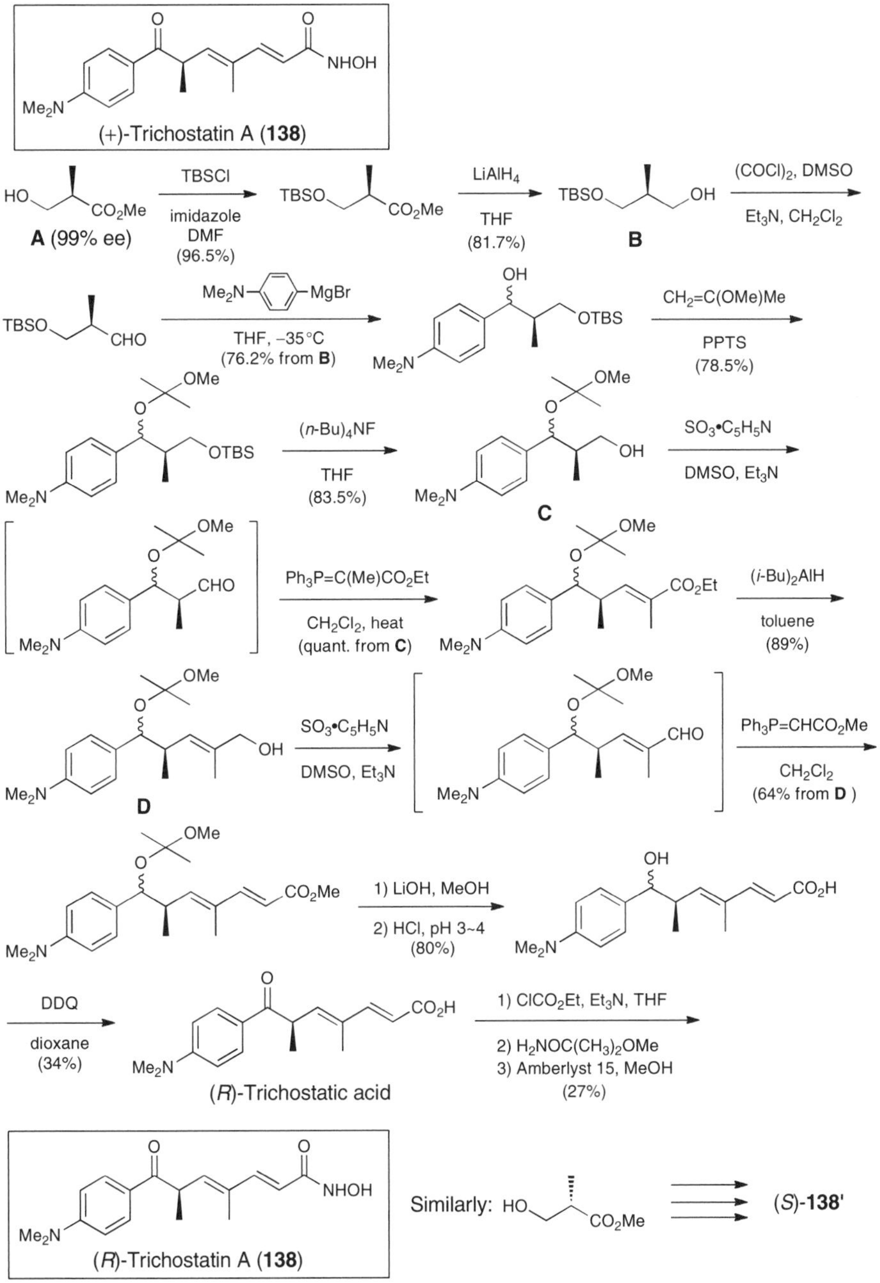

Figure 5.22 Synthesis of (+)-trichostatin A. Modified by permission of Shokabo Publishing Co., Ltd

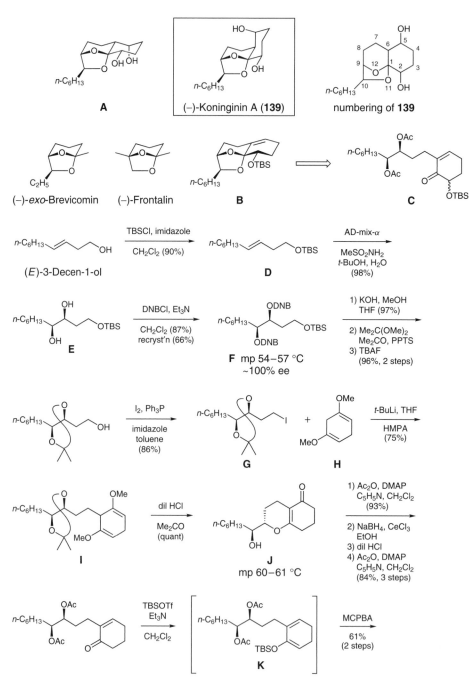

Figure 5.23   *Synthesis of (−)-koninginin A (1)*

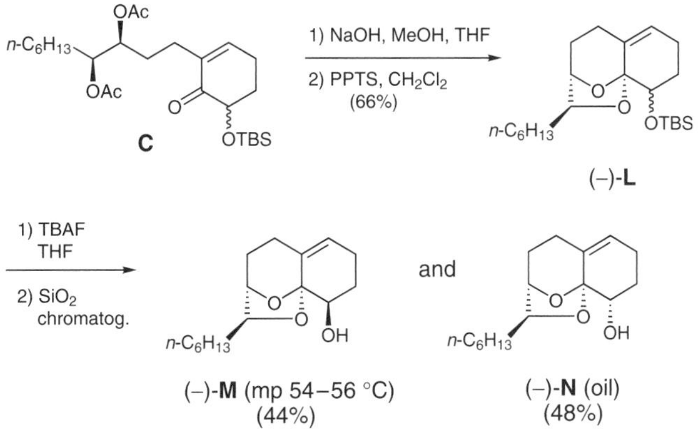

Figure 5.23 *(continued)*

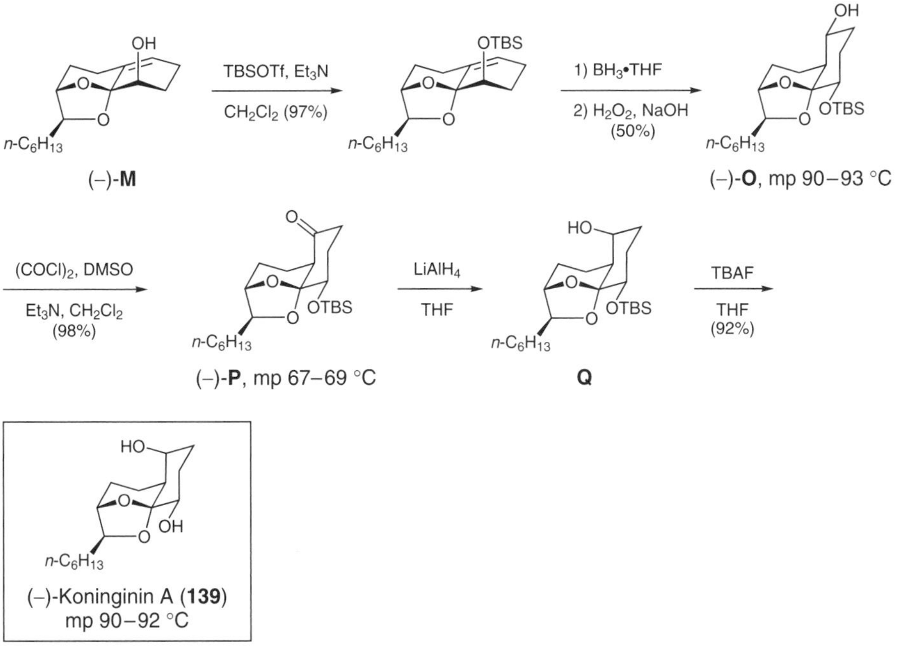

Figure 5.24   *Synthesis of (−)-koninginin A (2)*

Our 1995 synthesis of (−)-koninginin A (**139**) is summarized in Figure 5.23 and 5.24.[30] In the planning stage, I made an assumption that (−)-koninginin A must possess (9*S*,10*S*)-configuration, because pheromone acetals (−)-*exo*-brevicomin and (−)-frontalin were known to possess (*S*,*S*)-configuration. The structure **A**, then believed to be that of koninginin A, would be prepared from **B**, which would be derived from **C**. The 1,2-diol part of **C** were to be prepared by Sharpless asymmetric dihydroxylation.

The synthesis started from (*E*)-3-decen-1-ol, whose hydroxy group was protected as TBS ether to give **D**. Asymmetric dihydroxylation of **D** with AD-mix-α® furnished **E** (90.6% ee), which was purified by recrystallizing its *bis*-3,5-dinitrobenzoate **F**. The enantiomerically pure **F** was converted to iodide **G**. Alkylation of dihydroresorcinol dimethyl ether **H** with **G** yielded **I**. Treatment of **I** with dilute acid effected deprotection and cyclization to give crystalline **J**. Introduction of a hydroxy group at C-2 was executed by epoxidation of **K** and rearrangement of the resulting epoxide to give **C**. Removal of the acetyl group of **C** was followed by acid treatment to give levorotatory acetal **L** as a stereoisomeric mixture. When its TBS protective group was removed, the mixture could be separated by silica-gel chromatography to give crystalline (−)-**M** and oily (−)-**N**. In 1995, it was difficult to prove the crystalline isomer was (−)-**M**. This problem will be discussed later.

Figure 5.24 shows the conversion of (−)-**M** into (−)-koninginin A (**139**). The hydroxy group of **M** was protected as TBS ether. Hydroboration-oxidation of the product stereoselectively gave (−)-**O** by the α-attack of borane due to the steric hindrance caused by the bulky TBSO-group. Swern oxidation of **O** afforded ketone (−)-**P**, which was reduced to furnish **Q**. Finally, desilylation of **Q** gave (−)-koninginin A (**139**), whose spectral and chiroptical properties were in good accord with those of the natural product. Accordingly, the (1*S*,2*R*,5*S*,6*S*,9*S*,10*S*)-absolute configuration of (−)-koninginin A was established as depicted in **139**. The assigned stereochemistry was supported by X-ray analysis of our synthetic **139**.[32]

In 1995 when we finished the above-described synthesis of (−)-koninginin A, we assigned structure **A** (Figure 5.23) to our synthetic product in accordance with the structure proposed by Ghisalberti. In our synthesis, however, there was a stereochemically ambiguous step.

As shown in Figure 5.23 epoxidation of **K** was nonstereoselective, and two alcohols **M** and **N** were obtained. The ¹H-NMR spectral data of these two hydroxyacetals at 300 MHz were not sufficiently informative to allow the assignment of their stereostructures. For example, the crystalline isomer showed signals due to the proton at C-2 at $\delta = 3.90$ (dd, $J = 3.7, 11.1$ Hz), while the oily one exhibited them at $\delta = 3.84$ (dd, $J = 3.6, 10.6$ Hz). It was therefore difficult to decide which of them was **M** with an axial hydroxy group at C-2. In addition, the crystalline isomer provided only very fine needles that were unsuitable for an X-ray analysis. Nevertheless, we could convert the TBS ether of the crystalline hydroxyacetal to (−)-koninginin A, which was presumed to be (1*S*,2*S*,5*R*,6*R*,9*S*,10*S*)-**A**, by combining the then-accepted relative stereochemistry of (−)-koninginin proposed by Ghisalberti with the (9*S*,10*S*)-stereochemistry originating from the Sharpless AD reaction. The structure of the crystalline hydroxyacetal was therefore thought to be (−)-**N** in early 1995.[30]

In late 1995, Xu and Zhu reported another synthesis of (−)-koninginin A (**139**) and its diastereomer **A**.[31] They carefully analysed the 600 MHz ¹H-NMR spectra of **139** and **A**, and found subtle differences between the signals due to the proton at C-6 of **A** [$\delta = 1.70$ (1H, ddd, $J = 3.0, 5.8, 11.5$ Hz)] and that of **139** [$\delta = 1.58$ (1H, dd, $J = 2.9, 7.3$ Hz)]. Based on these observations, they concluded that (−)-koninginin A is (1*S*,2*R*,5*S*,6*S*,9*S*,10*S*)-**139**.

As I already published our synthesis of (−)-koninginin A assuming Ghisalberti's assignment of its relative configuration to be correct, Xu and Zhu's paper surprised me. So as to resolve the stereochemical problem unambiguously, we carried out an X-ray analysis of our synthetic (−)-koninginin A. The crystals were obtained as very fine needles, and therefore only 14 reflections (instead of the usual 25 reflections) could be used to determine the unit-cell parameters. Nonetheless, the structure of (−)-koninginin A could be resolved as (1*S*,2*R*,5*S*,6*S*,9*S*,10*S*)-**139**, which was in accord with the result of Xu and Zhu.

There were a number of reasons why we could not give, in 1995, **139** as the correct structure of (−)-koninginin A. Although our synthetic intermediates **M**, **O** and **P** as well as the final product **139** were all crystalline, their very fine needle-like crystals were difficult to be analysed by X-ray at the time (1994) when we carried out our synthetic work. Moreover, the ¹H-NMR of **O**, **P** and **139** could not be resolved sufficiently at 300 MHz to allow observation of the splitting patterns of the signals due to the proton at C-6, which were hidden inside the complicated signals at $\delta = 1.18-2.05$ (19H, m) in the case of **O**, $\delta = 1.17 - 1.7$ (11H, m) in the case of **P**, and $\delta = 1.18 - 2.33$ (19H, m) in the case of **139**. Only at 600 MHz the correct and precise observation was possible with regard to the signals due to the proton at C-6.

Our early preoccupation with the veracity of Ghisalberti's stereochemical assignment was the major reason for our erroneous conclusion in 1995. The lack of highly sophisticated analytical instruments such as a 600 MHz NMR spectrometer or a better X-ray diffractometer was the minor reason for our mistake.[32]

5.2.5 Cytoxazone

In 1998 Osada and coworkers isolated 140 mg of a novel cytokine modulator from 18 L of the culture broth of *Streptomyces* RK 95-31 found in a soil sample in Hiroshima Prefecture. That new immunosuppressant was named cytoxazone, and its structure (**140**, Figure 5.25) was elucidated on the basis of its NMR, CD and X-ray analyses. Cytoxazone inhibits the cytokine production via the signaling pathway of Th2 cells, but not Th1 cells. We envisaged that **140** with a 2-oxazolidinone ring might readily be synthesized from *p*-methoxycinnamyl alcohol by employing the Sharpless asymmetric dihydroxylation (AD) as the key reaction.

Figure 5.25 summarizes our synthesis of cytoxazone (**140**).[33] Commercially available *p*-methoxycinnamyl alcohol was converted to the corresponding TBS ether **A**, which was dihydroxylated with AD-mix-$\alpha^{\circledR}$ to give (2*S*, 3*S*)-**B** (99.6% ee). The corresponding cyclic sulfite **C** was cleaved with

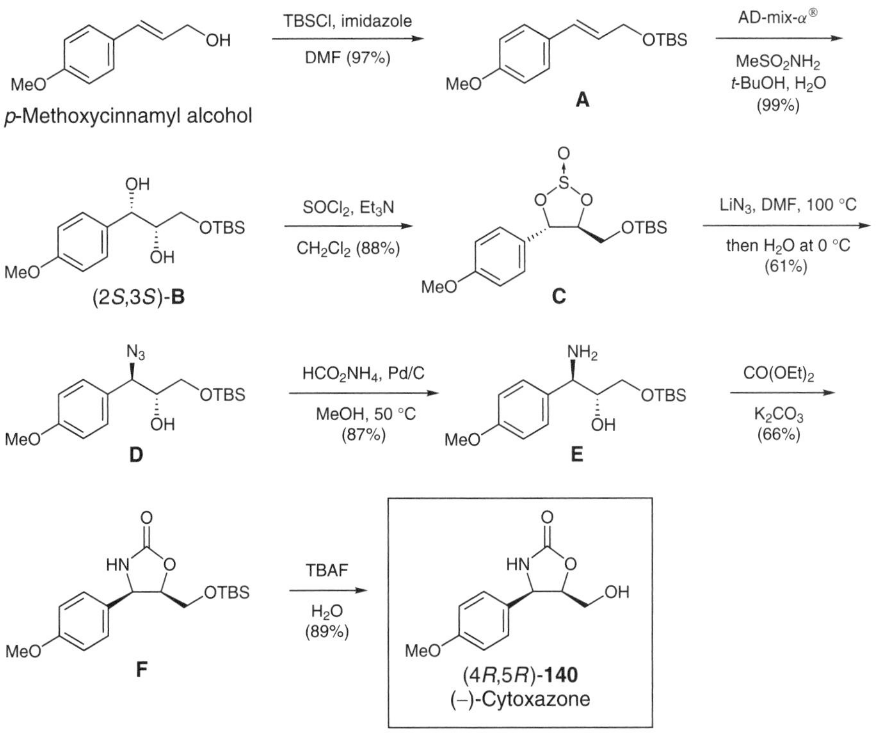

Figure 5.25 *Synthesis of (−)-cytoxazone*

lithium azide to give **D**, whose reduction afforded amino alcohol **E**. 2-Oxazolidinone ring was constructed by treatment of **E** with diethyl carbonate in the presence of potassium carbonate to give **F**. Finally, deprotection of the TBS group afforded (4*R*,5*R*)-(−)-cytoxazone (**140**). The overall yield of this simple synthesis of **140** was 26% (7 steps).

5.2.6 Neuchromenin

In 1996, Hayakawa *et al.* isolated 10 mg of (−)-neuchromenin from 2 L of the culture broth of *Eupenicillium javanicum* var. *meloforme* PF1181 as an inducer of neurite growth of PC12 cells at concentrations of 2.5–10 μg/mL. Its structure was deduced as **141** (Figure 5.26) by extensive spectral analysis, although its absolute configuration remained unknown. In order to establish its absolute configuration, we undertook the synthesis of the enantiomers of **141**.[34]

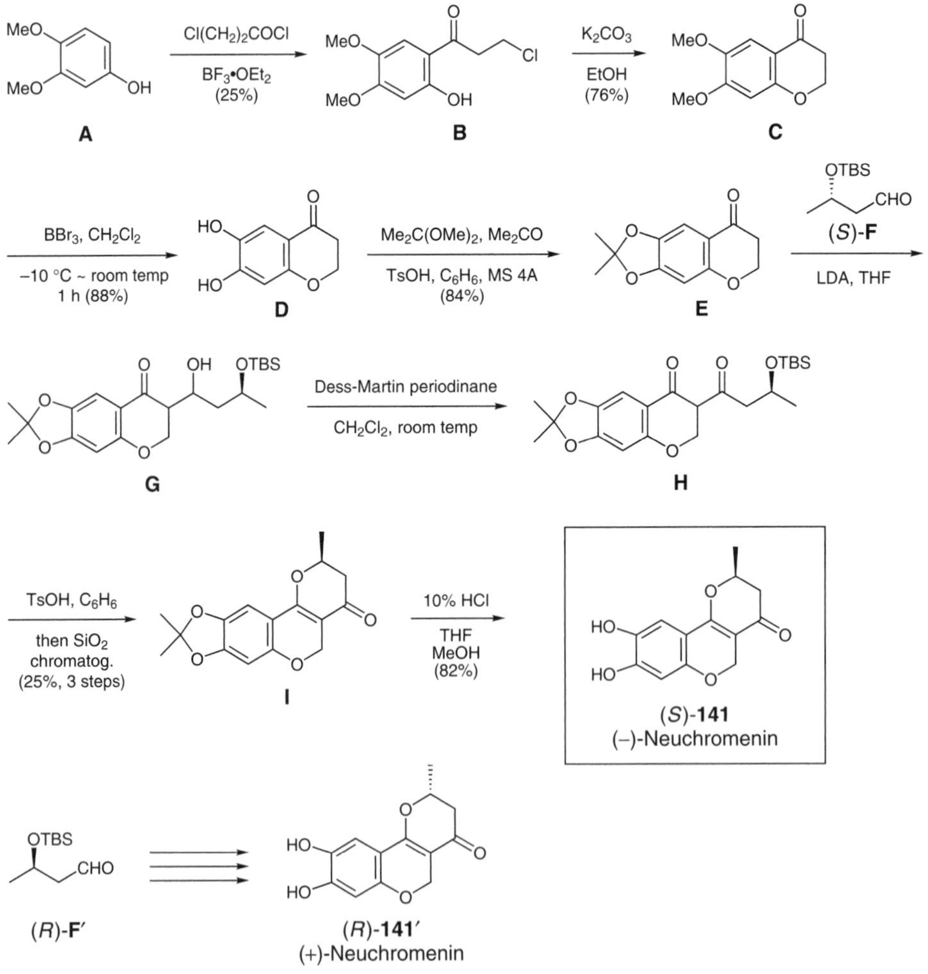

Figure 5.26 *Synthesis of the enantiomers of neuchromenin*

Figure 5.26 summarizes our synthesis of **141**. Friedel–Crafts acylation of neat **A** with 3-chloropropanoyl chloride and boron trifluoride etherate gave **B**. Ring closure of **B** to **C** was effected with potassium carbonate in ethanol. Then, **C** was converted to 1,3-benzodioxole **E** via **D**. The reason for this conversion was to make later deprotection easy.

The remaining and chiral four-carbon building block (*S*)-**F** was prepared from ethyl (*S*)-3-hydroxybutanoate by protection with TBS chloride, reduction with diisobutylaluminum hydride, and oxidation with PCC. Aldol reaction of **E** and (*S*)-**F** furnished **G**. The aldol **G** was oxidized with Dess–Martin periodinane, and the resulting diketone **H** was treated with acid to give **I**. Deprotection of **I** gave (*S*)-**141**, which was levorotatory. The absolute configuration of the naturally occurring (−)-neuchromenin was thus determined as *S*.

The enantiomeric purity of our synthetic (*S*)-**141**, however, was only 59% ee. Acid treatments (**H**→**I** and **I**→**141**) in the course of the synthesis caused partial racemization due to retro-aldol/aldol and/or retro-Michael/Michael reactions. Unfortunately, milder methods of deprotection such as hot dilute acetic acid or trifluoroacetic acid in dichloromethane were not effective enough to give **141** in appreciable yield. In enantioselective syntheses, retro-aldol/aldol and/or retro-Michael/Michael processes are most dangerous reactions to cause partial racemization.

Fortunately in this case, purer (*S*)-**141** could be obtained by further recrystallization of crude (*S*)-**141**, because (±)-**141** was less soluble than (*S*)-**141**. Consequently, 5.8 mg of (*S*)-(−)-neuchromenin (**141**, 91% ee), mp 195–200 °C, $[\alpha]_D^{20} = -491$ (*c* 0.11, MeOH), {cf. natural **141** : mp 195–200 °C, $[\alpha]_D^{20} = -520$ (*c* 0.1, MeOH)} could be obtained after five recrystallizations from ethyl acetate/hexane. Similarly, (*R*)-aldehyde **F′** yielded (*R*)-(+)-neuchromenin (**141′**), the unnatural enantiomer.

5.2.7   Nocardione A and B

In 2000, Otani *et al.* isolated two furano-*o*-naphthoquinones, (−)-nocardione A and (−)-nocardione B (**142** and **143**, Figure 5.27 and 5.28) as new tyrosine phosphatase inhibitors with moderate antifungal and cytotoxic activities. Due to the scarcity of the materials (only 8 mg of **142** and 0.3 mg of **143** were obtained from 4.5 L of the culture broth of *Nocardia* sp. TP-AO 248), their absolute configuration remained unknown. In order to solve this problem, we undertook a synthesis of optically active **142** and **143** with known absolute configuration.[35]

Figure 5.27 summarizes our synthesis of (*R*)-(+)-**143′**, the opposite enantiomer of the naturally occurring (−)-nocardione B. Commercially available 5-methoxy-1-tetralone (**A**) was treated with lithium hexamethyldisilazide (LiHMDS), followed by (*S*)-propylene oxide (**B**) in the presence of scandium triflate in dry toluene to give hydroxy ketone **C**. The hydroxy group of **C** was protected as 2,2,2-trichloroethoxycarbonate to give **D**, and its alicyclic ring was oxidized with selenium dioxide to give **E**.

Zinc and acetic acid reduced the 1,4-naphthoquinone **E** into hydroquinone **F** with concomitant removal of the protective group. Ring closure of **F** was effected under the Mitsunobu conditions to give **G**. Barton's benzeneseleninic anhydride [(PhSeO)₂O] smoothly oxidized **G** to furnish (*R*)-*ent*-nocardione B (**143′**) as dextrorotatory needles with orange color. Naturally occurring (−)-nocardione B (**143**) was therefore shown to possess (*S*)-configuration.[35]

Demethylation of **143′** with aluminum chloride was too harsh to give only (±)-nocardione A [(±)-**142**] as dark red needles. It became apparent that optically active **142** must be prepared by using a more readily removable protective group for the phenolic hydroxy group.

(*S*)-(−)-Nocardione A (**142**) could be synthesized, as shown in Figure 5.28, by employing a benzyl group as the protective group, which could be removed successfully by hydrogenolysis. The synthesis proceeded in the same manner as for **143′** via **C**, **D**, **E**, **F**, **G** and **H**. The final hydrogenolysis was successful only when a large amount (40–50 wt% of the substrate **H**) of palladium-charcoal was added to **H** in DMF under

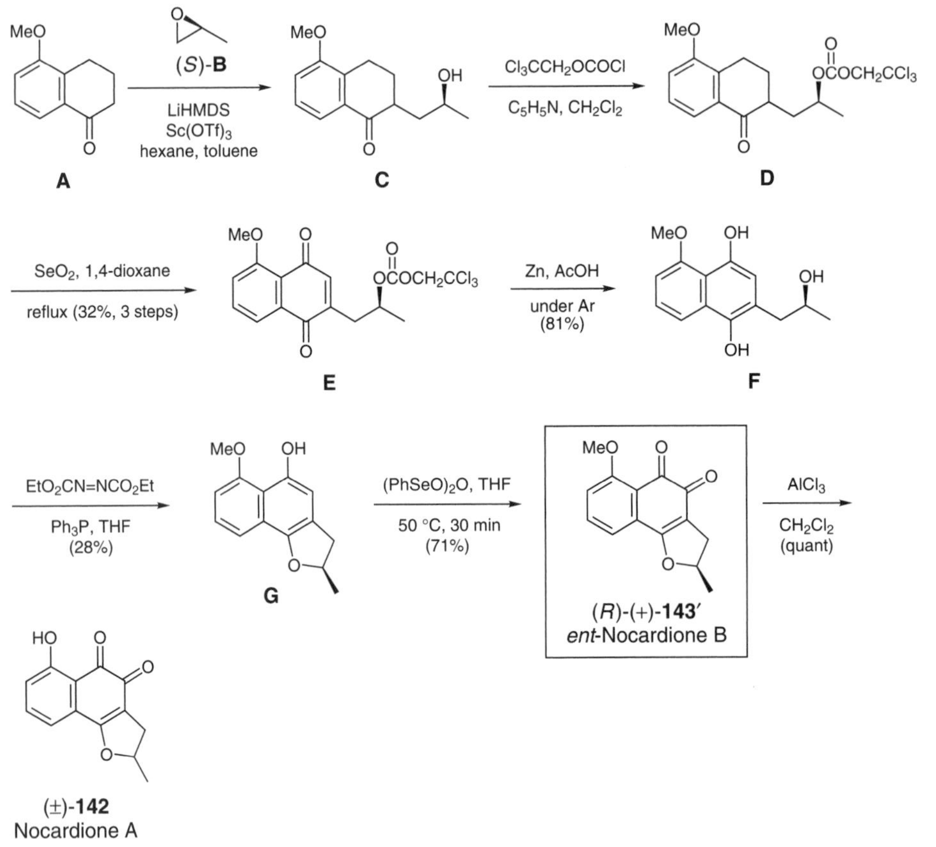

Figure 5.27   *Synthesis of (+)-ent-nocardione B*

hydrogen for a short period (5–10 min). (*S*)-(−)-Nocardione A (**142**) was obtained, after spontaneous air oxidation, as dark red needles, whose spectral and chiroptical properties were in accord with those of the natural product.[35]

5.2.8   Cytosporone E

In 2000, Clardy and coworkers isolated cytosporones A–E as new octaketide metabolites of the endophytic fungus CR 200 (*Cytospora* sp.). Cytosporones D and E (**144**, Figure 5.29) were reported to show strong antimicrobial activity, while the others were biologically inactive. Their structures were clarified mainly by NMR analysis.

In the case of cytosporone C (Figure 5.29), its structure was solved by X-ray analysis. According to Clardy *et al.*, the space group in which it crystallized required cytosporone C to be isolated as a racemic mixture. Although the specific rotations of cytosporones were not reported, it could be speculated that cytosporones in general might have been isolated as racemates. This speculation aroused my curiosity to scrutinize whether there would be any difference between the enantiomers of cytosporone E (**144**) with regard to their antimicrobial activity.

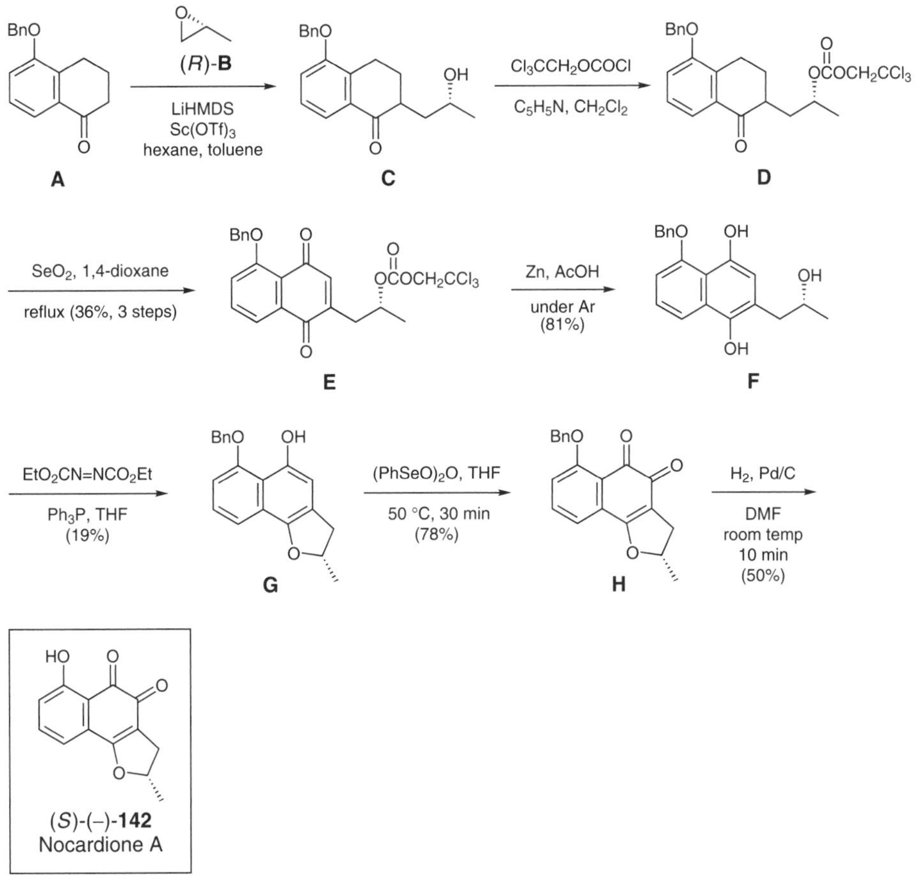

Figure 5.28   *Synthesis of (−)-nocardione A*

Figure 5.29 summarizes our synthesis of the enantiomers of cytosporone E (**144** and **144′**).[36] Commercially available methyl 3,4,5-trihydroxybenzoate (**A**) was chosen as the starting material. After benzylation and bromination, **A** yielded **B**, to which was attached the side-chain by Suzuki–Miyaura coupling to give **C**. Asymmetric dihydroxylation of **C** with AD-mix-β® proceeded sluggishly to furnish enantiomerically impure (3*R*,1′*R*)-**D** (45% ee as determined by HPLC analysis on Chiralcel® OG) in 54% yield after a week at 4 °C. Barton deoxygenation of thiocarbonyl derivative **E** gave (*S*)-**F**, which was hydrogenolysed to afford (*S*)-**144** of low ee.

We then attempted purification of impure (*S*)-**144** by enantioselective HPLC. Fortunately, TBS derivative **G** was found to be separable by preparative HPLC on Chiralcel® OD to give pure (*S*)-**G**. Deprotection of the TBS group of (*S*)-**G** under conventional conditions with TBAF caused partial racemization of (*S*)-**144**. However, treatment of (*S*)-**G** with dilute ethanolic hydrochloric acid at room temperature caused no appreciable racemization to give (*S*)-**144** (98.4% ee), $[\alpha]_D^{24} = -90.7$ (acetone), in 40% yield. Similarly, (*R*)-**144′** was also synthesized by employing AD-mix-α® instead of AD-mix-β®. (*R*)-Cytosporone E (**144′**), $[\alpha]_D^{25} = +91.3$ (acetone) could be obtained pure (>99% ee). As shown in the present case, use of preparative HPLC is becoming more and more important in the preparation of pure enantiomers.

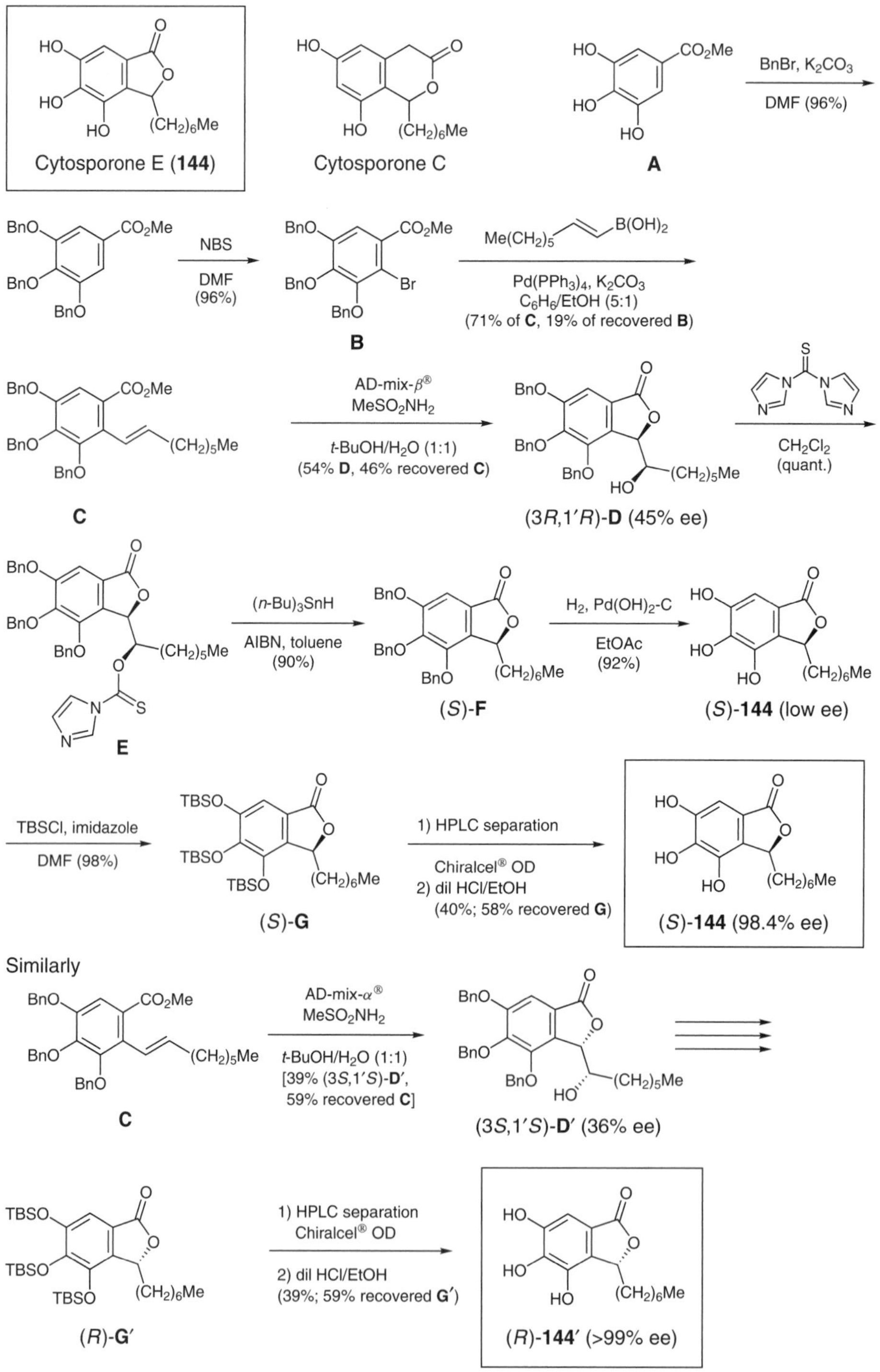

Figure 5.29 *Synthesis of the enantiomers of cytosporone E*

Antimicrobial activities of (*R*)-**144′** and (*S*)-**144** were examined at Sankyo Co. by employing twelve different micro-organisms. The enantiomers showed only very weak antimicrobial activity of the same degree. Their activity was far weaker than that of the practical antimicrobial agent, itraconazole® (Janssen). It sometimes happens that the synthetic samples of a natural product are better sources of exact and correct biological evaluation.

5.3 Other bioactive metabolites of micro-organisms

There are some microbial metabolites that show insecticidal or plant-growth regulating activities. Two examples will be given in this section. Both of them have interesting structures.

5.3.1 Monocerin

Monocerin (**145**, Figure 5.30) was first isolated in 1970 by Aldridge and Turner as an antifungal metabolite from culture filtrates of *Exserohilum monoceras* (=*Helminthosporium monoceras*), which protect wheat against powdery mildew (*Erysyphe graminis*). Subsequently, in 1979, Grove and Pople identified **145** as an insecticidal constituent of an entomogenous fungus *Fusarium larvarum*. Phytotoxic property of **145** was later reported in 1982 by Robeson and Strobel, who identified **145** as a phytotoxin produced by *Exserohilum turcicum*. The stereostructure of monocerin as depicted in **145** was proposed on the basis of its ^{1}H-NMR spectral and chiroptical studies.

We became interested in synthesizing the naturally occurring (+)-enantiomer (**145**) of monocerin, and accomplished the synthesis in 1989, as shown in Figure 5.30.[37]

Our strategy was to couple the aromatic part **A** with the aliphatic part **B** to give **C**. (*S*)-1,2-Epoxypentane served as the starting material to prepare (*S*)-**B**. Then, olefinic alcohol **D** was epoxidized with MCPBA, and the resulting epoxide was treated with boron trifluoride etherate to give **E** as an isomeric mixture. Treatment of **E** with 2.3 eq of *n*-butyllithium provided the corresponding dianion, which was quenched with carbon dioxide to give **F**. Lactonization of **F** under the Mitsunobu conditions furnished a mixture of two lactones **G** and **H**, which were separated by silica-gel chromatography. Treatment of **G** with 1.1 eq. of boron tribromide in dichloromethane for 30 min at −20 °C resulted in the removal of only one methyl group to give (+)-monocerin (**145**).[37]

5.3.2 Pinthunamide

Pinthunamide (**146**, Figure 5.31) is a sesquiterpene carboxamide isolated in 1989 from a fungus *Ampulliferina* sp. by Kimura *et al.*, and accelerates the root growth of lettuce seedlings by 150% at a dosage of 300 mg/L. Its unique structure with a bridged tricyclic ring system was determined by X-ray analysis together with other spectral methods, although its absolute configuration remained unknown. The unique structure **146** led us to synthesize it starting from a compound with known absolute configuration in order to determine the absolute configuration of pinthunamide.

Our synthesis started from hydroxy ketone **B** (Figure 5.31), which was obtained by asymmetric reduction of prochiral diketone **A** with fermenting baker's yeast.[38] The key step in the present synthesis was the ring formation by intramolecular alkylation of **C** to give **D**. To obtain **C**, the *endo*-hydroxy group of **B** was first epimerized via retro-aldol/aldol by treatment with *p*-toluenesulfonic acid in carbon tetrachloride. The tricyclic intermediate **D** was converted to (+)-pinthunamide (**146**), mp 187–189 °C, $[\alpha]_D^{21.5} = +60$ (EtOH), which was identical with the natural product. Its absolute configuration was thus determined as depicted in **146**.[39]

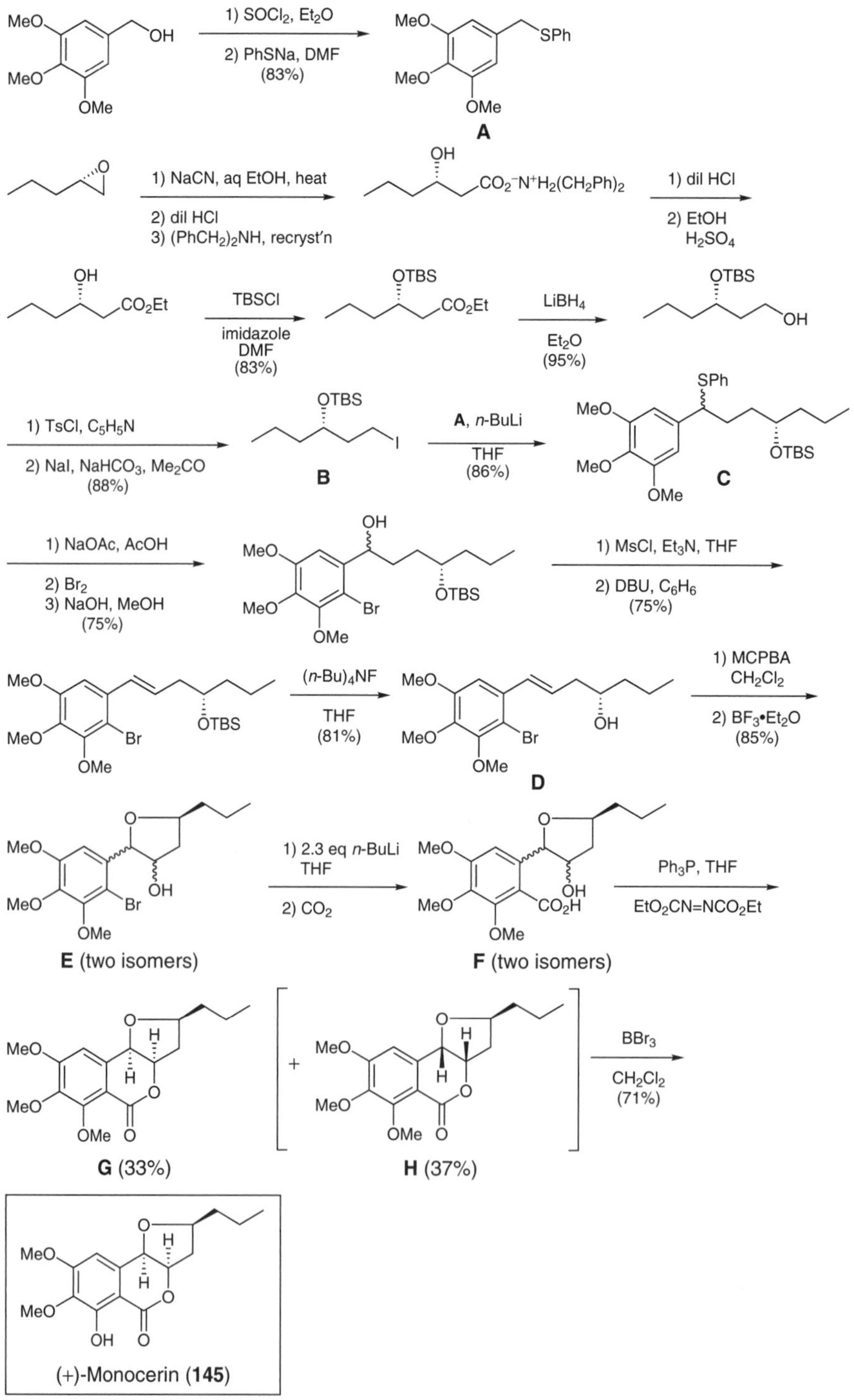

Figure 5.30 *Synthesis of (+)-monocerin. Modified by permission of Shokabo Publishing Co., Ltd*

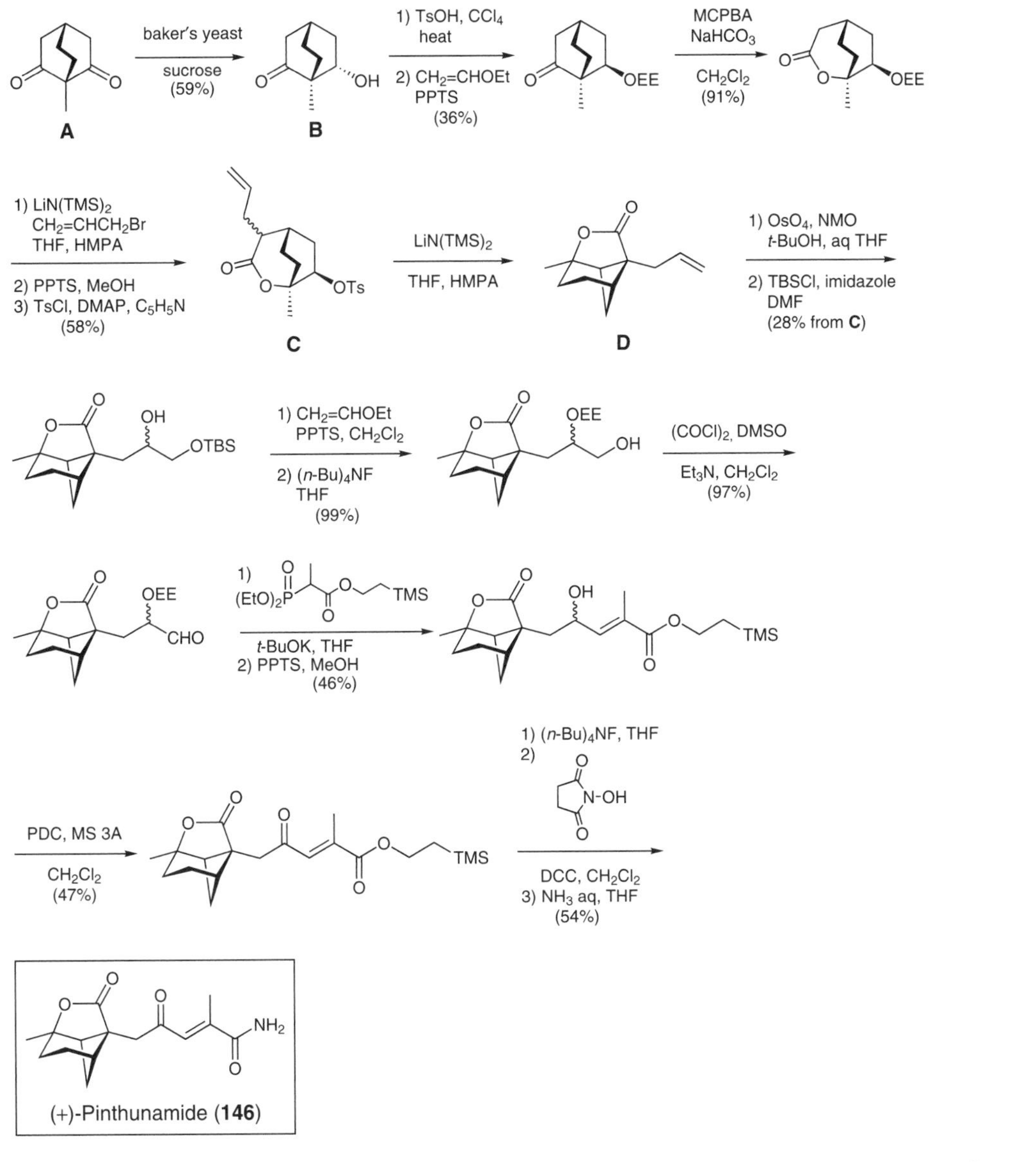

Figure 5.31   *Synthesis of (+)-pinthunamide. Modified by permission of Shokabo Publishing Co., Ltd*

Micro-organisms produce various compounds with unique structures. Many of them serve as interesting targets for total synthesis. When the microbial metabolites are produced in scarce amounts, organic synthesis can be an important tool to supply them sufficiently.

References

1. Mori, K. *Tetrahedron Lett*. **1981**, *22*, 3431–3432.
2. Mori, K.; Yamane, K. *Tetrahedron* **1982**, *38*, 2919–2921.

3. Mori, K. *Tetrahedron* **1983**, *39*, 3107–3109.
4. Hara, O.; Beppu, T. *J. Antibiot*. **1982**, *35*, 349–358.
5. Miyake, K.; Horinouchi, S.; Yoshida, M.; Chiba, N.; Mori, K.; Nogawa, N.; Morikawa, N.; Beppu, T. *J. Bacteriol*. **1989**, *171*, 4298–4302.
6. Mori, K.; Chiba, N. *Liebigs Ann. Chem*. **1989**, 957–962.
7. Mori, K.; Chiba, N. *Liebigs Ann. Chem*. **1990**, 31–37.
8. Horinouchi, S.; Beppu, T. *Proc. Jpn. Acad. Ser. B* **2007**, *83*, 277–295.
9. Mori, K.; Funaki, Y. *Tetrahedron Lett*. **1984**, *25*, 5291–5294.
10. Mori, K.; Funaki, Y. *Tetrahedron* **1985**, *41*, 2369–2377.
11. Mori, K.; Funaki, Y. *Tetrahedron* **1985**, *41*, 2379–2386.
12. Funaki, Y.; Kawai, G.; Mori, K. *Agric. Biol. Chem*. **1986**, *50*, 615–623.
13. Abe, T.; Mori, K. *Biosci. Biotechnol. Biochem*. **1994**, *58*, 1671–1674.
14. Mori, K.; Uenishi, K. *Liebigs Ann*. **1996**, 1–6.
15. Mori, K.; Kinsho, T. *Liebigs Ann. Chem*. **1991**, 1309–1315.
16. (a) Takikawa, H.; Hashimoto, T.; Matsuura, M.; Tashiro, T.; Kitahara, T.; Mori, K.; Sasaki, M. *Tetrahedron Lett*. **2008**, *49*, 2258–2261. (b) Hashimoto, T.; Tashiro, T.; Kitahara, T.; Mori, K.; Sasaki, M.; Takikawa, H. *Biosci. Biotechnol. Biochem*., **2009**, *73*, 2299–2302.
17. Okada, K.; Koseki, K.; Kitahara, T.; Mori, K. *Agric. Biol. Chem*. **1985**, *49*, 487–493.
18. Yabuta, G.; Ichikawa, Y.; Kitahara, T.; Mori, K. *Agric. Biol. Chem*. **1985**, *49*, 495–499.
19. Kitahara, T; Matsuoka, T.; Katayama, M.; Marumo, S.; Mori, K. *Tetrahedron Lett*. **1984**, *25*, 4685–4688.
20. Kitahara, T.; Kurata, H.; Matsuoka, T; Mori, K. *Tetrahedron* **1985**, *41*, 5475–5485.
21. Mori, K.; Tamura, H. *Liebigs Ann. Chem*. **1988**, 97–105.
22. Kitahara, T.; Kurata, H.; Mori, K. *Tetrahedron* **1988**, *44*, 4339–4349.
23. Mori, K.; Tomioka, H.; Fukuyo, E; Yanagi, K. *Liebigs Ann. Chem*. **1993**, 671–681.
24. Mori, K.; Sato, K. *Tetrahedron* **1982**, *38*, 1221–1225.
25. Mori, K.; Fujioka, T. *Tetrahedron* **1984**, *40*, 2711–2720.
26. Mori, K.; Takechi, S. *Tetrahedron* **1985**, *41*, 3049–3062.
27. Mori, K.; Koseki, K. *Tetrahedron* **1988**, *44*, 6013–6020.
28. Yoshida, M.; Hoshikawa, Y.; Koseki, K.; Mori, K.; Beppu, T. *J. Antibiot*. **1990**, *43*, 1101–1106.
29. Mori, K.; Abe, K. *Polish J. Chem*. **1994**, *68*, 2255–2263.
30. Mori, K.; Abe, K. *Liebigs Ann*. **1995**, 943–948.
31. Xu, X.-X.; Zhu, Y.-H. *Tetrahedron Lett*. **1995**, *36*, 9173–9176.
32. Mori, K.; Bando, M.; Abe, K. *Biosci. Biotechnol. Biochem*. **2002**, *66*, 1779–1781.
33. Seki, M.; Mori, K. *Eur. J. Org. Chem*. **1999**, 2965–2967.
34. Tanada, Y.; Mori, K. *Eur. J. Org. Chem*. **2001**, 1963–1966.
35. Tanada, Y.; Mori, K. *Eur. J. Org. Chem*. **2001**, 4313–4319.
36. Ohzeki, T.; Mori, K. *Biosci. Biotechnol. Biochem*. **2003**, *67*, 2584–2590.
37. Mori, K.; Takaishi, H. *Tetrahedron* **1989**, *45*, 1639–1646.
38. Mori, K.; Nagano, E. *Biocatalysis* **1990**, *3*, 25–36.
39. Mori, K.; Matsushima, Y. *Synthesis* **1993**, 406–410.

6

Synthesis of Marine Bioregulators, Medicinals and Related Compounds

Extensive human endeavor to elucidate the structures and functions of natural products of terrestrial organisms gave us substantial knowledge about terrestrial natural products. In contrast, modern human endeavor to clarify the structures and functions of marine natural products began in the 1950s after World War II. There remains many things to do. Studies on biofunctional molecules of marine origin will broaden our knowledge about their roles in marine ecological system, and also give us opportunities to design useful medicinals by modifying their structures. This chapter describes my synthetic works on marine antifeedants, medicinal candidates, and glycosphingolipids of medicinal interests.

6.1 Marine natural products of ecological importance such as antifeedants

In the marine ecological system, we can observe a subtle balance among the population of various different organisms. For example, seaweeds and seashells produce antifeedants against fishes so that they can avoid the attack by fishes. Just like terrestrial organisms, marine organisms employ hormones and pheromones to regulate their own lives. In this section I will give four examples of the syntheses of these ecologically important molecules.

6.1.1 Stypoldione

In 1979, Fenical and coworkers isolated (−)-stypoldione (**147**, Figure 6.1) as an ichthyotoxic and cytotoxic metabolite of the tropical alga, *Stypopodium zonale*, in the Western Caribbean Sea. Its structure was established as **147** by the X-ray analysis of its red crystal, without assignment of the absolute configuration. This unique diterpene with a spiro-*o*-benzoquinonefuran C_7 unit is extremely toxic to the reef-dwelling herbivorous fish *Eupomacentrus leucostictus* even at a dosage of $1.0\,\mu g/mL$, and probably functions as a chemical defense weapon of the alga. Stypoldione (**147**) is also an inhibitor of cell division in the fertilized sea-urchin egg assay at a $1.1\,\mu g/mL$ level of concentration. Gerwick and Whatley later isolated **147** from juvenile sea hare, *Aplysia dactylomela*, feeding on *S. zonale*, and thus proved the metabolic transfer between the brown alga and the sea hare.

The unique structure **147** of stypoldione attracted our attention, and we reported in 1992 an enantioselective synthesis of **147**, as shown in Figures 6.1 and 6.2.[1,2] As to the construction of the ring system

Chemical Synthesis of Hormones, Pheromones and Other Bioregulators Kenji Mori
© 2010 John Wiley & Sons, Ltd

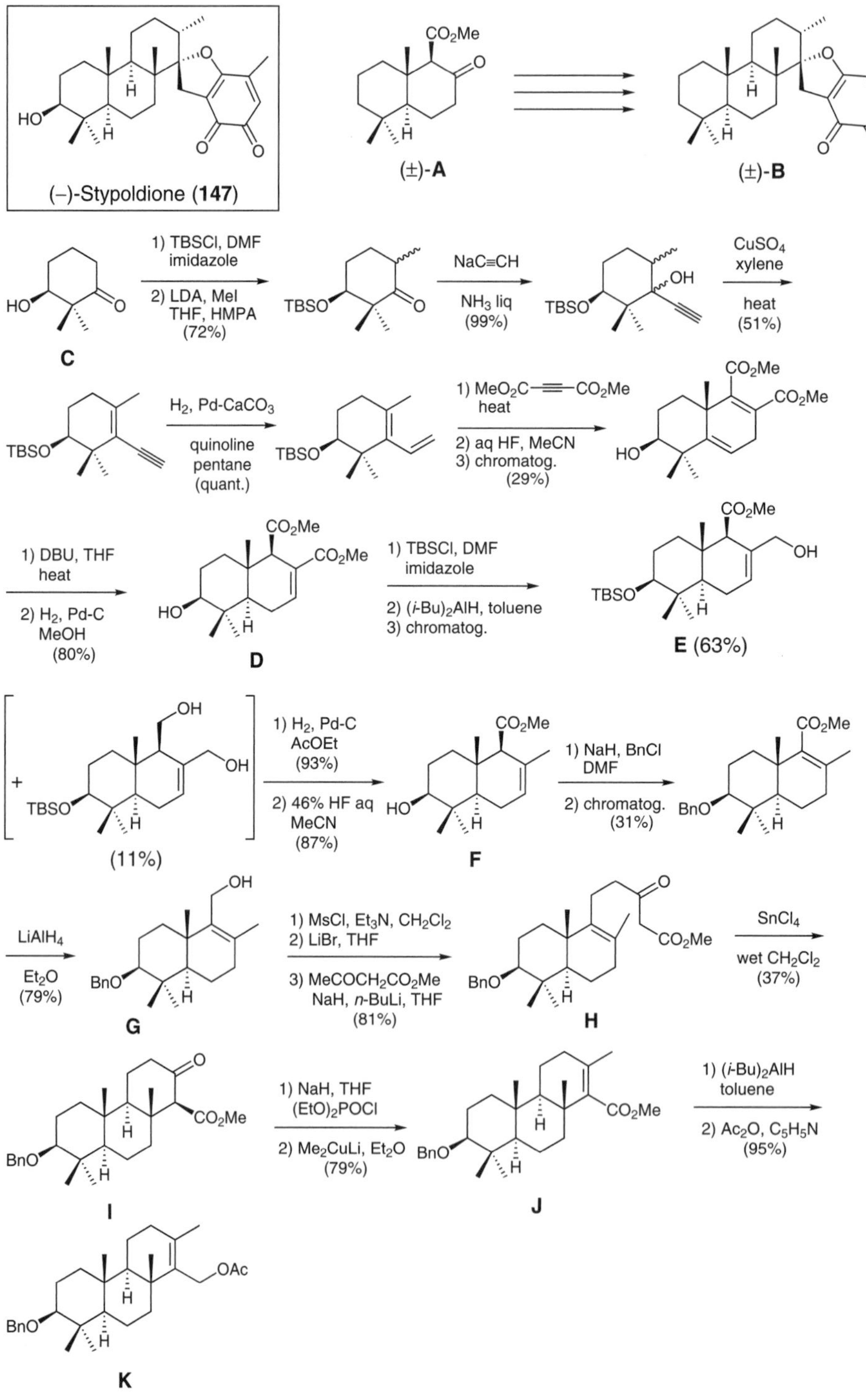

Figure 6.1 *Synthesis of stypoldione (1). Modified by permission of Shokabo Publishing Co., Ltd*

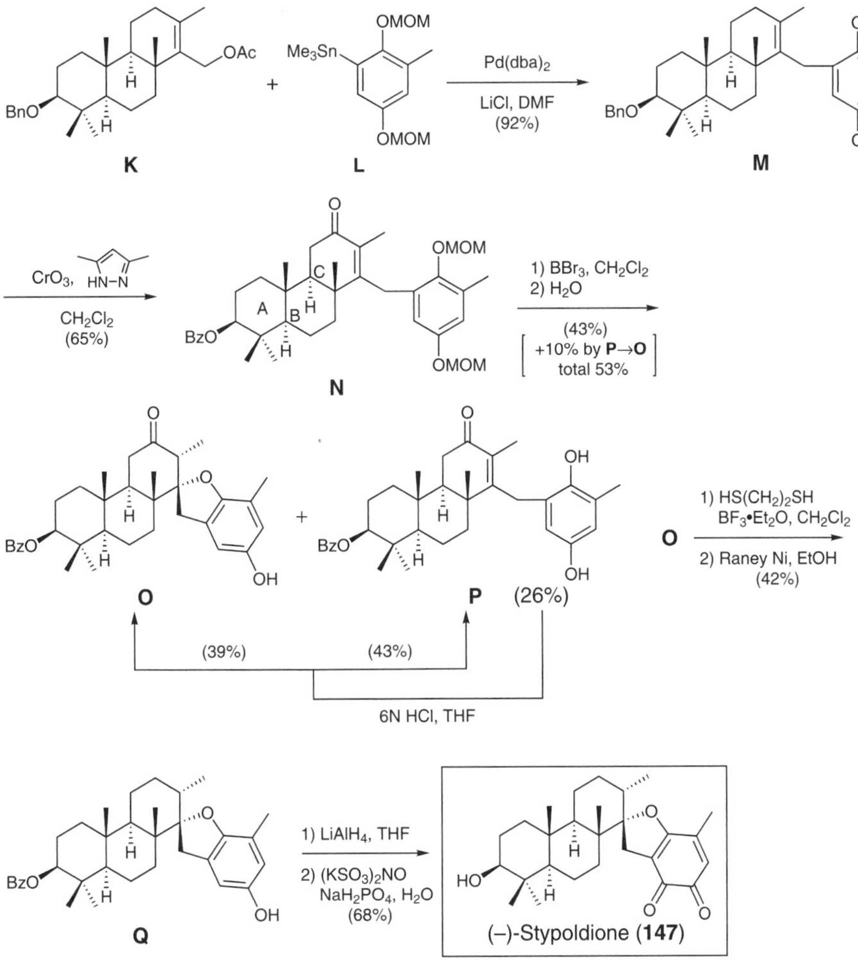

Figure 6.2   *Synthesis of stypoldione (2). Modified by permission of Shokabo Publishing Co., Ltd*

of **147**, we essentially followed our route employed in the synthesis of (±)-14-deoxystypoldione (**B**) from (±)-**A**.[3] Our starting material was the (*S*)-hydroxy ketone **C**, which was readily available by reducing the corresponding diketone with fermenting baker's yeast. Conversion of **C** to hydroxy diester **D** was already mentioned in connection with the synthesis of (−)-polygodial (**64**, Figure 3.13). Then, **D** was further manipulated to afford alcohol **E**.[4] Hydrogenolysis of the allylic hydroxy group of **E** gave **F**, which was converted to β-keto ester **H** via **G**.[5] Cationic cyclization of **H** with tin(IV) chloride gave tricyclic β-keto ester **I**. This ester **I** afforded the C_{20}-building block **K** via **J**.

Kosugi–Stille coupling of **K** with the stannane **L** gave **M** in 92% yield. Allylic oxidation of **M** furnished **N** with concomitant conversion of the benzyl protective group to benzoyl. Deprotection of the MOM protective groups of **N** with boron tribromide gave a mixture of **O** and **P** by partial Michael addition of the liberated phenolic hydroxy group of **P** to the conjugated ketone to give **O**. The noncyclized **P** gave an additional amount of **O** by acid treatment. Subsequently, the carbonyl group of **O** was reductively removed via the corresponding dithioacetal to give **Q**. Reduction of **Q** with lithium aluminum hydride removed the benzoyl protective group, and finally the product was oxidized with Frémy's salt to give $(-)$-stypoldione (**147**) as dark red needles, $[\alpha]_D^{25} = -62$ (CHCl$_3$).[2] Since the natural product was also levorotatory, its absolute configuration was established as depicted in **147**. Because our 1992 synthesis of **G** was inefficient (2.0% overall yield based on **C**, 16 steps),[1] another synthesis of **G** was described in our 1995 full paper (6.7% overall yield based on **C**, 12 steps).[2]

6.1.2   *meso-* and (±)-Limatulone

In 1985, Faulkner and coworkers isolated limatulone (**148a** and **148b**, Figure 6.3) from the intertidal limpet *Achmeia (Collisella) limatula* as feeding inhibitor against fish and crab. Indeed, it is the most potent fish-feeding inhibitor, and almost an order of magnitude more effective than polygodial (**64**, Figure 3.13), the well-known antifeedant. Food pellets containing limatulone at the level of 0.05% dry weight or more induces regurgitation in the intertidal fish *Gibbonsia elegans*, a known limpet predator.

As depicted in structures **148a** and **148b**, limatulone is a structurally unusual triterpene, consisting of two identical C$_{15}$ units. This unique structure caused a problem in the course of structure determination. Namely, it was difficult to decide whether the naturally occurring and optically inactive limatulone was **148a** or **148b**. We became interested in this stereochemical problem, and synthesized both *meso-* and (±)-limatulone in 1993 according to the retrosynthetic analysis as shown in Figure 6.3.[6,7]

We assumed that *meso-* and (±)-compounds such as **A** and **A′** may be separable at a certain stage of the synthesis. Intermediates like *meso*-**A** and (±)-**A′** will readily be generated by dimerization or its equivalent operation of the racemic key intermediate (±)-**B**. This route is simpler and more efficient than other routes that employ optically active intermediates. The intermediate (±)-**B** may be prepared from the

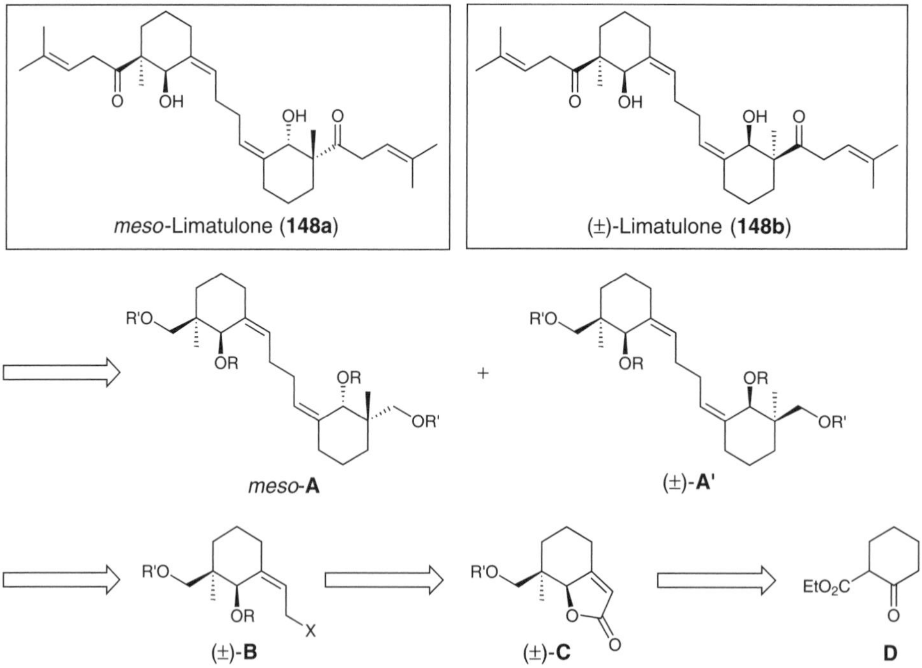

Figure 6.3   *Retrosynthetic analysis of limatulone. Modified by permission of Shokabo Publishing Co., Ltd*

known β-keto ester **D** via the lactone **C**, which possesses all of the necessary structural features in the cyclic moieties of limatulone. Since the conversion of *meso*-**A** and (±)-**A′** into *meso*-**148a** and (±)-**148b** may not be so difficult, the high efficiency of the separation of isomers **A** and **A′** will be the key to the success of the synthesis.

Figure 6.4 summarizes conversion of β-keto ester **D** to *meso*-**A** and (±)-**A′**. Keto carboxylic acid **E** was prepared from **D**, and heated with acetic anhydride in the presence of sodium acetate to give a separable mixture of lactones **F** and **G**. The undesired **F** afforded an additional amount of **G** by hydrolysis and relactonization. Further synthetic operations converted **G** to allylic alcohol **H**, which gave the two key C_{10}-building blocks **I** and **J**. Alkylation of sulfone **J** with bromide **I** was followed by reductive removal of the phenylsulfonyl group to give a mixture of **K** and **L**, which could not be separated. Fortunately, *meso*-**A** and (±)-**A′**, which were obtained by removing the EE protective group of **K** and **L**, respectively, were separable by silica-gel chromatography, and both were obtained as crystals. X-ray analysis of the crystal melting at 95–97 °C revealed its structure as *meso*-**A**. Another crystal melting at 89–91 °C must therefore be (±)-**A′**.

Further conversion of *meso*-**A** and (±)-**A′** to *meso*-**148a** and (±)-**148b**, respectively, is shown in Figure 6.5. Both **148a** and **148b** were oils. We were lucky to have *meso*-**A** as crystals of good quality, because this enabled us to solve its structure by X-ray analysis.

The ^{1}H- and ^{13}C-NMR spectra of our synthetic limatulone **148a** and **148b** were similar but slightly different from each other. These spectra were compared with the authentic spectra of natural limatulone kindly provided by Prof. Faulkner. The ^{1}H- and ^{13}C-NMR spectra of Faulkner's limatulone were identical with those of (±)-**148b**. Therefore, the natural limatulone reported in 1985 was the racemic one. To our surprise, however, the ^{1}H-NMR spectrum of another fraction from the HPLC separation of *Achmeia limatula* metabolite coincided with that of *meso*-**148a**. The presence of this fraction was not reported in the isolation paper, but Prof. Faulkner kindly provided us with a copy of the 360 MHz ^{1}H-NMR spectrum of that fraction. It therefore became clear that the limpet *Achmeia limatula* produces both *meso*- and racemic limatulones, **148a** and **148b**.[6,7] It may be of interest to study the biosynthesis of limatulone to clarify the reason for nonstereoselective cyclization of the squalene precursor.

6.1.3   Testudinariol A

In 1997, Spinella *et al.* isolated testudinariol A (**149**, Figure 6.6) as a metabolite of the marine mollusc *Pleurobrancus testudinarius*. This compound is a structurally unique triterpene alcohol, and thought to be a defensive allomone of *P. testudinarius*, because **149** was ichthyotoxic against a fish *Gambusia affinis*. The partially cyclized squalene skeleton present in **149** is unique and biosynthetically unusual as in the case of limatulone (**148**). The unique structure **149** of testudinariol A led us to achieve its synthesis in 2001.[8,9]

Figure 6.6 shows our synthetic plan for testudinariol A (**149**). Because the structural feature of target molecule **149** is its C_2-symmetry, **149** can be obtained by dimerization or its equivalent operation of **A**. The intermediate **A** may be prepared from **B** by (*Z*)-selective installation of the two-carbon appendage. For the stereoselective construction of the cyclopentane portion of **B**, an intramolecular ene reaction is appropriate employing **C** as the substrate. The intramolecular oxy-Michael-type cyclization of **D** has been adopted to prepare the tetrahydropyran ring of **C**. The intermediate **D** can be synthesized from **F** [(*R*)-glycidol] via the known diol **E**.

Our synthesis of (+)-testudinariol A (**149**) is shown in Figures 6.7 and 6.8. The starting (*R*)-glycidol was treated with allylmagnesium chloride to give diol **A**. Selective and stepwise protections of hydroxy groups of **A** were followed by oxidative cleavage of the terminal double bond to give **B**. The aldehyde **B** was subjected to the Horner–Wadsworth–Emmons reaction to afford **C** after removal of the TBS group.

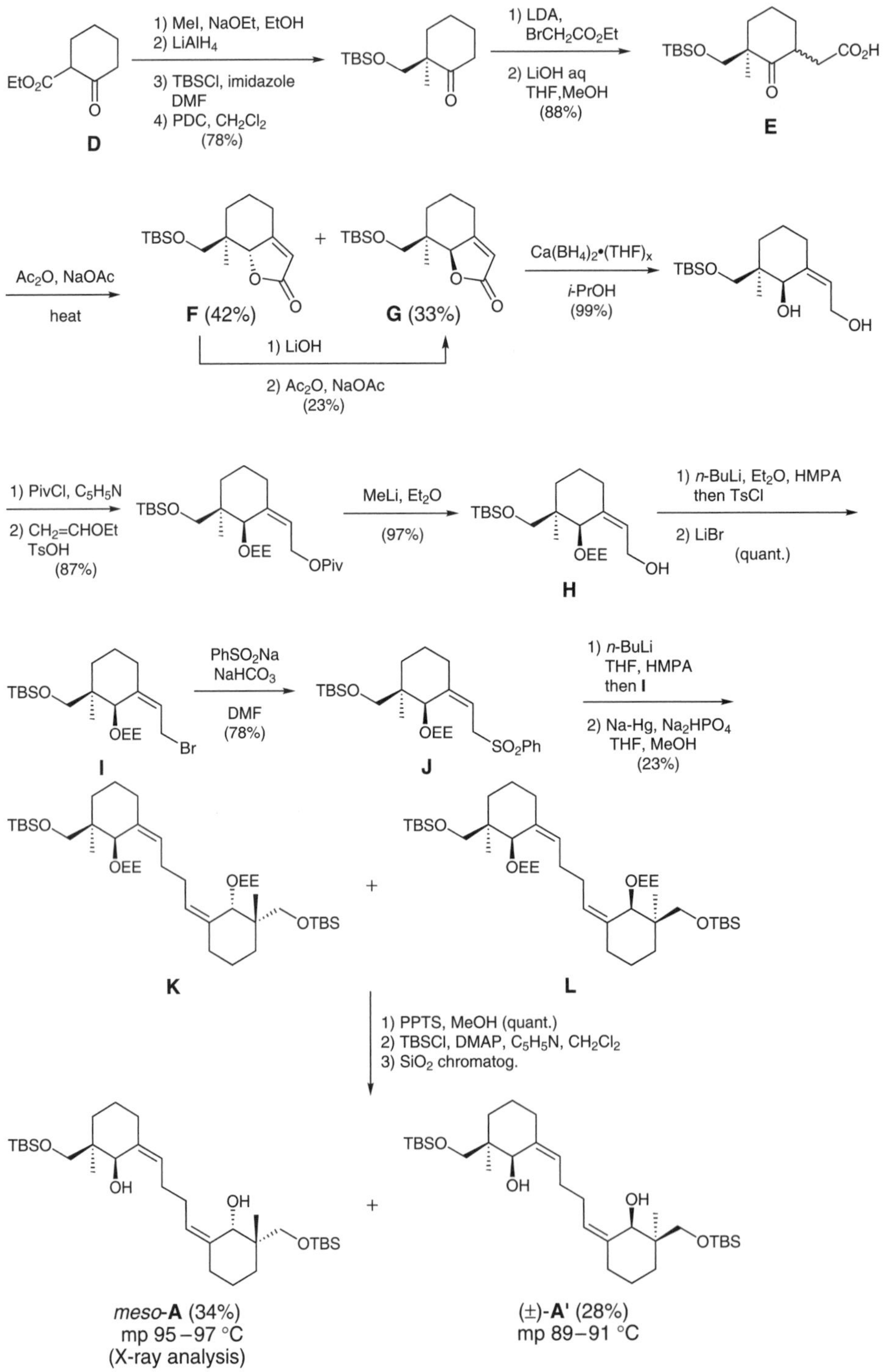

Figure 6.4   *Synthesis of limatulone (1). Modified by permission of Shokabo Publishing Co., Ltd*

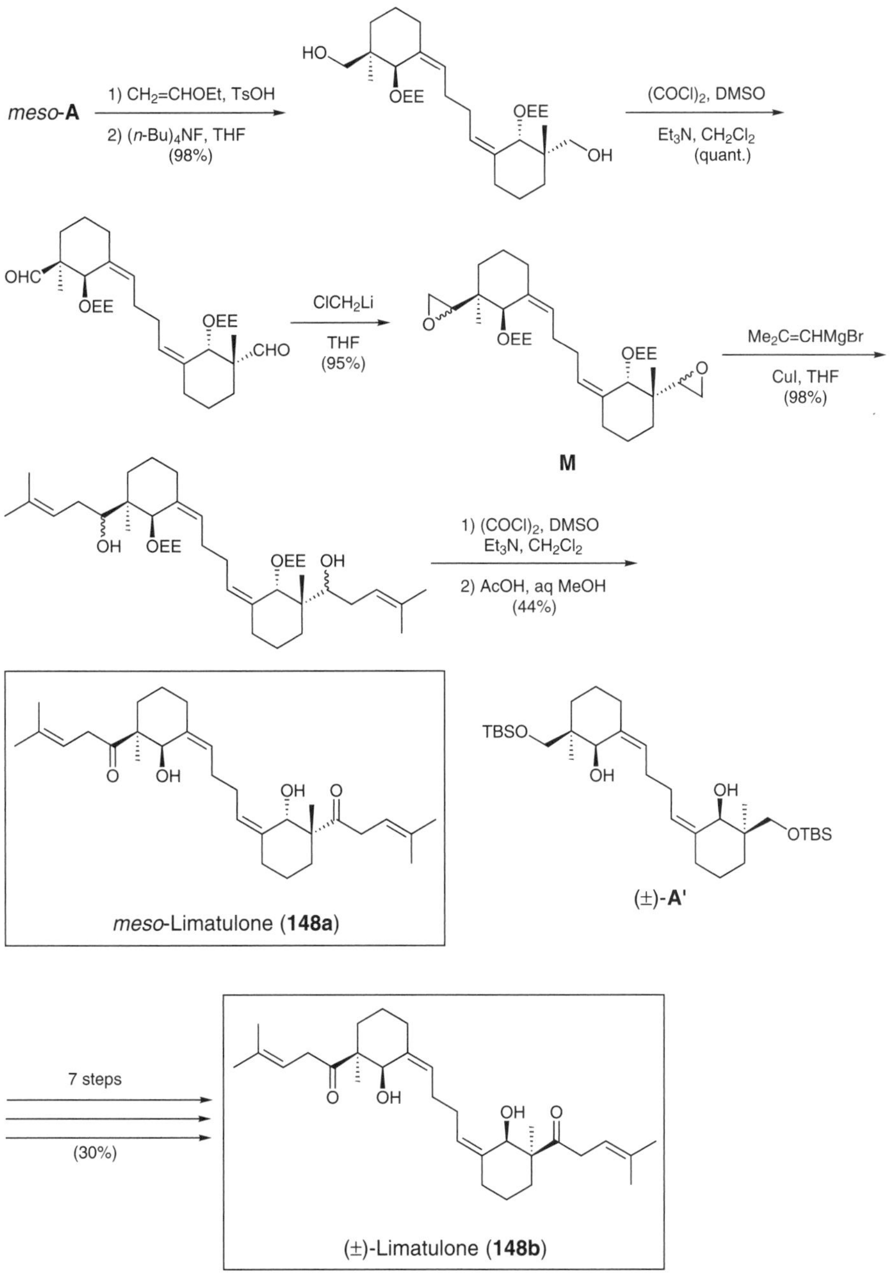

Figure 6.5 *Synthesis of limatulone (2). Modified by permission of Shokabo Publishing Co., Ltd*

The hydroxy ester **C** was treated with potassium *t*-butoxide (0.1 eq) in THF at −10 to 4 °C to give in 93% yield a mixture of the desired **D** and its three isomers in a 5:5:2:2 ratio. The unwanted three isomers were recycled to give an additional amount of **D**. By repeating this process for three times, **D** could be obtained in 68% yield.

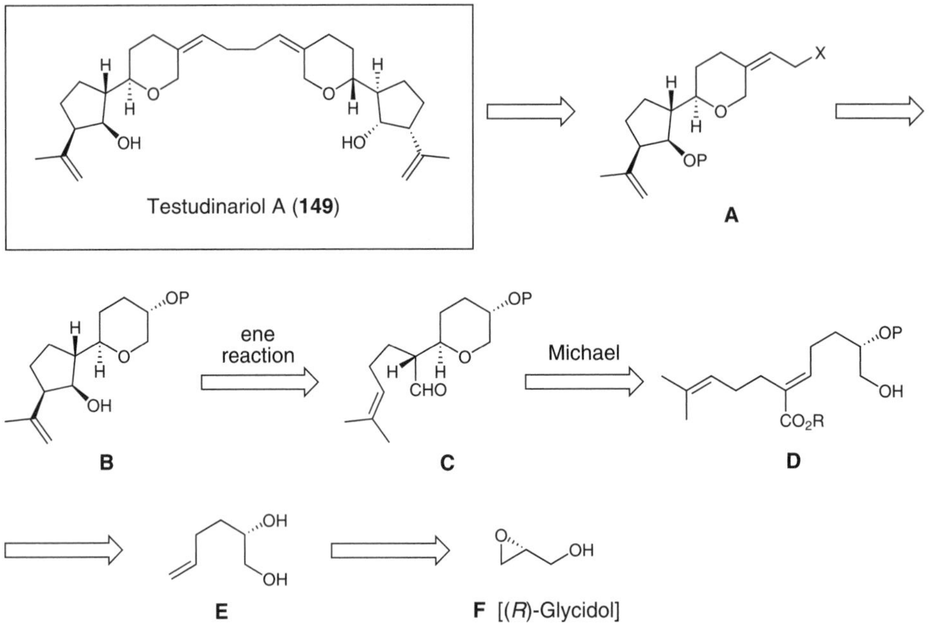

Figure 6.6 *Synthetic plan for testudinariol A*

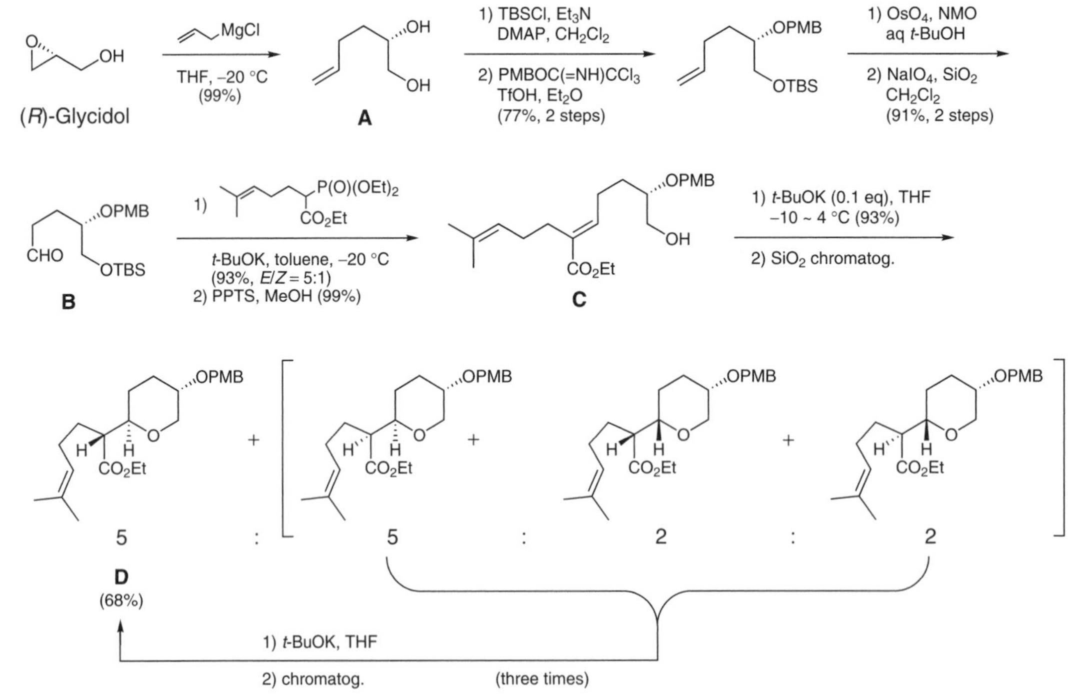

Figure 6.7 *Synthesis of testudinariol A (1)*

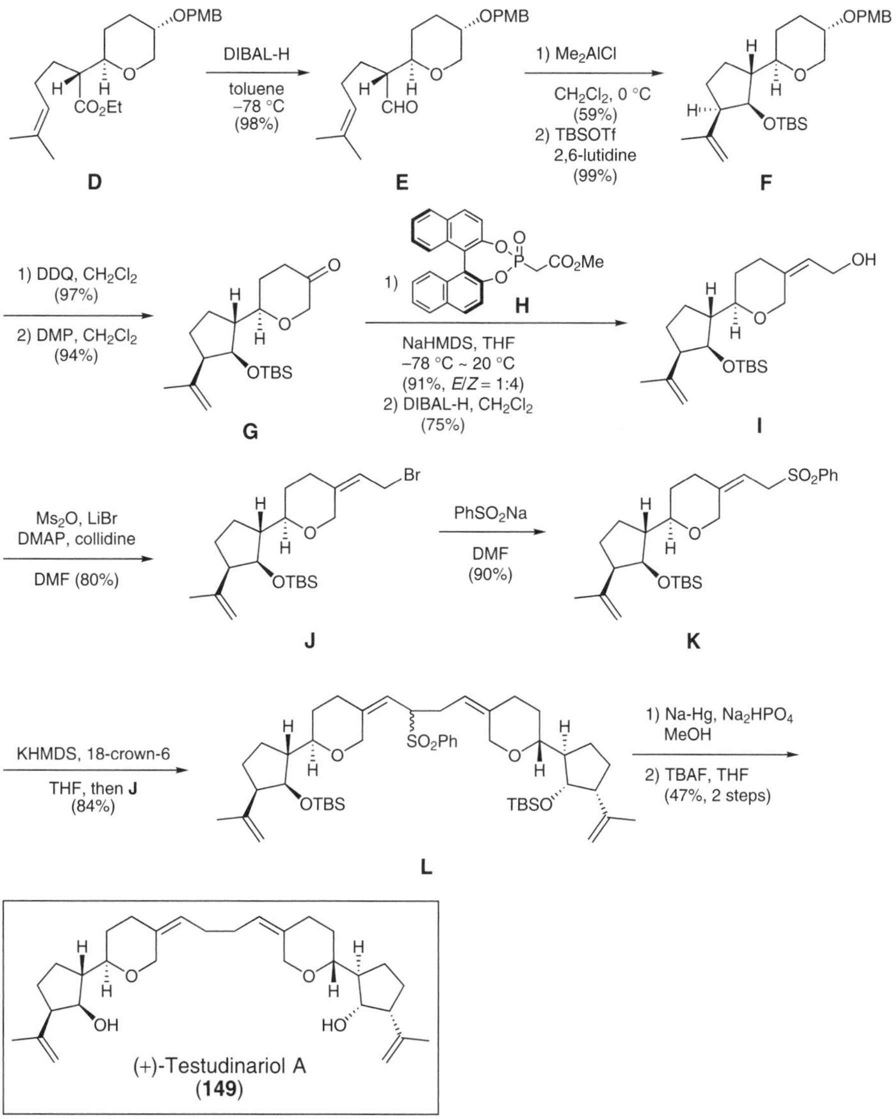

Figure 6.8   *Synthesis of testudinariol A (2)*

Conversion of **D** to (+)-testudinariol A (**149**) is summarized in Figure 6.8. After reduction of **D** with DIBAL-H, the resulting aldehyde **E** was treated with dimethylaluminum chloride in dichloromethane to give the cyclized product **F** after TBS protection. Ketone **G**, prepared from **F**, was then treated with a chiral phosphonoacetate **H** in the presence of NaHMDS to give the desired (*Z*)-ester as the major product. Reduction of the (*Z*)-ester with DIBAL-H furnished **I**. Alcohol **I** was converted to bromide **J** and sulfone **K**, respectively. Alkylation of **K** with **J** afforded **L**. Its reductive desulfonization and silyl deprotection yielded (+)-testudinariol A (**149**) in 4.4% overall yield based on **A** (19 steps). The spectroscopic and chiroptical properties of our **149** were in good accord with those reported for the natural product. Its absolute configuration was therefore determined as depicted in **149**.[8,9]

6.1.4 Stellettadine A

In 1996, Tsukamoto, Fusetani and their coworkers isolated stellettadine A (**150**, Figure 6.9) from a marine sponge, *Stelletta* sp., collected in Japan. This sponge metabolite induces at a concentration of 50 μM larval metamorphosis in ascidians, *Halocynthia roretzi*. Its structure is unique as bisguanidium alkaloid acylated with a chiral norsesquiterpene acid. Its *S* absolute configuration as depicted in **150′** was proposed on the basis of degradative studies. We became interested in the unique structure and bioactivity of stellettadine A, and synthesized its enantiomers **150** and **150′** in 2001.[10,11] Our work unambiguously established the *R* configuration of the naturally occurring stellettadine A (**150**).

Figure 6.9 summarizes our synthesis of (*S*)-(+)-stellettadine A (**150′**). (*S*)-(−)-Citronellal was treated with methoxymethylene triphenylphosphorane to give **A**, whose palladium-catalysed oxidation furnished **B**. Chain elongation of **B** by conventional Wittig chemistry was followed by functional group transformation to give acyl chloride **D** via **C**. For the guanidine moiety of **150′**, commercially available agmatine sulfate (**E**) was chosen as the starting material. Its nonguanidine primary amino group was protected as *t*-butoxycarbonyl(Boc) amide **F**. Acylation of **F** with **D** was successful under the conditions as depicted in Figure 6.9 to give *bis*-acylation product **G**. The Boc protective group of **G** was removed, and the resulting amine **H** was subjected to guanylation followed by alkaline hydrolysis to give (*S*)-(+)-stellettadine A (**150′**) as its dihydrochloride. In the same manner, (*R*)-(−)-**150** was also synthesized from (*R*)-(+)-citronellal.

Circular dichroism (CD) spectral comparison of (*R*)-**150**, (*S*)-**150′**, and the natural product definitely showed the absolute configuration of the natural isomer as *R*. Our synthetic work gave a more reliable result concerning the absolute configuration of stellettadine A than the initial degradative work.

6.2 Marine natural products of medicinal interest

There are a number of marine natural products whose bioactivities are of medicinal interest. We synthesized some of them. In this section, I will summarize several examples of our synthesis. Marine sphingolipids remained as the focus of our research interest because of their structural and biological diversities.

6.2.1 Punaglandin 4

Punaglandin 4 (PUG 4, **151**, Figure 6.10) is one of the chlorinated marine prostanoids isolated from a Hawaiian octocoral, *Telesto riisei*, by Scheuer and coworkers in 1985. Its remarkable antitumor activity coupled with its unique structure attracted the attention of chemists, and its synthesis was achieved by Yamada, Noyori, Shibasaki, and Mori's groups. Our synthesis relied on lipase-catalysed asymmetric process, and is summarized in Figures 6.10 and 6.11.[12]

Figure 6.10 illustrates the preparation of two key building blocks **E** and **I**. The former was synthesized by employing lipase-catalysed reaction, while the latter was derived from L-(+)-tartaric acid (**F**). The known (±)-4-hydroxy-2-cyclopentenone (**A**) was converted to **C** (as a mixture of four stereoisomers) via **B**. Treatment of **C** with pig-pancreatic lipase (PPL) afforded (−)-**D** in 25% yield. By this enzymatic hydrolysis of the acetate **C**, only the acetate corresponding to (−)-**D** was hydrolysed even in the presence of chlorine and silicone atoms in the molecule to give the desired (−)-**D**. The experimental simplicity of

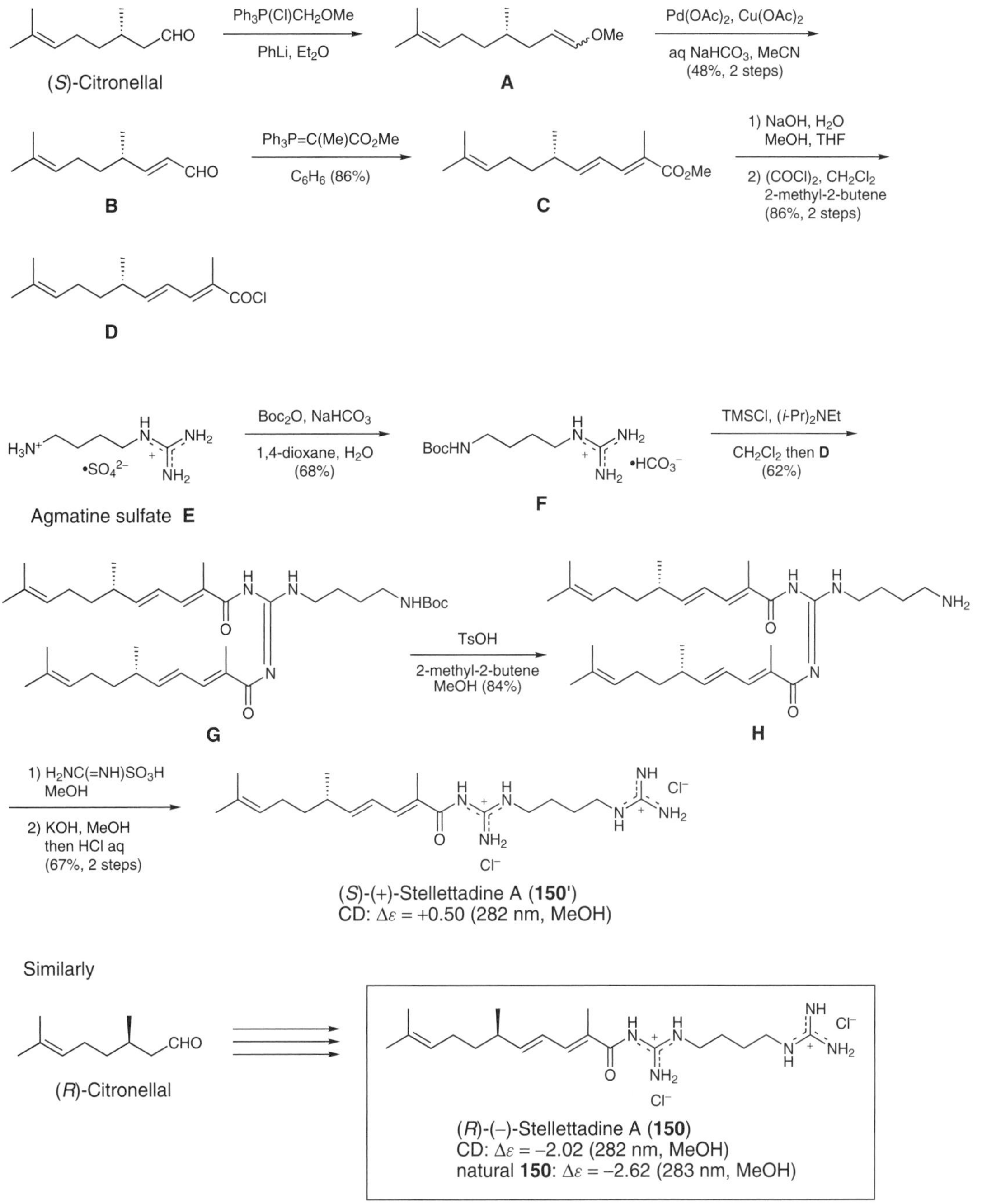

Figure 6.9 Synthesis of stellettadine A

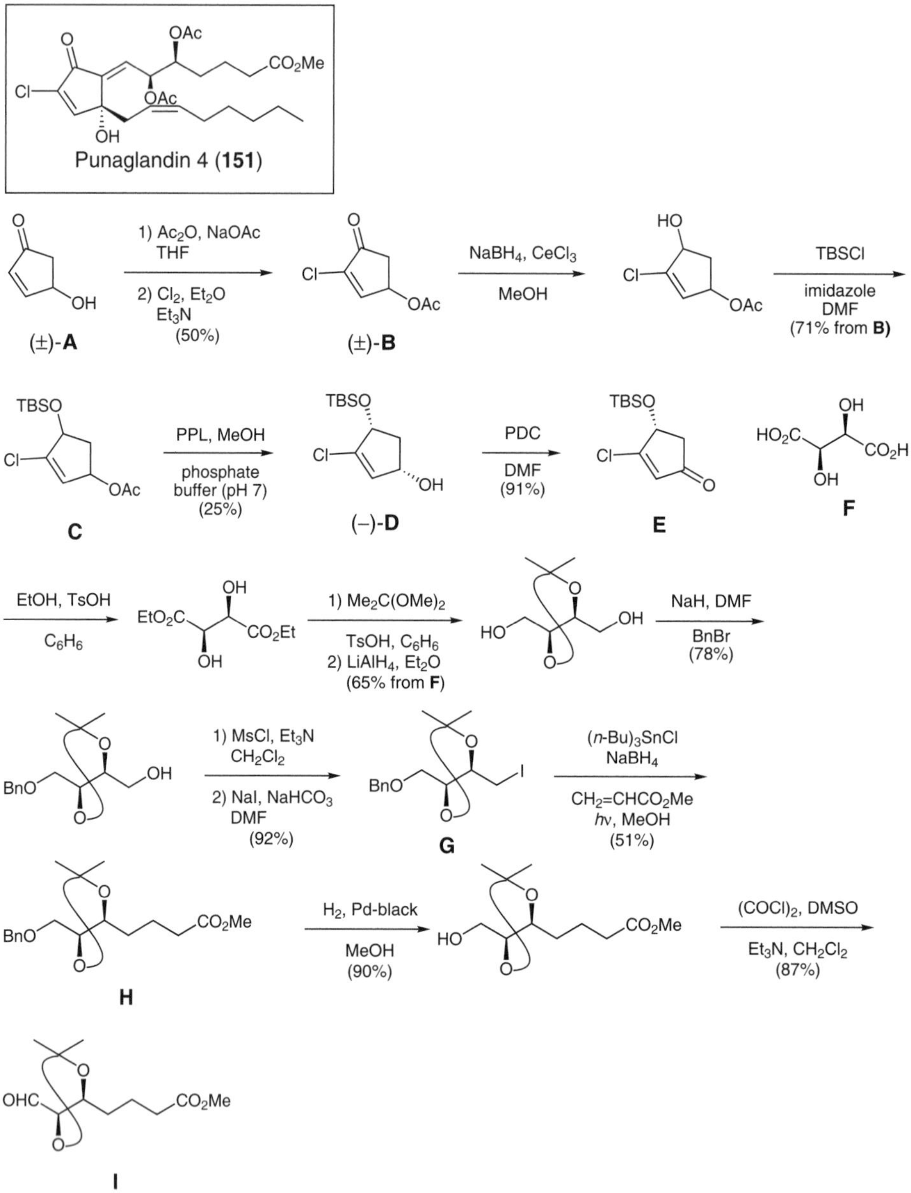

Figure 6.10 *Synthesis of punaglandin 4 (1). Modified by permission of Shokabo Publishing Co., Ltd*

this step was remarkable to secure (−)-**D** in pure form. Oxidation of (−)-**D** with pyridinium dichromate (PDC) gave the building block **E**.

L-(+)-Tartaric acid (**F**) was then converted to iodide **G**, whose addition to methyl acrylate under the standard conditions of radical reactions furnished **H**. Hydrogenolysis of **H** over palladium black yielded the corresponding alcohol, whose Swern oxidation afforded **I**, another building block.

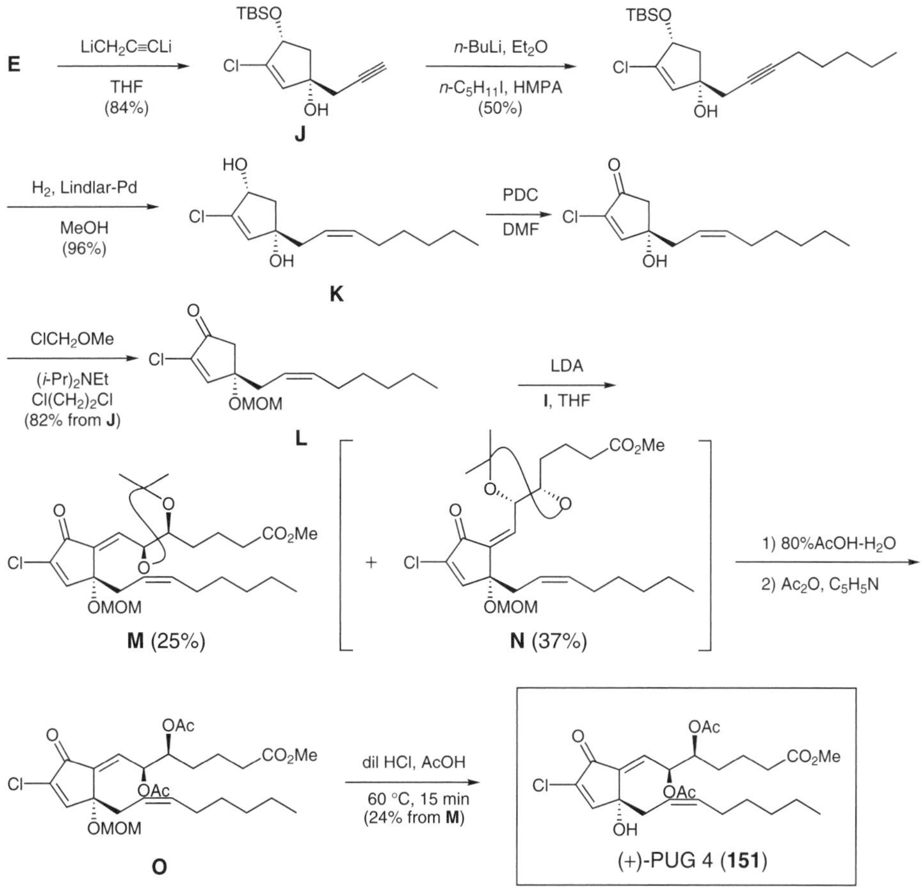

Figure 6.11   *Synthesis of punaglandin 4 (2). Modified by permission of Shokabo Publishing Co., Ltd*

Conversion of the chloro ketone **E** to (+)-punaglandin 4 (**151**) is summarized in Figure 6.11. Treatment of **E** with the dianion derived from propyne gave **J**, which was alkylated with 1-iodopentane. Semi-hydrogenation of the product afforded **K**, which was converted to **L**. The aldol reaction between **I** and **L** was problematic, and yielded the desired **M** as the minor product. The desired aldol product **M** was obtained in 25% yield after chromatographic purification. Punagladin 4 (**151**) was obtained in 24% yield based on **M**. The usefulness of lipase in enantioselective synthesis is well illustrated in the present synthesis of punaglandin 4 (**151**).[12]

6.2.2   Bifurcarenone

In 1980, Fenical and coworkers isolated bifurcarenone, an inhibitor of mitotic cell division, from the brown seaweed, *Bifurcaria galapagensis*, harvested in the Galapagos Islands. Its structure was proposed as **152A** (Figure 6.12) on the basis of chemical and spectroscopic studies, without assignment of the absolute configuration. Our 1989 synthesis of (±)-**152A**, however, revealed its spectroscopic properties to be different from those of natural bifurcarenone.[13] By synthesizing, (±)-**152**, we showed that the natural

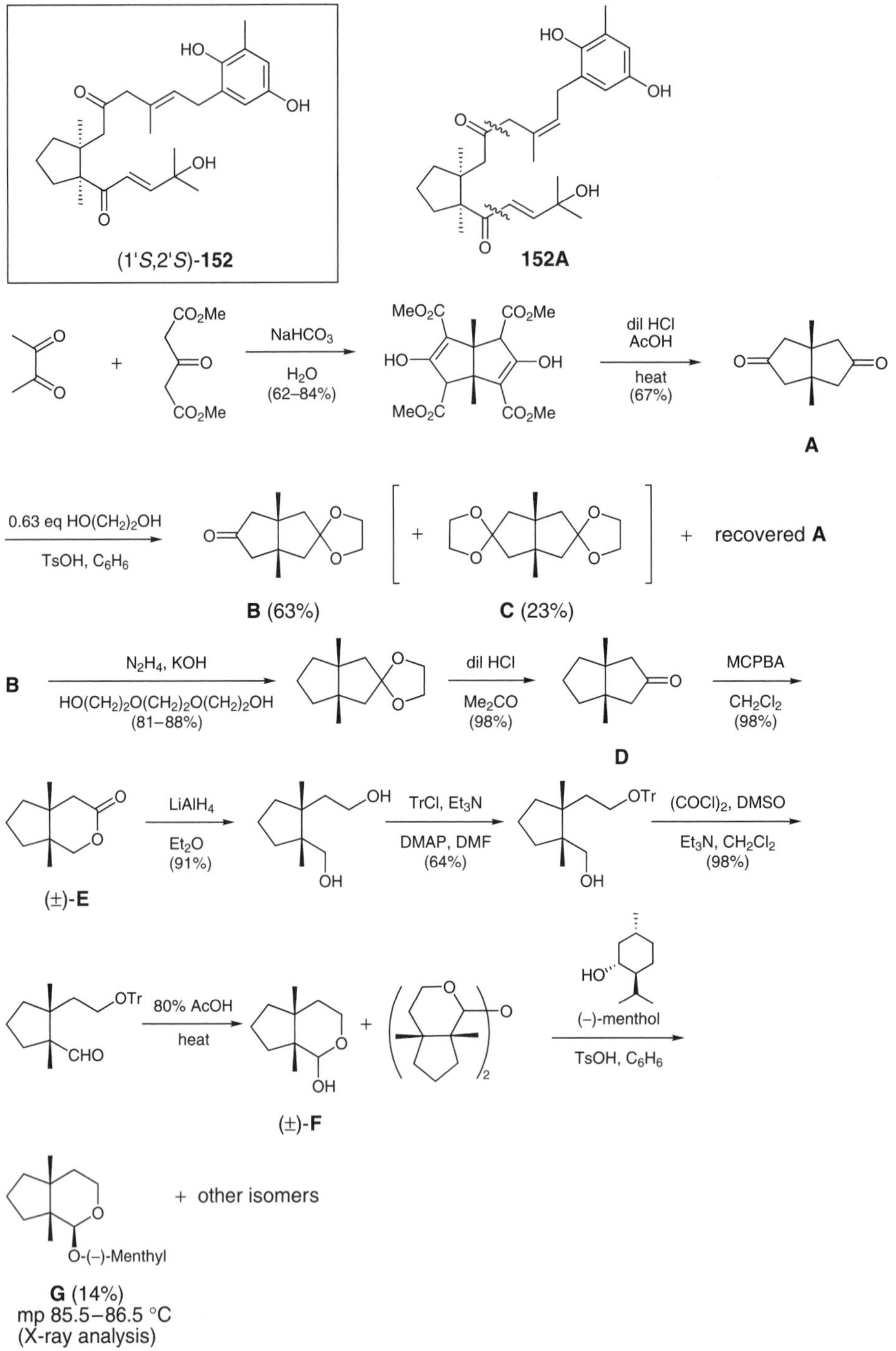

Figure 6.12 *Synthesis of (1'R,2'R)-bifurcarenone (1). Modified by permission of Shokabo Publishing Co., Ltd*

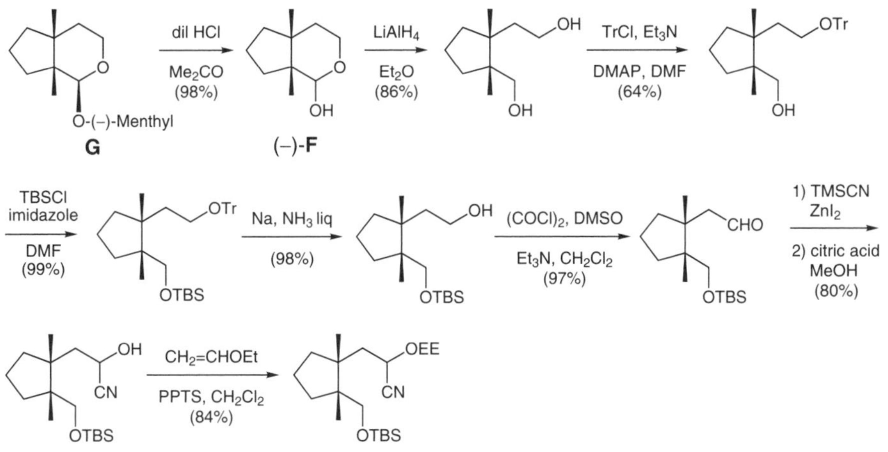

Figure 6.13　*Synthesis of (1′R,2′R)-bifurcarenone (2). Reprinted with permission of Shokabo Publishing Co., Ltd*

product is either (1′R,2′R)-**152′** or (1′S,2′S)-**152**. Our synthesis of (1′R,2′R)-**152′** (Figures 6.12–6.14) enabled us to assign (1′S,2′S)-**152** to the natural product.[14]

As shown in structure **152A**, we dissected the target molecule into three parts, the side-chain with a benzene ring, the stereogenic cyclopentane part, and the aliphatic side-chain. The synthesis of the cyclopentane part is summarized in Figures 6.12 and 6.13. The known *cis*-1,5-dimethylbicyclo[3.3.0]octane-3,7-dione (**A**) was synthesized according to the *Organic Syntheses* procedure. Acetalization of **A** with 0.63 equivalents of ethylene glycol and *p*-toluenesulfonic acid in benzene furnished the desired monoacetal **B** as the major product. Wolff–Kishner reduction of the ketone **B** under the Huang Minlon conditions was followed by acid treatment to give ketone **D**. Baeyer–Villiger oxidation of **D** afforded (±)-**E**, whose functional group transformations gave hemiacetal (±)-**F** contaminated with dimeric acetal. Treatment of this acetal mixture with (−)-menthol in the presence of *p*-toluenesulfonic acid afforded crystalline mixed acetal **G** together with other isomers. Fractional recrystallization of the crude mixture was repeated eight times to give pure **G**, whose structure could be solved by X-ray analysis, as depicted.

Conversion of the pure acetal **G** to the cyclopentane building block **H** is shown in Figure 6.13. Cyanohydrin-forming reaction was employed to attach the activating group for the subsequent side-chain elongation reaction in a later stage.

Figure 6.14 summarizes the preparation of the side-chain part with a benzene ring and the completion of the synthesis of (1′R,2′R)-bifurcarenone (**152′**). *o*-Cresol was our starting material, which yielded allylated **J** after five steps. Lemieux–Johnson oxidation of **J** afforded an aldehyde, whose Horner–Wadsworth–Emmons olefination was surprisingly nonselective to give (*E*)-**K** and (*Z*)-**L** in 25% and 21% yields, respectively, based on **J**. These two could be separated by silica-gel chromatography, and **K** was converted to allylic chloride **M**, which was coupled with the building block **H** to give **N**. The intermediate **N** afforded aldehyde **O** through six steps, and **O** was treated with dianion derived from 3-hydroxy-3-methyl-1-butyne to give **P**. (1′R,2′R)-Bifurcarenone (**152′**) was derived from **P** through five steps. A synthesis of **152A′** was also achieved by employing **L** as the key intermediate.

To establish the absolute configuration of the natural bifurcarenone, we measured the optical rotatory dispersion (ORD) spectra of both the synthetic (1′R,2′R)-bifurcarenone (**152′**) and the natural product, kindly provided by Prof. Fenical. Our synthetic sample showed the ORD spectrum antipodal to that of the natural product. The absolute configuration of bifurcarenone was therefore determined as 1′S,2′S.[14]

6.2.3 Elenic acid

In 1995, Scheuer and coworkers isolated and characterized elenic acid (**153**, Figure 6.15), an inhibitor of topoisomerase II, from the Indonesian sponge, *Plakinastrella* sp. Elenic acid has a unique structure, in which its phenol portion and β, γ-unsaturated carboxylic acid moiety are linked with a long polymethylene

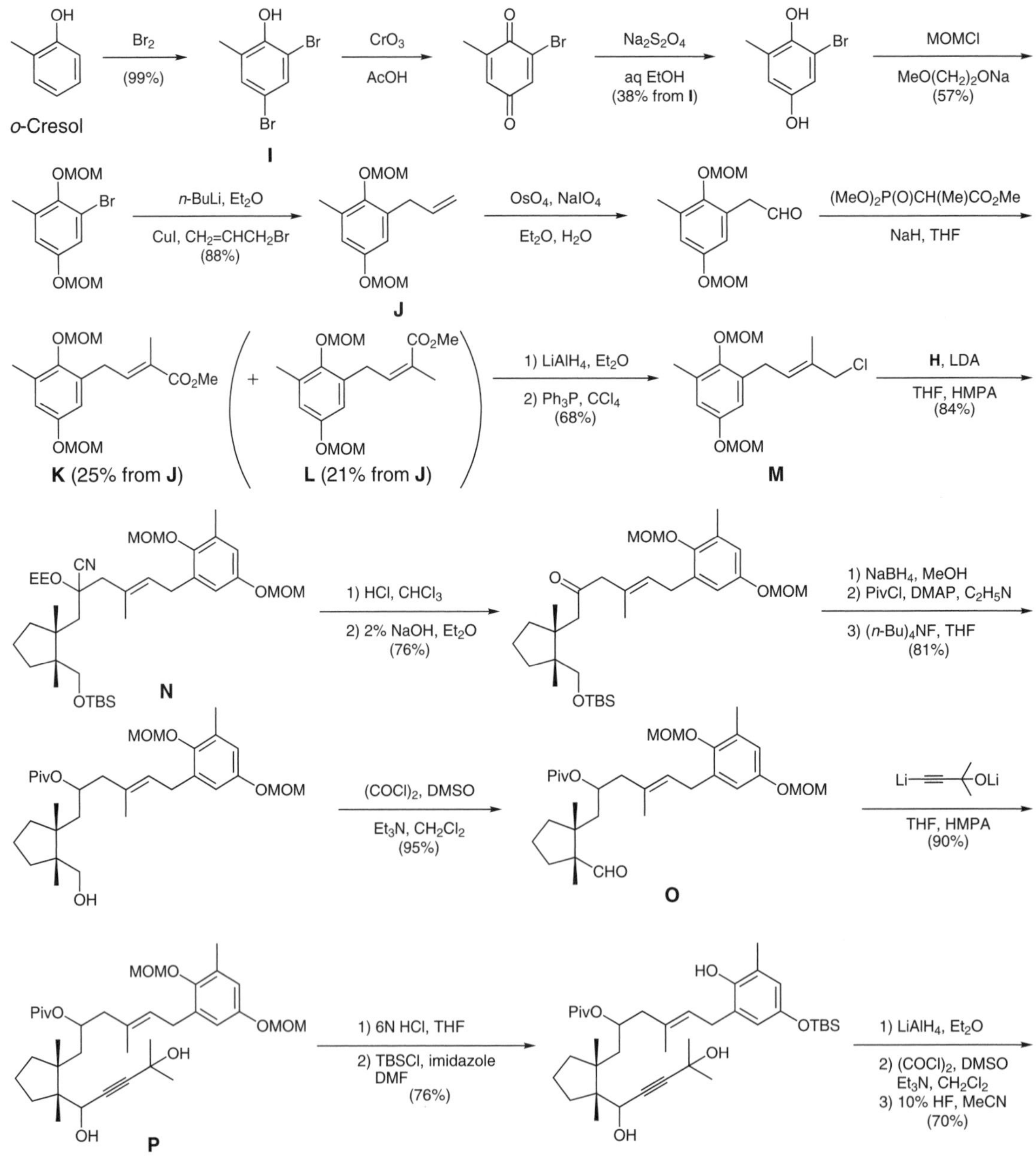

Figure 6.14 *Synthesis of (1′R,2′R)-bifurcarenone (3). Modified by permission of Shokabo Publishing Co., Ltd*

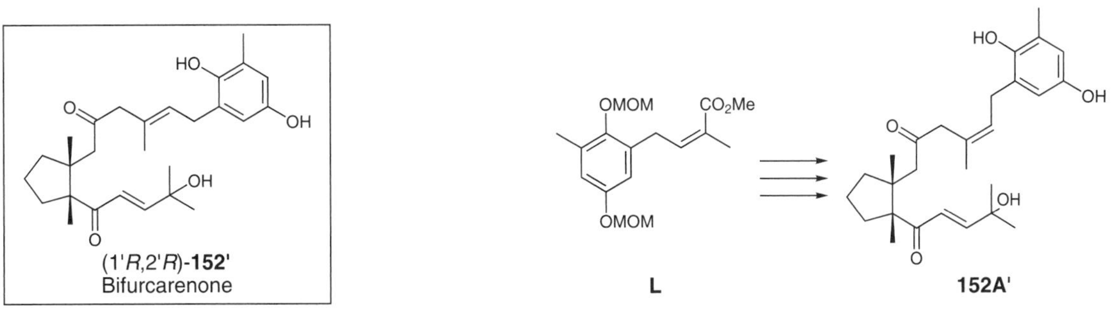

Figure 6.14 (continued)

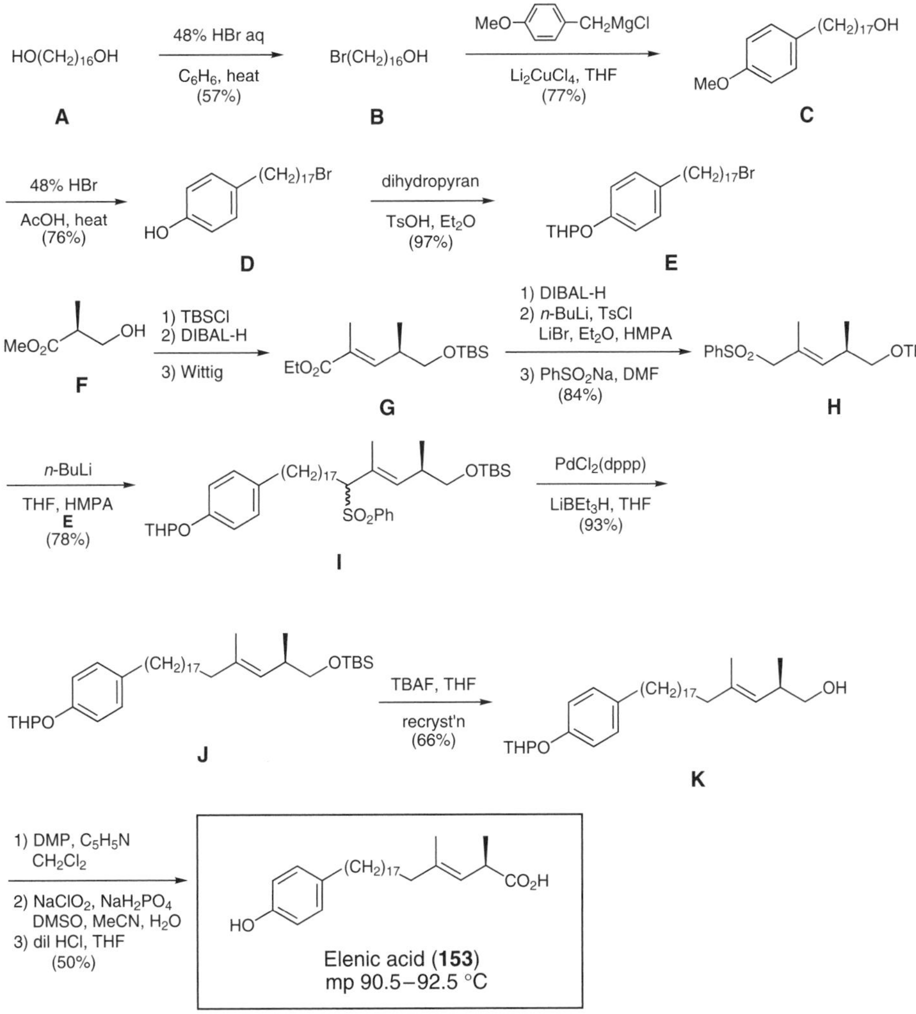

Figure 6.15 Synthesis of elenic acid

spacer. Scheuer's group determined the absolute configuration at C-2 to be *R*. We became interested in the unique structure and bioactivity of elenic acid, and synthesized it, as shown in Figure 6.15.[15]

Our synthesis started from 1,16-hexadecanediol (**A**), which was monobrominated to give **B**. Coupling of **B** with *p*-methoxybenzylmagnesium chloride afforded **C**. Treatment of **C** with hydrobromic acid in acetic acid gave **D**, which was converted to the key building block **E**. The chiral moiety of elenic acid was constructed from commercially available methyl (*S*)-3-hydroxy-2-methylpropanoate (**F**), which was converted to phenylsulfone **H** via **G**. Alkylation of the sulfone **H** with bromide **E** yielded **I**, whose phenylsulfonyl group was reductively removed under palladium catalysis to give **J**. The corresponding alcohol **K** was oxidized and deprotected to give (*R*)-elenic acid (**153**), whose ^{1}H- and ^{13}C-NMR spectra were in good accord with those of the natural product. The specific rotation of the synthetic **153** was $[\alpha]_D^{25} = -30$ (CHCl$_3$), while the natural product showed $[\alpha]_D = -27.2$ (CHCl$_3$). The *R* configuration of elenic acid was thus confirmed.[15] Elenic acid (**153**) inhibits the activity of both eukaryotic DNA polymerases and DNA topoisomerases.[16]

6.2.4   Symbioramide

In 1988, Kobayashi *et al.* isolated a novel ceramide symbioramide (**154**, Figure 6.16) from the laboratory-cultured dinoflagellate *Symbiodium* sp. obtained from the inside of gill cells of the Okinawan bivalve, *Fragum* sp. At 10^{-4} μM concentration, it increases the sarcoplasmic reticulum Ca^{2+}-ATPase activity by 30%. It also exhibits antileukemic activity against L1210 murine leukemia cells in vitro with an IC$_{50}$ value of 9.5 μg/mL. Kobayashi's work coupled with Nakagawa's synthetic work established the stereochemistry of symbioramide as (2*S*,3*R*,2′*R*,3′*E*)-**154**.

In continuation of our sphingolipid works (see 5.1.2), we achieved a synthesis of symbioramide as shown in Figure 6.16.[17] Synthesis of the α-hydroxy-β, γ-unsaturated carboxylic acid moiety **H** started from 1-bromopentadecane (**A**). Chain elongation of **A** with dianion of propargyl alcohol was followed by *E*-selective reduction of the triple bond to give allylic alcohol **B**. Sharpless asymmetric epoxidation of **B** with L-(+)-diethyl tartrate as the ligand was followed by TBS-protection of the hydroxy group to furnish **C**. Treatment of **C** with diphenyl diselenide and sodium borohydride yielded **D** as the major product. After oxidation and chromatographic purification, the desired **F** was obtained in 75% yield based on **C**. Protection of the secondary hydroxy group of **F** with *t*-butyldiphenylsilyl chloride gave **G**, whose TBS group was selectively removed with acetic acid, and the resulting alcohol was oxidized with Jones chromic acid to furnish the key building block **H**.

The synthesis of the dihydrosphingosine moiety **L** could be achieved readily, starting from L-serine (**I**). Garner aldehyde was treated with lithium pentadecynide to give **J**. Hydrogenation of **J** afforded **K**, which was deprotected to give C$_{18}$-dihydrosphingosine. Protection of its hydroxy groups furnished **L**, which was acylated with **H** to give protected ceramide **M**. Global deprotection of **M** in two steps afforded crystalline symbioramide (**154**), which was spectroscopically identical with the natural product.

Optical rotation values of symbioramide (**154**) taught us an interesting lesson. The temperature of the sample solution in a cell for rotation measurements influences the sign and magnitude of the specific rotation of **154**. The synthetic sample of ours showed $[\alpha]_D^{22} = -1.5$ (*c* 0.24, CHCl$_3$), while the natural **154** showed $[\alpha]_D^{19} = +5.8$ (*c* 0.1, CHCl$_3$), This discrepancy annoyed us. We measured the rotation of **154** very carefully using a polarimeter equipped with a cell in a constant temperature bath. On this apparatus, the specific rotation of our synthetic **154** was recorded as : $[\alpha]_D^{19} = +3.6$, $[\alpha]_D^{23} = +0.76$, $[\alpha]_D^{28} = -1.5$, $[\alpha]_D^{35} = -5.5$ (*c* 0.31, CHCl$_3$). In our previous measurement which gave the value $[\alpha]_D^{22} = -1.5$ (*c* 0.24, CHCl$_3$), the room temperature was 22 °C. During the measurement, however, the temperature of the sample solution was raised by the irradiation with the sodium D-line light. The temperature of the sample solution was therefore higher than 22 °C, and we recorded a negative rotation. The discrepancy was thus removed,

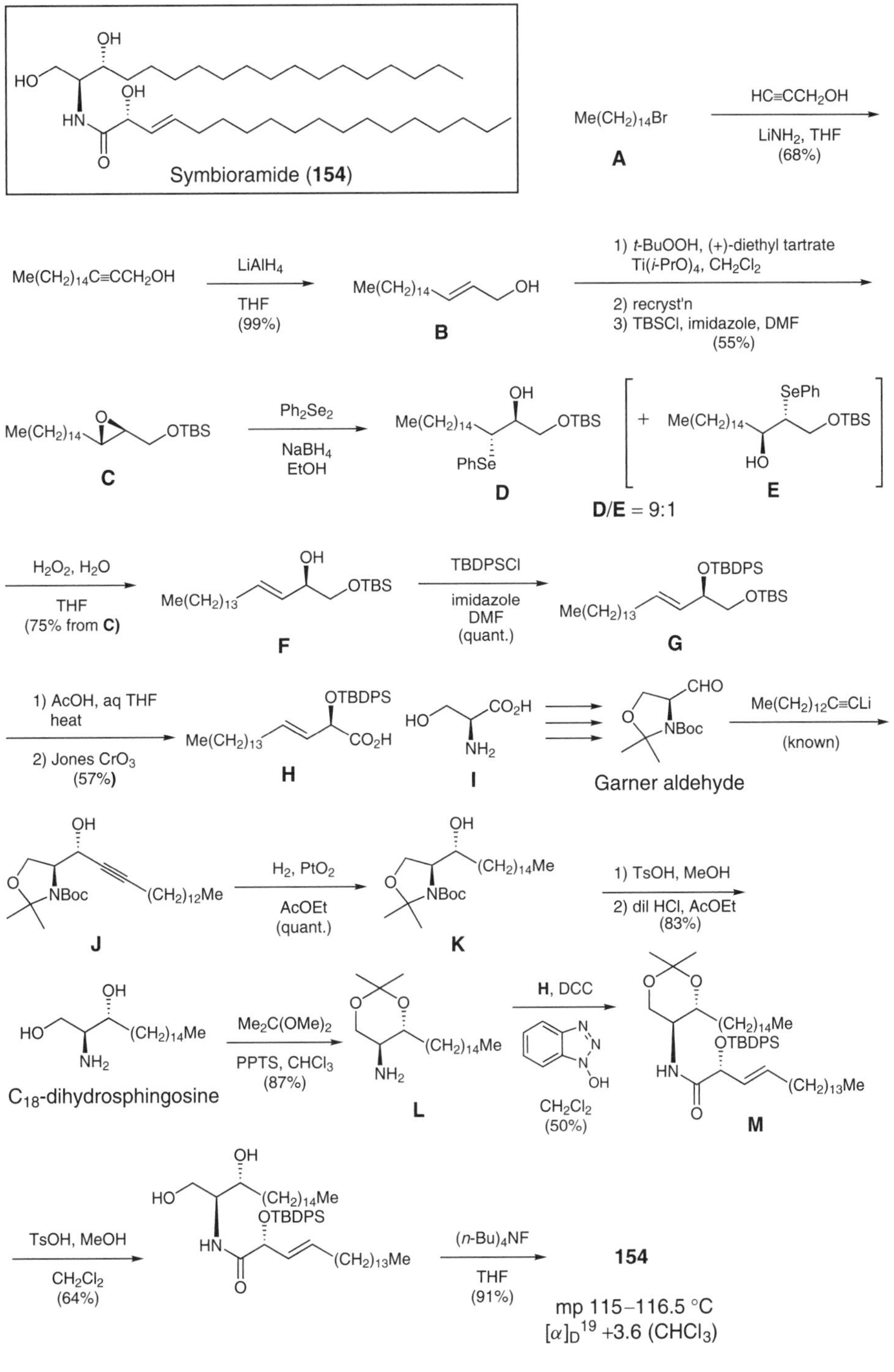

Figure 6.16   *Synthesis of symbioramide. Modified by permission of Shokabo Publishing Co., Ltd*

and the absolute configuration of symbioramide (**154**) was reconfirmed as depicted. The temperature dependence of the specific rotation value is a well-known but often overlooked phenomenon, which must be taken into account properly.

6.2.5   Penazetidine A

In 1994, Crews and coworkers isolated and identified penazetidine A (**155**, Figure 6.17), an inhibitor of protein kinase C, from the Indo-Pacific marine sponge, *Penares sollasi*. Its structure **155** was proposed on the basis of NMR and MS studies, although nothing was known about its absolute configuration. Assuming that its biosynthesis is in accord with that of other sphingosines, we synthesized both $(2S,3R,4S,12'R)$-**155** and $(2S,3R,4S,12'S)$-**155'**, as shown in Figure 6.17.[18,19]

For the synthesis of $(2S,3R,4S,12'R)$-penazetidine A (**155**), (S)-citronellol and (S)-Garner aldehyde (**F**) were employed as our starting materials. Chain elongation of (S)-citronellol under the Schlosser conditions gave **A**, whose double bond was cleaved to furnish aldehyde **B**. Treatment of **B** with lithium nonynide gave **C**, which was subjected to acetylene zipper reaction to afford **D**. Deoxygenation of **D** yielded (R)-**E**. Coupling of (R)-**E** with (S)-Garner aldehyde (**F**) gave **G** which was converted to epoxide **I** via **H**. Reduction of **I** with DIBAL-H was followed by mesylation to give mesylate **J**. This was treated with sodium hydride to give azetidine **K**. Removal of its tosyl and TBS-protective groups gave $(2S,3R,4S,12'R)$-penazetidine A (**155**). Similarly, (R)-citronellol was converted to $(2S,3R,4S,12'S)$-**155'**. The ^{1}H- and ^{13}C-NMR spectra of both **155** and **155'** were identical and in accord with the authentic spectra of the natural product provided by Prof. Crews. Both **155** and **155'** as well as natural penazetidine A were levorotatory, and it was impossible to decide which of **155** and **155'** is the natural product.

6.2.6   Penaresidin A and B

In 1991, penaresidin A (**156**, Figure 6.18) and B (**157**, Figure 6.18), actomyosin ATPase activators, were isolated by Kobayashi *et al.* from the Okinawan marine sponge *Penares* sp., and characterized as a mixture of the corresponding tetraacetyl derivatives. They were shown to be azetidine alkaloids related to phytosphingosines. Their absolute configuration, however, was not known. We assumed the absolute configuration of the azetidine moiety of penaresidins to be 2S,3R,4S, considering the possible biogenetic relationship between penaresidins and phytosphingosines, and started their synthesis.

Our synthesis of penaresidin A and B is summarized in Figure 6.18.[19–21] For the synthesis of penaresidin A (**156**), L-isoleucine was chosen as the starting material. Epoxide $(2S,3S)$-**A** was prepared from L-isoleucine in a manner similar to that illustrated in Figure 4.64. Treatment of **A** with lithium 1-decynide yielded **B**, which was subjected to acetylene zipper reaction to give terminal acetylene **C**. Mitsunobu inversion of the (R)-hydroxy group of **C** to furnish **D** was followed by hydrolysis and silylation to give **E**.

Alkynylation of Garner aldehyde **F** with alkyne **E** gave **G** with the required carbon skeleton. The triple bond of **G** was then reduced to give (E)-alkene **H**. The next epoxidation was unfortunately nonstereoselective to give the required epoxide **I** and its isomer. Reduction of the epoxide **I** was fortunately selective with DIBAL-H as the reductant, and **J** was obtained after mesylation.

Treatment of **J** with sodium hydride in THF closed the azetidine ring smoothly to give **K** in 81% yield. Then, **K** was reduced with sodium naphthalenide to remove the tosyl group. Penaresidin A [$(2S,3R,4S,11'S,12'S)$-**156**] was obtained after further deprotection of the TBS group. Peracetylation of **156** afforded the tetraacetyl derivative, whose spectroscopic and chiroptical properties were in accord with those of the tetraacetyl derivative of the natural product. Isomers of penaresidin A with $(2S,3R,4S,11'R,12'S)$- and $(2S,3R,4S,11'R,12'R)$-configurations were also synthesized, but their physical properties were different from those of the natural penaresidin A.

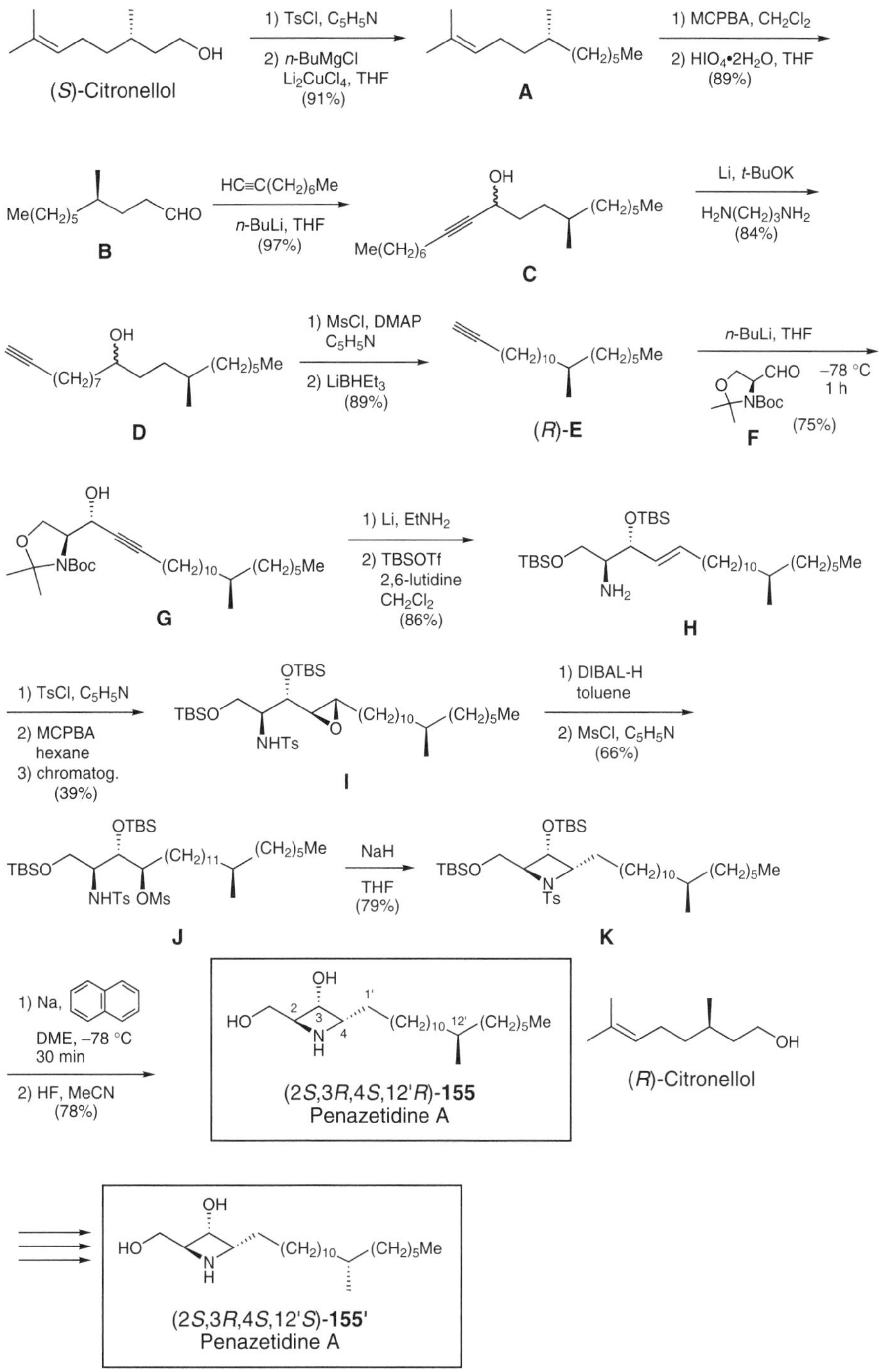

Figure 6.17 *Synthesis of penazetidine A*

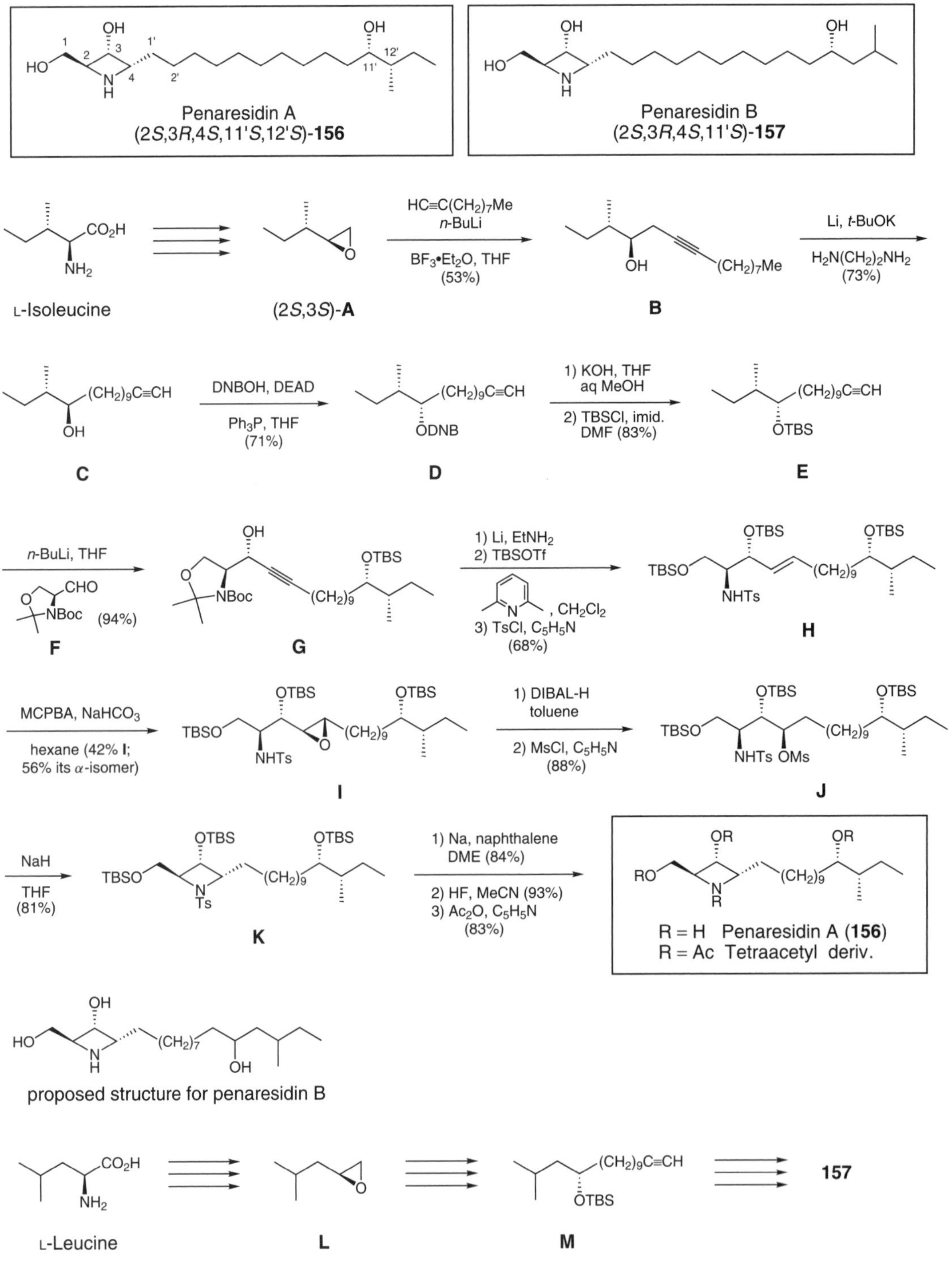

Figure 6.18   *Synthesis of penaresidin A and B*

Kobayashi *et al.* initially proposed a structure shown in the lower part of Figure 6.18 for penaresidin B. Our synthesis of **156** made it possible to critically analyse the ^{13}C-NMR spectrum of the tetraacetyl derivatives of the naturally occurring mixture of penaresidin A and B. Consequently, we proposed **157** as the structure of penaresidin B. Our proposal was confirmed by its synthesis. L-Leucine was converted to alkyne **M** via epoxide **L**. The alkyne **M** yielded **157**, whose spectral properties were in accord with those of penaresidin B.[21]

Absolute configuration at C-11′ of penaresidin A and B were determined as *S* also by Kobayashi *et al.* on the basis of ^{1}H-NMR data of their *tris-O*-MTPA esters.[22] As you have seen in this case, synthesis of **156** and **157** was the key to firmly establish the structures of penaresidin A and B.

6.2.7 Sulfobacin A, B and flavocristamide A

In 1995, sulfobacin A (**158**, Figure 6.19) and B (**159**), von Willebrand factor receptor antagonists, were isolated by Kamiyama *et al.* from the culture broth of a terrestrial bacterium *Chryseobacterium* sp. In the same year, the isolation of flavocristamide A (**160**) and B (=sulfobacin A, **158**), DNA polymerase α inhibitors, from the cultured mycelium of marine bacterium *Flavobacterium* sp. in Hokkaido was reported by Kobayashi *et al.* These are sulfonolipid, and unusual sphingosine derivatives. We became interested in synthesizing these three sulfur-containing compounds, and carried out their synthesis from L-cysteine.[23,24]

Sulfobacin B (**159**), the simplest member of these sulfonolipids, was synthesized as shown in Figure 6.19. L-Cysteine was converted to **C** via **A** and **B**. Attachment of a side-chain to **C** furnished **D**, which was hydrogenated over Adams' platinum oxide to give **E**. Sultine **F** was generated, when **E** was treated with hydrochloric acid. This conversion protected the hydroxy group at C-3. Acylation of **F** with **G** gave **H**, whose sultine ring was cleaved with ammonia to afford **I**. Oxidation of sulfinic acid **I** with hydrogen peroxide furnished sulfobacin B (**159**).[23,24]

For the synthesis of sulfobacin A (**158**) and flavocristamide A (**160**), TBS ether of (*R*)-3-hydroxy-15-methylhexadecanoic acid was necessary, which was synthesized from 10-bromo-1-decanol (**A**) as shown in Figure 6.20. Chain elongation of **A** under the Schlosser conditions gave **B**, which was oxidized with PCC to give aldehyde **C**. (±)-β-Hydroxy ester **D** was prepared from **C** by treatment with ethyl acetate and LDA. The corresponding (±)-acid was acetylated with vinyl acetate in the presence of lipase PS to give enantiomerically pure (*R*)-hydroxy acid and the acetylated (*S*)-acid. The former was converted to its TBS ether **E**.

For the synthesis of flavocristamide A (**160**), (*E*)-iodoalkene (**I**) was necessary. This was prepared from 9-decen-1-ol (**F**). Chain elongation of **F** yielded **G**, whose double bond was converted to a triple bond to give **H**. Subsequent hydroalumination of **H** was followed by quenching with iodine to give the desired iodoalkene **I**.

Sulfobacin A (**158**) was obtained by acylation of the amine **J** with **E** followed by deprotection and oxidation,[23,24] while flavocristamide A (**160**) was synthesized from **K** by alkenylation with **I** to give **L**. In the same manner as for sulfobacin A, **L** was converted to flavocristamide A (**160**) via **M**.[24]

6.2.8 Plakoside A

In 1997, Fattorusso and coworkers isolated plakoside A (**161**, Figure 6.21) as a metabolite of the Caribbean sponge, *Plakortis simplex*. It is structurally unique as a glycosphingolipid with a prenylated D-galactose moiety and cyclopropane-containing alkyl chains, and shows strong immunosuppressive activity without cytotoxicity. Only 5 mg of plakoside A could be secured from 57 g (dry weight) of the sponge. Since the absolute configuration at the stereogenic centers of the two cyclopropane moieties was unknown except

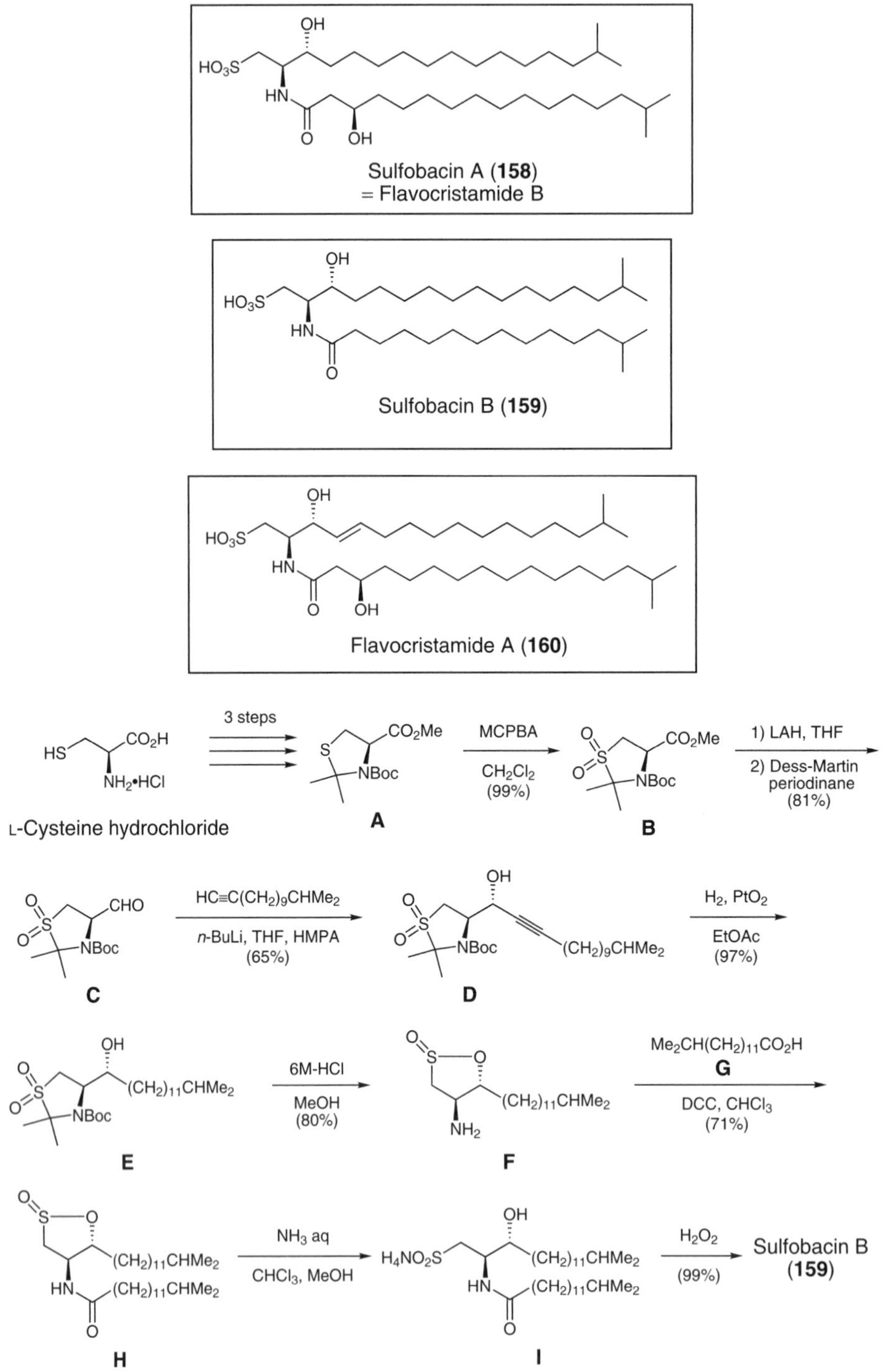

Figure 6.19   *Structures of sulfobacin A, B and flavocristamide A and synthesis of sulfobacin B*

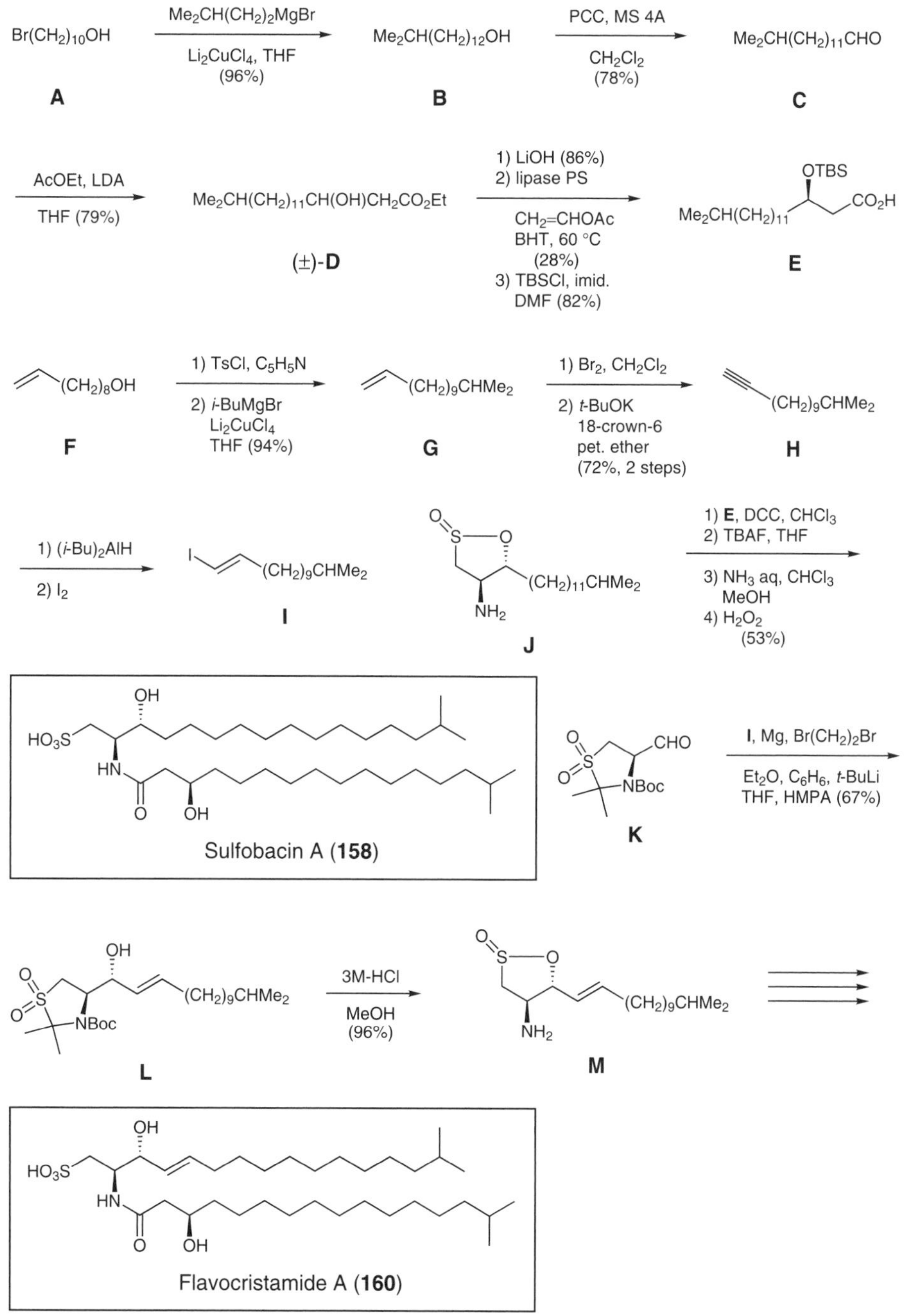

Figure 6.20 *Synthesis of sulfobacin A and flavocristamide A*

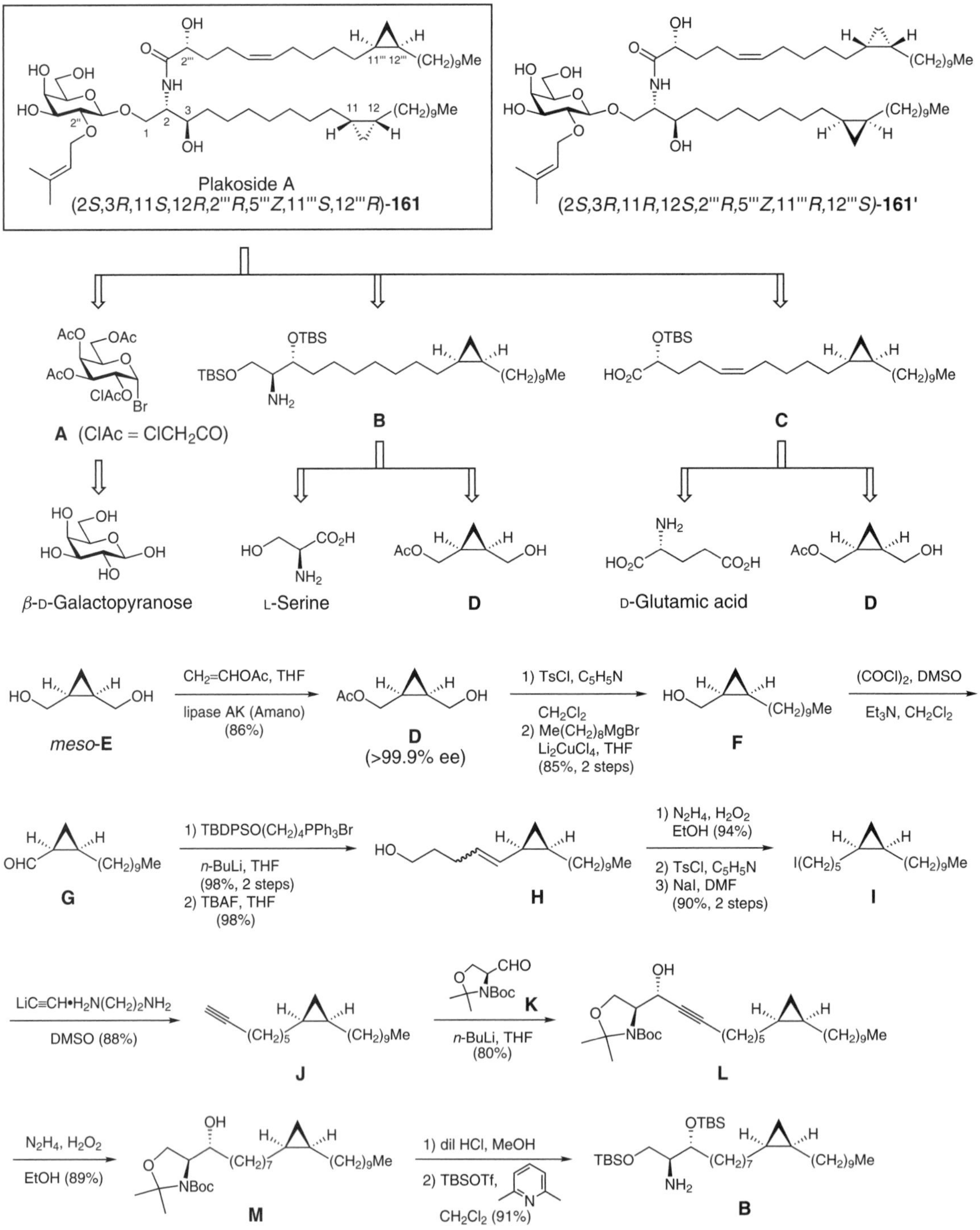

Figure 6.21 *Synthesis of plakoside A (1)*

Similarly

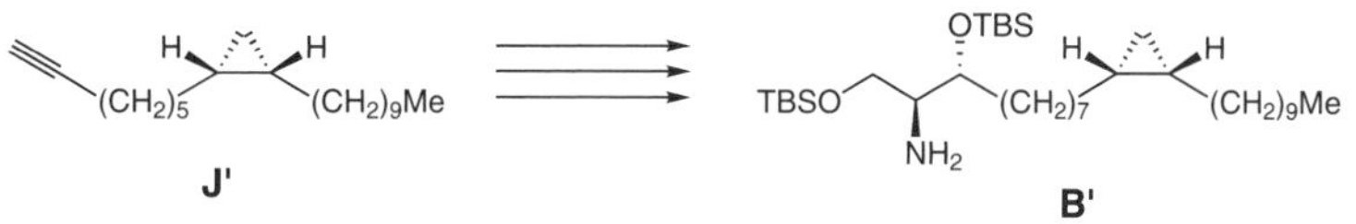

Figure 6.21 (continued)

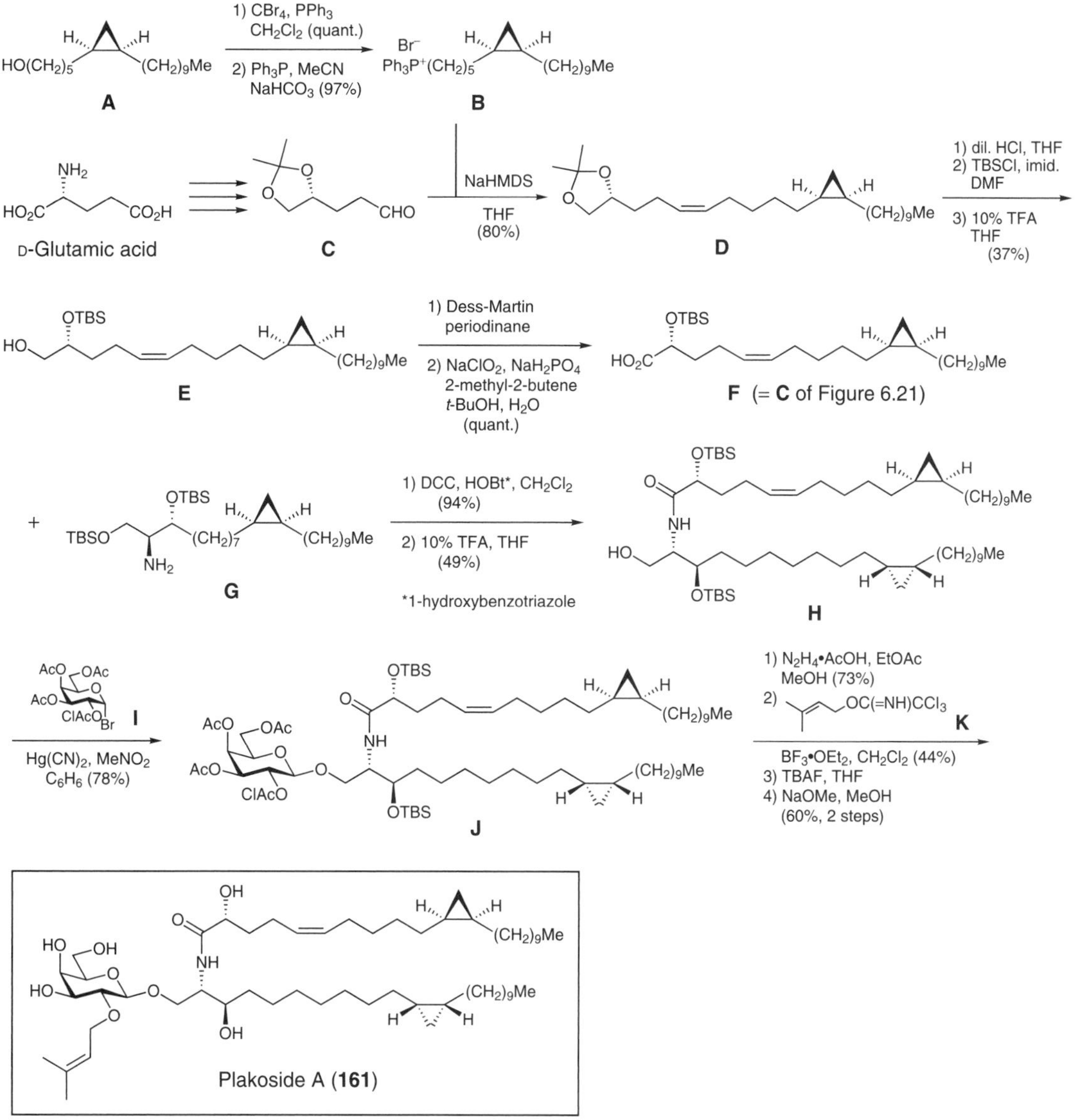

Figure 6.22 *Synthesis of plakoside A (2)*

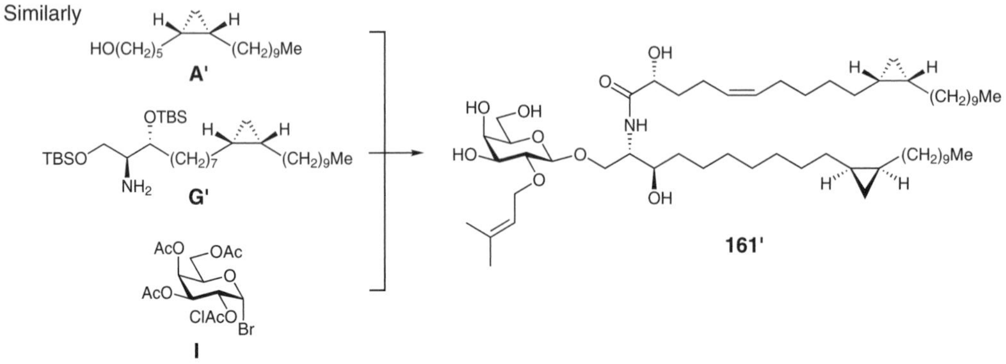

Figure 6.22   *(continued)*

that they were *cis*-disubstituted cyclopropanes, we decided to synthesize two diastereomers of plakoside A, (2*S*,3*R*,11*S*,12*R*,2′′′*R*,5′′′*Z*,11′′′*S*,12′′′*R*)-**161** and (2*S*,3*R*,11*R*,12*S*,2′′′*R*,5′′′*Z*,11′′′*R*,12′′′*S*)-**161′**, anticipating that one of them would be the natural product. We assumed that the two cyclopropane-containing side-chains in a given molecule have the same absolute configuration due to the enantioselective biocyclopropanation process.

Figures 6.21 and 6.22 summarize our synthesis of plakoside A (**161**) and its diastereomer **161′** in 2001.[25,26] As I will describe later in this section, plakoside A was proved to be **161**, not **161′**. Plakoside A (**161**) can be prepared from the three building blocks, **A, B** and **C**. D-Galactose will be the starting material for **A**, while **B** and **C** can be synthesized from cyclopropane alcohol **D** and L-serine or D-glutamic acid, respectively.

The monoacetate **D** was obtained by lipase-catalysed desymmetrization of *meso*-diol **E**. Chain elongation of **D** afforded **F**. The corresponding aldehyde **G** was subjected to a Wittig reaction to give **H**. Diimide reduction of **H** saturated the double bond, and further functional-group transformation gave iodide **I**. Treatment of **I** with lithium acetylide-ethylenediamine complex furnished **J**, which was coupled with Garner aldehyde **K** to give **L**. Diimide reduction of **L** smoothly furnished **M**, whose deprotection and silylation yielded the key building block **B**. Similarly, the diastereomeric alkyne **J′** afforded **B′**.

Figure 6.22 shows the completion of the synthesis. Alcohol **A** was converted to phosphonium salt **B**, which was coupled with **C** by the Wittig reaction to give **D**. Deprotection, bis-silylation, and mono-desilylation of **D** furnished alcohol **E**, which was oxidized to the key building block **F** (= **C** of Figure 6.21). Acylation of the sphingosine part **G** with the acid **F** was executed in the presence of *N*,*N*′-dicyclohexylcarbodiimide (DCC) and 1-hydroxybenzotriazole (HOBt). The product was partially desilylated to afford **H**. Glycosidation of the ceramide **H** with **I** gave **J**. Treatment of **J** with hydrazine acetate removed its monochloroacetyl (ClAc) protective group. Then, the product was treated with prenyl trichloroacetimidate (**K**) to attach a prenyl group at C- 2′ of the galactose moiety. Subsequent desilylation and deacetylation yielded plakoside A (**161**). Similarly, three building blocks **A′**, **G′** and **I** were assembled to give **161′**.[25,26]

In 2000, Nicolaou *et al.* synthesized **161′** by a different route, found its ¹H- and ¹³C-NMR spectra to be identical to those of plakoside A, and claimed the structure of plakoside A as **161′**.[27] We found that both **161** and **161′** exhibit entirely identical ¹H- and ¹³C-NMR spectra. Moreover, they were in complete agreement with those reported for the natural product. Although **161** and **161′** are diastereomeric, their stereogenic centers are separated by seven or eight carbon atoms, and they therefore showed identical spectroscopic properties. It was concluded that the absolute configuration of plakoside A could not be

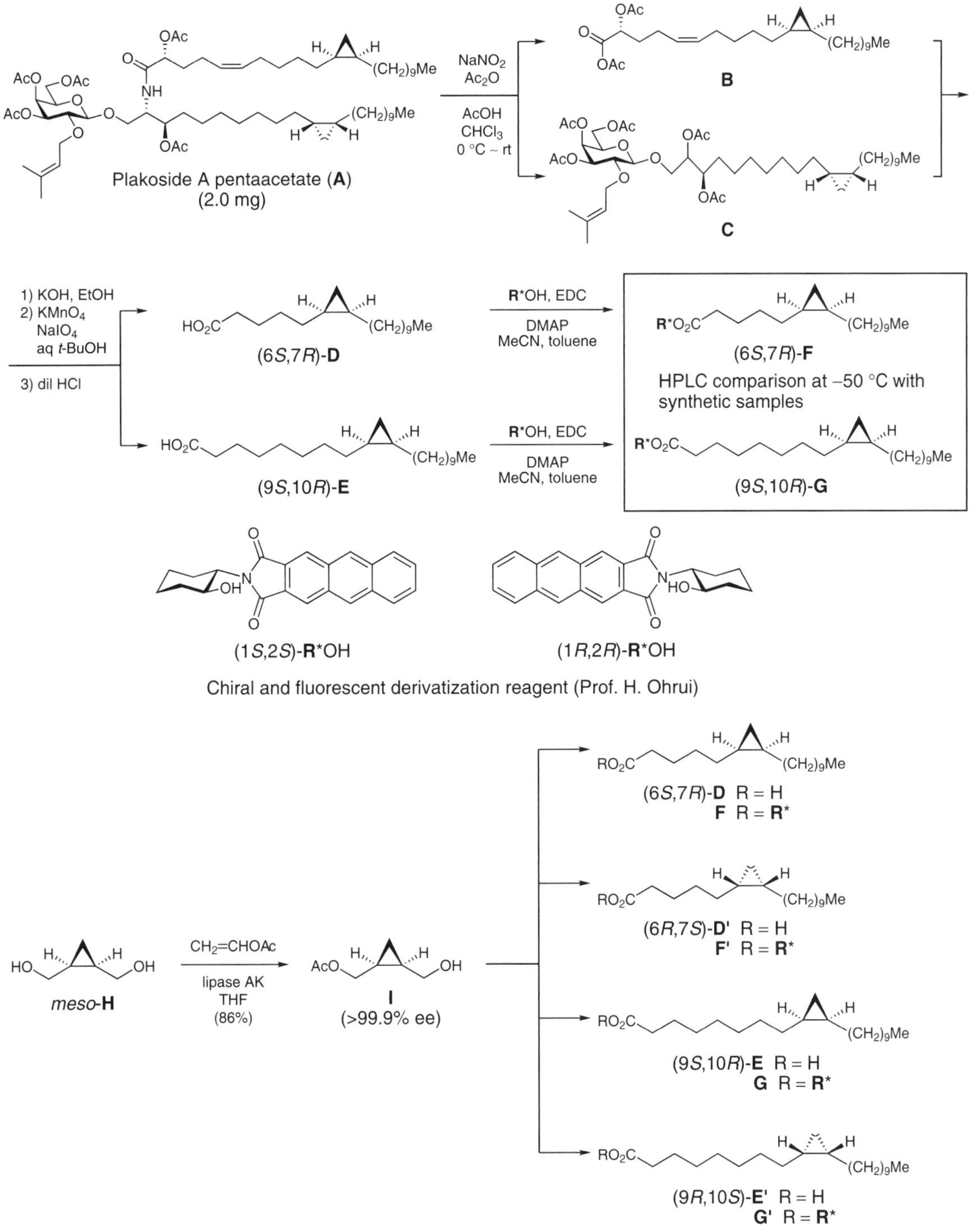

Figure 6.23 *Determination of the absolute configuration of plakoside A*

solved by its synthesis alone. Degradation and derivatization of plakoside A seemed necessary to solve the stereochemical problem.

Consequently, we decided to resume degradation studies on natural plakoside A. Professor Fattorusso reisolated 5 mg of plakoside A from the Caribbean sponge. Our strategy is shown in Figure 6.23. Degradation of plakoside A pentaacetate (**A**) will give two cyclopropane acids **D** and **E** resulting from the two carbon chains. By determining the absolute configuration of **D** and **E**, we will be able to determine the stereochemistry of plakoside A. Then, what kind of analytical method will give us the stereochemical assignment? HPLC analysis of esters **F** and **G** prepared from **D** and **E** after derivatization with Ohrui's chiral and fluorescent reagent (1*S*,2*S*)- and (1*R*,2*R*)-R*OH will meet the challenge.[28]

Degradation of plakoside A pentaacetate (**A**, 2.0 mg) was executed by first treating it with nitrous acid in acetic anhydride through *N*-nitrosation at the amide nitrogen of **A** to give **B** and **C**, which were further cleaved to give **D** and **E**, respectively.[29,30] A mixture of **D** and **E** was derivatized with Ohrui's reagent **R*OH**, and the products **F** and **G** were subjected to HPLC analysis at the column temperature of −50 °C. Owing to the presence of the anthracene ring in **F** and **G**, their picogram quantities were detectable by fluorescence, and therefore minute amounts of the degradation products could be analysed.

In order to determine the absolute configuration of **D** and **E**, it was necessary to prepare the synthetic reference samples of known absolute configuration. Conversion of *meso*-**H** to monoacetate **I** was followed by further synthetic steps to give **D**, **D′**, **E** and **E′**, all the possible candidates of **D** and **E** obtained by degradation. These were derivatized with Ohrui's reagent **R*OH**, and analysed by HPLC. The esters derived from plakoside A were (6*S*,7*R*)-**F** and (9*S*,10*R*)-**G**. Accordingly, the absolute configuration of plakoside A must be (2*S*,3*R*,11*S*,12*R*,2‴*R*,5‴*Z*,11‴*S*,12‴*R*)-**161**.[29,30]

A combination of enantioselective synthesis and HPLC analysis is a powerful method for the determination of the absolute configuration of a compound with stereogenic centers remote from other functionalities and stereogenic centers.

6.3   Glycosphingolipids and sphingolipids of medical interest

I have already shown several syntheses of sphingolipids as microbial metabolites or marine natural products. Sphingolipids are building blocks of the plasma membrane of eukaryotic cells. Their function is to anchor lipid-bound carbohydrates to cell surfaces, and to construct the epidermal water permeability barrier. The chemistry of sphingolipids is therefore closely related to dermatology or the science of skin. This section first treats sphingolipid in human epidermis.

Another topic in this section is the synthesis of glycosphingolipids as immunostimulating agents through natural killer T cells. This is our ongoing research subject at RIKEN (Institute of Physical and Chemical Research), and closely related to medical science. The initial discovery of α-galactosylsphingolipid with anticancer activity was brought about in the 1990s through screening of extracts of a marine sponge. I therefore put glycosphingolipid chemistry in this chapter dealing with marine natural products.

There are a number of reviews on sphingolipids. Biological and medical aspects of sphingolipid chemistry were reviewed by Kolter and Sandhoff.[31] Bio-organic chemistry of ceramide was reviewed by Kolter and coworkers.[32] Chemical aspects of sphingolipid research were also reviewed.[33,34]

6.3.1   Esterified cerebroside of human and pig epidermis

In 1989, Hamanaka and coworkers isolated a new esterified cerebroside from human epidermis, and found it to be a linoleic acid-containing acylglucosylceramide (**162**, Figure 6.24). She later named **162** "type I epidermoside," because it was specific to epidermis. Her works have been reviewed.[35] The same structure

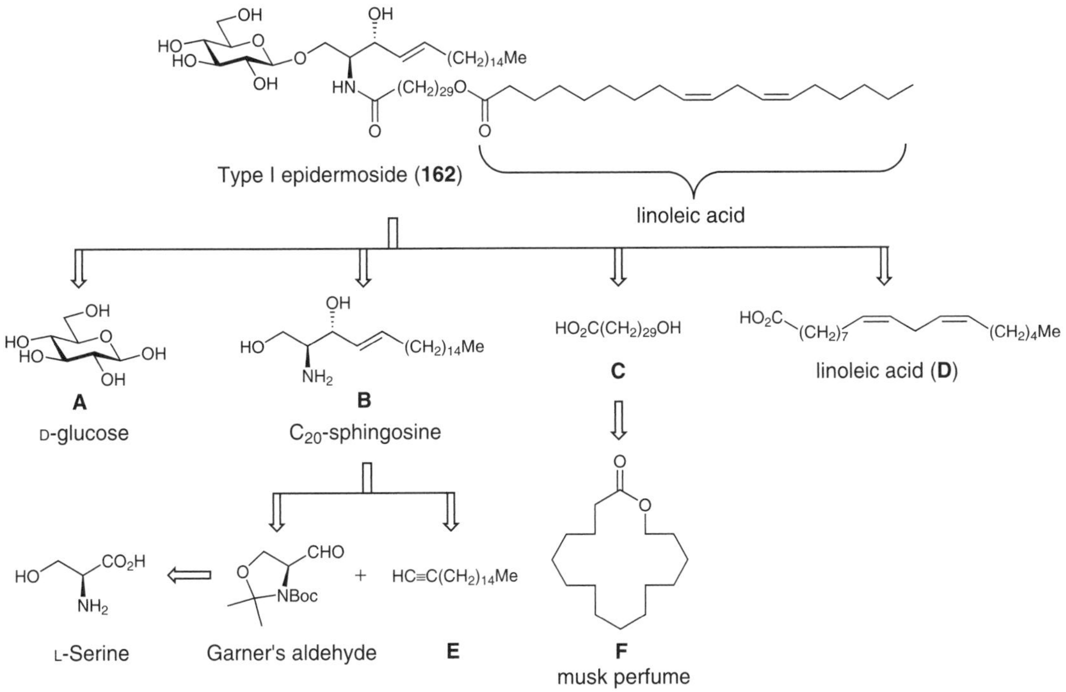

Figure 6.24   *Retrosynthetic analysis of type I epidermoside*

162 was also proposed by Downing and coworkers for the cerebroside which they isolated from pig epidermis. The esterified cerebroside **162** works as a functional water barrier in the skin.

We became interested in synthesizing **162** to confirm the proposed structure, and also to supply a sufficient amount of **162** to dermatologists for its further study. Figure 6.24 shows our retrosynthetic analysis of **162**.[36] Bond disconnection of **162** leads to D-glucose (**A**), (2S,3R,4E)-4-icosasphingenine (C$_{20}$-sphingosine, **B**), 30-hydroxytriacontanoic acid (**C**), and linoleic acid (**D**) as the necessary building blocks. C$_{20}$-Sphingosine (**B**) can be prepared from 1-heptadecyne (**E**) and L-serine, while **C** can be derived from 15-pentadecanolide (**F**), a musk perfume.

Our synthesis of **162** in 1991 is summarized in Figure 6.25.[36] Garner's aldehyde was converted to **D** via **A**, **B** and **C**. It was necessary to protect the amino group of **B** as trichloroacetamide, which could be removed readily in a later stage. Synthesis of the *p*-nitrophenyl ester **J** of 30-hydroxytriacontanoic acid was rather complicated, starting from 15-pentadecanolide. Its methanolysis followed by oxidation with pyridinium chlorochromate (PCC) furnished aldo ester **E**, one of the building blocks. In the route leading to the other building block, methyl 15-hydroxytriacontanoate was silylated to give **F**, which was reduced to furnish an alcohol. The alcohol then gave bromide **G**. A Wittig reagent derived from **G** was coupled with **E** to afford **H**, whose double bond was hydrogenated to give methyl 30-TBDPSoxytriacontanoate (**I**). After converting **I** to **J**, the hydroxy group of **J** was acylated with linoleyl chloride (**K**) to furnish the desired key building block, **L**.

Glycosylation of the sphingosine moiety **D** with acetobromo-D-glucose (**M**) smoothly gave β-D-glucopyranoside **N**. After removing the acetyl and trichloroacetyl groups of **N**, the free amino group was acylated with **L** to afford, after desilylation, type I epidermoside (**162**).[36] Thus, we obtained 74 mg of

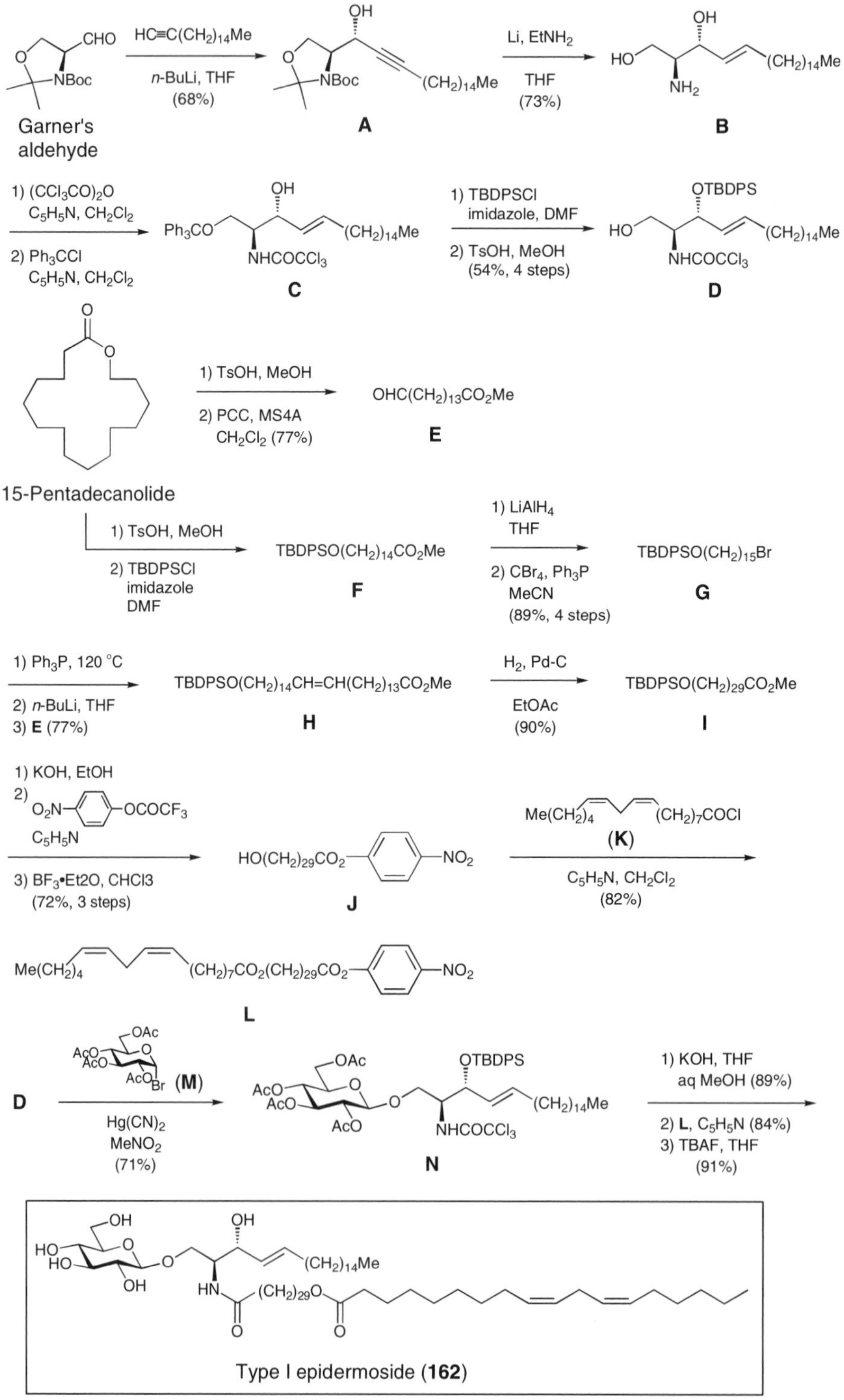

Figure 6.25 *Synthesis of type I epidermoside*

162 as a waxy solid melting at 122–125 °C. Its ^{1}H-NMR spectrum was identical with that of the natural product, and therefore the structure **162** of type I epidermoside was confirmed.

The extremely lengthy esterified side-chain of **162** may function as a connecting livet in the lipid bilayer system of epidermis. We also synthesized ceramide 1 (Figure 6.26), the free and extractable ceramide of human epidermis.[37]

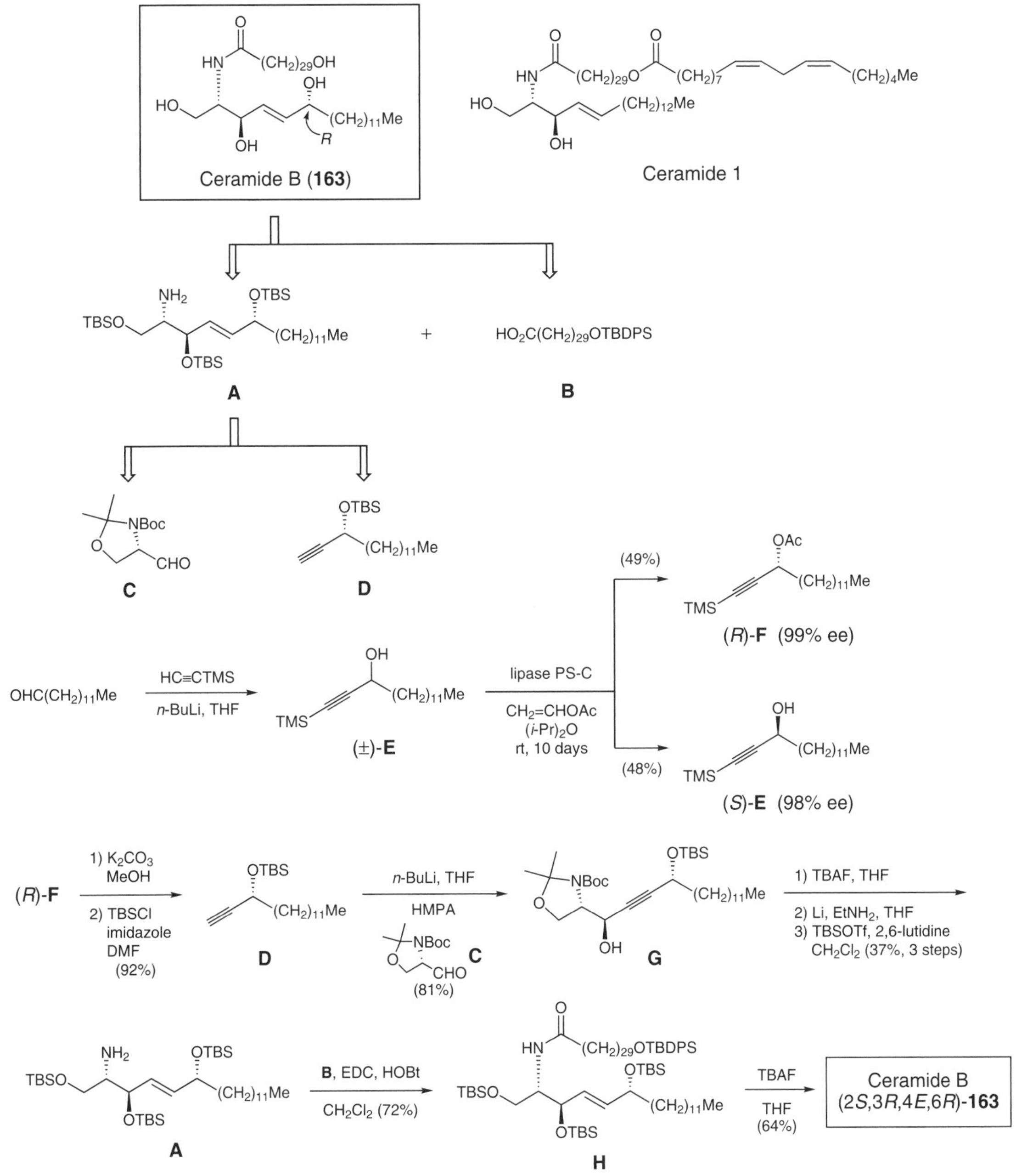

Figure 6.26   *Synthesis of ceramide B*

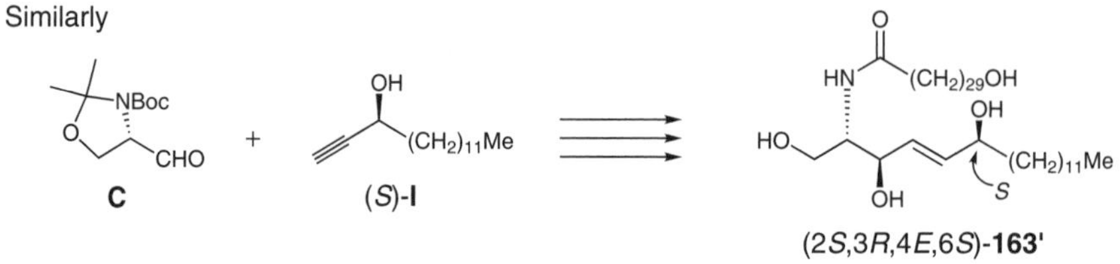

Similarly

(2S,3R,4E,6S)-**163'**

Figure 6.26   (continued)

6.3.2   Ceramide B, 6-hydroxylated ceramide in human epidermis

Ceramides are predominant lipids of human epidermis, acting as the water barrier to prevent loss of body water. They are classified into two groups, free ceramides and protein-bound ones. In the former case, the acyl side-chain of ceramides does not possess a terminal hydroxy group, while in the latter it is present to enable the binding of ceramides with proteins. In 1994, Downing and coworkers reported the isolation and identification of ceramide B (**163**, Figure 6.26), a protein-bound ceramide. It is a new 6-hydroxy-4-sphingenine-based ceramide. Its structure **163** was assigned by Downing *et al.* by extensive ¹H-NMR studies.

As to the stereochemistry of **163**, it almost certainly possesses (2S,3R,4E)-configuration, since all the known mammalian sphingosines possess that configuration. However, the absolute configuration at C-6 of **163** remained unknown. We became interested in solving this problem by synthesizing both 6*R*- and 6*S*-isomers of ceramide B as shown in Figure 6.26.[38,39]

Ceramide B (**163**) can be synthesized by connecting the sphingosine part **A** with the acyl part **B**. Preparation of **B** can be achieved as shown in Figure 6.25. It was an intermediate for the synthesis of type I epidermoside (**162**). The sphingosine part **A** is to be synthesized from Garner's aldehyde **C** and acetylene **D**. The problem was how to prepare both the enantiomers of 1-alkyn-3-ol like **D**.

1-Alkyn-3-ols are versatile intermediates in organic synthesis. In 1978, we reported asymmetric hydrolysis of the acetates of (±)-1-alkyn-3-ols with *Bacillus subtilis* esterase to give optically active acetates and alcohols.[40] Their enantiomeric purities, however, were mediocre. The present availability of a number of commercial lipases changed the situation completely. Treatment of (±)-acetylenic alcohol **E** with lipase PS-C (Amano Enzyme, Inc) and vinyl acetate in diisopropyl ether gave acetate (*R*)-**F** (99% ee) and the recovered alcohol (*S*)-**E** (98% ee) almost quantitatively after 10 days at room temperature. If both the enantiomers of 1-alkyn-3-ols are required, lipase-catalysed asymmetric acetylation leads to their successful preparation.

The acetate (*R*)-**F** was then converted to **D**, which was coupled with Garner's aldehyde **C** to give **G**. The building block **A** was prepared from **G**, and coupled with acid **B** to furnish **H**. Global desilylation of **H** afforded ceramide B (**163**) with 6*R*-configuration. Similarly, by employing (*S*)-**I** as an intermediate, (2S,3R,4E,6S)-**163'** was synthesized.

Both **163** and **163'** were then acetylated, and the ¹H-NMR spectra at 500 MHz of the two acetyl derivatives were compared with that of the acetyl derivative prepared from the natural ceramide B. By this comparison, it became clear that ceramide B is (2S,3R,4E,6R)-**163**. There were observed completely different signal patterns at $\delta = 5.0$–5.8, and the acetyl derivative prepared from **163** showed the identical ¹H-NMR spectrum with the authentic spectrum of the acetylated ceramide B. It is said that the content of ceramide B in old people is higher than that in young people. The hydroxylation at C-6 may reflect the aging of epidermis.

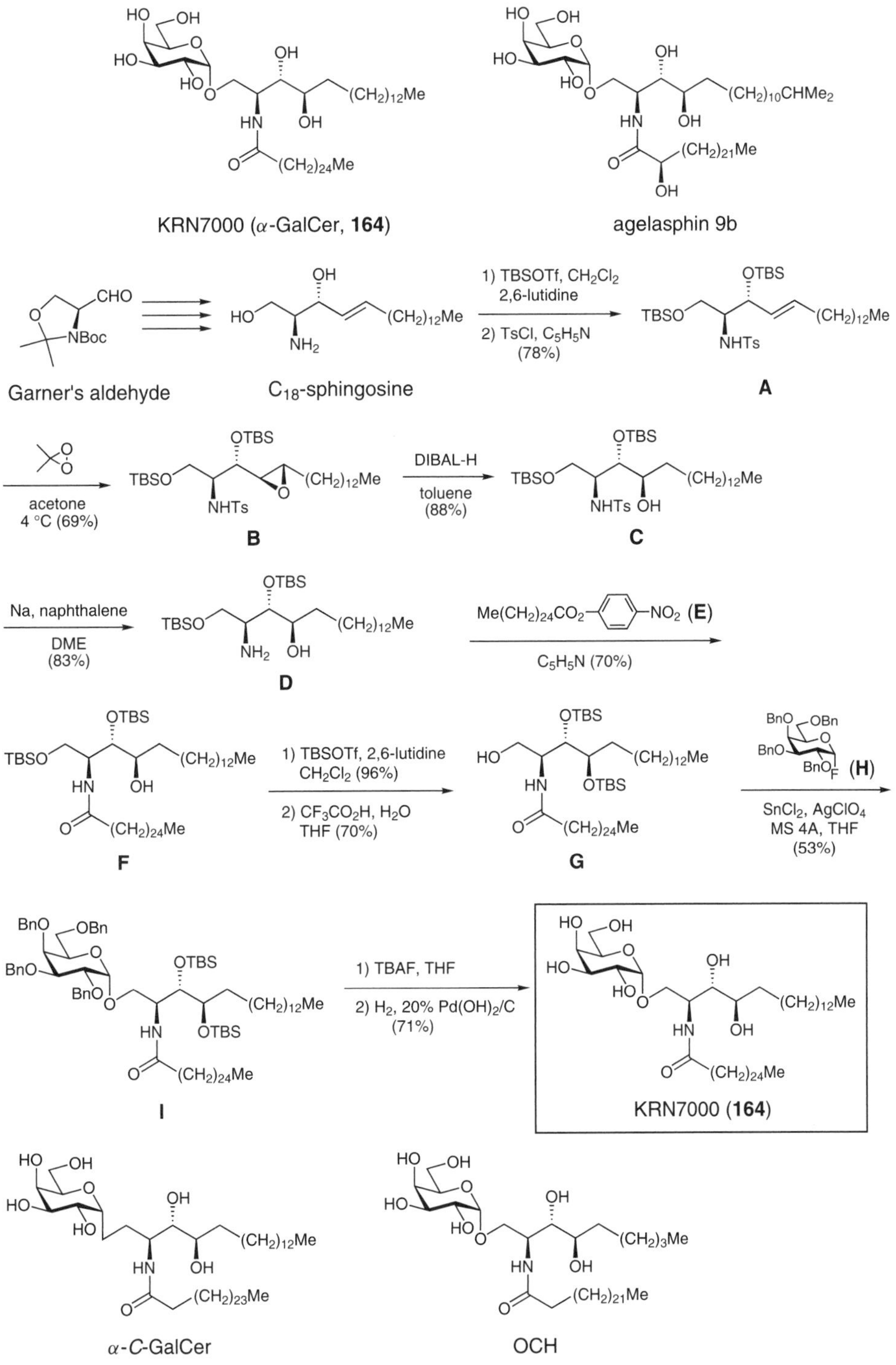

Figure 6.27   *Structures of KRN7000 and related compounds, and synthesis of KRN7000*

6.3.3   KRN7000, a glycosphingolipid that stimulates natural killer T cell

Most of my work described in this book was done at the University of Tokyo (until March, 1995) and at the Science University of Tokyo (April, 1995–December, 2001). In our Japanese system, we have a mandatory retirement at the age of 60 (the University of Tokyo) or 65 (Science University of Tokyo). Fortunately, I found my own laboratory bench at Fuji Flavor Company (a pheromone manufacturer) in 2002 to synthesize pheromones by myself. But I had no laboratory space to accommodate my coworkers to synthesize sphingolipids. Dr. S. Hamanaka thought that I should have a laboratory to continue my collaboration with medical scientists. Her effort gave me a chance to meet Prof. M. Taniguchi, an immunologist at Chiba University, on May 16, 2002. He was going to launch the Research Center for Allergy and Immunology of RIKEN as its Director. RIKEN (Institute of Physical and Chemical Research) is a big research organization supported by the Japanese government. I was told that Prof. Taniguchi was searching for capable synthetic chemists to synthesize analogs of KRN7000 (**164**, Figure 6.27).

KRN7000 (**164**) is an anticancer drug candidate developed by researchers at Kirin Brewery Company.[41] It was obtained through the modification of the structures of agelasphins (see Figure 6.27 for the structure of agelasphin 9b), which had been isolated in 1993 as anticancer glycosphingolipids from the extract of an Okinawan marine sponge, *Agelas mauritianus*. These glycosphingolipids exhibit anticancer activity in vivo in mice and humans, while they show no cytotoxicity at all in vitro. As some of my former students joined Kirin Brewery Co., and I too had a good relationship with Kirin, I knew the structure **164** of KRN7000. Indeed in 1998, we published a synthesis of KRN7000, as shown in Figure 6.27.[42]

Our synthesis of KRN7000 started from C_{18}-sphingosine, which was prepared from Garner's aldehyde. The key step was the stereoselective epoxidation of **A** to give **B** as the major product. This epoxidation was examined under various different conditions. MCPBA gave a mixture of epoxides, **B** having been the minor isomer [cf. Figures 6.17 (**H → I**) and 6.18 (**H → I**)]. Fortunately, dimethyldioxirane in acetone was found to give **B** as the major product (β-epoxide **B**/α-epoxide = 84:16 after 3 days at 4 °C). Reductive cleavage of the epoxy ring of **B** with DIBAL-H in toluene took place regioselectively to give the desired alcohol **C**.

Reductive removal of the *N*-tosyl group of **C** gave **D**, to which was added the acylating agent **E** to furnish **F**. Silylation of **F** was followed by partial desilylation of the persilylated product to provide **G**. Galactosylation of **G** with fluorosugar **H** gave the protected form **I** of the target molecule. Desilylation followed by debenzylation of **I** afforded KRN7000 (**164**). In 1998, synthesis of phytosphingosine derivative like **D** was rather complicated and cumbersome as shown here. At present, however, phytosphingosine is manufactured by fermentation, and commercially available.

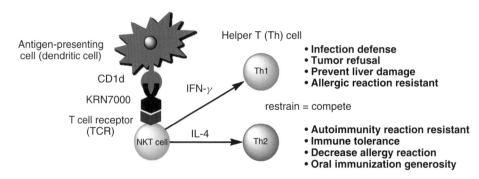

Figure 6.28   *Natural killer T (NKT) cells provide an innate-type immune response upon T cell receptor with CD1d-presented antigens such as KRN7000*

On May 16, 2002, Prof. Taniguchi taught us a lot about the biology of KRN7000. He and coworkers have shown that KRN7000 is a ligand to make a complex with CD1d (CD = cluster of differentiation) protein, a glycolipid-presenting protein on the surface of the antigen-presenting cells of the immune system.[43] Two lipid alkyl chains of **164** are bound in grooves in the interior of the CD1d protein, and the galactose head group of **164** is presented to the antigen receptors of natural killer (NK) T cells of the immune system (Figure 6.28).

Prof. Taniguchi actually showed us a docking model picture of **164** and CD1d protein. Indeed, the two alkyl chains of **164** seemed to be bound deeply in the grooves of the CD1d protein.[44]

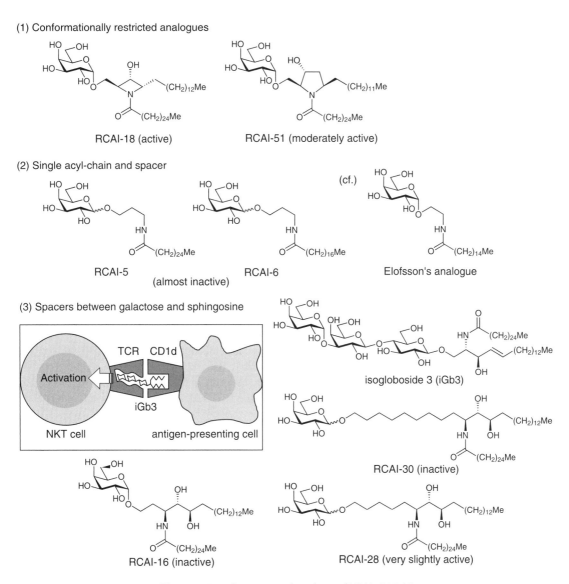

Figure 6.29   *Structures of analogs of KRN7000 (1)*

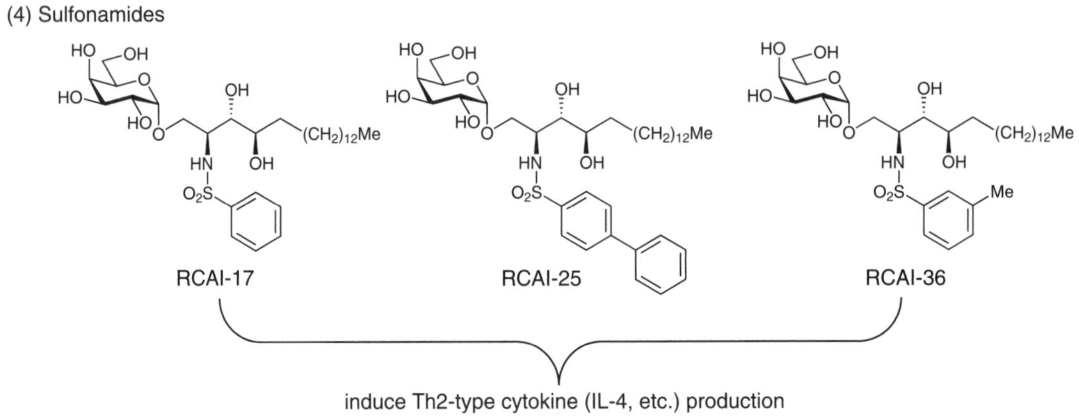

(4) Sulfonamides

induce Th2-type cytokine (IL-4, etc.) production

Figure 6.29   *(continued)*

After activation by recognition of CD1d-**164** complex, NKT cells release both helper T(Th)1 and Th2 types of cytokines at the same time in large quantities. Th1 type cytokines such as interferon (IFN)-γ mediate protective immune functions like tumor rejection, whereas Th2 type cytokines such as interleukin(IL)-4 mediate regulatory immune functions to ameliorate autoimmune diseases. Th1 and Th2 type cytokines can antagonize each other's biological actions. Because of this antagonism, use of KRN7000 for clinical therapy has not been successful yet. Prof. Taniguchi asked me to synthesize glycosphingolipids, which induce NKT cells to produce preferentially either Th1 or Th2 type cytokines, and convinced me to believe this project to be meaningful and important. He also told me that the natural ligand for CD1d protein was unknown.

Prof. Taniguchi's enthusiasm as the discoverer of NKT cells matched with my desire to have our own laboratory space, and we agreed to start the project. It began on April 7, 2003, at a small laboratory in the Research Institute of Seikagaku Kogyo Company, a carbohydrate-based pharmaceutical company in the suburb of Tokyo. In 2006, our laboratory was allowed to be in the Main Building of RIKEN in Wako-shi, Saitama.

In 2003, when we began our work, there were two remarkable analogs of KRN7000, as shown in Figure 6.27. Franck, Tsuji, and their coworkers reported that their synthetic α-D-*C*-galactosylceramide (α-*C*-GalCer, Figure 6.27) caused an enhanced Th1 type response in vivo in mice. On the other hand, Miyamoto *et al.* found that OCH (Figure 6.27), an analog of KRN7000 with a truncated sphingosine alkyl chain, caused NKT cells to produce IL-4 predominantly.

There are three excellent reviews available to understand the chemistry and biology of KRN7000 and related compounds.[45–47] In the next two sections, I will describe what we have done at RIKEN.

6.3.4   Analogs of KRN7000 prepared in 2003–2006

There are two options in choosing the strategy for drug discovery. One is to depend on high-throughput screening of a huge number of drug candidates produced by combinatorial chemistry. I simply did not like this. The other is to design candidate compounds by so-called "structure-based design" using the structural information obtained by X-ray crystallographic analysis of a lead compound and its receptor protein. This is a rational way. In our case, if we can have the structural information about CD1d protein-KRN7000-T cell receptor protein complex as revealed by X-ray crystallography, we can design possible drug candidates.

In 2003 when we started our works, there was no X-ray information available. Accordingly, we had to design our drug candidates empirically or by trial and error.

We first wanted to clarify the effect of conformational restriction in the ceramide part as caused by azetidine or pyrrolidine ring formation (Figure 6.29).[48] We synthesized many compounds, and gave them

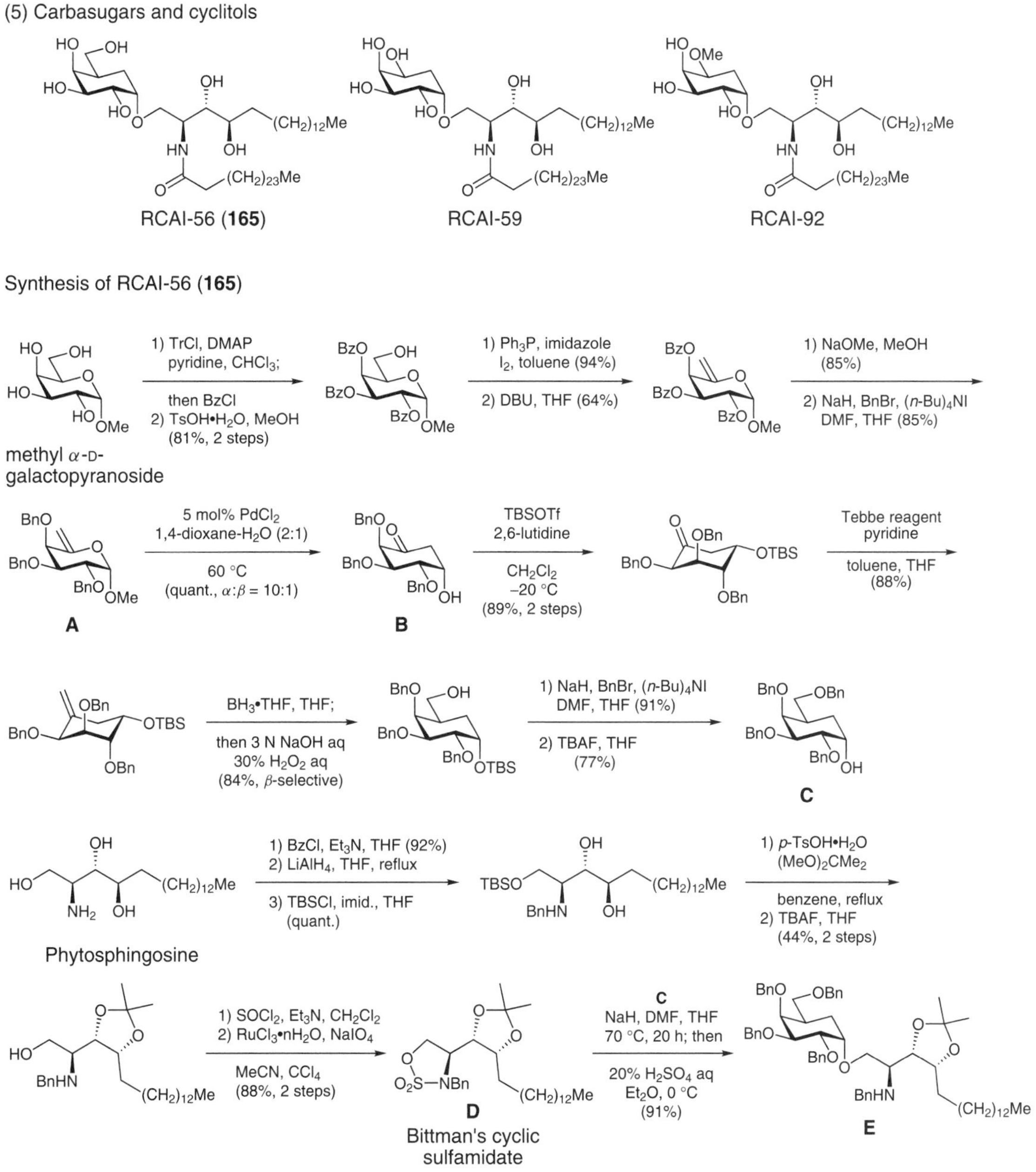

(5) Carbasugars and cyclitols

Synthesis of RCAI-56 (**165**)

Figure 6.30   *Structures of analogs of KRN7000 (2), and synthesis of RCA1-56*

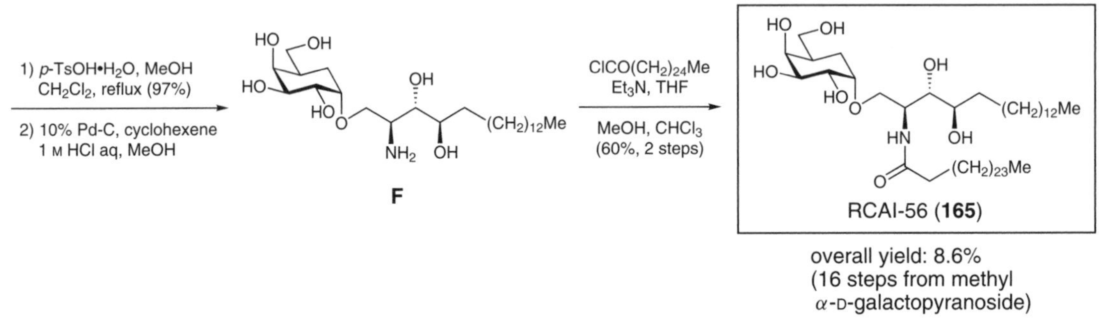

Figure 6.30   (continued)

code names, such as RCAI-1, RCAI-2, RCAI-3 and so on. RCAI stands for Research Center for Allergy and Immunology, RIKEN. Thanks to our experience in synthesizing azetidines such as penazetidine A (Figure 6.17) and penaresidin A and B (Figure 6.18), we quickly synthesized RCAI-18 and RCAI-51. They could induce the production of cytokines just like KRN7000. But their potencies could not surpass that of KRN7000, although RCAI-18 was almost as active as KRN7000. No preferential production of either Th1 or Th2 type of cytokines could be observed.[48]

We also synthesized compounds with a single acyl-chain and a spacer such as RCAI-5, because Elofsson and coworkers reported an analog that could be presented by CD1d protein.[49] However, both RCAI-5 and RCAI-6 were almost inactive.

In 2004, Zhou *et al.* proposed isogloboside 3 to be the natural ligand for CD1d protein.[50] Their claim might mean that the presence of a spacer between D-galactose and sphingosine could be allowed for the NKT cell activation. We synthesized RCAI-16, 28 and 30. All of them were inactive or only very slightly active.

Bioisosterism is an important and useful concept in drug design. Replacement of an atom in a bioactive compound by an atom of other elements of course gives a different compound. Sometimes, however, the new compound can also be bioactive. We therefore imagined that replacement of carboxamide in KRN7000 by sulfonamide linkage might give bioactive analogs. Accordingly, sulfonamide analogs such as RCAI-17, 25 and 36 were synthesized and bioassayed.[51] These analogs were shown to be the stimulants of mouse NKT cells to induce the production of Th2-biased cytokines such as IL-4 in vitro.

At the end of 2006, we were really waiting for the success of the X-ray crystallographic analysis of the CD1d protein-KRN7000-T cell receptor protein complex so that we would be able to achieve more meaningful work in this area.

6.3.5   Cyclitol, carbasugar and modified D-galactose analogs of KRN7000: RCAI-56 and RCAI-61

In 2005, an X-ray crystallographic analysis revealed the structure of the complex of human CD1d and KRN7000.[52] It was proved that the two lipid alkyl chains of KRN7000 are bound in the interior of the CD1d protein, and the galactose head group of KRN7000 is presented to the outside. However, we could not know anything about the roles of the oxygen functions of galactose in interaction with the T cell receptor. The awaited and decisive X-ray crystallographic structure of a human T cell receptor in complex with CD1d bound to KRN7000 was published in 2007 by McClusky, Rossjohn and their coworkers in Australia and the UK.[53]

It was revealed that human T cell receptor of NKT cell is interacting with the 2′-, 3′ and 4′-hydroxy groups of the galactose part of KRN7000, while 6′-hydroxy group is not involved in this hydrogen-bond network with any residues of human CD1d or T cell receptor. Moreover, it was found that the oxygen atom of the galactopyranose ring makes no hydrogen bonding with CD1d or T cell receptor, while the glycosidic

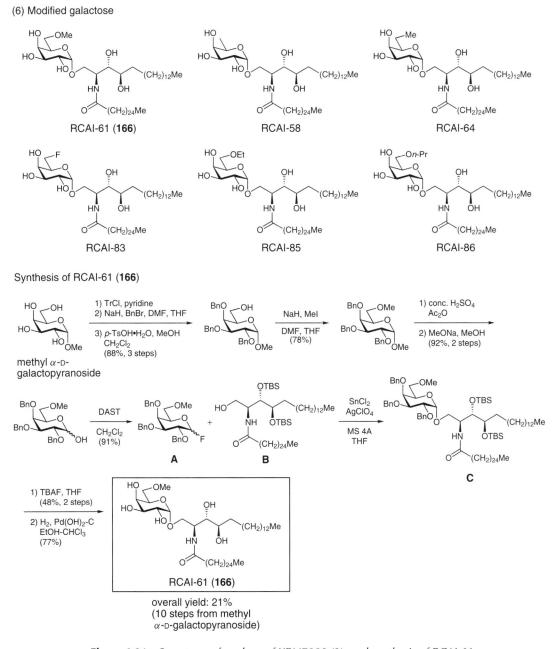

Figure 6.31 *Structures of analogs of KRN7000 (3), and synthesis of RCAI-61*

oxygen atom makes a hydrogen bond with CD1d. We therefore started our works to synthesize cyclitol or carbasugar analogs lacking the oxygen atom of the pyranose ring.[54,55] Modification at 6′-hydroxy group of galactose was also attempted.[56]

Among carbasugar and cyclitol analogs of KRN7000, RCAI-56 (**165**), RCAI-59 and RCAI-92 (Figure 6.30) were remarkably potent stimulants of mouse lymphocytes to produce Th1-biased cytokines such as interferon-γ in vivo. They were over 5 times more potent than KRN7000. Substitution of the pyranose oxygen with a methylene group turned out to be quite a successful way to enhance the bioactivity of the analogs of KRN7000.[54,55]

Synthesis of RCAI-56 (**165**) is summarized in Figure 6.30. So as to obtain D-carbagalactose, methyl α-D-galactopyranoside was converted to **A**, which was subjected to palladium-catalysed Ferrier rearrangement to give **B**. The keto cyclitol **B** served as the key intermediate to furnish 2,3,4,6-tetrabenzylated carbagalactose **C**. As to the sphingosine part, we were happy to purchase phytosphingosine manufactured by fermentation. It is commercially available now. Phytosphingosine was converted to Bittman's cyclic sulfamidate **D**, which was coupled with the carbasugar **C** to give **E**. Deprotection of **E** afforded **F**, which was acylated to furnish RCAI-56 (**165**). This was a lengthy synthesis, and the overall yield of **165** was 6.7% based on methyl α-D-galactopyranoside (16 steps).[54,55] Although its bioactivity is remarkable, RCAI-56 is rather difficult to prepare efficiently with a low cost.

Our subsequent attempts were to modify the primary hydroxy group (6′-OH) of KRN7000, and provided various analogs as shown in Figure 6.31. RCAI-61 (**166**) was shown to be the most promising analog of KRN7000 because of its synthetic availability and high bioactivity. RCAI-61 (**166**) brought about highly remarkable increase (×8.2 times stronger than KRN7000 as estimated by total amount of the produced interferon-γ at the dosage of 2 μg/mouse in vivo) in the production of interferon-γ even at a low concentration.[56]

Synthesis of RCAI-61 (**166**) is shown in Figure 6.31. Fluorosugar **A**, prepared from methyl α-D-galactopyranoside, was coupled with ceramide **B** to give **C**, whose deprotection furnished RCA-61 (**166**). The overall yield was 21% (10 steps) based on methyl α-D-galactopyranoside.[56] The short and efficient synthetic route leading to **166** makes it a promising drug candidate.

In conclusion, "structure-based design" was eventually fruitful. Reliable X-ray crystallographic structure determination evokes the imagination of chemists to design better drug candidates.

In this chapter, we have studied various syntheses of diverse marine natural products. Some of them are connected with drug-discovery programs. Molecular design in life science requires chemists' ingenuity to create better drugs than the lead compounds among natural products.

References

1. Mori, K.; Koga, Y. *Bioorg. Med. Chem. Lett.* **1992**, *2*, 391–394.
2. Mori, K.; Koga, Y. *Liebigs Ann.* **1995**, 1755–1763.
3. Mori, K.; Koga, Y. *Liebigs Ann. Chem.* **1991**, 769–774.
4. Mori, K.; Komatsu, M. *Liebigs Ann. Chem.* **1988**, 107–119.
5. Mori, K.; Komatsu, M. *Tetrahedron* **1987**, *43*, 3409–3412.
6. Mori, K.; Takikawa, H.; Kido, M; Albizati, K.F.; Faulkner, D.J. *Nat. Prod. Lett.* **1992**, *1*, 59–64.
7. Mori, K.; Takikawa, H.; Kido, M. *J. Chem. Soc., Perkin Trans. 1* **1993**, 169–179.
8. Takikawa, H.; Yoshida, M.; Mori, K. *Tetrahedron Lett.* **2001**, *42*, 1527–1530.
9. Yoshida, M.; Takikawa, H.; Mori, K. *J. Chem. Soc., Perkin Trans. 1* **2001**, 1007–1017.
10. Nozawa, D.; Takikawa, H.; Mori, K. *Bioorg. Med. Chem. Lett.* **2001**, *11*, 1481–1483.

11. Takikawa, H.; Nozawa, D.; Mori, K. *J. Chem. Soc., Perkin Trans. 1* **2001**, 657–661.
12. Mori, K.; Takeuchi, T. *Tetrahedron* **1988**, *44*, 333–342.
13. Mori, K.; Uno, T. *Tetrahedron* **1989**, *45*, 1945–1958.
14. Mori, K.; Uno, T.; Kido, M. *Tetrahedron* **1990**, *46*, 4193–4204.
15. Takanashi, S.; Takagi, M.; Takikawa, H.; Mori, K. *J. Chem. Soc., Perkin Trans. 1* **1998**, 1603–1606.
16. Mizushina, Y.; Murakami, C.; Ohta, K.; Takikawa, H.; Mori, K.; Yoshida, H.; Sugawara, F.; Sakaguchi, K. *Biochem. Pharmacol.* **2002**, *63*, 399–407.
17. Mori, K.; Uenishi, K. *Liebigs Ann. Chem.* **1994**, 41–48.
18. Yajima, A.; Takikawa, H.; Mori, K. *Liebigs Ann.* **1996**, 1083–1089.
19. Mori, K. *J. Heterocyclic Chem.* **1996**, *33*, 1497–1517.
20. Takikawa, H.; Maeda, T.; Mori, K. *Tetrahedron Lett.* **1995**, *36*, 7689–7692.
21. Takikawa, H.; Maeda, T.; Seki, M.; Koshino, H.; Mori, K. *J. Chem. Soc., Perkin Trans. 1* **1997**, 97–111.
22. Kobayashi, J.; Tsuda, M.; Cheng, J.-F.; Ishibashi, M.; Takikawa, H.; Mori, K. *Tetrahedron Lett.* **1996**, *37*, 6775–6776.
23. Takikawa, H.; Muto, S.; Nozawa, D.; Kayo, A.; Mori, K. *Tetrahedron Lett.* **1998**, *39*, 6931–6934.
24. Takikawa, H.; Nozawa, D.; Kayo, A.; Muto, S.; Mori, K. *J. Chem. Soc., Perkin Trans. 1* **1999**, 2467–2477.
25. Seki, M.; Kayo, A.; Mori, K. *Tetrahedron Lett.* **2001**, *42*, 2357–2360.
26. Seki, M.; Mori, K. *Eur. J. Org. Chem.* **2001**, 3797–3809.
27. Nicolaou, K.C.; Li, J.; Zenke, G. *Helv. Chim. Acta* **2000**, *83*, 1977–2006.
28. Ohrui, H. *Proc. Jpn. Acad. Ser. B* **2007**, *83*, 127–135.
29. Mori, K.; Tashiro, T.; Akasaka, K.; Ohrui, H.; Fattorusso, E. *Tetrahedron Lett.* **2002**, *43*, 3719–3722.
30. Tashiro, T.; Akasaka, K.; Ohrui, H.; Fattorusso, E.; Mori, K. *Eur. J. Org. Chem.* **2002**, 3659–3665.
31. Kolter, T.; Sandhoff, K. *Angew. Chem. Int. Ed.* **1999**, *38*, 1532–1568.
32. Brodesser, S.; Sawatzki, P.; Kolter, T. *Eur. J. Org. Chem.* **2003**, 2021–2034.
33. Tan, R.X.; Chen, J.H. *Nat. Prod. Rep.* **2003**, *20*, 509–534.
34. Liao, J.; Tao, J.; Lin, G; Liu, D. *Tetrahedron* **2005**, *61*, 4715–4733.
35. Hamanaka, S.; Suzuki, M.; Suzuki, A.; Yamakawa, T. *Proc. Jpn. Acad. Ser. B* **2001**, *77*, 51–56.
36. Mori, K.; Matsuda, H. *Liebigs Ann. Chem.* **1991**, 529–535.
37. Masuda, Y.; Mori, K. *J. Indian Chem. Soc.* **2003**, *80*, 1081–1083.
38. Mori, K.; Masuda, Y. *Tetrahedon Lett.* **2003**, *44*, 9197–9200.
39. Masuda, Y.; Mori, K. *Eur. J. Org. Chem.* **2005**, 4789–4800.
40. Mori, K.; Akao, H. *Tetrahedron Lett.* **1978**, *19*, 4127–4130.
41. Morita, M.; Motoki, K.; Akimoto, K.; Natori, T.; Sakai, T.; Sawa, E.; Yamaji, K.; Koezuka, Y.; Kobayashi, E.; Fukushima, H. *J. Med. Chem.* **1995**, *38*, 2176–2187.
42. Takikawa, H.; Muto, S.; Mori, K. *Tetrahedron* **1998**, *54*, 3141–3150.
43. Kawano, T.; Cui, J.; Koezuka, Y.; Toura, I.; Kaneko, Y.; Motoki, K.; Ueno, H.; Nakagawa, R.; Sato, H.; Kondo, E.; Koseki, H.; Taniguchi, M. *Science* **1997**, *278*, 1626–1629.
44. Kamada, N.; Iijima, H.; Kimura, K.; Harada, M.; Shimizu, E.; Motohashi, S.; Kawano, T.; Shinkai, H.; Nakayama, T.; Sakai, T.; Brossay, L.; Kronenberg, M.; Taniguchi, M. *Int. Immunol.* **2001**, *13*, 853–861.
45. Savage, P.B.; Teyton, L.; Bendelac, A. *Chem. Soc. Rev.* **2006**, *35*, 771–779.
46. Franck, R.W.; Tsuji, M. *Acc. Chem. Res.* **2006**, *39*, 692–701.
47. Wu, D.; Fujio, M.; Wong, C.-H. *Bioorg. Med. Chem.* **2008**, *16*, 1073–1083.
48. Fuhshuku, K.; Hongo, N.; Tashiro, T.; Masuda, Y.; Nakagawa, R.; Seino, K.; Taniguchi, M.; Mori, K. *Bioorg. Med. Chem.* **2008**, *16*, 950–964.
49. Wallner, F.K.; Chen, L.; Moliner, A.; Jondal, M.; Elofsson, M. *ChemBioChem* **2004**, *5*, 437–444.
50. Zhou, D.; Mattner, J.; Cantu III, C.; Schrantz, N.; Yin, N.; Gao, Y.; Sagiv, Y.; Hudspeth, K.; Wu, Y.-P.; Yamashita, T.; Teneberg, S.; Wang, D.; Proia, R.L.; Levery, S.B.; Savage, P.B.; Teyton, L.; Bendelac, A. *Science* **2004**, *306*, 1786–1789.

51. Tashiro, T.; Hongo, N.; Nakagawa, R.; Seino, K.; Watarai, H.; Ishii, Y.; Taniguchi, M.; Mori, K. *Bioorg. Med. Chem.* **2008**, *16*, 8896–8906.
52. Koch, M.; Stronge, V.S.; Shepherd, D.; Gadola, S.D.; Mathew, B.; Ritter, G.; Fersht, A.R.; Besra, G.S.; Schmidt, R.R.; Jones, E.Y.; Cerundolo, V. *Nat. Immunol.* **2005**, *6*, 819–826.
53. Borg, N.A.; Wun, K.S.; Kjer-Nielsen, L.; Wilce, M.C.J.; Pellicci, D.G.; Koh, R.; Besra, G.S.; Bharadwaj, M.; Godfrey, D.I.; McCluskey, J.; Rossjohn, J. *Nature* **2007**, *448*, 44–49.
54. Tashiro, T.; Nakagawa, R.; Hirokawa, T.; Inoue, S.; Watarai, H.; Taniguchi, M.; Mori, K. *Tetrahedron Lett.* **2007**, *48*, 3343–3347.
55. Tashiro, T.; Nakagawa, R.; Hirokawa, T.; Inoue, S.; Watarai, H.; Taniguchi, M.; Mori, K. *Bioorg. Med. Chem.* **2009**, *17*, 6360–6373.
56. Tashiro, T. Nakagawa, R.; Inoue, S.; Shiozaki, M.; Watarai, H.; Taniguchi, M.; Mori, K. *Tetrahedron Lett.* **2008**, *49*, 6827–6830.

7

Synthetic Examination of Incorrectly Proposed Structures of Biomolecules

Many incorrect structures of biomolecules have been proposed for natural products. Synthesis of compounds having the proposed structures often enables us to judge the correctness of the proposals. Especially in the cases of bioactive natural products, the synthetic compounds must be bioactive, if the proposed structures are correct. In some cases, we are able to revise the structures by synthesizing the biomolecules themselves after slightly modifying the proposed structures. In other cases, we are able to definitely disprove the proposed structures. In this chapter let us examine in depth the examples of incorrectly proposed structures of small biomolecules. We must admit that we humans are not omnipotent, while we may have a wrong desire to pretend ourselves to be omnipotent.

7.1 Origin of incorrect or obscure structures

These days chemistry students usually accept the structures of biomolecules shown in their textbooks as certain and unchangeable revelations of the unseen world of molecules. However, those who learned steroid chemistry in 1950s know that the initial structures proposed for cholesterol and deoxycholic acid by Wieland in 1928 (Figure 7.1) were challenged by the X-ray crystallographer Bernal in 1932, resulting in proposals of more plausible structures by others.[1,2] These historical episodes in steroid chemistry led us to believe that the truth about such matters would finally be revealed.

Since that time, a number of incorrect structures have been reported for small biomolecules. In this chapter, I will summarize my experience in examining incorrect or obscure structures by means of organic synthesis.

Incorrect or obscure structures are generally based on three categories:

(1) fabrication and/or falsification of facts, (2) use of inappropriate purification and/or analytical methods together with misinterpretation of the resulting analytical data, (3) use of inappropriate or unreliable bioassay methods. Examples will be given to illustrate these three categories.

Chemical Synthesis of Hormones, Pheromones and Other Bioregulators   Kenji Mori
© 2010 John Wiley & Sons, Ltd

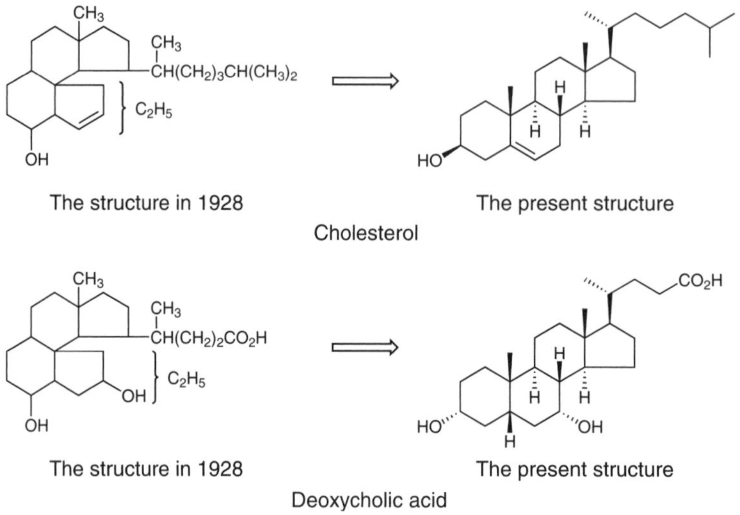

Figure 7.1   *Structures of cholesterol and deoxycholic acid*

7.2   Structure fabrications of historical interest

7.2.1   Kögl's auxin-a and -b, the plant-growth promoters

In 1933, Kögl *et al.* isolated auxin-a (**A**, Figure 7.2) and its lactone **B** as plant-growth promoters in human urine. This was indeed the first definite work to claim the isolation of an active principle responsible for the plant growth. One year later, they announced the isolation of auxin-a and auxin-b (**C**) from peanut oil and corn oil. The structures of these plant-growth promoters were proposed to be those as shown in Figure 7.2 by Kögl and Erxleben in 1934.

Auxin-a and b share the same trisubstituted cyclopentene ring in their structures. The presence of that ring system was deduced by permanganate oxidation of auxin-a and b, which gave auxin-glutaric acid (**D**) as crystals. Starting from (*S*)-3-methylpentanoic acid, Kögl and Erxleben synthesized three possible stereoisomers of **D**, and proposed one of them to be identical with the acid derived from the natural products.

The synthesis of these plant-growth promoters auxin-a and b was attempted by several groups. In 1966, Matsui (my former teacher) and Hwang synthesized a mixture of all of the possible stereoisomers of the δ-lactone **E** derived from auxin-b (**C**). The synthetic product, however, was biologically inactive.

In the same year of 1966, J.A. and J.F.G. Vliegenthart published a surprising paper entitled "Reinvestigation of authentic samples of auxin-a and auxin-b, and related products by mass spectrometry." They analysed the remnant samples of Kögl-labeled auxin-a, auxin-a lactone, auxin-b and auxin-glutaric acid, and found them not to be **A**–**D** but to be cholic acid, hydroquinone, thiosemicarbazide, and phthalic acid, respectively (Figure 7.2). It must be pointed out that the reported melting points of **A**–**D** are in good accord with those of cholic acid, hydroquinone, thiosemicarbazide, and phthalic anhydride.

These analytical results by Vliegenthart together with the fact that no one could reisolate auxin-a and b, in addition to the fact that the synthetic **E** was biologically inactive, made almost everyone to regard Kögl's auxin work as a typical scientific fraud.

Buffel in Belgium, however, published in 1985 a paper re-examining Kögl's reported data, and questioned the authenticity of the samples analysed by Vliegenthart. Buffel suggested some sample exchange

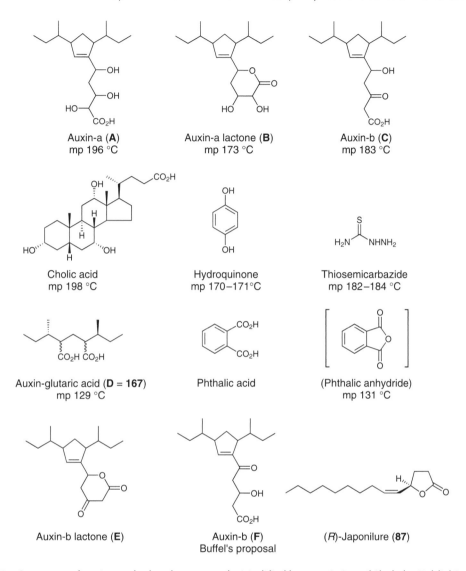

Figure 7.2 *Structures of auxins and related compounds. Modified by permission of Shokabo Publishing Co., Ltd*

between the late 1930s and early 1960s, probably during World War II. He thought the structure of auxin-a as **A** to be correct, but proposed a new structure **F** for auxin-b.

Three years after the publication of Buffel's paper, Matsui *et al.* announced the synthesis of **F** as a mixture of all of its possible stereoisomers, and found it to be biologically inactive. Thus, neither Matsui's 1966 synthesis of **E** nor his 1988 synthesis of **F**, both as stereoisomeric mixtures, yielded evidence to support the existence of auxin-b as a plant-growth promoter. We should, however, be careful enough to think about the possible effect of the wrong stereoisomers in the synthetic **E** and **F**. In pheromone perception, as we discussed in Chapter 4, there are some cases in which the opposite enantiomer of the natural pheromone strongly inhibits the bioactivity of the correct enantiomer. Especially in the case of the Japanese beetle pheromone (japonilure, **87**), its racemate lacks bioactivity due to the strong inhibition caused by the wrong enantiomer. No one can deny such a possibility also in the case of auxin-b.

Then, what kind of experiment can be rigorous enough to prove or disprove the correctness of Kögl–Erxleben's work? Is it necessary to synthesize all of the possible stereoisomers of **A** and **E** to separately bioassay the stereoisomers? In any scientific achievement, the most important part of the work is the correctness of the observed data such figures like melting points and specific rotations in the case of organic chemistry. In Kögl's auxin work, auxin-glutaric acid (**D**) was the most extensively studied compound. The acid derived from auxin-a and b and two other stereoisomers were synthesized, and their melting points and specific rotations were recorded precisely, although their absolute configuration remained unknown. I therefore decided to synthesize the three stereoisomers of auxin-glutaric acid (**D**) and to establish their absolute configurations. The melting points and specific rotations of our synthetic acids must carefully be compared with those reported by Kögl and Erxleben.

Figure 7.3 summarizes our synthesis of all three stereoisomers of auxin-glutaric acid (**167**).[3] Commercially available (*S*)-2-methyl-1-butanol (**A**) and L-isoleucine (**B**) were converted to (*S*)-3-methylpentanoic acid (**C**), which was previously employed by Kögl and Erxleben to synthesize auxin-glutaric acid. Then, **C** was converted to bromo ester **E**, whose intramolecular cyclization to give δ-lactone **F** was the key step. Chromic-acid oxidation of **F** furnished a stereoisomeric mixture of auxin-glutaric acid (**167**).

Prior to the separation of the stereoisomers of **167**, there was a need to decide what kind of analytical methods should be used in addition to chromatographic separation to check the purity of the separated isomers and also to assign the absolute configuration to each isomer. Among the three stereoisomers, only (2*S*,4*R*,5*S*,1′*S*)-**167** is not symmetrical, while the other two are symmetrical. In their ^{13}C-NMR spectra, therefore, (2*S*,4*R*,5*S*,1′*S*)-**167** should exhibit 13 signals, while (2*S*,4*S*,5*S*,1′*S*)- and (2*R*,4*R*,5*S*,1′*S*)-**167** should show only 7 signals. Accordingly, ^{13}C-NMR spectroscopy must be an ideal analytical method. As to the two isomers with 7 signals in their ^{13}C-NMR spectra, the X-ray crystallographic analysis of either of the isomers or its derivative will be the decisive way to establish their absolute configurations. The separation and identification of the three isomers of **167** could be achieved successfully, as shown in Figure 7.3. We were lucky to have (2*R*,4*R*,5*S*,1′*S*)-**I** as crystals suitable for X-ray analysis.

The melting points and specific rotations of our three stereoisomers of **167** were carefully measured and compared with the values reported previously by Kögl and Erxleben. As shown in Table 7.1, no correspondence was observed between our data and Kögl's. We therefore concluded that the Kögl–Erxleben work on auxin-a and b was a scientific fraud based on fabrication of the experimental data. The auxin work, which was once regarded as a marvellous piece of work in natural products chemistry, was just a fiction without rigorously checked scientific data. We should be careful to avoid such a mixture of poetry and truth.

Professor Frido J. Ritter, a pioneer of pheromone research in the Netherlands, wrote to me a letter after reading our paper on auxin-a and b. He was taught organic chemistry in 1945 by Kögl himself at Utrecht University. Kögl still firmly believed the auxin story at that time. Prof. Ritter continued that Kögl had an unlimited trust in his coworkers and failed to control their data adequately. In my opinion, those who are engaged in scientific research must be critical enough to examine the experimental data adequately. We must establish a good human relationship within our research team to avoid any scientific fraud.

7.2.2   Chemical communication system of the green flagellate, *Chlamydomonas*

During the 1930s and 1940s, Moewus and Kuhn (1938 Nobel Prize winner) extensively studied the biology and chemistry of the green flagellate, *Chlamydomonas eugametos*, and its sexuality. They claimed that a 3:1 mixture of (8*Z*)-crocetin dimethyl ester (**A**, Figure 7.4) and its (8*E*)-isomer **B** activated the female gametes of the flagellate, while a 1:3 mixture activated the males to enable copulation. Moewus and

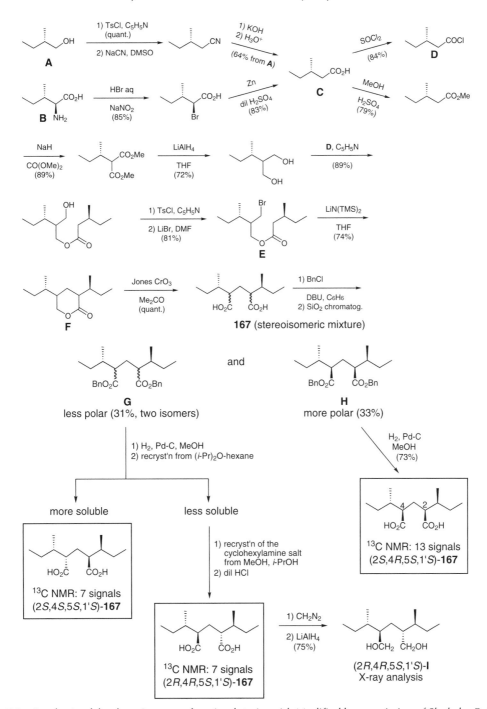

Figure 7.3 *Synthesis of the three isomers of auxin-glutaric acid. Modified by permission of Shokabo Publishing Co., Ltd*

Table 7.1 Comparison of the physical data of the three stereoisomers of **167** with those reported for the three stereoisomers of auxin-glutaric acid

Our work		
Compound	mp/°C	$[\alpha]_D^{22-24}$ (EtOH)
(2*S*,4*R*,5*S*,1'*S*)-**167**	73–74	+12.7
(2*S*,4*S*,5*S*,1'*S*)-**167**	86–89	−16.1
(2*R*,4*R*,5*S*,1'*S*)-**167**	88.5–89.5	+14.9

Kögl-Erxleben work		
Compound	mp/°C	$[\alpha]_D^{20}$ (EtOH)
α-acid	104.5	−4.9
β-acid	106–108	+9.35
γ-acid (= natural isomer)	129	−11.3

Kuhn's theory was revolutionary, because they claimed that chemical substances could regulate such a basic biological process as reproduction and also that a mixture of two isomers, not a single pure compound, was responsible for the bioactivity. Their work attracted attention of scientists in many countries, and in 1951 I myself read a review article on this work, and became fascinated by the beauty of the chemical communication system of the organism.

Their observations, however, could not be reproduced by others, or even by Moewus himself when he stayed at Columbia University. Therefore, in 1955, Professor Ryan, Moewus' host at Columbia, concluded that Moewus' work was incorrect. Incidentally, in 1956 when I was an undergraduate student, I attended Prof. Ryan's lecture on microbial genetics given at the University of Tokyo for a semester. He did not mention anything about *Chlamydomonas* on that occasion. In 1960 Kuhn and Löw also admitted that the culture filtrate of the alga provided by Hartmann in Tübingen did not show any spectroscopic evidence indicating the presence of crocetin dimethyl ester in it. The *Chlamydomonas* story by Moewus remained as an enigma among the pioneering works of chemical communications, while the name *Chlamydomonas* remained in the memory of many biologists and chemists including myself.

In 1995 Starr *et al.* in the USA reported the chemoattraction of male gametes by female gametes of a new species of the green flagellate, *Chlamydomonas allensworthii*. The female-produced attractant pheromone was isolated and identified as **168** by Jaenicke and Marner in Germany. They coined the name lurlenic acid to **168**, because they worked in Cologne near the Rock of Lorelei. The pheromone **168** attracts the male gametes at a concentration as low as 10^{-12} M.

Mainly because of its unique structure and bioactivity and partly because of my personal interest in *Chlamydomonas* chemistry, we undertook the synthesis of lurlenic acid (**168**) basing on the retrosynthetic analysis as shown in Figure 7.5. Lurlenic acid (**168**) can be constructed from three building blocks,

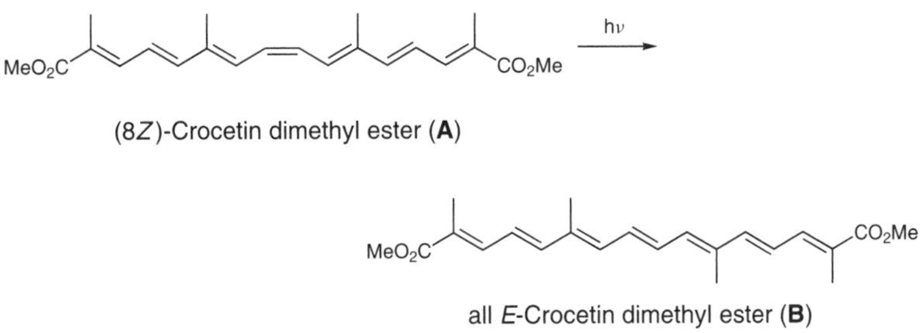
MeO2C
CO2Me
hν
(8Z)-Crocetin dimethyl ester (A)
MeO2C
CO2Me
all E-Crocetin dimethyl ester (B)

Pheromone precursor (**A**) shows no bioactivity.

↓ Irradiation for 24–25 min (blue violet light)

A / **B** = 3 : 1 This mixture activates the female gametes for copulation.

↓ Irradiation for 50 min (blue violet light)

A / **B** = 1 : 3 This mixture activates the male gametes for copulation.

↓ Further irradiation

All *E*-isomer **B** shows no bioactivity.

Figure 7.4 *Works of Moewus and Kuhn on the activation of the gametes of the green flagellate, Chlamydomonas eugametos, for copulation*

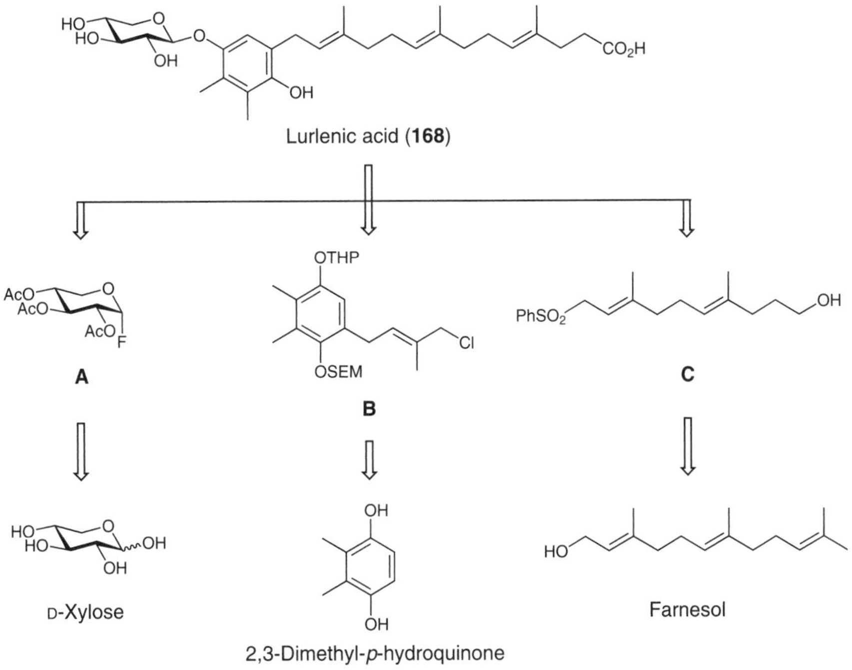
HO
HO
O
OH
O
CO2H
OH
Lurlenic acid (168)
AcO
AcO
O
AcO
F
A
OTHP
OSEM
Cl
B
PhSO2
OH
C
HO
HO
O
OH
OH
D-Xylose
OH
OH
2,3-Dimethyl-p-hydroquinone
HO
Farnesol

Figure 7.5 *Retrosynthetic analysis of lurlenic acid*

D-xylose part **A**, the aromatic part **B**, and the aliphatic part **C**, whose starting materials are to be D-xylose, 2,3-dimethyl-*p*-hydroquinone, and farnesol.

Our synthesis of lurlenic acid (**168**) is summarized in Figure 7.6.[4–6] The starting material **A** was half-protected and prenylated to give **B**, which was further protected and oxidized to furnish **C**. Wittig olefination

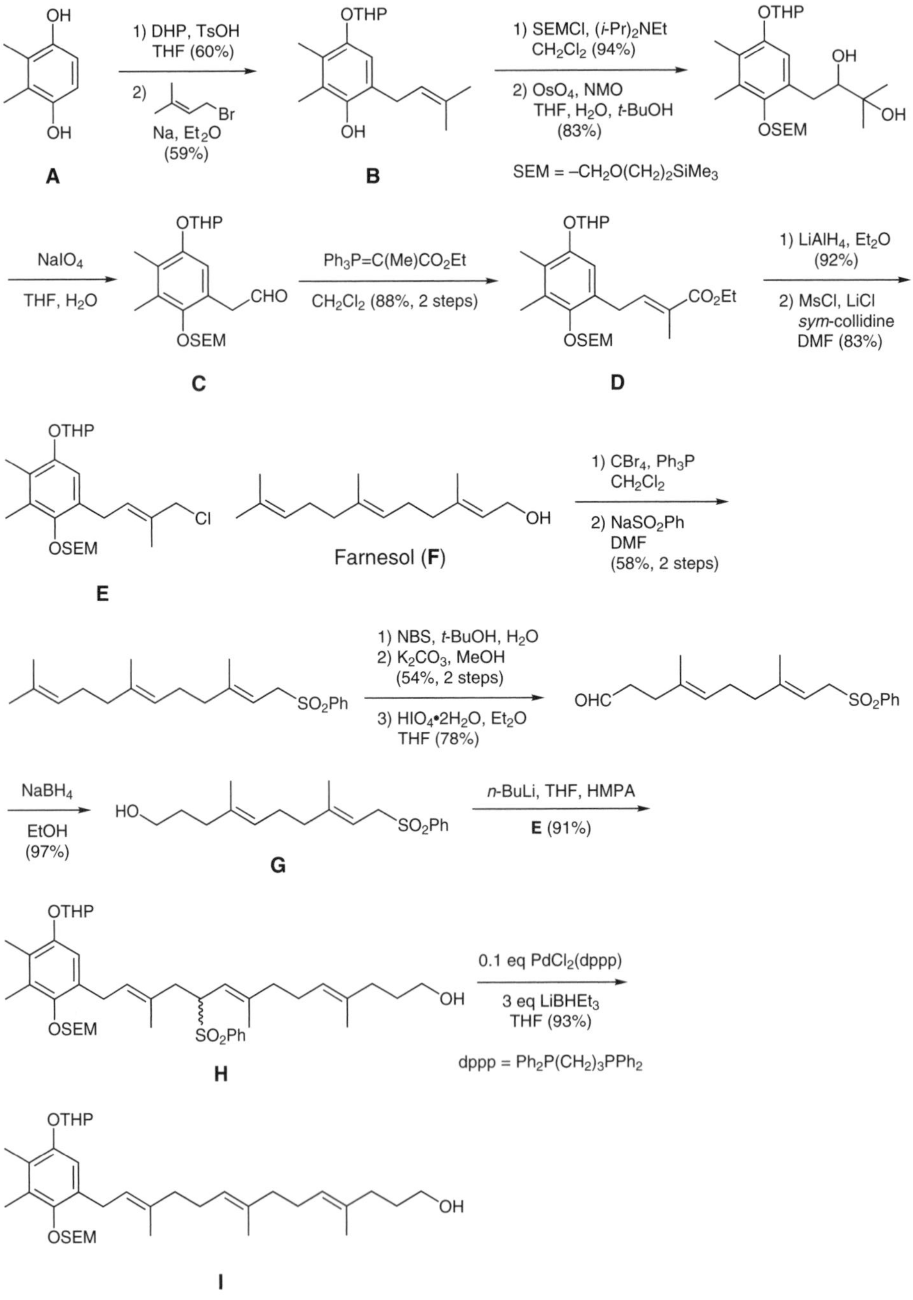

Figure 7.6   *Synthesis of lurlenic acid*

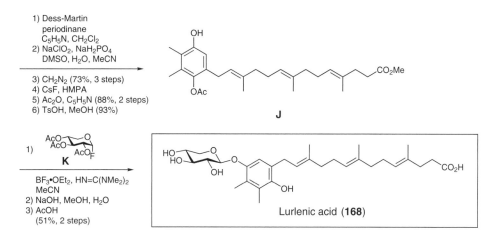

Figure 7.6 *(continued)*

of **C** afforded **D**, which was further converted to one of the key building blocks, **E**. Conversion of farnesol (**F**) to another key intermediate **G** proceeded without event. The dianion derived from **G** was alkylated with **E** to give **H**. Desulfonylation of **H** was executed with lithium triethylborohydride in the presence of a palladium catalyst to give **I**, which furnished **J**, the precursor for glycosylation. Glycosylation of **J** with 2,3,4-tri-*O*-acetyl-α-D-xylopyranosyl fluoride (**K**) afforded, after deprotection, lurlenic acid (**168**). The ^{1}H- and ^{13}C-NMR spectra of our synthetic **168** were completely identical with those of the natural product. Starr and Jaenicke bioassayed our sample, and the threshold concentration of **168** to attract the male gametes was about 10^{-13} M.[6]

Subsequent studies of ours showed the importance of D-xylose moiety for the bioactivity. Tetrahydropyranyl, β-L-xylopyranosyl, β-D-galactopyranosyl-, α-D-arabinopyranosyl- and β-D-arabinopyranosyl analogs of **168** were all inactive.[7] Structural requirements for the expression of pheromonal activity was further studied, indicating the importance of the phenolic hydroxy group, the appropriate length and the unsaturation of the side-chain part, and the presence of a polar group (CO_2H or OH) at the terminal position of the side-chain.[8]

In summary, as to *Chlamydomonas allensworthii*, its chemical communication system in connection with its sexuality was thus clarified. In contrast, as to Moewus' observations on *Chlamydomonas eugametos*, things remain obscure even now. The elusive work of Moewus and Kuhn played the role of a magic lure to attract scientists of the next generation to study *Chlamydomonas* by their revolutionary proposal of sex attractants as a mixture of small molecules.

7.2.3 Early fabrications of the structures of insect pheromones

In the 1960s, structure elucidation of insect pheromones was a very difficult task due to the scarcity of the sample available from the insects. As shown in Table 7.2, Jacobson *et al.* in the USA published three incorrect structures of pheromones.

The female pheromone of the gypsy moth was claimed to be gyptol (**A**, Table 7.2), while the true pheromone was disparlure (**85**). Although the female pheromone of the American cockroach was claimed to be a cyclopropane compound **B**, the genuine and major pheromone was periplanone-B. The female pheromone of the pink bollworm moth was proposed as propylure (**C**), while the genuine pheromone was gossyplure as a mixture of two isomers. In these three cases, the synthetic compounds with the false

Table 7.2 *Examples of incorrect and correct structures proposed for some insect pheromones*

Pheromone producers	Proposed structures	Correct structures
Gypsy moth (*Lymantria dispar*)	Gyptol (**A**) (Jacobson, 1961)	Disparlure (**85**) (Beroza, 1970)
American cockroach (*Periplaneta americana*)	**B** (Jacobson, 1963)	Periplanone-B (Persoons, 1976)
Pink bollworm moth (*Pectinophora gossypiella*)	Propylure (**C**) (Jacobson, 1966)	Gossyplure (1:1 mixture) (Hummel, 1973)

structures were pheromonally inactive, and therefore reinvestigations were initiated to clarify the true structures. Synthesis followed by bioassay was the key to clarify these structure fabrications.

7.3 Incorrect structures resulting from inappropriate use of purification or analytical methods

I have already discussed in previous chapters a number of examples belonging to this category. I will briefly summarize the examples already treated, and add two new examples.

Table 7.3 shows the already discussed examples of incorrect structures resulting from inappropriate use of purification or analytical methods. In the case of periplanone-A, the error was inevitable, because an efficient preparative HPLC system was not available yet at the time of that work.

Naurol A was isolated in 1991 by De Guzman and Schmitz as a metabolite of a Pacific sponge collected at Nauru Island. It showed weak cytotoxicity against murine lymphocytic leukemia cells, and its structure was proposed as **169** (Figure 7.7) based primarily on ^{1}H- and ^{13}C-NMR data. In 2000, we synthesized **169** as a racemic and diastereomeric mixture, starting from **A**.[9] The key step was the Stille–Kosugi coupling (**B** + **C**) to give **169**. Synthetic **169** showed ^{1}H- and ^{13}C-NMR spectra distinctly different from those of the natural naurol A. The correct structure of naurol A still remains unknown.

In 1998, Faulkner and coworkers isolated a new methyl ester as a constituent of the Philippine sponge, *Plakinastrella* sp., and proposed its structure as methyl (5Z,9Z)-17-methyl-5,9-nonadecadienoate (**170**). As I have my interest in elucidating the absolute configuration of the natural products with a *sec*-butyl terminal group, we synthesized its (*R*)-enantiomer **170** in 2001 by a Wittig reaction between **D** and **E**.[10] The synthetic (*R*)-**170** showed ^{1}H- and ^{13}C-NMR spectra different from those reported for the natural

Table 7.3 *Examples of incorrect structures resulting from inappropriate use of purification or analytical methods*

Proposed structures	Revised structures	Origin of the errors
Orobanchol (phytohormone)	53	Deduced by MS only. Insufficient NMR data.
Periplanone-A (insect pheromone)	79	GC purification instead of HPLC purification. (Pyrolysis took place.)
Acoradiene (insect pheromone)	80	Overlooking the previous work by others. Less-rigorous NMR analysis.
ar-Himachalene (insect pheromone)	84	Rotation measurement in an inproper solvent.
Koninginin A (antibiotic)	139	Less-rigorous NMR analysis.
Bifurcarenone (mitosis inhibitor)	152	Less-rigorous NMR analysis.

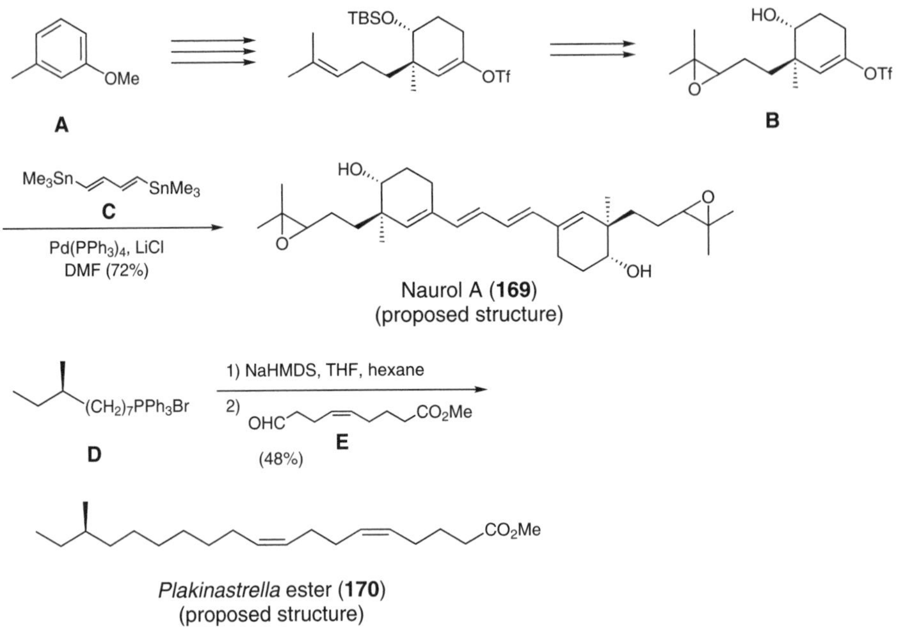

Figure 7.7   *Synthesis of the proposed structure of naurol A, and that of the proposed structure of a metabolite of a Philippine marine sponge*

product. After comparing the NMR spectra of the natural product with those of our synthetic sample, Faulkner told me that the natural material was a mixture of two compounds, neither of which had the proposed structure. He frankly told me that he had no idea what the correct structures were, and no material to work with. He admitted that there could be no excuse for mistakes such as this. I really like the honest way of the late Professor D.J. Faulkner, and miss him very much.

7.4   Inappropriate structural proposal caused by problems in bioassay methods

7.4.1   Blattellastanoside A and B, putative components of the aggregation pheromone of the German cockroach

The aggregation pheromone of the German cockroach (*Blattella germanica*) is excreted with its frass, and marks its harboring place. In 1990 Sakuma and Fukami at Kyoto University claimed the identification of the hydrochlorides of ammonia, methylamine, dimethylamine, trimethylamine and 1-dimethylamino-2-methyl-2-propanol as the attractant components of the aggregation pheromone. [In 2005, the sex attractant pheromone of *B. germanica* was identified as blattellaquinone (Figure 7.8), and confirmed by synthesis.[11]]

Sakuma and Fukami then demonstrated the presence of the arrestant components in the frass. Their work culminated in the isolation of the two arrestant components, which were shown to be chlorinated steroid glucosides, blattellastanoside A and B. The pheromone activity of the former was about 70 times stronger than that of the latter. Their structures were first proposed as 9α-chloro-11,12-epoxy-5α-stigmastan-3β-yl β-D-glucopyranoside (**A**, Figure 7.8) and 9α-chloro-11β-hydroxy-5α-stigmastan-3β-yl β-D-glucopyranoside (**B**), respectively, in April, 1992.

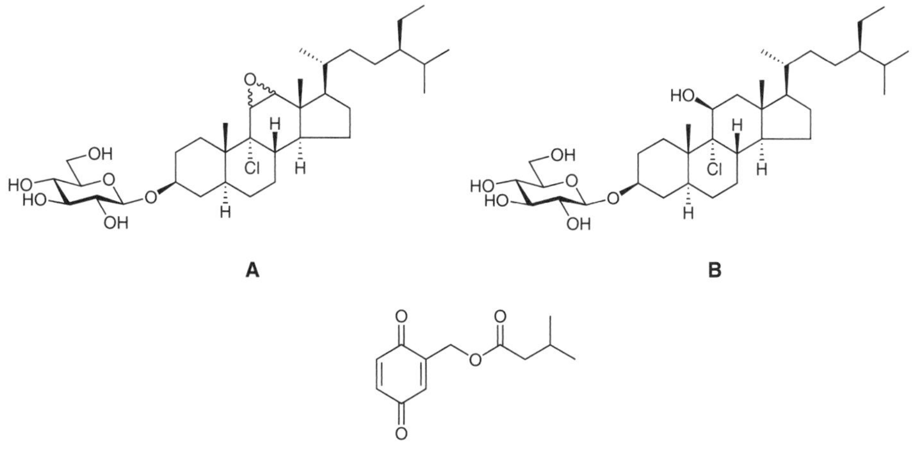

A   **B**

Blattellaquinone

Figure 7.8 *Structures proposed for blattellastanoside A and B in April, 1992, and the structure of blattel-laquinone. Modified by permission of Shokabo Publishing Co., Ltd*

I was intrigued by the unique chlorinated and C-ring-modified structures of **A** and **B**, and started the synthesis of **B**, whose stereochemistry had been proposed definitely as depicted in **B**. We first planned to synthesize the aglycone part of blattellastanoside B so as to verify the proposed structure **B** for balttel-lastanoside B. I thought that the ^{13}C-NMR spectrum of the aglycone part alone would be nearly identical with that of blattellastanoside B, except the signals due to β-D-glucopyranose moiety. Thus, if the proposed structure **B** were the correct one, the synthetic aglycone part would show the ^{13}C-NMR spectrum similar to that of the parent blattellastanoside B.

Figure 7.9 summarizes our synthesis of the aglycone part.[12] Because the target compound **I** possesses functional groups at C-9 and C-11, the $\Delta^{9(11)}$-steroid **D** can be envisaged as our intermediate, which was readily available by application of the radical-relay chlorination reaction directed by a template. By analogy with Breslow's work on 5α-cholestan-3α-ol, the desired 5α-stigmast-9(11)-en-3α-ol (**D**) was prepared from 5α-stigmastan-3β-ol (**A**). Treatment of **A** with *m*-iodobenzoic acid, triphenylphosphine and diethyl azodicarboxylate under the Mitsunobu conditions furnished **B** with the inverted α-configuration at C-3. Chlorination of **B** with sulfuryl chloride in the presence of a catalytic amount of benzoyl peroxide in carbon tetrachloride gave chloride **C**, which was dehydrochlorinated immediately with potassium hydroxide to afford **D**.

Jones chromic-acid oxidized **D** to ketone **E**, which was treated with *N*-bromosuccinimide and water to give bromohydrin **F**. Treatment of **F** with sodium methoxide afforded epoxide **G**, whose reduction gave epoxy alcohol **H**. Finally, cleavage of the epoxy ring of **H** with dry hydrogen chloride in chloroform afforded the proposed aglycone **I** as needles. The ^{13}C-NMR spectrum of **I**, however, was not in good accord with the ^{13}C-NMR signals due to the aglycone moiety of blattellastanoside B. The proposed structure **B** (Figure 7.8) was therefore incorrect.

In July 1992, Sakuma and Fukami reported new structures **A** and **B** (Figure 7.10) for blattellastanoside A and B in their poster presentation at 9th Annual Meeting of the International Society of Chemical Ecology in Kyoto. In the new structures, α-orientation of the epoxy ring of blattellastanoside A was proposed, and we decided to synthesize a compound with the proposed structure for blattellastanoside A.

Our synthesis of the proposed structure for blattellastanoside A is summarized in Figure 7.11.[12] The ketone **E** of Figure 7.9 was reduced with lithium tri(*t*-butoxy)aluminum hydride to give **A**, whose acetate **B**

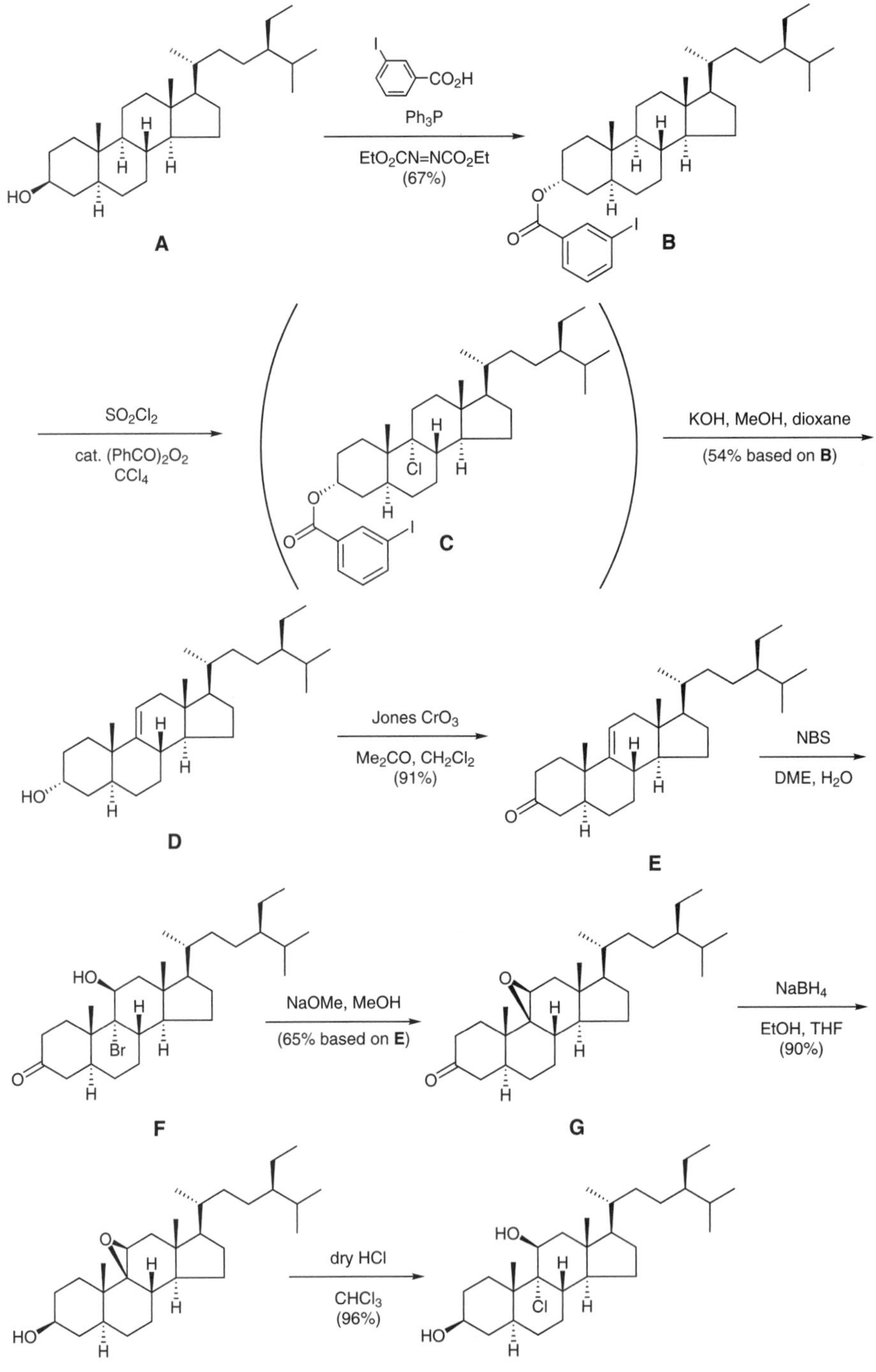

H

I (aglycone of the structure **B** of Figure 7.8)

Figure 7.9 *Synthesis of the aglycone part of the structure proposed for blattellastanoside B in April, 1992. Modified by permission of Shokabo Publishing Co., Ltd*

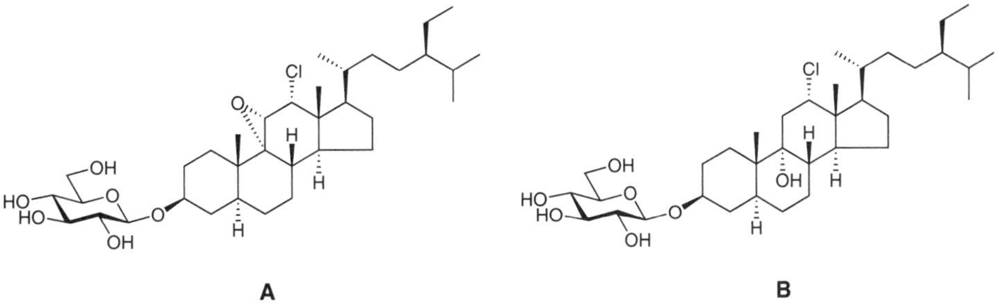

Figure 7.10   *Structures proposed for blattellastanoside A and B in July, 1992. Modified by permission of Shokabo Publishing Co., Ltd*

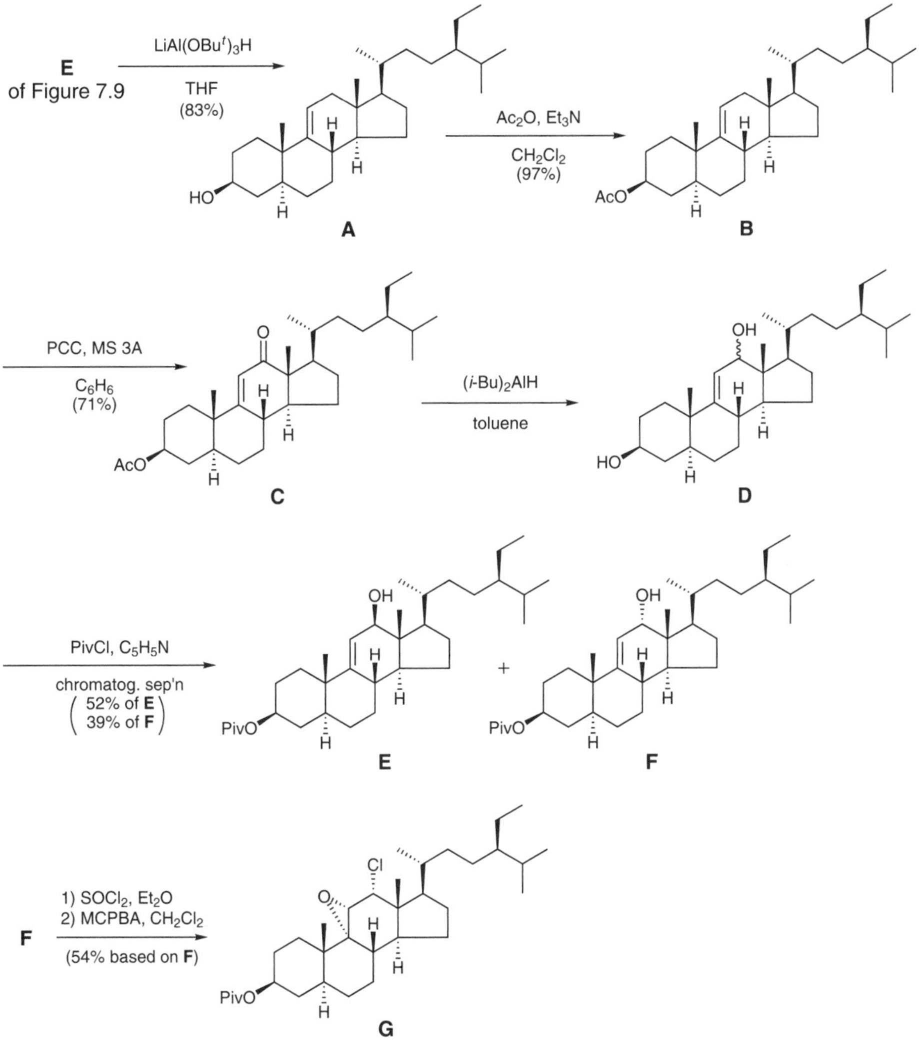

Figure 7.11   *Synthesis of the structure proposed for blattellastanoside A in July, 1992. Modified by permission of Shokabo Publishing Co., Ltd*

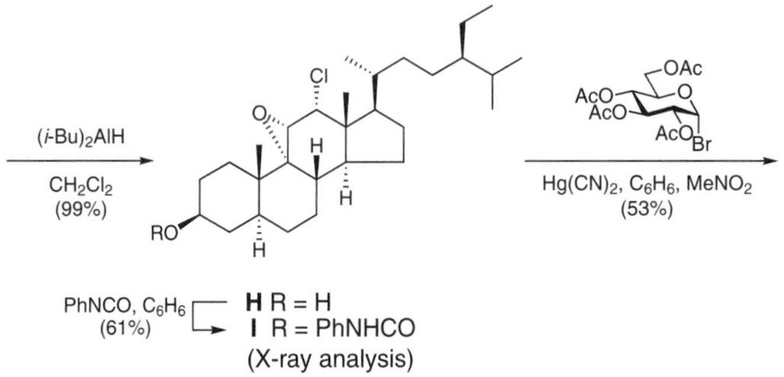

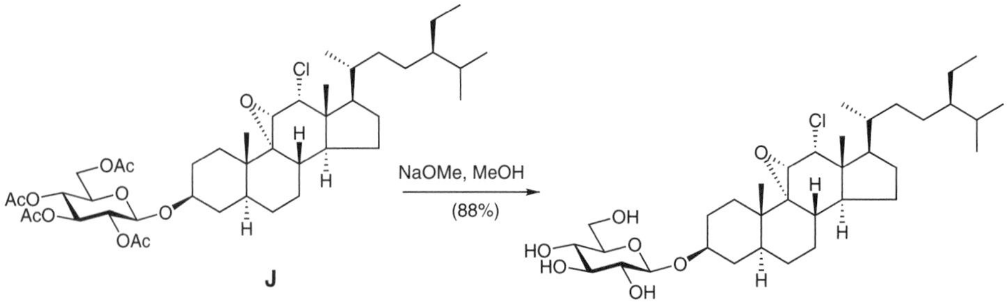

Figure 7.11   *(continued)*

was oxidized with pyridinium chlorochromate (PCC) to furnish acetoxy ketone **C**. Reduction of **C** with diisobutylaluminum hydride gave **D** as an inseparable mixture of stereoisomers. Monoacylation of **D** at C-3 with pivaloyl chloride yielded a mixture of **E** and **F**, which could be separated by silica-gel chromatography. Replacement of the hydroxy group of **F** with a chlorine atom was followed by epoxidation to give **G**, which was reduced to furnish **H**. The structure of **H** was confirmed by the X-ray analysis of its phenylcarbamate **I**. Conventional Königs–Knorr glycosylation of **H** with tetra-*O*-acetyl-α-D-glucopyranosyl bromide afforded **J**. Treatment of **J** with sodium methoxide in methanol furnished a crystalline glucoside with the structure **K**. Unfortunately, the ^{1}H- and ^{13}C-NMR spectra of **K** were different from those of blattellastanoside A. The proposed structure **K** (=**A** of Figure 7.10) was therefore incorrect.

The above results forced Sakuma and Fukami to propose new structures **A** (**171**) and **B** (**172**) for blattellastanoside A and B in 1993 as shown in Figure 7.12.[13,14] These structures are less unusual, because they have their functional groups only in rings A and B of the steroid nucleus. We immediately started our synthesis of **171** and **172**, as shown in Figures 7.13 and 7.14.[15]

Figure 7.13 summarizes our synthesis of blattellastanoside A (**171**).[15] Because **171** is a chlorinated steroidal glucoside with 5β-stigmastane skeleton, β-sitosterol **A** was chosen as the starting material. Accordingly, commercially available stigmasterol was converted to **A** in 70% overall yield by a known method. Epoxidation of **A** with *m*-chlorobenzoic acid gave epoxy alcohol **B** as a stereoisomeric mixture. Treatment of **B** with hydrogen chloride in chloroform was followed by oxidation of the product to give pure chloroketol **C** after recrystallization. Dehydration of **C** with hydrochloric acid caused epimerization at C-6 to give crystalline **D** with an axial hydrogen atom at C-6. The α, β-unsaturated ketone **D** was reduced with lithium tri(*t*-butoxy)aluminum hydride. The product was epoxidized with *m*-chloroperbenzoic acid, and

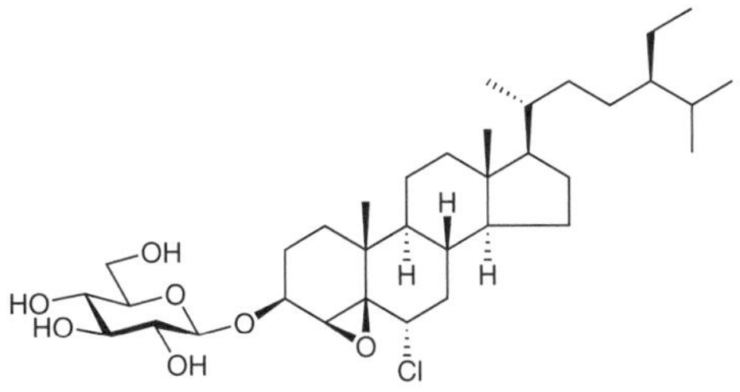

A

Blattellastanoside A (**171**)

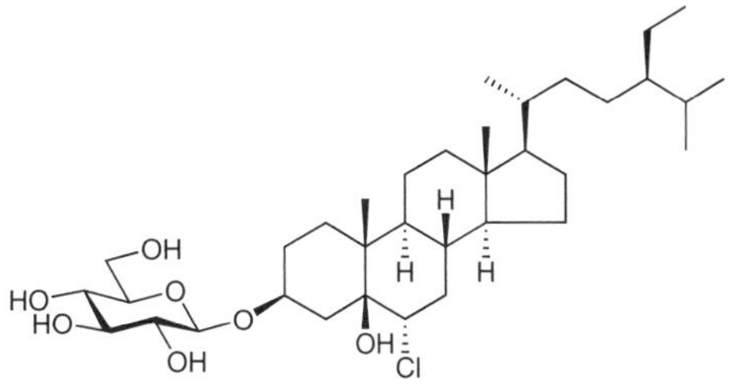

B

Blattellastanoside B (**172**)

Figure 7.12   *Structures proposed for blattellastanoside A and B in 1993. Modified by permission of Shokabo Publishing Co., Ltd*

then acetylalted to furnish **E**. Mild reduction of **E** with diisobutylaluminum hydride removed its acetyl group to give **F**, which was oxidized to the crystalline ketone **G**. The structure of **G** was confirmed by its X-ray analysis.

Treatment of **F** with acetobromo-D-glucose under the conventional Königs–Knorr conditions gave **H**, whose deacetylation afforded blattellastanoside A (**171**) as leaflets, mp 164–166 °C. The identity of our synthetic **171** with blattellastanoside A was confirmed by a comparison of their ^{1}H- and ^{13}C-NMR data and HPLC analysis.

Blattellastanoside B (**172**) was also synthesized as shown in Figure 7.14.[15] Reduction of the chloro epoxide **A** with diisobutylaluminum hydride was followed by acetylation to give **B** selectively, although the yield was only 29% with 68% of the recovered **A**. Treatment of **B** with sodium methoxide in methanol at 0 °C yielded **C**, which afforded epoxy alcohol **D** by further treatment with sodium methoxide at 60 °C. Glucosylation of **C** with acetobromo-D-glucose was followed by deacetylation to give blattellastanoside B (**172**) as microcrystalline powder, mp 158–160 °C. A comparison of the ^{1}H- and ^{13}C-NMR spectra of the synthetic **172** with those of blattellastanoside B proved the identity.

In 1993, we believed that our synthetic studies concluded the structural studies on the arrestant pheromone of the German cockroach, especially because Dr. M. Sakuma at Kyoto University told me that both **171** and **172** were pheromonally active. Surprisingly, in 1999, Scherkenbeck at Bayer AG and Wendler at University of Cologne in Germany together with their coworkers reported that blattellastanoside A (**171**) could not be detected in the feces of the German cockroach.[16] They synthesized **171** according to our procedure.[15] Their synthetic **171** showed no effect on the aggregation behavior of the German cockroach. Genuine arrestant pheromone of the German cockroach seems to be a mixture of carboxylic acids.[16] Their paper shocked me, because we had a joint publication with Dr. Sakuma to report the structure–pheromone activity relationship among the analogs of blattellastanoside A and B.[17] I immediately contacted Dr. Sakuma to hear his opinion about the German paper. He confessed to me that his bioassay method was not reproducible enough.

Sakuma and Fukami prepared the feces extract by extraction of feces-soaked filter paper. Filter paper may contain β-sitosterol, and contains plenty of D-glucose as cellulose, and these might have been the origin of blattellastanoside A and B. The exact reason for this erroneous result by Sakuma and Fukami still remains unclear.

Differolide (**135** and/or **135′**), which was discussed in Section 5.1.6, is another example of the structure based on nonreproducible bioassay.

7.4.2   2,2,4,4-Tetramethyl-*N,N*-bis(2,6-dimethylphenyl)cyclobutane-1,3-diimine as a putative antifeedant against the cotton boll weevil

In 1993, Miles *et al.* isolated a crystalline compound from the aerial part of the Thai plant, *Arundo donax*, and determined its structure by X-ray analysis as 2,2,4,4-tetramethyl-*N,N*-bis(2,6-dimethylphenyl)

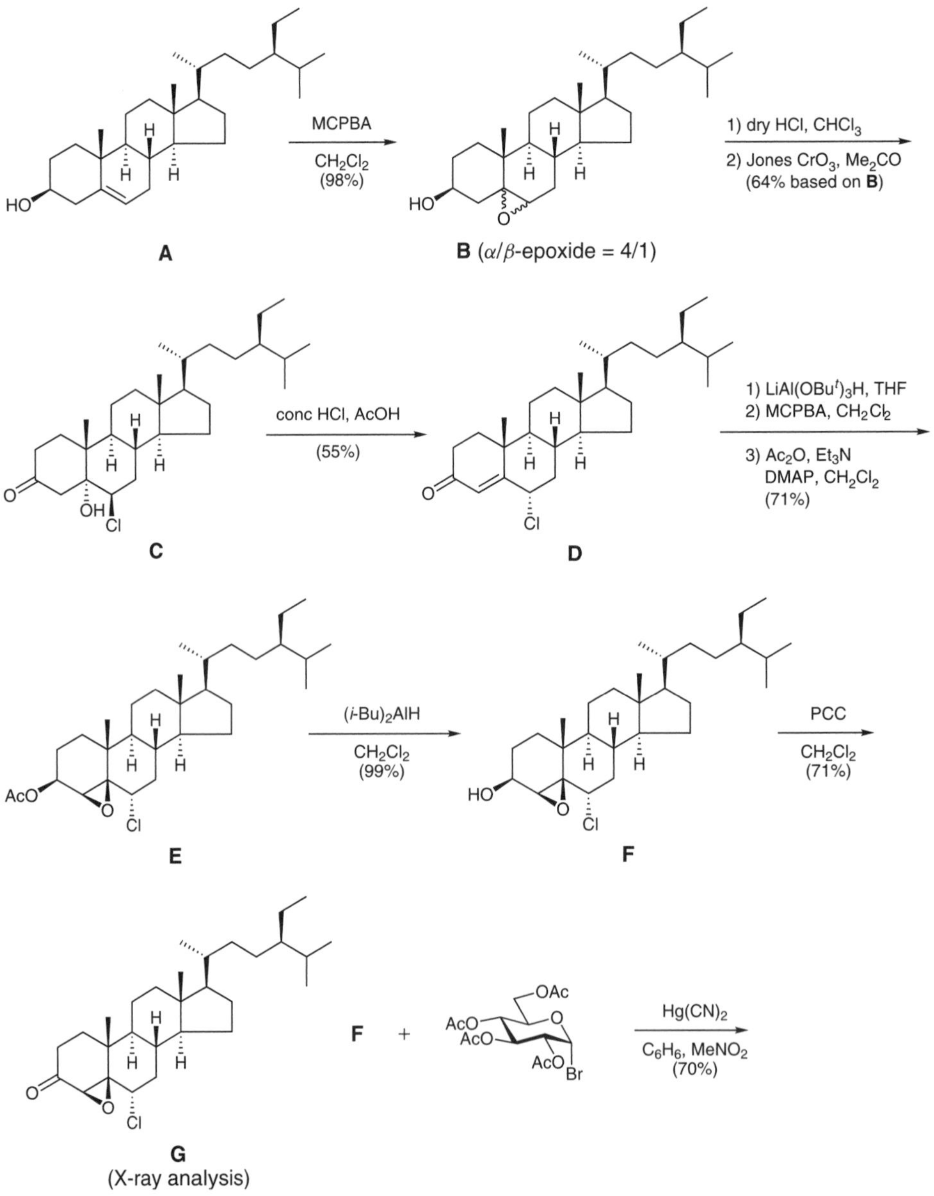

Figure 7.13   *Synthesis of blattellastanoside A. Modified by permission of Shokabo Publishing Co., Ltd*

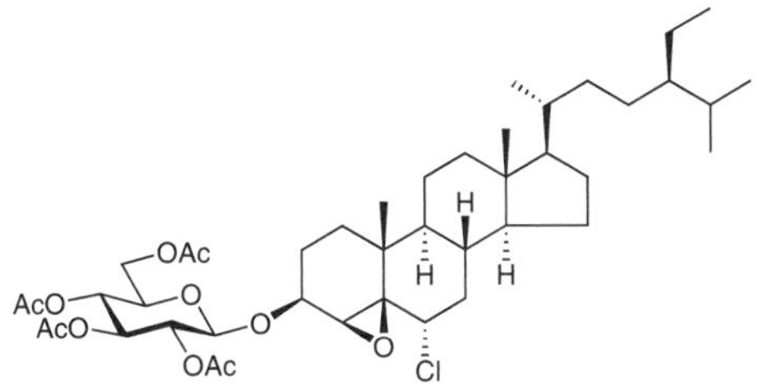

H

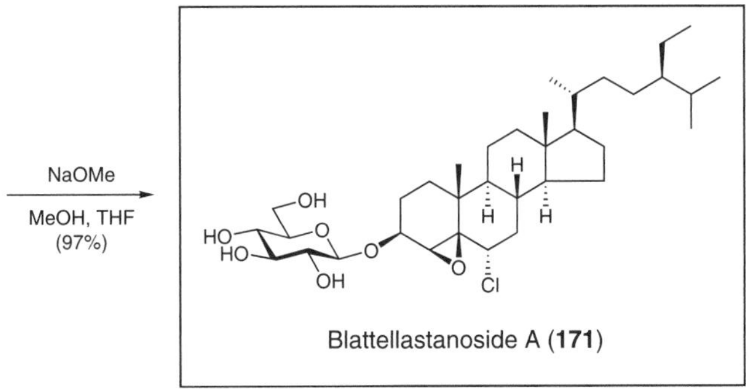

NaOMe
⎯⎯⎯⎯⎯⎯→
MeOH, THF
(97%)

Blattellastanoside A (**171**)

Figure 7.13 *(continued)*

cyclobutane-1,3-diimine (**173**, Figure 7.15). In addition, **173** was reported to show 54% inhibition of feeding against the notorious cotton pest, the boll weevil (*Anthonomus grandis*), at a dosage of 0.5 mg.

We synthesized **173** by titanium(IV) chloride-catalysed imine formation between **A** and **B**.[18] The product **173** showed no definite antifeedant activity against adult cotton boll weevils as tested by researchers at Sumitomo Chemical Co., Ltd. At the concentration of 500 ppm, it seemed even to stimulate feeding.[18] The structure **173** is too unusual to be a real natural product.

7.5 Human errors are inevitable in chemistry, too

In this chapter we have discussed some examples of incorrect structural proposals. Errors originated from (1) preoccupied and unusual assumptions about the possible structures, (2) inappropriate consideration of limitation of the available purification and analytical techniques, or (3) unreliable bioassay methods. In many cases, there was a human desire to reach a good result as soon as possible. In other words, there was pressure related to the well-known phrase, "publish or perish."

How can we avoid errors such as those discussed in this chapter? All researchers must maintain a calm and peaceful state of mind in order to judge their results correctly. A nonrigorous and too optimistic interpretation of the analytical or biological data is always a dangerous stumbling block. All research leaders must carefully examine the results obtained by their coworkers. Too competitive an atmosphere among coworkers may lead to misconduct.

Because science is a human achievement, errors will continue to occur. It is almost impossible to avoid errors, even if we work hard every day. The most important action for us to take is to admit our own errors as soon as we discover the correct answers.

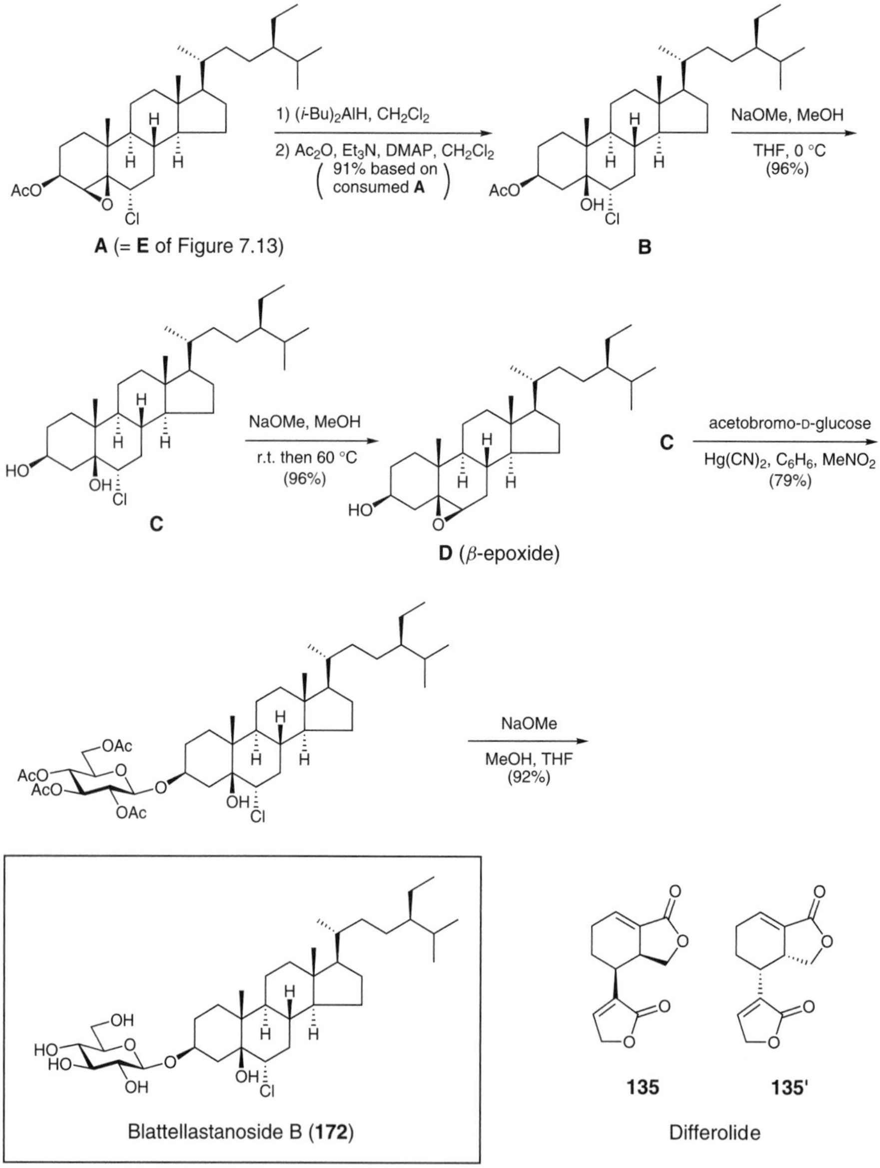

Figure 7.14 *Synthesis of blattellastanoside B and structure of differolide. Modified by permission of Shokabo Publishing Co., Ltd*

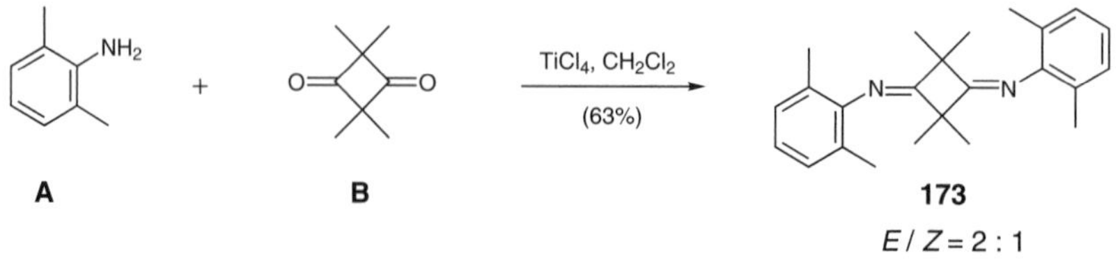

Figure 7.15 *Structure and synthesis of 2,2,4,4-tetramethyl-N,N-bis(2,6-dimethylphenyl)cyclobutane-1,3-diimine*

References

1. Strain, W. H. in *Organic Chemistry, an Advanced Treatise*, Vol II. Gilman, H., ed. Wiley: New York, 1943: pp. 1344–1349.
2. Fieser, L.F.; Fieser, M. *Steroids*, Reinhold: New York, 1959: pp. 70–74.
3. Mori, K.; Kamada, A.; Kido, M. *Liebigs Ann. Chem.* **1991**, 775–781.
4. Mori, K.; Takanashi, S. *Tetrahedron Lett.* **1996**, *37*, 1821–1824.
5. Mori, K.; Takanashi, S. *Proc. Jpn. Acad. Ser. B* **1996**, *72*, 174–177.
6. Takanashi, S.; Mori, K. *Liebigs Ann.* **1997**, 825–838.
7. Takanashi, S.; Mori, K. *Liebigs Ann.* **1997**, 1081–1084.
8. Takanashi, S.; Mori, K. *Eur. J. Org. Chem.* **1998**, 43–55.
9. Nozawa, D.; Takikawa. H.; Mori, K. *J. Chem. Soc., Perkin Trans. 1* **2000**, 2043–2046.
10. Takagi, M.; Takikawa, H.; Mori, K. *Biosci. Biotechnol. Biochem.* **2001**, *65*, 2065–2069.
11. Nojima, S.; Schal, C.; Webster, F. X.; Santangelo, R. G.; Roelofs, W. L. *Science* **2005**, 307, 1104–1106.
12. Mori, K.; Fukamatsu, K.; Kido, M. *Liebigs Ann. Chem.* **1993**, 657–663.
13. Sakuma, M.; Fukami, H. *Tetrahedron Lett.* **1993**, *34*, 6059–6062.
14. Sakuma, M.; Fukami, H. *J. Chem. Ecol.* **1993**, *19*, 2521–2541.
15. Mori, K.; Fukamatsu, K.; Kido, M. *Liebigs Ann. Chem.* **1993**, 665–670.
16. Scherkenbeck, J.; Nentwig, G.; Justus, K.; Lenz, J.; Gondol, D.; Wendler, G.; Dambach, M.; Nischk, F.; Graef, C. *J. Chem. Ecol.* **1999**, *25*, 1105–1119.
17. Mori, K.; Nakayama, T.; Sakuma, M. *Bioorg. Med. Chem.* **1996**, *4*, 401–408.
18. Mochizuki, K.; Takikawa, H.; Mori, K. *Biosci. Biotechnol. Biochem.* **2000**, *64*, 647–651.

8

Conclusion—Science as a Human Endeavor

I have discussed the syntheses of 173 small molecules with different bioactivities, and thereby summarized my research activities over half a century until 2009. In this final chapter, I will describe what I have encountered in my career as a chemist. My hope as a scientist will conclude this book.

8.1 Small molecules are also beautiful

I love synthesis of small molecules that can be completed by a single person. During my professorship, I always gave a single target compound to each of my students. My students had to finish their syntheses within two, three, or five years. Otherwise, they failed to obtain their Master's or Doctor's degrees. The target molecules were therefore not too complicated to prepare. Successful students felt confident of their capability and skill as chemists, and could start fruitful careers in either industries or in academia.

In our Japanese literature, we have "haiku" or short poem consisting of only seventeen Japanese phonetic characters. Although it is short, it can express much according to the sensitivity and creativity of its author. Synthesis of small molecules is just like "haiku". As you have seen in the previous chapters, small molecules sometimes play pivotal roles as hormones, pheromones, and other bioregulators. I love simplicity and regard it as an element of real beauty. Someone may prefer complexity to simplicity. All of us have different personal tastes. In my case, I want to love simplicity just because it is my own personal desire.

Auguries of Innocence

To see a World in a Grain of Sand
And a Heaven in a Wild Flower,
Hold Infinity in the palm of your hand
And Eternity in an hour.

(William Blake)

Chemical Synthesis of Hormones, Pheromones and Other Bioregulators   Kenji Mori
© 2010 John Wiley & Sons, Ltd

8.2   Continuous efforts may bring something meaningful

Execution of a difficult synthetic work is often tedious, and requires days or months of unsuccessful attempts. I shook my separatory funnels for tens of thousands of times. We have to endure and tolerate such repetition. In August 1977, I was invited to give a lecture at a terpenoid conference in Varenna, Lake Como, Italy. Professor G. Jommi at the University of Milan was the organizer of the conference. He said to me after a concert during the conference, "Oriental music is monotonous and full of repetition." I commented, "Yes, it is. It is just like our own lives. Lives are monotonous and full of repetition." Such a monotonous life, however, may eventually turn out to be a meaningful one.

In April 1965, when I was working on the synthesis of the gibberellins and related diterpenoids, I was given a prize by the Japan Society for Agricultural Chemistry (now the Japan Society for Bioscience, Biotechnology, and Agrochemistry). On the day of the Award Ceremony, my wrist watch was unfortunately out of order. As a young and poor academic chemist, I had no money to buy a new one. I therefore borrowed a wrist watch of my friend T. Ogawa (who is now a Professor Emeritus of the University of Tokyo, and the Director of RIKEN Wako Institute), and finished my lecture within the given time. After the ceremony, Professor T. Yabuta, then 76-year-old discoverer of the gibberellins, told me, "Dr. Mori, you don't have a watch. I will give it to you." But he did not give it to me soon. I almost forgot Prof. Yabuta's words. Then, in July 1965, I received a nice wrist watch from him with a letter as follows. "When we met together in the Award Ceremony, I told you that I will give you the same type of watch as I have. You may now think that I am a liar. Here it is! Please use it, and develop your synthetic works on the gibberellins as time passes." With this encouragement of the old Professor Yabuta, I could finish the gibberellin synthesis at the end of 1967. Encouragement by others will turn the monotonous days into a happy moment. Those who have been encouraged by others will learn to encourage others in their future lives.

In July 1973, Professor T. Mukaiyama organized a US−Japan Seminar on Organic Synthesis in Tokyo. I became acquainted with many American chemists on that occasion. Professor Barry Sharpless was among them. He gave a talk on organoselenium chemistry. I invited him and his wife to visit my apartment flat in the suburb of Tokyo. After twenty two years, at the time of my retirement from the University of Tokyo, Barry wrote to me: "I still remember fondly how you befriended a young American assistant professor on his first trip to Japan. My wife and I will never forget the evening at your home. You and I drank *a lot of* beer and instantly became good friends while discussing the future of organic chemistry as if we had an inside track on knowing where it was going." We talked on the future of name reactions and enantioselective synthesis in that evening. Friendship will give us courage to endure the monotonous days. It should be added that among about twenty chemists who attended the 1973 US−Japan Seminar, I was the only person to talk on the synthesis of optically active compounds. All the others gave talks on racemic synthesis.

I was in Italy in August 1977 as I already told you. In that terpenoid conference, I talked on enantiose-lective synthesis of pheromones. Some of my colleagues in Japan discouraged me by saying that I stopped working on the challenging task of the synthesis of diterpenoids with a number of stereogenic centers and began working on the far less difficult synthesis of pheromones with only one or two stereogenic centers. Of course, I had my own reason. What I was doing in pheromone works was enantioselective synthesis, and what I did in diterpenoids was racemic synthesis. In 1977 there was neither chiral GC analysis nor reliable asymmetric reactions. I was frustrated by the improper opinion of some of my colleagues in Japan, and talked with Professor E. Wenkert, an expert in diterpenoid synthesis, at the calm and beautiful shore of Lake Como. I told him that I love my pheromone synthesis and regard it as a meaningful work in spite of the cold attitude of some of my Japanese colleagues. He said to me, "Well, if you think that you are working in the most meaningful field, you would better continue it. If your field is really interesting, then others will also follow you." Through monotonous and repeating days, we may be able to find out what is really interesting. My experience tells me that our continuous efforts will bring us something meaningful, if the direction of the efforts is the correct one.

> *Always be cheerful.*
> *Pray unceasingly.*
> *Under all circumstances give thanks.*

(I Thessalonians 5:16–18)

8.3 Can a scientist eventually have a hope in the future?

I have worked for half a century as a synthetic chemist. I have synthesized many hormones, pheromones, and other bioregulators. But I still have a big question. Are we allowed to know something perfectly? Recent development in visualization of molecular events in organisms is quite often due to the advances in molecular probe or fluorescent tag methodology. If we want to know the molecular events in organisms, we have to add something to that molecule so as to detect its location. Such a molecular probe may influence the course of the event. Accordingly, what we can see by using a molecular probe may not be perfectly the same as what we want to see. I call this kind of thought my "**biological uncertainty principle**". Even after sacrificing my whole life, if I cannot know something perfectly, what an unhappy man I am! This thought brings me to my future prospect as follows.

> *What we see now is like the dim image in a mirror; then we shall see face to face.*
> *What I know now is only partial, then it will be complete.*

(1 Corinthians 13:12)

Acknowledgements

I thank my past and present coworkers in chemistry and biology. My thanks are due to Dr. Takuya Tashiro (RIKEN) and my wife Keiko Mori for their help in preparing the illustrations and typing the text, respectively. I am thankful to Shokabo Publisher, Tokyo, for permitting me to use some of the Schemes in my book "Seibutsu-Kassei Tennenbutsu no Kagakugosei (Chemical Synthesis of Bioactive Natural Products)" published in 1995.

Gloria Deo et memento Mori!

Index

Chemical Synthesis of Hormones, Pheromones and Other Bioregulators   Kenji Mori
© 2010 John Wiley & Sons, Ltd

neuchromenin, 214–15
NKT cells *see* natural killer T cells
NMR spectroscopy, 11–14, 109–10
Nocardia sp., 215
nocardione, 215–16, 217
nyctinasty, 45

OCH, 260
octant rule, 75
olean, 178–80, 199
olive fruit fly *see Bactrocera oleae*
Orgyia
  detrita, 184
  postica, 151
Orobanche minor, 63, 64
Orobanche spp., 61–3
orobanchol, 63, 64, 67
orobanchyl acetate *see* alectrol
oryzalexins, 53–6
Oryza sativa (rice plant),1, 53, 56, 118

palaearctic bee *see Andrena wilkella*
"paper factor", 82
paraconic acid, 190–1
Pectinophora gossypiella, 275
Pen II, 195, 196
Pen III, 195
penaresidin A, 242–5
penaresidin B, 242–5
Penares spp., 242
penazetidine A, 242, 243
penicillin, 201, 204
Penicillium
  funiculosum, 195, 196
  notatum, 201
Periplaneta americana, 126–8, 275–6
periplanone, 126–8, 275, 276
Pharaoh's ant *see Monomorium pharaonis*
Phaseolus vulgaris, 43, 44, 60
PHB *see* poly *β*-hydroxybutyrate
pheromones
  categories, 107
  clarification of structure, 124–33
  definition, 2, 107
  determination of absolute configuration,
    111–22
  enantioselective synthesis, 107–11
  incorrect structures, 275–6, 278–83
  preparation of pure sample for bioassay,
    133–43

stereochemistry–bioactivity relationships, 110,
  122–4, 143–84
phyllanthrinolactone, 45–9, 50
Phyllanthus urinaria, 45, 47
phytoalexins, 49–59
phytocassanes, 56–9
phytohormones, 19–20
  abscisic acid, 38–41
  brassinosteroids, 41–5
  diterpenes related to gibberellins, 30–7
  gibberellins, 19–30
  phyllanthrinolactone, 45–9
Phytophtora infestans, 56
pink bollworm moth *see Pectinophora gossypiella*
pinthunamide, 219–21
pisatin, 49–52
Pisum sativum, 49
Plakinastrella sp., 238, 276
Plakortis simplex, 245
plakoside A, 245–52
Platynereis dumerilii, 184
Pleurobrancus testudinarius, 227
polygodial, 95–7, 98
Polygonum hydropiper, 95
poly-*β*-hydroxybutyrate (PHB), 141, 159
Polyporus arcularius, 192, 195
polyzonimine, 102–3
Polyzonium rosalbum, 102
Popillia japonica, 137
posticlure, 151
powdery mildew, 219
propylure, 275
prothoracicotropic hormone (PTTH), 81
PTTH *see* prothoracicotropic hormone
PUG 4 *see* punaglandin 4
punaglandin 4, 232–5
purification methods, 126–8, 267, 276–8
Pyricularia oryzae see Magnaporthe grisea
Pyrrhocoris apterus, 82, 84

quorum sensing, 3

rape plant *see Brassica napus*
RCAI-56, 261–2, 264
RCAI-61, 262–4
red clover *see Trifolium pretense*
red-flour beetle *see Tribolium castaneum*
red imported fire ant *see Solenopsis invicta*
retrosynthetic analysis, 6
Rhizoctonia solani, 56